Jacek Zaucha • Kira Gee
Editors

Maritime Spatial Planning

past, present, future

Editors
Jacek Zaucha
Institute for Development and Maritime Institute
University of Gdańsk
Gdańsk, Poland

Kira Gee
Human Dimensions of Coastal Areas
Helmholtz Zentrum Geesthacht
Geesthacht, Schleswig-Holstein, Germany

https://doi.org/10.1007/978-3-319-98696-8

To our beloved: Irena, Justyna, Stanisław, Dave
Jacek & Kira

Foreword

Over the last 20 years, marine/maritime spatial planning (MSP)[1] has gained a strong political presence in Europe and elsewhere. Before 2006, only a handful of countries had begun to spatially plan sea areas, such as China, where marine functional zoning was first proposed by government in 1998. In Europe, efforts began in 2002 as part of the EU-funded BaltCoast project involving Germany, Sweden, Estonia, Poland, Latvia, Denmark and Finland. Belgium, Germany and the Netherlands then became forerunners of MSP in Europe, approving integrated management plans for their waters in 2005. By 2017, the number of countries with MSP initiatives of some type had grown to about 60, the majority of which are in Europe but also some in Central America, Africa and Asia (Ehler 2017; Santos et al. 2019).[2]

Given the growing interest in applying MSP around the world, it is time to take stock of what has been achieved by MSP and where future challenges might lie. Research interest in MSP has grown exponentially in recent years, and scholars are analysing various dimensions of MSP including its rationale, methods and outcomes in subject areas such as geography, planning, political and social sciences and ecology.

Based on the growing body of practical experience with MSP, critical questions have emerged regarding the MSP process and also the overall objectives

[1] We use the terms marine and maritime spatial planning in this book. The term "marine" is arguably more strongly associated with the marine environment and "maritime" with marine activities and uses, although in planning practice, "marine spatial planning", "maritime spatial planning" and "marine planning" are all used to describe similar processes. See also Section 3 of Chap. 1.

[2] Ehler, C. (2017) "World-Wide Status and Trends of Maritime/Marine Spatial Planning" presented at the 2nd International Conference on Marine/Maritime Spatial Planning, UNESCO, Paris.

of MSP. For example, what is the relationship between MSP and the ecosystem approach to marine management? Is MSP contributing to the sustainable development of the ocean? How inclusive are MSP processes in practice? Other challenges are emerging at the intersection of theory and practice, such as how to account for the social benefits of MSP or various understandings of spatiality. Questions have also been raised regarding MSP as a form of "ocean grabbing".

This book seeks to comprehensively address these and other issues related to MSP. At a critical juncture in the EU, where countries are required to develop maritime spatial plans by 2021,[3] it is the first comprehensive outlook on MSP seen through the lenses of different scientific disciplines. It brings together the perspectives of authors at the edge of research and practice: people who have been practically involved in MSP in their countries, regions or sea basins, either running MSP-related projects or assisting the elaboration of maritime spatial plans, and people who have considered MSP from a theoretical perspective. The authors represent disciplines as diverse as macro-spatial planning, oceanography, land-use planning, ecology, political and social sciences, as well as geography and economics. Rather than a definitive treatment of MSP, it is an attempt to capture the various roles of MSP in managing social-ecological systems, its intended and unintended socio-economic and ecological outcomes, the uncertainties that surround its conceptualisation as well as some critical questions concerning the concept of MSP as a whole. The book project was supported by an international interdisciplinary conference on MSP that took place in autumn 2017 in Poland.

The book begins with a presentation of the essence, origin and interdisciplinary character of MSP (Chap. 1). This is complemented by a general analysis of the ocean as the subject of planning in Chap. 2. What key ocean perspectives have developed over time, how have we come to see the ocean and what contradictions are emerging from these views?

Ecological perspectives are presented in the following two chapters. Chapter 3 focuses on the challenges and opportunities for ecosystem-based management and MSP in the Irish Sea, drawing on recent projects and experiences. Chapter 4 takes readers outside of Europe, outlining the role of systematic conservation planning in MSP and explaining how this has been applied in the Benguela Current Large Marine Ecosystem in South Africa and Namibia.

The ecological perspective is followed by an economic perspective, starting with Chap. 5 on classical location theory. In discussing its applicability to

[3] See the EU's Maritime Spatial Planning Directive (EC 2014), https://eur-lex.europa.eu/legal-content/EN/TXT/PDF/?uri=CELEX:32014L0089&from=EN.

marine space, it shows that while MSP cannot neglect market forces, the market itself is unable to allocate space to issues considered important from a societal point of view, requiring the public process of MSP as a corrective. Chapter 6 continues the economic theme by outlining current thinking on "Blue Growth" and its relationship with MSP. It points out that MSP can play an important role in supporting Blue Growth but only if intertwined with other measures of Integrated Maritime Policy and territorial development.

Chapters 7, 8, 9 and 10 present a socio-cultural perspective. Chapter 7 begins by interrogating the concept of "socio-cultural", examining how it is being defined and applied across the MSP landscape. Cultural ecosystem services, seascape and well-being are examined as central concepts. Chapter 8 continues this focus by conceptualising social sustainability in MSP, articulating how social sustainability could be conceived in MSP and describing how this framework could be applied to analyse MSP practice. Key features of social sustainability elaborated are deepening democratic decision-making, inclusion of socio-cultural values and knowledge, and equitable distribution and social cohesion. Chapter 9 argues that to recapture its democratising potential, MSP requires explicit engagement with politics and power. It highlights the use of the boundary object lens and citizen science as two potential avenues to facilitate this engagement. Chapter 10 presents an analytical framework to characterise participation in MSP, including a participation ladder emphasising power sharing, roles, functions and learning in transboundary contexts.

All of the above perspectives are interrelated, which is further emphasised by the next section of the book that has a practical perspective. Chapter 11 reflects on MSP from a land-sea interaction (LSI) perspective. It raises questions about the role and limitations of MSP in addressing sustainable development of the world's oceans as many of the issues it is concerned with are inextricably linked to activity on the land. It ends with an exploration of how LSI matters might inform future directions for MSP and may be heralding a new era of Territorial Spatial Planning, which spans both land and sea. Chapter 12 is dedicated to the Mediterranean and illustrates the policy framework supporting MSP implementation in the Mediterranean Sea. Following on from Chap. 11, it discusses the importance of linking MSP with integrated coastal management (ICM/ICZM) given the high relevance of LSIs. Chapter 13 is a critical analysis of stakeholder processes in MSP. Varying participatory models of MSP are currently developing for MSP, but delivering multi-sector participatory MSP processes is faced with many challenges. Various disconnects are identified between the conceptual underpinnings of MSP and the reality of recent stakeholder processes.

Chapter 14 looks at the use of scenarios to inform the development of MSPs, presenting an in-depth example of scenario-building for the Celtic Seas. Scenarios are revealed as important in understanding the aspirations of different stakeholders towards integration within the MSP process, as well as the realities of encouraging co-location between sea uses.

Chapters 15, 16 and 17 present a dedicated governance perspective on MSP. Chapter 15 places MSP in the context of risk governance, relating risk to the process of developing a maritime spatial plan. Chapter 16 considers the role of the law of the sea in MSP, highlighting that the United Nations Convention on the Law of the Sea (UNCLOS) has also provided the framework for the further development of, inter alia, global ocean governance. Chapter 17 goes beyond the confines of territoriality and considers MSP in areas beyond national jurisdiction (ABNJ). The chapter highlights the existing legal framework and describes which organisations could foster MSP in ABNJ. Furthermore, it assesses whether existing MSP tools are transferable to ABNJ.

The book concludes with an outlook towards the next generation of MSPs and MSP as a practice. Chapter 18 considers approaches to evaluating MSP and presents a methodology for designing a flexible and context-specific evaluation of MSP. Chapter 19 considers what basic skills are needed to achieve a successful professional practice of MSP, drawing on the visions and insights of consultants, maritime sectors, policymakers, scientists and teachers of MSP.

As the editors of the book and MSP researchers and practitioners, we would like to explicitly acknowledge those who have paved the way for MSP development in the EU, in particular, in its initial stages. Among others, key pioneers of MSP in Europe are Bernhard Heinrichs (propagator of the concept at the VASAB forum and creator of the first marine spatial development plan in the EU), Charles Ehler (propagator of the concept throughout the world through the structures of IOC-UNESCO), Nico Nolte (creator of the first maritime spatial plan for the exclusive economic zone in the EU), Haitze Siemers (forward-looking director within DG MARE and responsible for the EU MSP Directive) as well as Angela Schultz-Zehden (creator of the EU MSP platform and coordinator of numerous MSP pilot projects and various MSP innovations).

We are also grateful to all the authors in this book for contributing their thoughts, know-how and experience and for devoting their time to highlight the complexity of MSP, its interrelated risks and challenges as well as its opportunities.

Gdańsk, Poland — Jacek Zaucha
Geesthacht, Germany — Kira Gee

References

EC. (2014). Directive 2014/89/EU of the European Parliament and of the Council of 23 July 2014 Establishing a Framework for Maritime Spatial Planning. Official Journal of the European Union, L 257/135.

Ehler, C. (2017). World-Wide Status and Trends of Maritime/Marine Spatial Planning. Presented at the 2nd International Conference on Marine/Maritime Spatial Planning, UNESCO, Paris.

Santos, C. F., Ehler, C., Agardy, T., Andrad, F., Orbach, M. K., & Crowder, L. B. (2019). Marine Spatial Planning. In C. Sheppard (Ed.), *World Seas: An Environmental Evaluation, Second Edition, Volume Three: Ecological Issues and Environmental Impacts* (pp. 571–592). London, San Diego, Cambridge MA, Oxford: Academic Press.

Acknowledgements

The book resulted from intensive discussions between the editors and authors of the chapters facilitated under the project (research grant) "Economy of maritime space" financed by the Polish National Science Centre—decision no. 2015/17/B/HS4/00918 and executed by the Institute for Development in Sopot.

Contents

1 **Maritime/Marine Spatial Planning at the Interface of Research and Practice** 1
Charles Ehler, Jacek Zaucha, and Kira Gee

2 **The Ocean Perspective** 23
Kira Gee

3 **Challenges and Opportunities for Ecosystem-Based Management and Marine Spatial Planning in the Irish Sea** 47
Tim O'Higgins, Linda O'Higgins, Anne Marie O'Hagan, and Joseph Onwona Ansong

4 **Systematic Conservation Planning as a Tool to Advance Ecologically or Biologically Significant Area and Marine Spatial Planning Processes** 71
Linda R. Harris, Stephen Holness, Gunnar Finke, Stephen Kirkman, and Kerry Sink

5 **Can Classical Location Theory Apply to Sea Space?** 97
Jacek Zaucha

6 **Maritime Spatial Planning and the EU's Blue Growth Policy: Past, Present and Future Perspectives** 121
Angela Schultz-Zehden, Barbara Weig, and Ivana Lukic

7 Socio-cultural Dimensions of Marine Spatial Planning 151
Emma McKinley, Tim Acott, and Tim Stojanovic

8 Adding People to the Sea: Conceptualizing Social Sustainability in Maritime Spatial Planning 175
Fred P. Saunders, Michael Gilek, and Ralph Tafon

9 Politics and Power in Marine Spatial Planning 201
Wesley Flannery, Jane Clarke, and Benedict McAteer

10 Towards a Ladder of Marine/Maritime Spatial Planning Participation 219
Andrea Morf, Michael Kull, Joanna Piwowarczyk, and Kira Gee

11 Taking Account of Land-Sea Interactions in Marine Spatial Planning 245
Sue Kidd, Hannah Jones, and Stephen Jay

12 Linking Integrated Coastal Zone Management to Maritime Spatial Planning: The Mediterranean Experience 271
Emiliano Ramieri, Martina Bocci, and Marina Markovic

13 Stakeholder Processes in Marine Spatial Planning: Ambitions and Realities from the European Atlantic Experience 295
Sarah Twomey and Cathal O'Mahony

14 Scenario-Building for Marine Spatial Planning 327
Lynne McGowan, Stephen Jay, and Sue Kidd

15 Managing Risk Through Marine Spatial Planning 353
Roland Cormier and Andreas Kannen

16 The Role of the Law of the Sea in Marine Spatial Planning 375
Dorota Pyć

17 **The Need for Marine Spatial Planning in Areas Beyond National Jurisdiction** 397
Susanne Altvater, Ruth Fletcher, and Cristian Passarello

18 **Evaluation of Marine Spatial Planning: Valuing the Process, Knowing the Impacts** 417
Riku Varjopuro

19 **Education and Training for Maritime Spatial Planners** 441
Helena Calado, Catarina Fonseca, Joseph Onwona Ansong, Manuel Frias, and Marta Vergílio

Index 469

Notes on Contributors

Tim Acott is Reader in Human Geography at the University of Greenwich. He is Director of the Greenwich Maritime Centre and is the Chair of the Coastal and Marine Research Group at the Royal Geographic Society. Over the last 8 years, he has worked extensively on understanding the social and cultural importance of fisheries through sense of place and cultural ecosystem services. He has co-edited books on "Social Issues in Sustainable Fisheries Management" and "Social Wellbeing and the Values of Small-Scale Fisheries" and has published numerous articles. His most recent research is leading a project exploring the socio-natural values of lowland wetlands in England from a co-constructionist perspective.

Susanne Altvater is a lawyer and planner working as a consultant on marine issues for s.Pro (sustainable projects). Her project work primarily addresses marine policy and regulatory issues, mainly related to maritime spatial planning (MSP), for example, by supporting the EU MSP Platform designed to offer support to all EU Member States in their efforts to implement MSP in the years to come. Additionally, she is working on international law related to the Areas Beyond National Jurisdiction (ABNJ), mainly asking the question how concepts of marine biodiversity protections in these areas could be integrated into the sectoral use expectations or already ongoing uses. The existing regional organisations and stakeholders play an important part of her research.

Joseph Onwona Ansong is a consultant involved in projects including the Multi-use in European Seas (MUSES), EU MSP Platform and UNESCO Large Marine Ecosystem (LME) Learn MSP Toolkit. Before joining s.Pro, he worked with the Centre for Marine and Renewable Energy Ireland (MaREI), UCC on the SIMCelt Project: Supporting Implementation of MSP in the Celtic Seas to support practical cross-border cooperation in the implementation of MSP within the Celtic Seas. He had interned with Marine Policy and Regional Coordination Section of

UNESCO-IOC[4], Paris. Ansong holds an Erasmus Mundus Master in MSP and a Bachelor in Human Settlement Planning.

Martina Bocci holds a Master's degree in Biology and a PhD in Ecology. She has more than 15 years' experience in the fields of environment and sustainability through participation in coastal and marine environmental studies in Italy and abroad, public consultations and stakeholder engagement, monitoring programmes, planning, design and management projects. She is involved as senior expert in the activities of the European MSP Platform. She works on Integrated Coastal Zone Management (ICZM), land-sea interactions (LSIs), multi-use of the sea, marine and coastal economic development initiatives (Blue Growth Strategy, EUSAIR) and provides technical support to institutions implementing marine and coastal policies including the Water Framework Directive (WFD) and the Marine Strategy Framework Directive (MSFD).

Helena Calado holds a degree in Geography and Regional Planning and a PhD in Land-Use Planning and Management. She is a researcher at Marine and Environmental Sciences Centre (MARE) and professor at the Biology Department of University of the Azores, where she teaches Spatial Planning, Legislation and Environmental Management, and MSP. With over 25 years of experience in land-use planning, coastal zone management plans, and environmental impact assessments, she focuses her research on climate change impacts and mitigation measures, marine protected areas, and MSP for small islands, particularly the Azores.

Jane Clarke is a PhD candidate in the School of Natural and Built Environment at Queen's University Belfast. Her key research interests are marine spatial planning (MSP) and climate change, with a particular emphasis on the decision-making process relating to the transition to a low-carbon economy. Her research is funded by the Northern Ireland Department for Economy and subsidised by funding from the Marine Institute of Ireland.

Roland Cormier is a guest scientist at Helmholtz Zentrum Geesthacht and an associate of the Institute of Estuarine and Coastal Studies at the University of Hull. He is also a member of the International Council for the Exploration of Sea (ICES) Working Group on Marine Planning and Coastal Zone Management (WGMPCZM) and the Group of Experts on Risk Management in Regulatory Systems of the United Nations Economic Commission for Europe. His research is on risk-based approaches to legislative and policies in marine management and conservation. He is active as a consultant in environmental risk management in Europe, Canada and the United States, as well as a lecturer in universities in Canada and Europe.

Charles Ehler is an MSP consultant to UNESCO's IOC. He was a senior executive in the National Oceanic and Atmospheric Administration and United States Environmental Protection Agency for 32 years. In 2007, he received an award from

[4] International Oceanographic Commission of the United Nations Educational, Scientific and Cultural Organization.

the Intergovernmental Panel on Climate Change for his contribution to its award of the Nobel Peace Prize. He is the author of over 100 publications including the influential 2009 UNESCO guide to MSP for IOC. From 1968 to 1973, he taught regional planning at the University of Michigan, the University of California, Los Angeles (UCLA), and Stony Brook University.

Gunnar Finke is a German national based in Namibia as an advisor for the German Development Cooperation (GIZ). He provides technical and policy advice to Angola, Namibia and South Africa as Parties to the Benguela Current Convention (BCC) in implementing MSP. Prior to this assignment, he worked for the Blue Solutions Initiative, a global cooperation project to collate, share and generate knowledge and capacity for sustainable marine ecosystem management. He also advised the German Federal Ministry for Economic Cooperation and Development (BMZ) on marine biodiversity. Gunnar is a member of the World Commission on Protected Areas (WCPA) of the International Union for the Conservation of Nature (IUCN).

Wesley Flannery is a senior lecturer in the School of the Natural and Built Environment at Queen's University Belfast. His key research interests are in MSP, integrated coastal zone planning and stakeholder participation in environmental decision-making. He has conducted research on behalf of the Marine Institute of Ireland and the OSPAR Commission (Commission managing Convention for the Protection of the Marine Environment of the North-East Atlantic), the Irish government and local-level environmental NGOs. He is an editor of the *ICES Journal of Marine Science*.

Ruth Fletcher is a senior programme officer supporting the delivery of the Healthy Ocean Strategy at UN Environment World Conservation Monitoring Centre (UNEP-WCMC). As an interdisciplinary scientist, Ruth has expertise in ecology, economics, marine governance and social aspects of marine biodiversity issues. She is leading a project aiming to inform spatial planning in ABNJ. Her area-based planning expertise is also being used in the work on how to support Sustainable Development Goal delivery. Previous experience includes industry engagement and the assessment of social and cultural values in the marine environment ensuring she has an understanding of a wide range of diverse stakeholders.

Catarina Fonseca holds a Master's degree in Environmental Science and Technology from the University of Lisbon. She was a member of several research projects, including the Transboundary Planning in the European Atlantic (TPEA) project and has been involved in the several editions of the Erasmus Mundus Master Course on Maritime Spatial Planning as an invited speaker. Her research interests are mainly focused on spatial planning, governance, environmental management and public participation.

Manuel Frias holds a Master's degree in Geography from the University of Seville. He found his call in Utrecht (the Netherlands) when he studied Geographical Information Systems (GIS) as an Erasmus student. After some years working in

Madrid for a telecommunication company, he moved to Finland. He has been working at the Helsinki Commission (HELCOM) Secretariat in Helsinki since 2009 mainly with data management, GIS and MSP. He has worked in many projects related to MSP and is involved in a project which will create the first Baltic Marine Spatial Data Infrastructure prototype to access distributed MSP data.

Kira Gee has 20 years of experience in ICZM and MSP in the UK and Germany. She has been a government consultant in Germany developing a national ICZM strategy and prepared advice on transboundary MSP. She has been involved in numerous MSP projects and has co-authored a number of guiding documents such as the PlanCoast MSP Handbook, the BaltSeaPlan Vision 2030 and the PartiSEApate recommendations on transboundary MSP governance. In her academic work at Helmholtz Zentrum Geesthacht and the University of Liverpool she has pursued a particular interest in perceptions and cultural values related to the sea. She is also associated with the EU MSP platform and s.Pro GmbH.

Michael Gilek is Professor of Environmental Science at Södertörn University, Sweden. His research focuses on marine governance and MSP and how environmental and sustainability ambitions are addressed in various contexts in the Baltic Sea and beyond. Additional research interests include science-policy interactions and challenges of delay and science denial in environmental policy. He has coordinated numerous research projects, many of which have been interdisciplinary and transdisciplinary in their set-up. He has also published widely in international scientific journals and books.

Linda R. Harris is a researcher in the Institute for Coastal and Marine Research, and Department of Zoology at Nelson Mandela University in South Africa. She is a coastal ecologist specialising in sandy beaches, spatial ecology, and spatial prioritisation and planning. Her focus is on quantitative, hypothesis-driven research that contributes to real-world conservation and management initiatives (largely in South Africa). She has keen interest in developing new methods in ecology and conservation science, understanding biodiversity patterns and ecological processes and analysing animal movement patterns. She has also been involved in sea turtle conservation and research for the last decade.

Stephen Holness is a research associate at the Centre for African Conservation Ecology and the Institute for Coastal and Marine Research at Nelson Mandela University in Port Elizabeth, South Africa. He is a systematic conservation planner who focuses on protected area and land and sea spatial planning. He has particular interest in applying rapid but robust systematic assessments in areas which previously have not taken a rational evidence-based spatial approach to conservation and resource allocation issues. His background is in protected area management.

Stephen Jay is a specialist in MSP based in the University of Liverpool's Department of Geography and Planning. He has led a number of research projects investigating the uptake of marine planning, especially within Europe. He has recently been appointed as Director of the Liverpool Institute for Sustainable Coasts and Oceans (LISCO), which combines the marine and maritime expertise of the National Oceanography Centre, the University of Liverpool, and Liverpool John Moores University. He also is a key contributor to the work of the EU MSP Platform.

Hannah Jones is a member of the Atlantic area team for the European MSP Platform. She is a Research Project Manager at the University of Liverpool and has worked on numerous European-funded projects including the Celtic Seas Partnership project, the Erasmus Strategic Partnership for MSP and the Horizon 2020 funded AQUACROSS project, which aims to support EU efforts to protect aquatic biodiversity and ensure the provision of aquatic ecosystem services. She is also working on an ecosystem services evidence matrix for the European Environment Agency (EEA) European Topic Centre (ETC) for Inland Coastal and Marine waters.

Andreas Kannen is based at the Department of Human Dimensions of Coastal Areas at Helmholtz Zentrum Geesthacht. His work involves approaches for institutional and social assessments of environmental and economic changes in coastal areas, with particular focus on changes in marine use patterns, MSP and institutional aspects of marine governance. He has been involved in several international research networks and advisory groups and has served as chair of the ICES WGMPCZM from 2011 to 2016. He has also been co-chair of several related ICES workshops such as Quality Assurance of MSP and Cultural Dimensions of Ecosystem Services.

Sue Kidd is an academic and chartered town planner from the University of Liverpool's Department of Geography and Planning. She has acted as an advisor to the EU, government departments, regional and local authorities and non-governmental organisations. Sue has an interest in integrated approaches to planning, and much of her work has focused on sustainable development in coastal and marine areas. She has been at the forefront of the theory and practice of MSP and has been engaged in a variety of ways in assisting the roll-out of MSP in the Celtic Seas and wider European Seas.

Stephen Kirkman is a marine scientist for the Oceans and Coasts branch of South Africa's Department of Environmental Affairs. He has conducted considerable research in the Benguela Current LME regarding conservation status and trends of marine top predators and their use as indicators of ecosystem changes, understanding effects of climate variability and other changes on biological communities and populations, and identifying areas of conservation importance.

Michael Kull is a senior research fellow at Nordregio and holds a PhD (Political Science) from the University of Helsinki, where he also serves as an adjunct professor. He was a visiting professor at Tallinn University of Technology (public administration), at Getulio Vargas Foundation in Rio de Janeiro (environmental governance), and at Strasbourg University (political sociology). His research interests and publications include multilevel governance, rural development, and the governance of marine spaces. He coordinated/coordinates the Baltic SCOPE and Pan Baltic Scope project tasks at Nordregio. In the BONUS BASMATI project, he works on governance models/tools and stakeholder participation in MSP.

Ivana Lukic is a consultant at s.Pro and at the SUBMARINER Network for Blue Growth. She works on projects and studies related to MSP, in particular, stakeholder engagement and Blue Growth. Her professional background includes work on regional offshore renewable energy in the context of the Rhode Island Ocean Special Area Management Plan. As the leader of the World Ocean Council MSP Working Group, she also has extensive international experience with industry engagement for MSP. She holds an International Master Diploma in MSP and research experience from the Research Center in Biodiversity and Genetic Resources (Portugal).

Marina Markovic is based at UNEP/MAP PAP/RAC, Croatia. She holds two Master's degrees in Coastal Management and experience of working with different international organisations. In 2009, she joined the Priority Actions Programme/Regional Activity Centre (PAP/RAC) team, where she has been involved in a number of initiatives related to managing coastal and marine areas in the Mediterranean. In her recent work, she was part of the team developing a methodology for vulnerability/cumulative effect assessment for marine areas as well as providing technical support to national authorities in developing coastal and marine strategies and plans. She is involved in a number of PAP/RAC projects and initiatives dealing with MSP and its links to ICZM.

Benedict McAteer is a PhD researcher in the School of the Natural and Built Environment at Queen's University Belfast. His key research interests lie in the social and political realms of marine governance, as well as participative research in environmental decision-making. His research is funded by the Northern Ireland Department for Economy.

Lynne McGowan holds a Master's degree in Town Planning and a PhD in ICZM. McGowan is an honorary research associate, having previously worked at the University of Liverpool as a post-doctoral researcher. Her work has largely focused on LCIs, the transboundary dimensions of MSP and opportunities for Blue Growth. She has contributed to projects such as the Celtic Seas Partnership, and SIMCelt and is a member of the Secretariat for the Irish Sea Maritime Forum.

Emma McKinley is a Ser Cymru Research Fellow based at Cardiff University. She has an interdisciplinary background and is an experienced mixed methods researcher, applying a range of techniques across a diverse spectrum of topics including MSP, ecosystem services, public perceptions, coastal community resilience and sustainability and Blue Growth. She is the Vice-Chair of the Royal Geographic Society (with the Institute of British Geographers) Coastal and Marine Research Group, Chair of the newly formed Marine Social Sciences Network, and sits on the committee of the Society of Conservation Biology's Conservation Marketing and Engagement group.

Andrea Morf (PhD human ecology, MSc Env. Sciences) is a senior research fellow in Nordregio and scientific analyst and coordinator at the Swedish Institute for the Marine Environment (SIME). She also works as a researcher and teacher at the School of Global Studies at Gothenburg University, Sweden. Her experience includes public administration, teaching, consultancy, and scientific research and knowledge synthesis in the areas of environmental management, integrative coastal and marine management, MSP, participation, conflict and risk management, and environmental pedagogy. Geographically, her focus so far has mainly been on Sweden, the Nordic context, and Europe's coasts and marine areas.

Anne Marie O'Hagan is a senior research fellow at the Marine and Renewable Energy Ireland, Environmental Research Institute (ERI), UCC. Her background is in environmental science and law. She is currently the principal investigator on the Marine Institute-funded Navigate project on Ocean Law and Marine Governance. Previously she has worked on the legal aspects of ocean energy development, planning and management systems for offshore energy devices and their environmental effects. Her research is focused on aquaculture, MSP, environmental assessment and risk-based consenting. She represents MaREI on a number of working groups including the International Energy Agency's (IEA) Ocean Energy Systems Agreement Annex IV, ICES Working Group on Marine Renewable Energy and the all-Ireland Marine Renewables Industry Association (MRIA).

Linda O'Higgins is a senior post-doctoral researcher at the ERI, UCC, Ireland. As a marine microbiologist, her primary research focus is on lower trophic level functioning and resilience to anthropogenic pressures in natural shelf ecosystems. As a coordinator for the SIMCelt project, she works to develop public understanding of scientifically sound ecosystems-based approaches to, and their centralisation within, Ireland's MSP initiatives at local and regional scales.

Tim O'Higgins is a multidisciplinary scientist working at the interface between environmental science and policy. He is a research fellow at the Marine and Renewable Energy Ireland Research Centre in UCC. He specialises in science communication and outreach with specific experience in developing and promoting the use of practical guidelines at the science policy interface and has been a guest editor for the Journal *Ecology and Society*. He has published on a range of topics on environmental management and has written a textbook on EBM entitled *It's Not Easy Being Green: A Systems Approach to Environmental Management*. Tim has over ten years' experience in mapping and the analysis of GIS data and 15 years of experience in marine environmental management.

Cathal O'Mahony is a senior researcher in the MaREI Centre for Marine and Renewable Energy, UCC. His research has focused on integrated approaches to coastal and marine management, participatory mechanisms in coastal and marine planning. Over the last decade, he has contributed to projects which examined different aspects of MSP processes, linkages between approaches to coastal management and MSP, and science-policy interaction for marine management. He is a member of the Coastal and Marine Governance research group within the MaREI Centre and has collaborated with institutes from Europe's different regional seas.

Cristian Passarello has an interdisciplinary background and expertise in coastal and marine policy, marine biodiversity and ecosystem services. He studied across different European countries and holds an MSc in Environment and Resource Management from the Institute of Environmental Studies (IVM) of Amsterdam. During his studies, he specialised in ecosystem services and biodiversity and focused on legitimacy theory within marine protected areas as well as on policy measures for the reduction of marine plastic pollution. He currently works for s.Pro (sustainable projects) in Berlin.

Joanna Piwowarczyk (Institute of Oceanology, Polish Academy of Sciences, Sopot, Poland) is a researcher in the Marine Ecology Department of the Institute of Oceanology of the Polish Academy of Sciences. Her research focuses on participatory approaches to marine governance, stakeholders' roles in decision-making processes, and behavioural change to support sustainability in marine and coastal areas. She has been involved in several EU-funded projects and so far co-authored 15 scientific papers on interactions between people and the sea.

Dorota Pyć (PhD) is Associate Professor of Law and Head of the Maritime Law Department at the University of Gdańsk (Poland), where she specialises in the law of the sea, maritime law and marine environmental law. She chairs the Maritime Law Commission of the Polish Academy of Science. From 2013 to 2015, she served as the Undersecretary of State at the Ministry of Transport Construction and Maritime Economy and then at the Ministry of Infrastructure and Development of the Republic of Poland. She is Project Manager of SEAPLANSPACE—*Marine spatial planning instruments for sustainable marine governance*, which is funded through Interreg South Baltic 2014–2020.

Emiliano Ramieri holds a Master's Degree in Environmental Sciences. He is senior consultant at Thetis and has more than 15 years' experience in coastal and marine policies, planning and studies. He is involved in projects and initiatives dealing with MSP, ICZM, LSI and multi-use of the sea space, with specific focus on the contexts of the Mediterranean Sea and the Adriatic-Ionian Region. Competences on ICZM and MSP are coupled with a long experience on climate change vulnerability assessment and adaptation in coastal and marine systems. He has been a member of the European Topic Centre on Climate Change Impact, Vulnerability and Adaptation (ETC-CCA) of the EEA since 2011.

Fred P. Saunders works as an associate professor and researcher in the School of Natural Sciences, Technology and Environmental Studies at Södertörn University. He has held numerous academic positions and has also worked professionally for a national parks agency in Australia. His work is focused on understanding the politics and power relations of resource use and conservation in many parts of the world.

Angela Schultz-Zehden (MBA, MSc, BSc), founder and manager-owner of 's.Pro sustainable-projects GmbH' and the 'SUBMARINER network for Blue Growth EEIG', is one of the core experts in MSP and Blue Growth in Europe and has been at the forefront of developing both concepts. s.Pro sustainable-projects GmbH is one of the leading companies for MSP projects around the world. Since 2001, she has led a string of key projects (including BaltCoast, PlanCoast, BaltSeaPlan, PartiSEApate) and has authored numerous key guiding documents such as the MSP Handbook (2008), BaltSeaPlan Findings (2013), Baltic Blue Growth Studies (2013, 2017) and the LME: Learn MSP Toolkit (2018). Since 2016, she has been acting as Lead Manager of the EU MSP Assistance Mechanism and a European Commission service run by s.Pro to support EU Member States in the implementation of the MSP Directive. In addition to her work with s.Pro, she is also founder and Managing Director of the not-for-profit 'SUBMARINER network for Blue Growth EEIG', which brings together a wide range of public and private actors from all Baltic Sea Region countries to promote and implement activities necessary for using marine resources innovatively and sustainably. For her work, she has received the European Commissions' award 'Woman of the Year in the Blue Economy 2016'.

Kerry Sink is a principal scientist and the Marine Programme Manager at the South African National Biodiversity Institute. She works at the science-policy interface and is passionate about building the offshore biodiversity knowledge base in Africa. She has led the marine component of the National Biodiversity Assessment and a technical team to implement a new network of Marine Protected Areas in South Africa. She is a member of South Africa's MSP Working Group convened by the Department of Environmental Affairs.

Tim Stojanovic is Lecturer in Sustainable Development and Geography at the University of St Andrews, UK, on the council of the Scottish Oceans Institute and leads the Marine and Coastal Environment research team. His research interests include the governance, planning and management of coasts and oceans, and the social implications of change in the world's oceans.

Ralph Tafon is a PhD candidate at the School of Natural Sciences, Technology and Environmental Studies, Södertörn University, Sweden. He studies how politics, power and discourse are challenging and/or stabilising key MSP logics and decisions on the one hand, and how the same processes and mechanisms may both empower differently positioned stakeholders and transform oceans governance into a non-domination process, on the other hand.

Sarah Twomey is a human geographer specialising in Integrated Coastal and Marine Governance, based in the MaREI Centre for Marine and Renewable Energy, UCC. She is an international marine consultant and experienced group facilitator focused on the implementation of marine policy with specific expertise: developing capacity and supporting governments in Northwest Europe and Africa in statutory-based MSP, designing and applying cross-sectoral stakeholder engagement strategies.

Sarah is a PhD candidate and is in the final stages of developing and testing a meta-analytical framework to produce recommendations to resolve sectoral conflicts and optimise transboundary marine governance in disputed ecosystems.

Riku Varjopuro is Head of unit in Finnish Environment Institute, Marine Research Centre, Sustainable Use of the Marine Areas. He is a social scientist whose projects have addressed coastal management, environmental regulation of marine aquaculture, interactions between environment and fisheries and, most recently, EU marine protection policies and MSP. His research focuses on practices of decision-making and participation, and he has also taken part in and coordinated policy evaluation processes. He is coordinating a project that supports cross-border MSP in Estonian and Finnish waters.

Marta Vergílio holds a Master's degree in Environmental Engineering from the University of Aveiro and a PhD in Biology from the University of the Azores, working on the ecological structure in spatial plans for small islands. She was a member of several research projects and is currently working in the Multi-use in European seas (MUSES) and the Planning in a liquid world with tropical stakes (PADDLE) projects. Her main scientific interests are environmental planning and management, protected areas, green infrastructures, ecosystem functions and services and MSP.

Barbara Weig is a part-time consultant working for s.Pro sustainable-projects GmbH and the SUBMARINER Network for Blue Growth in several EU projects related to MSP and Blue Growth. As an external consultant, she works hand in hand with the Ministry of Economic Affairs, Transport, Employment, Technology and Tourism Schleswig-Holstein, where she is also based. Additionally, she is experienced in university teaching. She holds a diploma in Economic Geography (University of Mainz) and a PhD from Hamburg University. Her PhD thesis analysed the complexity of regional development in coastal areas.

Jacek Zaucha is Professor of Economics at the University of Gdańsk, Faculty of Economics. He is affiliated with the Maritime Institute in Gdańsk and the EU MSP Platform, and is a founder of the Institute for Development in Sopot. He is also an acting Vice-President of the Committee for Spatial Economy and Regional Planning of the Polish Academy of Sciences. His research focuses on regional and spatial development, territorialisation of development policy, MSP, spatial development and integration in the Baltic Sea Region, and territorial cohesion and EU Cohesion Policy. He is a leader of the team preparing the first maritime spatial plan for Polish marine waters.

List of Figures

Fig. 2.1	The shifting sea? Photo: Kira Gee	26
Fig. 2.2	The living sea? Photo: Kira Gee	34
Fig. 2.3	The beautiful sea? Photo: Kira Gee	40
Fig. 2.4	The romantic sea? Etching: Kira Gee	42
Fig. 3.1	Main geographic and oceanographic features of the Irish Sea. Data sources: Background bathymetry from http://www.emodnet-bathymetry.eu/. EEZ and territorial seas and boundaries from http://marineregions.org/. Location of oceanographic features re-drawn from information contained in Simpson 1974 and Simpson and Hunter 1976	48
Fig. 3.2	Map of Dublin Bay showing human modification. The hatched area shows the extent of the Bull Island prior to 1913. Effluent data and Pollution Load Index from O'Higgins and Wilson (2005) and O'Higgins (2006)	50
Fig. 3.3	Map of sectors explicitly addressed in the MSP Directive. Fishing pressure is expressed as swept area ratio. Renewable energy is shown in pale green. Main roll-on roll-off shipping routes are shown in yellow. Fishing pressure is shown in blue to purple. Designated bathing waters are shown in red. Aquaculture sites are shown as black circles. Data sources: Background bathymetry from http://www.emodnet-bathymetry.eu/. Aquaculture sites—https://atlas.marine.ie/#?c=53.9043:-15.8972:6. Shipping density and roll-on roll-off are based on data from https://data.gov.uk/. Fishing Intensity: OSPAR, https://odims.ospar.org/. Offshore wind farms: http://www.emodnet-humanactivities.eu/search-results.php?dataname=Wind+Farms+%28Polygons%29	57
Fig. 3.4	Examples of crowd-sourced and open data. (**a**) Photography User Days based on the InVEST model (Adamowicz et al. 2011).	

(**b**) Data from Dublin's traffic monitoring system showing temporal patterns in beach use for Dollymount Strand. Data source: https://data.gov.ie/dataset/volume-data-for-dublin-city-from-dublin-city-council-traffic-departments-scats-system 59

Fig. 4.1 The Benguela Current Large Marine Ecosystem (shaded grey) includes the EEZs of Angola, Namibia and western South Africa in the south-east of the Atlantic Ocean. Note that Cabinda is an exclave of Angola. World EEZ (version 6) boundaries available from http://marineregions.org/. For BCLME region, see BCC (2014) 81

Fig. 4.2 Original set and delineation of EBSAs adopted for the BCLME and SIO portion of mainland South Africa within the respective countries' EEZs. EBSAs in the surrounding high seas are excluded, except for the Benguela Upwelling System EBSA (light grey) that falls mostly within the BCLME. World EEZ (version 6) boundaries available from http://marineregions.org/. For EBSA boundaries, see https://www.cbd.int/ebsa 83

Fig. 4.3 Illustration of the advances in EBSA and MSP processes that can be achieved by SCP. (**a**) Draft revision of existing EBSAs and proposed EBSAs and ESAs in the Namibian EEZ designed using SCP; (**b**) existing and proposed MPAs as gazetted (Republic of South Africa 2016) relative to the revised and proposed EBSAs in South Africa (note that the two EBSAs in the adjacent high seas are also shown in light grey); and (**c**) example of the site-level interrogation of SCP inputs and outputs that can guide both MPA regulations and spatial management of activities in the rest of the EBSA and surrounding areas. In this case, the SCP accounted for existing protection provided by the Aliwal Shoal MPA by including that area in the new delineation of proposed MPAs (and proposed revision of EBSA boundaries). The SCP data supported fine-scale planning of the proposed zonation of the proposed MPA, with the different zones allowing only those activities that are compatible with the underlying biodiversity features. This application could be extended through all EBSAs in an MSP to minimise user-environment conflicts; beyond MPAs and EBSAs, the MSP could focus on resolving only user-user conflicts. Refer to the Government Gazette (Republic of South Africa 2016) for full details on the proposed MPAs, draft regulations and allowed activities per proposed zone. World EEZ (version 6) boundaries available from http://marineregions.org/. Proposed MPAs from Republic of South Africa (2016) 90

Fig. 5.1 Location semicircles (in sea areas) around the sea gateways of different importance (Zaucha 2018) 103

Fig. 5.2	Location zones in sea areas taking into account the phenomenon of the non-homogeneity of marine space and locations of ecosystem services and services exploiting sea abiotic assets (Zaucha 2018)	104
Fig. 5.3	Location zones in sea areas along with linear structures (Zaucha 2018)	105
Fig. 5.4	Influence of public choice on the market processes that together shape maritime spatial development (Zaucha 2018)	112
Fig. 6.1	Largest MEAs in terms of GVP and employment	129
Fig. 6.2	The fastest-growing MEAs	130
Fig. 6.3	The most promising MEAs	130
Fig. 10.1	The split ladder of participation (Source: Hurlbert and Gupta 2015, p. 105, adapted)	226
Fig. 11.1	ESaTDOR (European Seas Territorial Development and Risks) LSI typology of European maritime regions. Source: Based on University of Liverpool (2013, p. 6)	250
Fig. 11.2	Framework for addressing LSI. Source: Based on European MSP Platform (2017a)	252
Fig. 11.3	Land-sea pressure impact matrix. Source: Based on University of Liverpool (2016a, p. 16)	254
Fig. 11.4	Example of LSI value chain analysis. Source: Based on Ecorys (2012, pp. 32, 55)	256
Fig. 12.1	Links among EcAp, MSP and ICZM principles (source: UNEP(DEPI)/MED IG.23/23)	280
Fig. 12.2	Recommendations for marine and coastal planning in Boka Kotorska Bay (Montenegro) deriving from environmental vulnerability assessment (source: PAP/RAC and MSDT 2017)	285
Fig. 13.1	Widely accepted definition of stakeholders and their categories in MSP and marine governance (based on Pomeroy and Douvere 2008; Long 2012; Roxburgh et al. 2012; Flannery et al. 2015; Jay 2015, Jay et al. 2016)	297
Fig. 13.2	The Continuum of stakeholder participation (using the categories of industry, civil society and government—the latter can include different levels of authority from local, regional to national) in European MSP with various stages ranging from information provision to collaboration between all categories of stakeholders. The arrows represent the flow of information and the direction of interactions between stakeholders (Adapted from Arnstein 1969)	299
Fig. 13.3	Map of EU Member States bordering the Atlantic Ocean and the extent of their respective Exclusive Economic Zones (EEZs). Data sources: EEA and EMODNET	304
Fig. 13.4	Map of the island of Ireland. Data sources: EEA and EMODNET	305

Fig. 13.5 Map of Spain and Portugal with the Iberian coast to the west. Data sources: EEA and EMODNET 306
Fig. 13.6 Map illustrating the location of the two MSP pilot areas within the European Atlantic 307
Fig. 13.7 Phases of the MSP planning cycle illustrating how the participation of stakeholders informed the entire process of the TPEA project. (SW=Stakeholder Workshop) (Twomey and O'Mahony 2014) 308
Fig. 14.1 Types of scenario 332
Fig. 14.2 The SIMCelt possibility space 336
Fig. 14.3 Example of mapping cooperation and spatial impacts 337
Fig. 14.4 Mapping individual drivers onto the possibility space 338
Fig. 14.5 The four scenarios 342
Fig. 14.6 Future directions for selected sectors 343
Fig. 14.7 Key outcomes of the scenarios workshop 346
Fig. 15.1 Activities and outputs of the planning process steps in relation to the ten tenets 366
Fig. 18.1 Key questions in an evaluation of effectiveness and SEA. Linkages to indicators 418
Fig. 18.2 Evaluation approaches in relation to the degree and reasons of complexity of the planning contexts (Terryn et al. 2016, 1087) 422
Fig. 18.3 Different foci of evaluation in relation to steps of the spatial planning process (Carneiro 2013, 215) 425
Fig. 18.4 A scheme of a theory of change (Coryn et al. 2011, 201) 429
Fig. 18.5 Theory of change, considering factors that influence a logical sequence of events. Modified from Coryn et al. (2011) and Mayne (2012) 431
Fig. 19.1 Number of students in the Erasmus Mundus Master Course on Maritime Spatial Planning per background and edition 447
Fig. 19.2 Number of theses from the Erasmus Mundus Master Course on Maritime Spatial Planning per theme and edition 448
Fig. 19.3 Number of theses from the Erasmus Mundus Master Course on Maritime Spatial Planning per topic and edition 449
Fig. 19.4 Number of theses per MSP step and edition (MSP steps from the UNESCO "A Step-by-Step Approach toward Ecosystem-based Management" (Ehler and Douvere 2009)) 450
Fig. 19.5 Professional categories included in the pool of professionals 455
Fig. 19.6 (**a**) Main types of training attended by professionals in the last five years and (**b**) level of benefit of the training to the professionals 456
Fig. 19.7 Ranking of the importance of skills needed by an MSP practitioner, according to the expert respondents; highest values indicate more important skills 456

Fig. 19.8 Ranking of importance of backgrounds needed by an MSP practitioner, according to the expert respondents; highest values indicate more important backgrounds 458
Fig. 19.9 Levels of importance of specific training for MSP practitioners 459
Fig. 19.10 Areas of knowledge that the respondent professionals considered should be covered by specific MSP training 460
Fig. 19.11 Most effective types of MSP training according to the respondent professionals 460

List of Tables

Table 1.1	A comparison of MSP principles put forward by VASAB, HELCOM-VASAB and the European Union (Zaucha 2018)	9
Table 1.2	Types of MSP and their corresponding tasks (Zaucha 2018)	12
Table 4.1	SCP elements that link to each of the seven EBSA criteria	80
Table 7.1	Key socio-cultural concepts and their potential application in marine spatial planning	154
Table 8.1	A conceptual approach to examine social sustainability in MSP	191
Table 10.1	Participation ladders and related metaphors and important dimensions raised	223
Table 10.2	A ladder or stairway of MSP participation. Steps build on each other and do not reflect the "dark" manipulative and technocratic sides of participation	236
Table 11.1	LSI issues and arrangements in the European Atlantic	262
Table 11.2	LSI issues and arrangements in the Baltic Sea	263
Table 11.3	LSI issues and arrangements in the Black Sea	264
Table 11.4	LSI issues and arrangements in the Black Sea	264
Table 11.5	LSI issues and arrangements in the Baltic Sea	265
Table 13.1	List of key international and European instruments relevant to the European Atlantic that require stakeholder participation	300
Table 13.2	Rationale for actively involving stakeholders in MSP (NOAA Coastal Services Centre 2007; EC 2008; Ehler and Douvere 2009; EC 2014)	302
Table 13.3	Participatory mechanisms proposed by the TPEA stakeholders and justifications for their use in an engagement process	309
Table 13.4	High-level summary of the stakeholder mechanisms employed and trends in representation across different categories of stakeholders at the TPEA multi-sector workshops (2012–2014) in the northern and southern European Atlantic pilot areas	310

Table 15.1	Hoyle's quality management principles adapted to marine spatial planning from Cormier et al. 2015	360
Table 18.1	Topics and criteria of two MSP evaluation frameworks (Carneiro 2013; TPEA 2014)	427
Table 18.2	An example of a plausible theory of change in MSP	430
Table 18.3	Steps of an evaluation process	436
Table 19.1	Competences of a marine planner/marine planning team based on the MSP process	452
Table 19.2	MSP competences and knowledge and related skills	462

List of Boxes

Box 3.1	The Bull Lagoon, Dublin Bay: An Historical, Social-Ecological Perspective	50
Box 17.1	Explanation of the Two Main Concepts of the Chapter	398
Box 17.2	Processes of the BBNJ Working Group	400

1

Maritime/Marine Spatial Planning at the Interface of Research and Practice

Charles Ehler, Jacek Zaucha, and Kira Gee

1 Introduction to the Growing Practice of MSP

Marine/maritime spatial planning (MSP) is about managing the distribution of human activities in space and time to achieve ecological, economic and social objectives and outcomes. It is a political and social process informed by both the natural and social sciences. Over the last 20 years, MSP has matured from a concept to a practical approach to moving towards sustainable development in the oceans. Integrated marine spatial plans have been implemented by about 20 countries, and it is expected that by 2030, at least a third of the surface area of the world's exclusive economic zones will have government-approved marine spatial plans (Ehler 2017).

Academic interest in MSP has grown exponentially over the past decade. A November 2017 search of the "Web of Knowledge" of the Institute for Scientific Information (ISI) found over 900 scientific papers on MSP pub-

C. Ehler (✉)
Intergovernmental Oceanographic Commission (IOC) of the United Nations Educational, Scientific and Cultural Organisation (UNESCO), Paris, France
e-mail: charles.ehler@mac.com

J. Zaucha
Institute for Development and Maritime Institute, University of Gdańsk, Gdańsk, Poland

K. Gee
Human Dimensions of Coastal Areas, Helmholtz Zentrum Geesthacht, Geesthacht, Schleswig-Holstein, Germany

J. Zaucha, K. Gee (eds.), *Maritime Spatial Planning*,
https://doi.org/10.1007/978-3-319-98696-8_1

lished in international peer-reviewed journals and almost 10,000 articles in Google Scholar when searching for "marine spatial planning" alone (Santos et al. in press). According to Merrie and Olssen (2014), much of this increase in academic interest in MSP seems to have been derived from the first international workshop on MSP organised by the Intergovernmental Oceanographic Commission (IOC) of the United Nations Educational, Scientific and Cultural Organization's (UNESCO) in 2006. The 2nd International Conference on Marine/Maritime Spatial Planning, organised by IOC-UNESCO and the European Union (EU) and held in Paris in 2017, attracted over 700 applicants, reflecting the growing interest in the topic from a practical perspective.[1] Many scientific conferences such as the Annual International Council for the Exploration of the Sea (ICES) Sciences Conference are now regularly offering sessions on MSP, and a dedicated MSP Research Network has emerged, founded at the University of Liverpool with numerous subgroups focusing on specific topics.

2 The Imperative for a Multidisciplinary Approach in MSP

The imperative for employing a multidisciplinary approach stems from the nature of marine space as a multi-dimensional concept requiring insight from many scientific disciplines and types of knowledge (Ansong et al. 2018). Space is the subject of research and investigation by physicists, biologists, geographers, economists, political scientists, spatial planners, sociologists, philosophers and scholars of culture. As Faludi notes (2013, 8), "Territory is not necessarily a fixed entity enveloping all major aspects of social and political life within its boundaries. Rather, it is the object of negotiation and compromise, open to multiple interpretations." Space—and with this, marine space—must therefore be seen as a dynamic entity composed of a multitude of interrelations.

A key premise is that there is no single maritime space and that each of its delimitations is arbitrary. We are dealing with a number of overlapping sea spaces, each of which has its very own constituting relationships. For instance, many decisions concerning MSP are made in metropolitan centres far away from the coast, which is why maritime space in the regulatory dimension can have a discontinuous, network-like character. This is also the case with economic issues, illustrated by the fact that many of the economic benefits generated in the sea are realised far inland. At the same time, many traditional boundaries in the sea are currently dissolving. These include boundaries

[1] www.msp2017.paris/.

between state and economic actors, for example, or boundaries of perception, and even national and other administrative boundaries and borders. In various senses of the word, marine space is losing its traditional role as a frontier, instead becoming a contact point and boundary object for a variety of political, economic and environmental interests and views.

The definition of what exactly constitutes maritime space, and therefore the object of planning efforts, is a key challenge for spatial approaches to management. Where does the use of land affect the sea and the use of the sea affect the land? As each affects the other to some extent, is their separation in MSP not merely an artificial exercise? Similarly, ecological or cultural marine spaces may easily extend across land and water, giving rise to complex administrative and political questions. Some scholars have taken these notions even further, moving away from the consideration of "being" in the context of the sea and concentrating instead on the processual "becoming", understanding oceans as a mobile and processional entity in line with their constantly changing nature (Anderson and Peters 2014). Given the changeable character of the space, ecosystems and societies that constitute the ocean, it is therefore all the more important to consider the dynamic and process-oriented nature of MSP rather than any static outcomes it may produce. Analysing and shaping MSP in this specific context—dealing with multiple concepts of space and associated actors and stakeholders, dynamic yet also seeking stability as part of administrative processes and legislative frameworks—requires a multidisciplinary approach, drawing on the knowledge of different disciplines.

Another reason for MSP to take a multidisciplinary approach is its link to the sustainability discourse (see also Chap. 8 in this book). Seas and oceans are vulnerable ecosystems that consist of interrelated biological, chemical and physical processes. They provide humankind with numerous ecosystem goods and services, as well as abiotic benefits such as wind for offshore wind farming or navigation routes for shipping. Preserving them and securing their proper development is therefore of key importance to humankind. The ecosystem-based approach to marine spatial governance has been proposed as a central tool for achieving this overarching goal (see e.g. Carneiro 2013; Jay 2012; Douvere 2008; Gilliland and Laffoley 2008), but this poses new questions as to how to combine sustainable use of natural resources and the preservation of ecologically valuable species and habitats (Hassler et al. 2017). The central dilemma of sustainable development—how to simultaneously preserve and exploit ecosystems—also applies to the sea, perhaps even more so because in many respects, the oceans are still poorly understood.

A new sense of the ocean is also manifesting itself in the context of economics. This is exemplified in the popularity of the concept of "blue growth" most recently introduced by the Organisation for Economic Co-operation and Development (OECD) and European Commission (see also Chaps. 5 and 6 in this book). As stated by the European Commission (EC 2016, 2), for the EU and many nations around the world, oceans hold a key to the future. According to the OECD, in 2010 the blue economy resulted in global products and services worth US $1.5 trillion, or 2.5% of the world gross value added, providing 31 million jobs (OECD 2016, 13). New ways of reaping the benefits of ocean space are emerging (EC 2014b), and current uses are undergoing profound transformations (Zaucha 2009). Many emerging activities have strong transboundary dimensions, with traditional uses also facing increasing pressure to transnationalise. As a result, ever more international maritime networks are emerging—of sea basin transmission grids, shipping routes or transnational oil pipelines, in particular, in enclosed seas such as the Baltic or the Mediterranean. But the resulting pressures are rarely restricted to particular areas or sea spaces either. Even physically constrained maritime activities can have profound impacts on the surrounding maritime space—in the case of pollution or underwater noise, sometimes across very long distances. Transboundary approaches are therefore called for not only in economic development but also in resource management and protection (see also Chaps. 3 and 4 in this book).

From a research perspective, a key question is thus whether maritime development is simply the next stage in our emancipation from the geographical determinism first proposed by Ratzel (1882). The environment has long since ceased to determine human activities on land, but will we witness the same development in the sea? And what does this imply for marine governance including MSP? What will be the guiding principles of these developments, and what priorities will we set for the ocean? There are some indications that a new social awareness is emerging of the seas, driven by recent issues such as marine pollution and the powerful imagery that has now become available on life in the ocean (see also Chap. 7 in this book). Societies are reassessing the value they are placing on the ocean and are becoming more aware of its role in well-being and quality of life. Just like governance itself, the shifting values and beliefs about the ocean are also a topic for research, as these will guide our management approaches of the future—linking back to the sustainability discourse referred to earlier.

3 Origins and Development of MSP

The emergence of MSP is usually ascribed to the increasing intensity of maritime use, exceeding the capacity of marine areas to meet all demands simultaneously. Access to marine space is usually not restricted, potentially leading to overuse and conflicts. As many marine goods and services are not priced in the market, conflicts often cannot be resolved through economic analysis alone (Ehler 2017). MSP has so far developed as a new governance regime under the so-called public choice mechanism. Public choice is important as ocean space is not (yet) traded in the market, therefore requiring democratic decision-making in order to avoid risks of overexploitation (the tragedy of the commons). Usually in public choice, selected representatives are expected to make specific decisions; in this case, decisions on how marine space should be used. Public choice decision-making also entails consideration of important societal values such as biodiversity or social justice. For public choice mechanisms to work well, proper process and the involvement of all stakes are crucial (see also Chaps. 9, 10 and 13 in this book).

The idea that became MSP was initially proposed in 1976 by international and national interests in developing marine protected areas as a response to the environmental degradation of marine areas caused by human activities (Olsson et al. 2008). In the early 1980s, zoning plans were created for the Great Barrier Reef in Australia (Day 2002), although in Europe at least this did not lead to a more comprehensive debate concerning the essence of MSP. The Great Barrier Reef zoning plans also had a primary goal of marine conservation—a very different character and scope to the multiple-objective marine/maritime spatial plans currently being created in Europe and elsewhere.

The European discussion surrounding the possibility of spatial planning in the sea began in earnest around 2000, with the first mention of the term MSP in 2001 (VASAB 2001). A veritable explosion of publications occurred in the years 2007–2009, mostly composed of policy documents and handbooks indicative of a more practical engagement with MSP (EC 2007a, b, 2008b; Ehler and Douvere 2007; Acker and Hodgson 2008; Ekebom et al. 2008; Schultz-Zehden et al. 2008; Zaucha 2008; Ehler and Douvere 2009). The first academic papers concerned with the concept and practical implementation of MSP also appeared at this time (Douvere and Ehler 2008). At this point, the first maritime spatial plan in the EU had been elaborated, namely by the German federal state of Mecklenburg-Vorpommern for its territorial sea, which was approved in 2005 (Heinrichs et al. 2005).

The real breakthrough, however, came with the EU integrated maritime policy, as outlined in the Green Book (EC 2006) and Blue Book (EC 2007a) and presented, in detail, in the EU Action Plan (EC 2007b). This was followed by the publication of the "Roadmap for Maritime Spatial Planning: Achieving Common Principles in the EU" (EC 2008b), which describes MSP as "providing a framework for arbitrating between competing human activities and managing their impact on the marine environment". Its objective is described as "balancing sectoral interests and achieve sustainable use of marine resources in line with the EU Sustainable Development Strategy". It also stresses that MSP is a process involving data collection, stakeholder consultation and participatory development of a plan, including a process of monitoring and review.

The driving force of this debate were studies conducted by UNESCO,[2] VASAB[3] and the European Commission, as well as a broad range of EU-funded pilot projects on MSP in European sea basins.[4] In 2014 the MSP Directive (EC 2014a) was officially adopted. It obliges coastal EU member states to prepare maritime spatial plans by March 2021 and sets out a range of minimum requirements for these plans, such as giving consideration to land-sea interactions, considering environmental, economic, social and safety aspects, ensuring coherence between MSP and other processes such as integrated coastal management, ensuring the involvement of stakeholders and transboundary cooperation between member states and with third countries.

The exact nature of MSP, and what it can achieve as part of a legal process, continues to be contentious. This is illustrated by contrasting two views. According to a widely-used definition by UNESCO (Ehler and Douvere 2007), "MSP is a public process of analysing and allocating the spatial and temporal distribution of human activities in marine areas to achieve ecological, economic, and social objectives that are usually specified through a political process". VASAB, on the other hand, argues that MSP should be treated as a legally defined hierarchical process that aims to find a compromise between competing user needs (on the surface of the sea, in its waters and on the sea floor) in accordance with the values and objectives of a given community. These values and objectives are set out in international and state priorities and agreements. Furthermore, spatial development of marine areas should be

---

[2] Ehler, C., Douvere, F. (2009). Marine Spatial Planning: a step-by-step approach towards ecosystem-based management. Paris: Intergovernmental Oceanographic Commission, UNESCO. IOC Manual & Guides No. 53, IOCAM Dossier No. 6.

[3] Visions and strategy related to the Baltic Sea, cooperation of the Baltic ministers of spatial planning—cf. Zaucha (2013).

[4] See EU projects and initiatives on the MSP Platform website, www.msp-platform.eu.

shaped by using proper instruments, including visions and strategies (Zaucha 2008, 2). The first difference thus concerns the broader framework of the planning process itself: According to UNESCO, it is a public process, while for VASAB it is a hierarchical and legally- defined process, although the element of axiological choice—that is, the framing of objectives—is placed outside the framework of MSP. UNESCO advocates the paradigm of sustainable development; VASAB, on the other hand, does not specify this directly. UNESCO limits MSP to the allocation of marine space, and VASAB focuses on the ensuing consequences of such actions, although the UNESCO documents also indicate a similar belief in this respect (Ehler 2014).

Currently, the most frequently- used definition of MSP in Europe is the definition derived from the EU MSP Directive. This refers to "spatial planning of sea" areas as a "process through which appropriate organs of member states analyse and organise human activity in sea areas in order to achieve ecological, economic and social objectives" (EC 2014a, 140). This definition is brief and general in terms of the methods that are to be employed, and it narrows down the process to one conducted by public administrations—which seems restrictive, although it makes sense from the perspective of the Directive which persuades member states to engage in this kind of planning.

Not least through the many pilot projects and initiatives on MSP that have taken place in Europe, but also from applying MSP in national and subnational contexts, different approaches have emerged for initiating and carrying out MSP. These are dependent on the respective definition and rationale of MSP and generally vary between a more environmental or economic focus and a more strategic or conflict resolution focus. Language is an indication of these differences, expressed in the name of MSP as either *maritime spatial planning* (EC 2007a, 2014a; Acker and Hodgson 2008, 1; Schultz-Zehden et al. 2008, 11) or *marine spatial planning* (Ekebom et al. 2008, 4; Ehler and Douvere 2009, 7; Tyldesley 2004, 1; MSPP 2006, 1; IOPTF 2010, 47; SWAM 2014; Ehler 2014; Blasbjerg et al. 2009; HM Government 2011). While some use the terms interchangeably, they do seem to reflect a slightly different understanding of the significance and role of MSP. The tradition of the European Commission (which uses the term *maritime*) translates into minimising conflicts between maritime sectors, while the approach of UNESCO (which uses the term *marine*) focuses on the ecological and environmental issues encapsulated within such planning. OECD (2016, 21) proposes the following differentiation of those terms: The term[5] *maritime* should be understood as "being connected with the sea, especially in relation to seafaring, commercial

[5] "Maritime" will be understood as "being connected with the sea, especially in relation to seafaring, commercial or military activity", while "marine" will be understood as "of, found in, or produced by the sea, 'marine plants'; 'marine biology'".

or military activity", while "marine" should be understood as "of, found in, or produced by the sea, 'marine plants'; 'marine biology'". Cormier et al. (2015, 1) apply the term *maritime* in relation to economic connotations and *marine* in relation to ecological ones. The practice of planning, however, does not always confirm this semantic dichotomy as, for example, in England the spatial planning of sea areas is oriented towards economy despite the fact that the term *marine* is used (see also Douvere and Ehler 2009a, b; Jay et al. 2013).

Table 1.1 collects the most important MSP principles set out by international decision-making bodies around the same time during the initial, constituting phase of MSP: VASAB (Zaucha 2008, 4), VASAB along with Helsinki Commission (HELCOM)[6] (cf. Zaucha 2014) as well as the European Commission (EC 2008b, 10–13). Despite striking similarities, these also highlight the differing underlying values of the proposing organisations. VASAB, for example, in conjunction with HELCOM, an advocate of ecology-related issues, initiated a catalogue of principles that starts with sustainable development and the ecosystem approach. The European Commission, in contrast, as well as the original concept put forward by VASAB prioritises spatial efficiency, that is, the role of MSP in minimising of spatial conflicts.

4 Common Denominators for MSP

Despite the various differences highlighted above, there are many common denominators for MSP in Europe and also beyond. It is beyond doubt that spatial planning of sea areas:

- concerns four-dimensional maritime space (the sea surface and the lower part of troposphere above it, the water column, the sea bottom and the subsoil beneath it);
- encompasses both space and time;
- aggregates individual human preferences in relation to marine space by a process of public choice (although at times this choice is deficient in terms of uneven balance of power, see also Chap. 9 in this volume);
- concerns human activity and its consequences;

[6] HELCOM is the Helsinki Commission created as an executive body of the Convention for the Protection of the Marine Environment of the region of the Baltic Sea, drawn up in Helsinki on March 22, 1974. In 1992, the previous international agreement was replaced with a Convention for the Protection of the Marine Environment of the region of the Baltic Sea, drawn up in Helsinki on April 9, 1992.

Table 1.1 A comparison of MSP principles put forward by VASAB, HELCOM-VASAB and the European Union (Zaucha 2018)

VASAB Principles (2008)	VASAB-HELCOM Principles (2010)	EU Principles (2008)
1. MSP should have a pro-active and long-term character. Its foundation should be a vision and strategic objectives agreed upon by the Baltic countries. 2. MSP should be conducted by an institution which is independent from industry or sectoral interests. 3. MSP should be conducted on the basis of the principle of diversity, transparency and dialogue (co-participation) with stakeholders. 4. MSP should respect the ecosystem approach. 5. MSP should encompass all elements of multi-dimensional maritime space and should take into consideration the most important spatial changes which are seasonal in nature—related to time (annual cycle). 6. MSP should have adaptive and continuous character. Despite the fact that the planning process may vary in respective countries—the most important thing is preserving its continuous nature. 7. MSP should take advantage of the results of research (*evidence-based spatial planning*). 8. MSP should be coordinated transnationally—certain areas need to be planned together with neighbouring countries. 9. MSP should be constructed on the basis of the principle of a hierarchy of plans (*nested approach*). 10. It is necessary to achieve a compliance of maritime and land spatial planning. 11. MSP should be precautionary in nature. 12. In the early stages, MSP should take into consideration experiences, recommendations and information acquired from pan-Baltic organisations as well as CEMAT (Council of Europe Conference of Ministers responsible for spatial/regional planning). 13. In case there is no MSP, the decision-making process concerning management of sea areas needs to have an integrated (horizontal and vertical coordination) and transparent nature—one which encompasses dialogue with stakeholders.	1. Sustainable management 2. Ecosystem approach 3. Long-term objectives and perspective 4. The principle of caution 5. Participation and transparency 6. High-quality data and sources of information 7. International coordination and consultations 8. Coherent planning of land and sea areas 9. Planning suited to characteristic features and specific conditions of various areas 10. Continuous nature of planning	1. Employing spatial planning of sea areas depending on the area and activity type 2. Objectives in spatial planning of sea areas enabling settling disputes 3. Transparent manner of developing spatial planning of sea areas 4. Participation of the interested parties 5. Coordination in member states—simplifying the decision-making process 6. Ensuring legal effects of domestic spatial planning of sea areas 7. Cross-border cooperation and consultations 8. Monitoring and assessing components of the planning process 9. Achieving coherence between spatial management plans concerning the land and the sea—relations with integrated Coastal Zone Management (ICZM) 10. Reliable databases and scientific foundations

- is integrated (at least by definition);
- refers to the sea as a functional ecosystem;
- influences market processes at sea;
- requires transnational coordination within sea basins;
- requires coordination and connection with spatial planning on land;
- is conducted in a continuous and adaptive manner encompassing monitoring and evaluation; and
- employs—as best as it can—available research and information.

The most important differences of opinions concerning the essence and methods of MSP relate to:

- the degree of generality/specificity and the stage during which the aggregation of individual preferences into collective preferences occurs;
- relations between planning and management (in literature on the subject, e.g. Griffin (2006), planning is perceived as a function of management; however, some identify planning with management, for example, Tyldesley (2004, 4) or Ehler and Douvere (2009), who directly speak of a marine spatial management plan—hence the numerous discussions concerning relations of MSP and coastal zone management);
- the degree of specificity, integration and legal power of maritime spatial plans (e.g. in Norway marine spatial plans do not constitute a binding law but fulfil regulatory functions through the existing responsibilities of competent authorities);
- the degree of specificity and scope of data and information required for MSP; and
- the scope and methods of mobilising (engaging) stakeholders in the MSP process.

This last point relates to the fact that in research at least, the view has come to dominate that the process can be more important than the output of MSP (Payne et al. 2011). While some still regard MSP as simple regulatory plans that form a framework for administrative decision-making, it is increasingly evident that modern governance processes require more sophisticated techniques than administrative solutions and top-down directives as evidenced by terrestrial experiences (Faludi 2010, 21–23; 2015, 17; Healey 2000, 112–113; 2010, 226–227; Dühr et al. 2010, 102–111). Emphasising the process-based dimension means to emphasise (changing) social preferences regarding spatial management—a view that is also prevalent in the community of MSP practitioners. For example, Tomas Andersson, spatial planner and pioneer of MSP in Sweden,[7] describes MSP as "*a process to prepare society to meet an uncertain future and try to guide the development of space (and the use of resources) in a desirable direction*". Process-based planning often makes use of vision-based tools or employs scenarios, and uses the degree of mobilisation as a measure of success. Planning with an emphasis on the process, however, requires time and human resources as well as patience (e.g. Morf et al. forthcoming), not least because an iterative, adaptive approach might sometimes be interpreted as a lack of progress. In addition, it also requires intuition and experience on the part of planners—and a fine sense of timing, for example, when to present results to political decision-makers or when to begin a new planning cycle (see also Chap. 19 in this book).

Table 1.2 is a general typology of different MSP approaches, corresponding objectives and the types of planning documents that might be produced as a result. It highlights that MSP has numerous other methods of implementing collective choices for marine space apart from regulatory plans—including a scenario (see also Chap. 14 in this book), a vision or another form of capturing spatial arrangements over time.

5 Ten Common Misunderstandings and Key Areas for Future Research

Based on the above, and reflecting many conversations and discussions with researchers and practitioners on MSP, we end by listing some common misunderstandings with regard to MSP. We argue that these are also critical fields of research on MSP, although many other research topics could probably be added.

First, despite its origins within the field of conservation, MSP is not an exclusive domain of environmental protection. The ecosystem approach, one of the principles of MSP in the EU, emphasises the importance of achieving good ecological status for the sea. But the ecosystem approach also seeks to secure "permanent use of sea resources and services by present and future generations" (EC 2008a, art. 1). Humans constitute an integral part of the ecosystem; therefore, it is necessary to integrate protection and use. This comes back to issues related to the guiding principle of MSP—what are the core values it is attempting to promote, and how can sustainability be translated into practice?

[7] *Speech during the 2016 Baltic Days in St. Petersburg, pers. comm.*

Table 1.2 Types of MSP and their corresponding tasks (Zaucha 2018)

Type of planning	Key objective	Types of planning documents
Information-based planning	The principal objective of this type of MSP is to identify marine resources (including their mapping), and to assess their robustness (sensitivity analysis) along with associated pressures and demands. Conflict analysis is also part of the objectives, as well as risk analysis (planning actions in the event of threats).	• Studies • Spatial analyses • Reports • Conflicts matrixes • Risk maps
Strategic and vision-based (indicative) planning	The principal objective is to inspire other actors that are shaping spatial development through their actions. Spatial planning has no or insufficient authority over these actors and is therefore unable to enforce desirable actions.	• Pilot plans • Scenarios • Visions • Strategies • Other policy-related documents of the indicative character
Regulatory planning	Planners have causative power resulting from legal regulations or economic instruments and can enforce desirable actions. The plan is a means of implementing publicly agreed objectives and priorities—for example, in relation to using marine resources, environmental protection or limiting conflicts.	• Plans in the character of a local law • Other binding documents of this type

Second, continuing this line of thinking, MSP is not a universal remedy. Some challenges require solutions other than spatial planning. For example, in relation to eutrophication, spatial planning can only contribute to solutions but not resolve the original issue. Interactions and cooperation with other processes and policy areas are therefore essential if MSP is to play its part in the greater scheme of ocean governance. This also relates to the transboundary nature of MSP and its reliance on various forms of integration, resulting in a coherent approach to management supported by multiple policies (see later in this chapter).

Third, MSP does not—and should not—replace sectoral planning and programming. Space is an integrative concept, but like its terrestrial counterpart, MSP requires understanding of the various sectoral interests in this space. Understanding sectoral interests also extends into the social sphere and the preferences of society at large, such as aesthetic preferences with respect to landscapes. Non-material values and preferences need to be revealed and understood in order to be placed alongside the more commercial interests—requiring the engagement of social sciences and researchers to ensure they can be fully integrated in the MSP process.

Fourth, MSP should not be confused with licensing, permitting, or similar processes of granting permission to use marine space. Even where a maritime spatial plan exists, permits and the associated processes of environmental impact assessment are a key requirement for using that space responsibly. This is related to the fact that even today, the marine environment and human impacts of use are poorly understood. Even where human interference with marine ecosystems seems acceptable, this should be verified through detailed analyses and research.

Fifth, MSP is not a one-time choice. Social preferences regarding maritime space are subject to dynamic changes, suggesting the need for continuous reinterpretation. Drawing up and approving a plan is merely a precursor for another plan that builds on the experiences of the first. Planning processes are therefore also learning processes, and MSP institutions must see themselves as learning institutions. It is encouraging that this fact is being recognised by countries engaged in revisions of their first marine/maritime spatial plans, although it must also be noted that the general thrust of MSP is unlikely to change in revisions of a plan, suggesting a degree of path dependency. Germany, for example, commencing its revision of the 2009 maritime spatial plan for the Exclusive Economic Zone (EEZ) in 2019, is determined to improve on the first planning process, which to a large extent was still experimental. In this context, the lack of a definitive plan is not a failure when a diligent planning process has taken place that changed stakeholder awareness and revealed tacit knowledge about the sea.

Sixth, MSP is more than drawing lines on a map. As stated in the beginning of this chapter, planning encompasses all the various actions that lead to rational spatial development. Spatial allocation is an important activity but only one among many. A vital element of MSP, for example, is defining the rules and principles to be used in location processes, as well as the rules for the negotiation processes that constitute MSP. A distinguishing feature of MSP in this context is its integrated approach, that is, its ability to analyse correlations, such as the mutual interactions of various (future) sea uses, cumulative pressures and their impacts on the functioning of the marine ecosystem. This in turn requires good collaboration with science and research to enable MSP to be evidence-led, as well as to recognise existing uncertainties and deal with them accordingly (e.g. through the precautionary principle or by making them transparent).

Seventh, MSP is conducted at multiple scales, encompassing both horizontal and vertical dimensions. A distinguishing feature of the sea is its higher geographical continuity of ecosystems, requiring a coherent approach to management across administrative boundaries. This may be achieved by a set of shared principles concerning the organisation of the MSP process, or by agreeing on shared objectives for ocean management, or a nested hierarchy of plans, for example. Yet achieving coherence is no easy feat, especially when maritime spatial plans are also influenced by regional and local plans or other strategic documents. Challenges related to integration have recently been evaluated by the BaltSpace project, focusing on policy and sector, stakeholder and knowledge integration (Saunders et al. 2016). Achieving integration in all these dimensions requires collaboration and coordination—and with this, understanding of the specific enablers and barriers to both. This is particularly important in international contexts as planning cultures may differ across borders but also within countries where institutional cultures and the respective value bases may also diverge.

Eighth, MSP is not an ideal process. It is a social process, and as such its benefits may diverge from what is expected. MSP might be a source of considerable drawbacks—for example, if it is appropriated by well-organised powerful interest groups or if risks associated with the MSP process are not taken into account (see also Chap. 15 in this book). The social dimensions of MSP are currently the subject of one of the most intense and heated scientific debates in MSP research, and it has been pointed out that rather than a "rational" process, MSP is in fact a highly politicised process. Rather than the ecological or economic results of the plan (see Chap. 9 in this book; Boucquey et al. 2016), the focus is therefore increasingly shifting to the sociopolitical results of the MSP process. This involves aspects such as power and the distribution of benefits achieved by the plan (Flannery et al. 2016). There also appears a postulate of phronetic evaluation of MSP (Kidd and Ellis 2012; Flyvbjerg 2004; Kidd and Shaw 2014). Therefore, MSP cannot be treated uncritically as an unquestioningly positive process. Continuous monitoring and evaluation are required not only of economic and ecological results but also of the social effects of this process. In countries only beginning MSP efforts, it is essential to develop awareness and build capacity among stakeholders to prevent MSP from being dominated by the strongest interest groups.

Ninth, there is an issue regarding the efficiency of MSP. Generally, efficiency is brought down to the design and implementation of the planning process which is supposed to lead to "balanced" outcomes (cf. e.g. Saunders et al. 2016). Mistakes in process design and implementation can lead to social resistance and a lack of legitimacy of the planning process, thereby reducing process efficiency. At the same time, mistakes and extra time spent on a process can be instrumental in promoting learning, and failures can ultimately act to improve relations between planners and stakeholders—forcing both sides to approach each other, forcing compromise and forcing both sides to engage with each other's viewpoints. This takes time and continuity—for example, in terms of staffing, in order to build the required level of trust. Engaging with efficiency also takes evaluation and a critical assessment of the process (see Chap. 18 in this book).

And last not least, there is the issue of working with stakeholders in a meaningful way. Many experiences have shown that tokenistic involvement will not lead to the desired results, but that long-term and honest commitment is necessary. Research that contributes to understanding stakeholders, their core values and motivations, as well as mechanisms for successful process design, is therefore of key importance for MSP in the future.

6 Conclusion

This chapter has attempted to sketch the development of MSP from its initial conceptualisation as a zoning tool for marine conservation to a multi-dimensional approach to spatial marine governance. MSP is continuing to develop as a practice around the world, although the number of initiatives that have reached the implementation stage is still comparatively low: Out of a total of 60 MSP initiatives in 2017, 37% were at the pre-planning stage,

33% at the plan preparation stage and 19% had an approved plan. Eleven per cent have gone as far as revising their plans, with some countries like the Netherlands now in their third cycle of MSP (Ehler 2017). Many plans cover the EEZ, sometimes encompassing large sea areas as a result. Conflicts among uses still constitutes the most common driving force for MSP, closely followed by the need for a more integrated approach and concerns about marine conservation and new and emerging uses, indicating that strategic use of MSP as part of targeted development planning for the sea is still less well developed. The greater proportion of maritime spatial plans is also advisory rather than regulatory, although many rely on other authorities for the implementation of management plans (Ehler 2017).

Given the current level of interest in MSP and the political support it has in many regions of the world, the number of countries engaging with MSP is set to increase. It has been estimated that by 2030, a third of the world's EEZs will be covered by government-approved maritime spatial plans (Ehler 2017)—with the possibility of extending even further into areas beyond national jurisdiction (see also Chap. 17 in this book). Views on whether this is desirable or not are likely to vary, not least in line with different interpretations of MSP as a concept.

What is certain is that MSP will continue to face challenges. At a practical implementation level, a key challenge is that MSP requires authority in order to be effective, which takes time to establish. Added to this is the fact that MSP is rarely free, but requires the allocation of (often scarce) government funds. Moreover, MSP usually requires painful decisions related to various trade-offs and this might decrease its acceptance. Win-win situations are rare in contemporary MSP. Methodological challenges are likely to arise from different practices of MSP, not least from evaluating them in order to assess the actual benefits of MSP. Also planning culture and experience varies among countries. Is MSP worth the effort, and what kind of MSP yields which benefits to whom, how and when? Is it possible to generalise or is effective MSP always context-specific? Another challenge is that MSP does not occur in isolation but requires transnational cooperation—which may not be an easy feat in times of increasing international strife and competition. Climate change is likely to pose its own challenges, related for example to adaptiveness of marine/maritime spatial plans but also linked to geostrategic issues, such as exploitation of the Arctic. Interdisciplinary and transdisciplinary efforts are required for successfully addressing these and other issues, requiring the expertise of a wide range of scientists and practitioners today and in the next generation.

Acknowledgements The authors are grateful to the project "Economy of maritime space" funded by the Polish National Science Centre for contributing the Open Access fee for this chapter and facilitating our discussions and preparation of the book. The Maritime Institute in Gdansk also partially paid the Open Access fee.

References

Acker, H., & Hodgson, S. (2008). *European Commission Legal Aspects of Maritime Spatial Planning*. Final Report to DG Maritime Affairs & Fisheries. MRAG, Framework Service Contract, No. FISH/2006/09—LOT2.

Anderson, J., & Peters, J. K. (2014). *Water Worlds: Human Geographies of the Ocean*. Farnham, Surrey and Burlington, VT: Ashgate Publishing, 214 pp.

Ansong, J., Calado, H., & Gilliland, P. (2018). A Multifaceted Approach to Building Capacity for Marine/Maritime Spatial Planning: Review and Lessons from Recent Initiatives. *Marine Policy* (Submitted).

Blasbjerg, M., Pawlak, J. F., Sørensen, T. K., & Vestergaard, O. (2009). *Marine Spatial Planning in the Nordic Region Principles, Perspectives and Opportunities*. Copenhagen: Nordic Council of Ministers.

Boucquey, N., Fairbanks, L., Martin, K. S., Campbell, L. M., & McCay, B. (2016). The Ontological Politics of Marine Spatial Planning: Assembling the Ocean and Shaping the Capacities of 'Community' and 'Environment'. *Geoforum, 75*, 1–11.

Carneiro, G. (2013). Evaluation of Marine Spatial Planning. *Marine Policy, 37*, 214–229.

Cormier, R., Kannen, A., Elliott, M., & Hall, P. (2015). *Marine Spatial Planning Quality Management System*. ICES Cooperative Research Report 327. Retrieved December 31, 2016, from http://www.ices.dk/sites/pub/publication%20reports/cooperative%20research%20report%20(crr)/crr327/marine%20spatial%20planning%20quality%20management%20system%20crr%20327.pdf.

Day, J. C. (2002). Zoning-Lessons from the Great Barrier Reef Marine Park. *Ocean & Coastal Management, 45*(2–3), 139–156.

Douvere, F. (2008). The Importance of Marine Spatial Planning in Advancing Ecosystem-Based Sea Use Management. *Marine Policy, 32*, 762–767.

Douvere, F., & Ehler, C. (Eds.). (2008). Special Issue on the Role of Marine Spatial Planning in Implementing Ecosystem-Based Sea Use Management. *Marine Policy, 32*(5), 759–843.

Douvere, F., & Ehler, C. (2009a). Ecosystem-based Marine Spatial Management: An Evolving Paradigm for the Management of Coastal and Marine Places. *Ocean Yearbook, 23*, 1–26.

Douvere, F., & Ehler, C. (2009b). New Perspectives on Sea Use Management: Initial Findings from European Experience with Marine Spatial Planning. *Journal of Environmental Management, 90*, 77–88.

Dühr, S., Colomb, C., & Nadin, F. (2010). *European Spatial Planning and Territorial Cooperation*. London and New York: Routledge, 452 pp.

EC. (2006). Green Paper Towards a Future Maritime Policy for the Union: A European Vision for the Oceans and Seas. Brussels, 7.6.2006 COM(2006), 275 final, Vol. II.

EC. (2007a). Communication from the Commission to the European Parliament, the Council, the European Economic and Social Committee and the Committee of the Regions. An Integrated Maritime Policy for the European Union. Brussels, 10.10.2007, COM(2007), 575 final.

EC. (2007b). Accompanying Document to the Communication from the Commission to the European Parliament, the Council, the European Economic and Social Committee and the Committee of the Regions. An Integrated Maritime Policy for the European Union. Brussels, 10.10.2007, SEC(2007) 1278.

EC. (2008a). Directive 2008/56/EC of the European Parliament and of the Council of 17 June 2008 Establishing a Framework for Community Action in the Field of Marine Environmental Policy Marine Strategy Framework Directive. Official Journal of the European Union, L 164.

EC. (2008b). Roadmap for Maritime Spatial Planning: Achieving Common Principles in the EU. Brussels, 25.11.2008, COM(208) 791.

EC. (2014a). Directive 2014/89/EU of the European Parliament and of the Council of 23 July 2014 Establishing a Framework for Maritime Spatial Planning. Official Journal of the European Union, L 257/135.

EC. (2014b). Commission Staff Working Document A Sustainable Blue Growth Agenda for the Baltic Sea Region. Brussels, 16.5.2014 SWD(2014) 167 final.

EC. (2016). Communication from the Commission to the European Parliament, the Council, the European Economic and Social Committee and the Committee of the Region. International Ocean Governance: An Agenda for the Future of Our Oceans. Brussels, 10.11.2016, JOIN(2016) 49 final.

Ehler, C. (2014). *A Guide to Evaluating Marine Spatial Plans.* IOC Manuals and Guides No. 70, ICAM Dossier 8. Paris: UNESCO, Intergovernmental Oceanographic Commission UNESCO IOC, 96 pp.

Ehler, C. (2017). World-Wide Status and Trends of Maritime/Marine Spatial Planning. Presented at the 2nd International Conference on Marine/Maritime Spatial Planning, UNESCO, Paris.

Ehler, C., & Douvere, F. (2007). *Visions for a Sea Change.* Technical Report of the International Workshop on Marine Spatial Planning, November 8–10, 2006. Paris: Intergovernmental Oceanographic Commission, UNESCO. IOC Manual & Guides No. 46, ICAM Dossier 3.

Ehler, C., & Douvere, F. (2009). *Maritime Spatial Planning. A Step-by Step Approach. Toward Ecosystem-based Management.* Manual and Guides No 153, ICAM Dossier No. 6. Paris: Intergovernmental Oceanographic Commission UNESCO IOC, 99 pp.

Ekebom, J., Jäänheimo, J., & Reker, J. (Eds.). (2008). Towards Marine Spatial Planning in the Baltic Sea BALANCE WP4 Final Report.

Faludi, A. (2010). Beyond Lisbon: Soft European Spatial Planning. *disP—The Planning Review, 46*(182), 14–24.

Faludi, A. (2013). Territory: An Unknown Quantity in Debates on Territorial Cohesion. *European Journal of Spatial Development*. Refereed Article No. 51.

Faludi, A. (2015). Place Is a No-Man's Land. *Geographia Polonica, 88*(1), 5–20.

Flannery, W., Ellis, G., Ellis, G., Flannery, W., Nursey-Bray, M., van Tatenhove, J. P., Kelly, C., Coffen-Smout, S., Fairgrieve, R., Knol, M., & Jentoft, S. (2016). Exploring the Winners and Losers of Marine Environmental Governance/Marine Spatial Planning: Cui bono?/"More than Fishy Business": Epistemology, Integration and Conflict in Marine Spatial Planning/Marine Spatial Planning: Power and Scaping/Surely Not All Planning Is Evil?/Marine Spatial Planning: A Canadian Perspective/Maritime Spatial Planning—"ad utilitatem omnium"/ Marine Spatial Planning: "It Is Better to Be on the Train than Being Hit by It"/ Reflections from the Perspective of Recreational Anglers.... *Planning Theory and Practice, 17*, 121–151.

Flyvbjerg, B. (2004). Phronetic Planning Research: Theoretical and Methodological Reflections. *Planning Theory and Practice, 5*(3), 283–306.

Gilliland, P. M., & Laffoley, D. (2008). Key Elements and Steps in the Process of Developing Ecosystem-Based Marine Spatial Planning. *Marine Policy, 32*, 787–796.

Griffin, R. W. (2006). *Fundamentals of Management: Core Concepts and Applications*. Boston: Houghton Mifflin, 559 pp.

Hassler, B., Blažauskas, N., Gee, K., Gilek, M., Janßen, H., Luttmann, A., Morf, A., Piwowarczyk, J., Saunders, F., Stalmokaite, I., Strand, H., & Zaucha, J. (2017). *BONUS BALTSPACE D 2:2: Ambitions and Realities in Baltic Sea Marine Spatial Planning and the Ecosystem Approach: Policy and Sector Coordination in Promotion of Regional Integration*. Huddinge: Södertörn University.

Healey, P. (2000). Institutionalist Analysis, Communicative Planning, and Shaping Places. *Journal of Planning Education and Research, 19*(2), 111–121.

Healey, P. (2010). *Making Better Places—The Planning Project in the Twenty-First Century*. London: Palgrave.

Heinrichs, B., Schultz-Zehden, A., & Toben, S. (2005). The Interreg III B Balt Coast Project. A Pilot Initiative on Integrated Coastal Zone Management in the Baltic Sea (2002–2005), Coastline Reports 2005 no. 5.

HM Government. (2011). *UK Marine Policy Statement*. London: The Stationery Office Limited.

IOPTF. (2010). *Final Recommendations of The Interagency Ocean Policy Task Force*. Washington, DC: The White House Council on Environmental Quality.

Jay, S. (2012). Marine Space: Manoeuvring Towards a Relational Understanding. *Journal of Environmental Policy and Planning, 14*(1), 81–96.

Jay, S., Flannery, W., Vince, J., Liu, W. H., Xue, J., Matczak, M., Zaucha, J., Janssen, H., van Tatenhove, J., Toonen, H., Morf, A., Olsen, E., Vivero, J., Mateos, J., Calado, H., Duff, J., & Dean, H. (2013). International Progress in Marine Spatial Planning. In A. Chircop, S. Coffen-Smout, & M. McConnel (Eds.), *Ocean Yearbook 27* (pp. 171–212). Leiden: Martinus Nijhoff Publishers.

Kidd, S., & Ellis, G. (2012). From the Land to Sea and Back Again? Using Terrestrial Planning to Understand the Process of Marine Spatial Planning. *Journal of Environmental Policy and Planning, 14*(1), 49–66.

Kidd, S., & Shaw, D. (2014). The Social and Political Realities of Marine Spatial Planning: Some Land-Based Reflections. *ICES Journal of Marine Science, 71*(7), 1535–1541.

Merrie, A., & Olssen, P. (2014). An Innovation and Agency Perspective on the Emergence and Spread of Marine Spatial Planning. *Marine Policy, 44*, 366–374.

Morf, A., Strand, H., Gee, K., Gilek, M., Janssen, H., Hassler, B., Luttman, A., Piwowarzyk, J., Saunders, F., Stalmokaite, I., & Zaucha, J. (forthcoming). Peer Reviewed Version of BONUS BALTSPACE Deliverable 2.3 Report, Swedish Institute for the Marine Environment Report Series, Swedish Institute for the Marine Environment, University of Gothenburg, Gothenburg, Sweden.

MSPP. (2006). Marine Spatial Planning Pilot. Final Report. MSPP Consortium. Retrieved May 15, 2018, from http://www.abpmer.net/mspp/docs/finals/MSPFinal_report.pdf.

OECD. (2016). *The Ocean Economy in 2030*. Paris: OECD Publishing.

Olsson, P., Folke, C., & Hughes, T. P. (2008). Navigating the Transition to Ecosystem-Based Management of the Great Barrier Reef, Australia. *Proceedings of the National Academy of Science of the U S A, 105*(28), 9489–9494.

Payne, I., Tindall, C., Hodgson, S., & Harris, C. (2011). Comparison of National Maritime Spatial Planning (MSP) Regimes Across EU. IN: Comparative Analysis of Maritime Spatial Planning (MSP) Regimes, Barriers and Obstacles, Good Practices and National Policy Recommendations. "Seanergy 2020". Retrieved December 29, 2016, from http://www.seanergy2020.eu/wp-content/uploads/2011/07/110707_final-deliverable-d.2.31.pdf.

Ratzel, F. (1882). *Antropho-Geographie*. Stuttgart: J. Engelhorn, 526 pp.

Santos, C. F., Ehler, C. N., Agardy, T., Andrade, F., Orbach, M. K., & Crowder, L. B. (in press). Marine Spatial Planning. In C. Sheppard (Ed.), *World Seas: An Environmental Evaluation* (Vol. III). Elsevier.

Saunders, F., Gilek, M., Gee, K., Göke, C., Hassler, B., Lenninger, P., Luttmann, A., Morf, A., Piwowarczyk, J., Schiele, K., Stalmokaite, I., Strand, H., Tafon, R., & Zaucha, J. (2016). *Exploring Possibilities and Challenges for MSP Integration in the Baltic Sea*. Stockholm: Bonus BaltSpace. Retrieved December 7, 2017, from http://www.baltspace.eu/index.php/published-reports.

Schultz-Zehden, A., Gee, K., & Scibior, K. (2008). *Handbook on Integrated Maritime Spatial Planning*. Berlin: S.PRO, 98 pp.

SWAM. (2014). *Marine Spatial Planning—Current Status 2014. National Planning in Sweden's Territorial Waters and Exclusive Economic Zone (EEZ)*. Gothenburg: The Swedish Agency for Marine and Water Management.

Tyldesley, D. (2004). *Coastal and Marine Spatial Planning Framework for the Irish Sea Pilot Project*. London: Defra. Retrieved from http://jncc.defra.gov.uk/pdf/Tyldesley%20Marine%20spatial%20planning.pdf.

VASAB. (2001). *Wismar Declaration and VASAB 2010+ Spatial Development Action Programme*. Wismar: VASAB 2010.

Zaucha, J. (2008). *Sea Use Planning and ICZM Input to the Long Term Spatial Development Perspective*. Final Report from Working Group 3. Riga: Vision and Strategies Around the Baltic Sea. Retrieved May 27, 2018, from http://www.vasab.org/east-west-window/documents.html.

Zaucha, J. (2009). The Marine Economy in the Face of New Development Trends (Spatial Aspects). In T. Markowski (Ed.), *The Polish Spatial Development Concept Versus European Vision of Spatial Development Perspectives* (pp. 134–156). Warsaw: Polish Academy of Science, Committee for Spatial Economy and Regional Planning.

Zaucha, J. (2013). Programming Development of the Baltic Sea Region. In T. Kudłacz & D. Woźniak (Eds.), *Programming Regional Development in Poland. Theory and Practice*. Studies of Polish Academy of Science, Committee for Spatial Economy and Regional Planning, *35*: 177–190.

Zaucha, J. (2014). *The Key to Governing the Fragile Baltic Sea. Maritime Spatial Planning in the Baltic Sea Region and Way Forward*. Riga: VASAB. 110 pp.

Zaucha, J. (2018). *Gospodarowanie przestrzenią morską*. Warszawa: Instytut Rozowju i Sedno, 408 pp.

2

The Ocean Perspective

Kira Gee

1 Introduction

> Who has known the ocean? Neither you nor I, with our earth-bound senses, know the foam and surge of the tide that beats over the crab hiding under the seaweed of his tide pool home; or the lilt of the long, slow swells of mid-ocean, where shoals of wandering fish prey and are preyed upon, and the dolphin breaks the waves to breathe the upper atmosphere. (…) To sense this world of waters known to the creatures of the sea we must shed our human perceptions of length and breadth and time and place, and enter vicariously into a universe of all-pervading water. For to the sea's children nothing is so important as the fluidity of their world. (Carson 1937)

In the Western world, as elsewhere, our human history is closely interwoven with the sea. Human relationships with the sea have been considered from angles as different as philosophy, geography, military studies, navigation and seafaring, natural sciences, political sciences, and social sciences and have featured in the various fields of art, literature, and music for centuries if not millennia. Planning is a relative newcomer in this long list of disciplines, bringing its very own perspectives and epistemologies. These in turn are driven—in part at least—by established notions such as the ability to delineate administrative

K. Gee (✉)
Human Dimensions of Coastal Areas, Helmholtz Zentrum Geesthacht, Geesthacht, Schleswig-Holstein, Germany
e-mail: kira.gee@gmx.de

J. Zaucha, K. Gee (eds.), *Maritime Spatial Planning*,
https://doi.org/10.1007/978-3-319-98696-8_2

boundaries in the sea, as well as other perspectives that enable the sea to be subjected to a planning rationale in the first place.

How we think of the sea, and how we came to think of the sea in spatial planning terms, is the main focus of this chapter. It does not seek to present a comprehensive overview of man's relationship with the sea—this would be the subject of a book in its own right. Rather, it is selective in highlighting key perspectives that have developed over time and that still determine how we think of the ocean in our Western world—and I do emphasise that this is a Western perspective. Our ways of thinking about the sea influence how we choose to manage the ocean and what limits current approaches to management, and they are also important for understanding some of the conflicts this causes in marine management and governance today.

So what do we see when we look out to the sea? What do we mean when we say "ocean", and how are we in the Western world currently conceptualising the ocean? This chapter aims to draw out some fundamental lines of thought and show how these have shifted over time in response to certain driving forces. One perspective is that of differing attempts at understanding, delineating, and ultimately exploiting the ocean, leading to the duality between an industrial, exploitative perspective (often labelled "blue growth") on the one hand and the environmental perspective on the other. But oceans are also social spaces, communication spaces, and cultural spaces—and they play an important role in how we as humans understand ourselves as communities and individuals. The sections are in no particular order of importance. Section 2 outlines some of the fundamental challenges we have as humans in understanding a watery world so very different from our own. Section 3 discusses endeavours to enclose the ocean as part of nation's territory. Section 4 moves on to scientific attempts at making the ocean more amenable to exploitation, leading on to a discussion of some current policy lines within the European Union (EU) related to the oceans. Section 5 considers the ocean as an aesthetic and affective space. The chapter closes by offering some thoughts on what this might imply for maritime spatial planning (MSP).

2 Grasping the Ungraspable

Water is the cradle of philosophy, and according to Thales of Milet (around 600 BC), water is the cradle of all things. He considered the earth to float on water, and also saw water as the *arche*, the element and the first principle of existing things—in other words, the origin of all things to which all things must return. It has been suggested that Thales' philosophy may have been influenced by his life on the coast and first-hand observations of the ocean

(Scholtz 2016), a fact which may also be true of another early philosopher and pupil of Thales, Anaximander, who came to consider water as the origin of life. In his philosophy, which was still founded on the idea of a first and all-encompassing principle, water and earth produced fish through heat, and independent humans initially developed in these fish-like beings—a transition that took place in the sea. Heraclitus (about 535–475 BC) was first to speak more specifically of the transition between the elements, encapsulated in the principle of "panta rhei", or everything flows, describing the idea that the cosmos itself is engaged in a permanent circular movement: earth becomes water, water becomes earth, and in this permanent transition and change, everything is in fact one. Heraclitus' world is like the sea, a world that never stands still yet one that is indestructible, often encapsulated in the phrase "no-one steps in the same river twice". In some way, Heraclitus could be said to pre-empt a more modern take on the geography of the oceans, a view of the oceans as a "dynamic system that is perpetually being remade" (Steinberg 2014), or a system that is less an object but a constant state of becoming (Ryan 2012). Everyday language has also taken up metaphors of the sea to symbolise change, such as stemming the tide of something, a wave of innovation, or a flood of new ideas.

Building on these philosophical considerations, is it possible to approach the nature of the ocean more closely, from within so to speak? How can this exceptionally ungraspable space (Steinberg 2014, p. xvi) be grasped after all? In the same piece, Steinberg (2014) summarises some of the inherent difficulties we humans face when encountering and describing the ocean. One is that human ocean experiences are always indirect, requiring mediation by a range of tools, not least to enable some form of immersion in the water. As a result of these physical barriers, we can never truly be "of the ocean". The ocean must therefore be regarded as the "other", something that is not terra firma and something that is always to some degree unknowable. Seas and oceans thus become an object, "a substance, a surface of difference" in a land-ocean binary. Much of our human perception and representation, including artistic representation, has reproduced that difference—in science, for example, by restricting ourselves to analysing particular ocean uses, or the mobilities of species, or experiences of those gazing at the ocean, rather than looking at the entirety of experiences and the co-construction of the ocean by humans and the water itself. Another difficulty, also argued by Steinberg (2014), is that locations in the ocean are difficult to grasp. Maps and planning documents suggest a false sense of the static, obscuring the continuous movement of the water that makes it impossible to truly locate a point in the ocean as a permanent material place. Returning to Heraclitus at this point, the ocean is constant

Fig. 2.1 The shifting sea? Photo: Kira Gee

flux—requiring us to re-think conceptions of ocean space in terms of both geophysical and social processes. This may also have implications for MSP: What are we actually able to locate and own in the sea? What kind of map do we require for doing so? And for what purpose? (Fig. 2.1).

3 The Territorial Perspective

3.1 Mare Liberum and Mare Clausum

The question of what can be localised in the sea is one that also arose around 1600, in the context of a dispute concerning ownership in the sea. At that time, physical and moral perspectives of the sea were still interlinked (Scholtz 2016), and so the question of whether property was possible in the sea was inevitably also a moral one, touching upon the morality of law and freedom and early expressions of international law (see also Chap. 16 in this volume).

The discussion began with a dispute over who ruled the sea routes from Europe to India. Exploration had become common at the time and mainly served the acquisition of property, both in terms of tradable goods such as spices and in terms of territory which was colonised and appropriated by subjugating the indigenous peoples. This gave rise to new forms of conflict and competition, and when the Portuguese claimed sole user rights of the trade routes to India, the Netherlands protested. *Mare Liberum*, or "The Free Sea", was written in 1609 by Hugo Grotius, a Dutch jurist and philosopher, in defence of the idea that the sea belonged to all. Grotius is known as one of the fathers of international law, but his argument is based on natural law, which he considered of universal validity (Scholtz 2016).

Natural law is based on the assumption that God, or nature, has given Earth to all of humankind as common property. Private property is considered necessary as some things can be consumed, or only be used by one person at a time, and are therefore no longer available to others. Early property rights to land and livestock, for example, are based on this idea, as is collective or public property, in the sense that something can become the property of a particular community at the exclusion of other people. Both are distinct from common property which belongs to all of humankind. In order to find recognition as private or public property, certain conditions must be met. Movable goods, for example, need to be explicitly appropriated, and non-moveable goods such as land must be delineated, built on, or guarded to indicate their appropriation.

Grotius argues that private or public ownership of the sea is impossible as well as immoral. In a distinction that is carried over in today's international law of the sea, he contends that resources such as fish may be appropriated but that the sea itself as an immeasurable good does not allow its possession. Four arguments stand out in making this case. Firstly, Grotius contends that private property is only possible for things in which one has a personal interest. Fishers, for example, are interested in fish and might want to protect their catch from rivals, but they do not need to protect the sea itself as there is always enough of the sea for others to also fish (Tuck 1999). Secondly, he argues, it is a fundamental right of all private individuals to acquire goods and protect them, as long as this does not take away the legitimate goods of another person—another argument related to the boundless nature of the sea. Thirdly, trade is an essential means of sharing wealth and facilitating the just distribution of goods in the world. Oceans and winds enable trade over long distances, but free trade demands free seas, and anyone enclosing the sea for the purpose of owning it, and thereby restricting the freedom of others, would commit an injustice (Scholtz 2016). In the dispute over ownership of the trade routes' access to East India, Grotius therefore contended that the Dutch had a fundamental right to seek trade in the East Indies and that Portuguese attempts at preventing this could only be legitimate if they could claim ownership of the seas—which clearly they could not (Tuck 1999). The fourth, and perhaps most interesting, argument is that the innate nature of the sea itself prevents it from being privately or publicly owned. Because it is fluid, it cannot be possessed in the sense of being demarcated as an object or property. It cannot be bought or sold or divided up through contracts. The sea "wants" to serve everyone, and it can do just that because it is apparently inexhaustible and not used up by any particular activities or—at the time of Grotius at least—not damaged by human use.

> All that which has been so constituted by nature that though serving some one person it still suffices for the common use of all other persons, is to day and ought to remain in the same condition as when it was first created by nature. (…) The air belongs to this class of things for two reasons. First, it is not susceptible of occupation; and second its common use is destined for all men. For the same reasons the sea is common to all, because it is so limitless that it cannot become a possession of any one, and because it is adapted for the use of all, whether we consider it from the point of view of navigation or of fisheries. (Grotius 1609/1916 translation)

The particular nature of the sea is encapsulated in the following paragraph:

> (...) the question at issue is the outer sea, the ocean that expanse of water which antiquity describes as the immense, the infinite, bounded only by the heavens, parent of all things; the ocean which the ancients believed was perpetually supplied with water not only by fountains, rivers, and seas, but by the clouds, and by the very stars of heaven themselves; the ocean which, although surrounding this earth, the home of the human race, with the ebb and flow of its tides, can be neither seized nor inclosed; nay, which rather possesses the earth than is by it possessed. (ibid.)

This also points to a distinction between the seashore and inner sea and the outer sea. Grotius contends that the shore and inner seas can be occupied and used and therefore considered public property, but even there the sea is an agent that cannot be contained:

> [The shore] becomes therefore the property of the occupier, but his ownership lasts no longer than his occupation lasts, inasmuch as the sea seems by nature to resist ownership. For just as a wild animal, if it shall have escaped and thus recovered its natural liberty, is no longer the property of its captor, so also the sea may recover its possession of the shore. (ibid.)

But freedom of the seas also requires that the freedom of everyone else is respected. Thus, Grotius also acknowledges that there must be laws and limits to what is permitted at sea. Protection and jurisdiction, however, are set apart from ownership, and there is a clear statement that fleets maintained for the protection of navigation or the punishment of pirates under a certain jurisdiction do not then lead to ownership of the sea:

> We recognize, however, that certain peoples have agreed that pirates captured in this or in that part of the sea should come under the jurisdiction of this state or of that, and further that certain convenient limits of distinct jurisdiction have been apportioned on the sea. Now, this agreement does bind those who are parties to it, but it has no binding force on other nations, nor does it make the delimited area of the sea the private property of any one. It merely constitutes a personal right between contracting parties. (ibid.)

This last aspect is important as it recognises realities such as piracy, which did threaten this rather idyllic picture. The sea, of course, was a stage for nations to compete for influence and territory, and England was only able to become a global power thanks to its ability to control the sea. In line with the desire for hegemony, in 1635, the Englishman John Selden developed the opposing doctrine of "*mare clausum*", reaffirming what had become standard practice based on the accepted notion that states have jurisdiction over their neighbouring waters. Although jurisdiction did not allow them to ban fishing and sailing in these waters, it did permit them to introduce regulations that effectively resulted in the same (Tuck 1999). Mare clausum thus amounted to a division of the sea into national spheres of interest, to the exclusion of other states (Ratter 2018). To some degree, this argument was based on the rights of states to national security and their ability to restrict a certain sea area to other states. The point of contention then became what stretch of water could reasonably be controlled by a coastal state. Arguably, the range of the most advanced cannon at around 1700 was three nautical miles, the birth of the 3 nm zone (Ratter 2018).

3.2 UNCLOS: A History of Enclosure?

International maritime law evolved in an ongoing compromise between the principle of freedom of the seas for navigation and resource management for the allocation of exploitation rights (Portman 2016). In the past, distances and the limited ability to travel had effectively created spatial monopolies over resources. In the early twentieth century, some nations expressed their desire to extend national claims, for example, to include mineral resources, to protect fish stocks, and to enforce pollution controls. In 1945, in an interpretation of the principle of a nation's customary right to protect its natural resources, President Truman extended US-American control to all the natural resources of the US continental shelf (Ratter 2018; Portman 2016). Similar claims quickly became standard practice. By 1967, only 25 nations still used the old 3-mile limit; many more had set a 12-nautical-mile territorial limit and eight had even set a 200-nautical-mile limit. By that point, national sovereignty was no longer a question of expressing and exercising power as a physical presence: Annexing maritime areas became a matter of simply staking a national claim, either in line with or even disregarding international guidelines (Ratter 2018).

The United Nations Convention on the Law of the Sea (UNCLOS) began to be negotiated in the 1950s. The first round of negotiations led to the Convention on the Territorial Sea and the Contiguous Zone, the Convention on the Continental Shelf, and the Convention on the High Seas (all 1958). Nevertheless, nations continued to make varying claims of territorial waters, and so the Third United Nations Conference on the Law of the Sea was convened in New York in 1973 to set limits, to agree on navigation, archipelagic status, and transit regimes, as well as set out exclusive economic zones (EEZs) and continental shelf jurisdiction. Countries now have sovereignty over their internal waters and territorial seas up to 12 sm, sovereign rights in the EEZs to conduct certain activities, and the rights to exploit certain resources of the continental shelf. But apart from defining ocean boundaries and associated rights, the convention also establishes general obligations for safeguarding the marine environment and protecting freedom of scientific research on the high seas (Portman 2016).

In areas beyond national jurisdiction, the principle of a "common heritage of mankind" was introduced, ensuring that no state is able to claim or exercise sovereignty or sovereign rights over any part of these sea areas (see also Chap. 16 in this volume). Consciously or unconsciously, this reaffirms Grotius' legal and moral notion of the sea as common property: "All rights in the resources of the Area are vested in mankind as a whole" (UNCLOS Art. 137). Buoyed perhaps by the spirit of the times and the idea of a more equitable distribution of global wealth, Article 140 goes on to specify that "Activities in the Area[1]

[1] Area beyond national jurisdiction.

shall, as specifically provided for in this Part, be carried out for the benefit of mankind as a whole, irrespective of the geographical location of States".

The self-interest of countries to have exclusive fishing rights and rights over other resources, however, soon trumped any burgeoning international ideals. Countries began to demarcate areas according to the newly agreed extensions, in some cases resulting in huge territorial gains. Countries are continuing to extend their jurisdictional authority seawards, mainly by invoking an UNCLOS provision which allows coastal states to establish the outer edge of the continental margin up to 350 nm wherever the margin extends beyond 200 nautical miles. At the same time, special transit rights apply in the case of straits and also for land-locked states.

So where will it all end? In the face of ongoing and prospective disputes over marine resources such as the Arctic, the question of who can claim property in the sea remains a highly pertinent one. Despite the noble intentions encapsulated in the idea of a common heritage of mankind, UNCLOS has effectively condoned a veritable race between coastal states to carve up the ocean—racing to secure resources and therefore also political and economic power in a rapidly changing world. It has created wholly new maps of the world and led to new theatres of conflict (Ratter 2018). The trend to increasing territorialisation of the sea is inextricably linked to the increasing industrialisation of marine resource exploitation (Vitzthum 1981), enabling countries to go faster and deeper and becoming ever more efficient in extracting resources from the sea. Some authors have compared this to the colonisation of continents in earlier periods—with a clear advantage to those countries that have a coast or islands and are wealthy already and can afford the expensive technology.

The key question—not least for MSP—is whether some degree of ownership, or at least custodianship of sea areas, is able to prevent a tragedy of the commons (see also Chap. 5 in this volume). This will depend on whether exercising jurisdiction over natural resources is also taken to mean responsibility for their conservation, leading to prudent utilisation. The alternative development path may be unlimited "ocean grabbing" in an environment that still lacks a comprehensive approach to governance (Portman 2016). Are the current rules that guide exploitation sufficient, and who polices them? And what really is our attitude to the conservation of the ocean: Does this offer an alternative trajectory?

4 Scientific Discovery and Ocean Resources

4.1 Changing Relations with the Ocean in the Wake of Discovery

Despite the many dangers associated with it, the obvious "otherness" of the ocean has not stopped humans from being curious about it. Until the last quarter of the eighteenth century, European understanding of the ocean's depth derived mostly from the imagination, based on stories recorded in ancient literature and the Bible. In the late eighteenth and nineteenth century, this changed as a result of technological advances and a growing interest in natural sciences, as well as a burgeoning interest in the ocean as a place. An important shift took place at this time, a re-interpretation of the ocean as a desirable place rather than a barrier to overcome (Rozwadowski 2005).

The changing Western relations with the ocean during this period are down to a confluence of factors. Of particular importance is the coming together of the expansionist tendencies of the great maritime nations of the time, Britain and the USA, and the growing interest in and capability of scientific exploration. Ocean resources, and the economic benefits associated with them, were highly desirable in a time of international competition, and in particular in the USA there was a strong link between commercial maritime interests and early scientific institutions such as the Coast Survey (Rozwadowski 2005). Discovering and understanding ocean resources required systematic scientific investigation of ocean places, and locating them in the vastness of the ocean required new kinds of maps. But although perceptions of the sea were still driven by ambitions for using its resources, the ocean increasingly became an object of investigation in itself.

In terms of getting to know the ocean, two approaches became predominant in the eighteenth century: hydrography and natural sciences. Both were preoccupied with the deep sea, albeit for different reasons. Charting the ocean floor and arriving at a bathymetric chart of ocean areas was related to navigational safety but also to early commercial endeavours such as attempts to lay the first submarine cable between the USA and Britain. But there was also growing interest in a physical geography of the sea as such, in order to understand the physical phenomena of the sea. The first bathymetric charts and vertical elevation profiles of large parts of the Atlantic appeared in the mid-nineteenth century based on deep-sea sounding programmes. Hydrographic exploration and exploitation stimulated each other, bringing together the natural curiosity and spirit of the early oceanographers, the financial might of investors and companies, and the political interest and naval capability of seafaring nations. As oceanography developed, early, more holistic approaches were gradually replaced by mathematical analysis. Thus the oceans "came to be seen not as trackless wastes (the view of ancient and classical authors), nor as part of a great interlinked cosmic machine (Humboldt's view in the early nineteenth century) but as physical phenomena subject to mathematical analysis" (Mills 2009, p. 10)—a view that still predominates today.

The natural sciences were also interested in the deep sea, driven mainly by the question of what creatures existed in the ocean and whether life was possible at all at great depth. As was the case for hydrography, two interests combined. Commercial interests mainly related to fish and whales as key resources—whose exploitation had grown exponentially due to better equipment and economic interest—but there was also the innate desire of science to learn more about marine life and its interconnections. Technological advances such as the advent of microscopes had led to recognition of the abundance and variety of life in the sea, and faced with this great and infinite life force, the response at the time was one of wonder. In his famous "Cosmos", published in 1845, the great polymath Alexander von Humboldt gives rather poetic descriptions of the ocean (Fig. 2.2):

> The application of the microscope increases, in the most striking manner, our impression of the rich luxuriance of animal life in the ocean, and reveals to the astonished senses a consciousness of the universality of life. In the oceanic depths, far exceeding the height of our loftiest mountain chains, every stratum of water is animated with polygastric sea-worms, Cyclidiæ and Ophrydinæ. The waters swarm with countless hosts of small luminiferous animalcules, Mammaria (of the order of Acalephæ), Crustacea, Peridinea, and circling Nereides, which when attracted to the surface by peculiar meteorological conditions, convert every wave into a foaming band of flashing light. (Cosmos 1845, p. 305)

Although a more mechanistic perspective also began to emerge, there was no initial contradiction between the desire to understand, collect, and classify individual species and a holistic view of nature. This interest in collecting specimen was not restricted to scientists but extended to the population at large; shell collecting and marine aquaria for example became favourite Victorian pastimes in Britain (Rozwadowski 2005). As the available knowledge grew, sea monsters were gradually replaced with scientific evidence of life in the ocean, and blank areas on ocean charts were gradually filled, often resorting to the local knowledge of sailors and whalers regarding the distribution and geographical range of species.

Fig. 2.2 The living sea? Photo: Kira Gee

Overall, the scientific approach has thus been one of mapping and structuring the ocean. Gradually, ocean space was placed, delimited, and sounded, and its material and spatial properties began to be understood in ever greater detail. This led to an expansion of the utilitarian relationship with the sea, not only through fishing and whaling but also, for example, seaside holidays. As technology improved, greater attention could be placed on ocean resources (Laloë 2016), a trend which is still ongoing today, for example, with oil exploration in the Arctic. The means available to exploration today have changed dramatically: Two-dimensional perspectives of the ocean have come to be replaced with three-, four- and even five-dimensional approaches (including the air above the sea and the substrate) made possible through filming and diving, opening up entirely new perspectives of the sea. Understanding connections has brought much greater awareness of the significance of the ocean to humankind as a whole, not least in the recent context of climate change. Last but not least, new technologies of exploring the deep have also brought about a new sense of wonder at the diversity and beauty of life in the ocean, evidenced for example by the popularity of documentaries such as the BBC's "Blue planet" series.

4.2 Non-utilitarian Perspectives

At the same time, there is also greater awareness of the fragility of the ocean—pushed in recent times by issues such as overfishing, pollution, invasive species, and lately microplastics. The seemingly inexhaustible ocean resources first began to appear finite in the nineteenth century in the wake of more efficient and larger-scale exploitation. A new type of ownership of ocean resources had arisen contrary to Grotius' ideas, and it was clearly damaging to the interests of others who would also have a legitimate interest in these same resources. Social and moral criticism had also begun of industrialisation in general, as it became apparent that it not only produces human poverty but also impoverished, damaged nature (Scholtz 2016). The more human intervention changed nature, the greater the need became to account for and guide human action. Also, the more the knowledge was gained of the sea, and the greater the understanding of its diversity and wonder, the more pronounced the application of non-utilitarian thought in conceptions of the sea.

Bioethics—understood here to describe the relationship between the biosphere and a growing human population (Potter 1971)—is concerned with a responsible human relationship with nature and arose out of an expectation of nature that goes beyond economic benefits and utilitarian value. Two perspectives come together here. The first is the notion of nature's intrinsic value, in other words, the value possessed by things or organisms in and of themselves. The ocean is mostly valued instrumentally, that is, for the benefits associated with it—recreation, traditional fishing, an aesthetically pleasing view—but it could equally be valued as an entity all in itself, a carrier of value independent of any human observer. The second perspective recognises the need for rules in order to limit human intervention and reverse further damage. This can arise from a utilitarian argument and the idea that the sea gives pleasure and contributes to our welfare, but also from the idea of intrinsic value of nature, protecting nature for its own sake. The conviction that nature conservation is morally good, and that untouched nature should continue to exist, is quite prevalent in Western countries, in particular Germany (Gee 2013), first emerging in the 1970s when resource shortages and environmental degradation led to the rise of environmentalism (Jepson and Canney 2003). Certain moral values serve as action guides here, i.e. a shared feeling that we (as individuals in a society) "ought" to behave in a certain way (Rokeach 1973).

Environmental protection has become an important focus of international coastal and ocean policy, driven by the many transboundary and global threats to the marine environment (Portman 2016). UNCLOS is one of the first such international agreements, although it could be argued that nations are more concerned with pushing through their national interests and are forgetting the obligations towards marine resources that also come with it. Despite the rise of sustainability as a unifying concept, there is still a divide between the desire to protect ocean resources on the one hand and facilitating their exploitation on the other (Portman 2016). Contradictory policy objectives are making it difficult to come to a unified guide to human action. This is amply illustrated in the EU's maritime policy (see also Chap. 6 in this volume). Although integration has become a central theme in maritime policy, discussions are ongoing in 2018 on how the demands of blue growth can best be reconciled with environmental protection and conservation.

4.3 Contradictory Policy Goals: Can They Be Reconciled?

In the early 2000s, the EU's incipient maritime policy was strongly influenced by the global economic crisis, the need for the EU to position itself against other powers such as China, and the difficult socio-economic situation in many EU member states. One of the first steps was the 2006 Green Paper "A Future Maritime Policy for the Union: A European Vision of the Oceans and Seas" (EC 2006) which describes the importance of oceans for innovation and cites geography as a reason for Europe's special relationship with the sea. The purpose of the paper, however, is clearly economic, asserting that Europe must revitalise its economy and emphasising the role already played by European oceans. Holistic ocean management is seen as a new approach, designed to overcome the largely sectoral and fragmented policy-making of the past. In its strategic objectives for 2005–2009 the European Commission thus declares "the particular need for an all-embracing maritime policy aimed at developing a thriving maritime economy, in an environmentally sustainable manner" (EC 2005).

One of the key problems is that the new approach to holistic ocean governance was underpinned by two pillars: the Lisbon strategy for growth and better jobs (European Council 2000) and maintaining and improving the status of the resource on which all maritime activities depend. In 2007, the

Commission followed up on the Green Paper by publishing the so-called Blue Book on Integrated Maritime Policy (IMP) (EC 2007). Essentially an economic policy, this seeks to coordinate relevant sectoral policies by promoting cross-cutting issues, including blue growth, marine data and knowledge, and MSP. Although it is also anchored in the Gothenburg agenda for sustainability, the IMP predominantly rests on the Lisbon agenda for growth and jobs and can therefore be understood as a bifurcation point, a point in time when the economic "blue growth" rationale becomes a dominant discourse and branches off from the concurrent development of the environmental pillar and its central paradigm of Good Environmental Status (GES). Still, the IMP does point to the need to achieve the full economic potential of the seas in harmony with the marine environment, and thus also offers an anchor for environmental policy, in the sense that the IMP cannot be fully achieved without also achieving environmental objectives.

The Commission's economic priorities have since been reaffirmed repeatedly, such as the communication on Blue Growth Opportunities for Marine and Maritime Sustainable Growth (EC 2012) and Innovation in the Blue Economy: Realising the Potential of Our Seas and Oceans for Jobs and Growth (EC 2014) (see Chap. 6 in this volume). In parallel, the Marine Strategy Framework Directive (MSFD), adopted in 2008 (EC 2008), presents a "comprehensive and integrated approach to the protection of all European coasts and marine waters".[2] The main aim of the MSFD is to achieve GES of the EU's marine waters. The main reason for this is instrumental, as the main purpose for doing so is to protect the resource base upon which economic and social activities depend. The ecosystem approach is presented as the guiding principle, although there is no clear definition of how this should be understood and implemented. Numerous other communications followed, mostly concerned with the implementation of the MSFD.

The concept of ecosystem-based management—along with its complementary principle, the precautionary approach—continues to be ill-defined and thus a struggle for marine managers and policymakers. Can MSP act as a bridge between the environmental and economic policy objectives? And what of the social dimension, the all-important third pillar of sustainability (see also Chap. 8 in this volume)? Early indications are that European countries are taking rather different approaches to the ecosystem approach and that it is a strong guiding principle in MSP in only some countries.

[2] http://ec.europa.eu/environment/marine/eu-coast-and-marine-policy/index_en.htm, accessed 8 March 2018.

5 Human Dimensions of the Ocean: The Ocean as a Place of Attachment

So far, this chapter has sought to draw out some of the prevalent ocean perspectives and their historical roots. Different attempts at perceiving, mapping, and categorising the ocean have been traced, starting with Hugo Grotius and UNCLOS as an approach to territorialise the ocean and following on with science and exploration as ways of understanding the physicality of the ocean and the ocean as an environment and resource. We have also seen that current maritime policy is divided into economic and environmental strands and that reconciliation seems difficult. A pervading theme throughout has been the presence of apparently opposing views—the enclosed versus the free sea or the utilitarian and non-utilitarian perspective of the ocean. It has also become clear that views of the ocean always reflect the general mood and world view of the time—such as the spirit of exploration in the nineteenth century or the era of discovery and trade in the seventeenth century.

This last section gives another ocean perspective, namely that of the ocean as a place as experienced and cherished by people. This builds on the idea that from a geographical perspective, there are fundamentally different ways of seeing the ocean. The first is the practice of regarding the ocean as a collection of material, tangible entities, resulting in particular spaces composed of physical-material facts—such as ocean currents, water depth, water temperature, and flora and fauna. The second is the understanding of the ocean as a visual phenomenon, referring to the appearance of the ocean as we see it. The third—and the focus of this section—is the sea not as a space but as a place—moreover, a place that can generate deep-seated attachment and with this, care. Moving away from the idea of ocean space as an extension of terrestrial space and its associated "protocols of measuring and distributing surfaces" (Laloë 2016, p. 2), this perspective is perhaps least amenable to governance and potentially conflicting with a purely spatial and rational perspective reliant solely on physical data and scientific evidence.

5.1 The Ocean as a Place

The ocean as a place refers to deeper meanings and symbolisms, attachments and internal pictures of the sea we may hold. Perceptions of the sea have changed over the centuries in response to greater technological control, giving rise to an ambiguous image of the sea, appearing cold, inapproachable, and dangerous on the one hand yet representing summer, sun, and beach life on the other. Sea bathing and ocean going for pleasure became popular during the nineteenth century (Fischer and Hasse 2001), a "brief period of time when the sea held enough romance and mystery to fire the imagination but less threat than in previous centuries" (Rozwadowski 2005, p. 21). Maritime novels gained large followings and helped to create a rich popular imagination of the ocean, and there was increasing aesthetic appreciation of the ocean inspired by paintings of seascapes. The sea also became a place of reflection and transcendence, inspiring new experiences of the sublime—engendering a new sense of place of the ocean related to its purported health benefits, leisure, and aesthetic interests.

But is the sea in itself also a place or just an object to be gazed at from a distance? Relph (1976) noted that "every identifiable place has unique content and patterns of relationship that are expressed and endure in the spirit of place" (p. 76). Sense of place represents a combination of what could be termed "intrinsic personality" of the environment and the "emotional attachment to localities developed by individuals and communities in the course of living and growing within the setting of home" (Muir 1999, p. 273; Tuan 1975). In order to understand the values assigned to a place, it is therefore important to explore this emotional relationship of people with places. The greater the emotional involvement in a place, and the greater the meaning assigned to it, the greater the likelihood of strong attachment to the place and therefore value.

The relationship between sea and place is not an easy one to resolve. Since the sea is not dwelled in in the usual sense of the word, it is theoretically conceivable that the capacity of the sea to turn into a place is inherently limited. A more likely conclusion is that notions of place arise differently in the context of the sea. Global communication has arguably contributed to the demise of "space" in the sense that everywhere has long since become somewhere. But there is also a different, inherent sense of belonging to a home place, which extends to the sea just as much as it does to the land. A strong sense of belonging to the sea has been found in Irish and Scottish Gaelic fishing communities, described as "not so much a landscape, not a sense of geography alone, nor of history alone, but a formal order of experience in which all these are merged" (MacKinnon and Brennan 2012, p. 7). Those working with the sea carry a deeper way of knowing the sea which is distinct from more formal ways of knowing. MacKinnon & Brennan find this reflected in the place names given to the sea by fishermen, indicative of a unique way of knowing the marine environment. This knowledge, they argue, represents a more complete way of knowing the sea than the objective precision of the natural sciences alone can deliver. This is because it also encompasses emotional energy as an indicator of "home" and a sense of responsibility for that home which is “place”. Similar descriptions can be drawn from sailors describing emotional experiences of being on the water or other professional users of the sea or leisure users. The sea is thus just as much a place as the land, with subjectivity of place not only arising from direct use of the marine environment but also imagery and traditional knowledge. Especially in the context of immaterial or experiential conceptions of the ocean, there is no universal, tangible, physical reality but multiple ocean realities which can be appreciated for many different reasons (Fig. 2.3).

But how is meaning derived from the ocean? Essentially, this is a question of perception, understood here as different ways of experiencing and interpreting the ocean. Perception in turn is linked to the general values a person holds, as well as their general beliefs about the world at large and what is important in this world, which comes back to notions such as bioethics raised in the previous section.

Fig. 2.3 The beautiful sea? Photo: Kira Gee

5.2 The Ocean as a Cultural Landscape?

An interesting perspective with relevance for the perception of the ocean is a duality that is better known from landscape research, namely the dualism between natural and cultural landscapes. The common understanding is that natural landscapes are those uninfluenced by man, having grown from natural processes and still determined by natural processes, while cultural landscapes in the broadest sense are those that are shaped by man. Fischer (2007) describes the so-called dialectic of the Enlightenment whose opposing trends led to the conception and importantly, valuation of both types of landscape. The first of these trends is the re-evaluation of nature and a re-interpretation of wild and threatening landscapes as something pleasing and beautiful. The coast—and imaginably also the ocean—is a prime example of a place which was no longer seen as dangerous and a location of divine retribution but came to be regarded as "wilderness" and a sublime place. Wilderness, often defined as extreme landscape formations, was re-interpreted as something of great value; much later, this was to become the founding idea of National Parks. Today, the "natural" is still a by-word for that which is inherently good, desirable, and pure; it has become all the more desired the less immediate our connection to nature. "The longing of tourists for 'beautiful' or even 'wild' nature is fed by the unconscious assumption that the 'natural' is needed as a cure for the over-civilization of the world in which we live" (Fischer 2007, p. 3). The idea of wilderness is one that influences perceptions of the ocean and does appear to be the opposite to the idea of the ocean as an increasingly "industrial" landscape. The second trend that began during the Enlightenment is the transformation and re-interpretation of inhospitable terrain and "badlands" (e.g. heathland, floodplains) into something that represented progress, therefore also becoming inherently good but for different reasons. Cultural landscapes became appreciated for the fact that they were man-made; as an added benefit (which may or may not be transferrable to the sea) they were regarded as aesthetically pleasing. Natural and cultural landscapes are therefore both valued in their own right but for different reasons and based on different value sets—an analogy that could readily apply to the ocean (Fig. 2.4).

Gazing out to the sea can give the impression that the sea is still very much a natural landscape, untouched as it seems by any human influences and nothing but an infinite expanse of water stretching to the horizon. Although it is no "dwelling place" in the usual sense of the word, the sea does have long-standing links to cultural practices such as fishing or trading. In recent years, cultural practices have become markedly more intense, expressed for example

Fig. 2.4 The romantic sea? Etching: Kira Gee

in the growing numbers of vessels in the sea and the growing numbers of structures such as bridges, platforms, and off-shore wind farms. The visual alteration of the sea, the appearance of fixed structures in its infinite expanse, may suggest the sea is indeed becoming a "cultural seascape", shaped by man just like cultural landscapes on land. At the same time, it is unclear what would constitute a "natural seascape": A seascape that appears unaltered visually? An unpolluted sea? An ocean untouched by any human influence? An open question is also whether there is similar romanticism as far as the "untouched" sea is concerned, or whether some form of attachment may have developed to the new cultural seascape, regarding bridges or other structures—a symbol of development and progress for example. Could the dialectic of the Enlightenment be in the process of being repeated for the sea?

6 Conclusion

This chapter is a somewhat eclectic collection of ocean perspectives that have shaped our Western views of the ocean. I have attempted to trace the origins of some of our thinking and the conditions during which the pervading views first emerged. I have also attempted to highlight how these perspectives are reflected in ocean policy—or sometimes lack appropriate reflection.

Naturally, the chapter cannot hope to capture the entire range of ocean perspectives and can rightly be criticised for being selective. For example, I have ignored some of the darker current perspectives, such as the role of the ocean as a barrier in the context of "fortress Europe". I have also ignored the role of the sea in military expansion, the changing role of the sea in times of climate change, or the fact that the sea can seem remote to people living a long distance from the coast. Many more perspectives could be listed here. What does become apparent, however, is the fact that ocean perspectives are as diverse and changeable as the ocean itself—and just as iridescent and fascinating in all their diversity. Oceans emerge as spatial metaphors and a way of structuring the perception of the social (Luutz 2007), much like regions that have come to be understood as produced by collective action (Paasi 1986). This implies the contingency of spatial entities and their disappearance when they are no longer reproduced by society.

The latter aspect in particular has implications for MSP. In the face of the many parallel constructs of the ocean and the many diverging roles the ocean plays in our society and subconscious, which ocean can and should we attempt to manage? Can diverging constructs of the sea as a transport space, fishing grounds, recreational space, natural habitat, or aesthetic place, plus the associated value sets and power relations, ever be brought together in a cohesive approach? Or will the ocean continue to remind us that attempts at management are temporary at best, that "panta rhei", everything, including philosophies of management, is in constant flux?

Acknowledgements This work was supported by the BONUS BALTSPACE project which has received funding from BONUS (Art 185) funded jointly from the EU's Seventh Programme for research, technological development, and demonstration and from Baltic Sea national funding institutions. The Open Access fee of this chapter was provided by the same project.

References

Carson, R. (1937). Undersea. *Atlantic Monthly, 78*, 55–67.

EC. (2005). Strategic Objectives 2005 – 2009: Europe 2010: A Partnership for European Renewal Prosperity, Solidarity and Security. Communication from the President in agreement with Vice-President Wallström. Brussels, 26.1.2005, COM(2005) 12 final.

EC. (2006). Commission Green Paper: Towards a Future Maritime Policy for the Union: A European Vision for the Oceans and Seas. COM (2006) 275 final.

EC. (2007). An Integrated Maritime Policy for the European Union. Brussels, 10.10.2007, COM(2007) 575 final.

EC. (2008). Directive 2008/56/EC of the European Parliament and of the Council, of 17 June 2008, Establishing a Framework for Community Action in the Field of Marine Environmental Policy (Marine Strategy Framework Directive).

EC. (2012). Blue Growth Opportunities for Marine and Maritime Sustainable Growth. Brussels, 13.9.2012, COM(2012) 494 final.

EC. (2014). Innovation in the Blue Economy: Realising the Potential of Our Seas and Oceans for Jobs and Growth. Brussels, 13.05.2014, COM/2014/0254 final/2*.

European Council. (2000). Lisbon European Council, March, 23–24. Presidency Conclusions.

Fischer, L. (2007). *Cultural Landscape and a Natural Landscape—Notes with Regard to the Wadden Sea Region*. Presentation at the LanceWadPlan Final Conference, Wilhelmshaven, June 19. Retrieved from http://www.lancewad.org.

Fischer, L., & Hasse, J. (2001). Historical and Current Perceptions of the Landscapes in the Wadden Sea Region. In: Vollmer, M., Guldberg, M., Maluck, M., Marrewijk, D. & Schlicksbier, G. (eds). *Landscape and Cultural Heritage in the Wadden Sea Region – Project Report*, pp. 72–97.

Gee, K. (2013). *Trade-Offs Between Seascape and Offshore Wind Farming Values: An Analysis of Local Opinions Based on a Cognitive Belief Framework*. PhD dissertation, Department of Geography, University of Göttingen.

Grotius, H. (1916). *The Free Sea* (R. van Deman Magoffin, Trans.). New York: Oxford University Press.

Humboldt, A. (1845). *Cosmos: A Sketch of the Physical Description of the Universe*, Vol. 1 by Alexander von Humboldt, translated by E. C. Otte. from the 1858 Harper & Brothers edition of Cosmos, volume 1. Retrieved from http://www.gutenberg.org/ebooks/14565.

Jepson, P., & Canney, S. (2003). Values-Led Conservation. *Global Ecology & Biogeography, 12*, 271–274.

Laloë, A. F. (2016). *The Geography of the Ocean. Knowing the Ocean as a Space*. Routledge.

Luutz, W. (2007). Vom "Containerraum" zur "entgrenzten" Welt—Raumbilder als sozialwissenschaftliche Leitbilder? *Social Geography, 2*(1), 29–45.

MacKinnon, I., & Brennan, R. (2012). *Belonging to the Sea. Exploring the Cultural Roots of Maritime Conflict on Gaelic Speaking Islands in Scotland and Ireland.* Scottish Crofting Federation and Scottish Association for Marine Science.

Mills, E. (2009). *The Fluid Envelope of Our Planet: How the Study of Ocean Currents Became a Science*. Toronto: University of Toronto Press.

Muir, R. (1999). *Approaches to Landscape*. London: Macmillan Press.

Paasi, A. (1986). The Institutionalization of Regions: A Theoretical Framework for Understanding the Emergence of Regions and the Constitution of Regional Identity. *Fennia, 46*, 105–146.

Portman, N. (2016). *Environmental Planning for Oceans and Coasts: Methods, Tools, and Technologies*. Springer International Publishing.

Potter, V. R. (1971). *Bioethics: Bridge to the Future*. Prentice-Hall.

Ratter, B. M. W. (2018). *Geography of Small Islands. Outposts of Globalisation*. Springer International Publishing.

Relph, E. (1976). *Place and Placelessness*. London: Pion.

Rokeach, M. (1973). *The Nature of Human Values*. The Free Press, Macmillan Publishing.

Rozwadowski, H. (2005). *Fathoming the Ocean. The Discovery and Exploration of the Deep Sea*. Cambridge, MA and London: The Belknap Press of Harvard University Press.

Ryan, A. (2012). *Where Land Meets Sea. Coastal Explorations of Landscape, Representation and Spatial Experience*. Routledge.

Scholtz, G. (2016). Die Philosophie des Meeres. mareverlag, Hamburg.

Steinberg, P. (2014). Foreword: On Thalassography. In J. Anderson & K. Peters (Eds.), *Water Worlds: Human Geographies of the Ocean* (pp. xiii–xvii). Farnham: Ashgate.

Tuan, Y.-F. (1975). Place: An Experiential Perspective. *Geographical Review, 65*, 151–165.

Tuck, R. (1999). *Hugo Grotius*. Reprinted in L. May & M. McGill (Eds.). 2014. *Grotius and Law*. Routledge, pp. 37–68.

Vitzthum, W. G. (Ed.). (1981). *Die Plünderung des Meeres—Ein gemeinsames Erbe wird zerstückelt*. Fischer Taschenbuch Verlag, Band 4248.

3

Challenges and Opportunities for Ecosystem-Based Management and Marine Spatial Planning in the Irish Sea

Tim O'Higgins, Linda O'Higgins, Anne Marie O'Hagan, and Joseph Onwona Ansong

1 Introduction

The Ecosystem Approach to Management, synonymous with Ecosystem-Based Management (EBM), is "*an approach which integrates the connections between land, air water and all living things including human beings and their institutions*" (Mee et al. 2015). This approach is enshrined in both Directives of the European Union's (EU) Integrated Maritime Policy, the Marine Strategy Framework Directive (MSFD) and the Directive on Maritime Spatial Planning, and both mandate a regional approach. EBM fundamentally applies a place-based approach (Olesen et al. 2011; McLeod et al. 2005), where an ecosystem represents the place and effective ecosystem-based marine management and planning must incorporate spatial considerations to manage human uses at a scale that encompasses its impacts (Lackey 1998).

The theory of EBM is now at least 50 years old and builds on the early insights of Hardin (1968) in recognising the tragedy of the commons and the

T. O'Higgins (✉) • L. O'Higgins • A. M. O'Hagan • J. O. Ansong
MaREI Centre for Marine and Renewable Energy, Environmental Research Institute (ERI), University College Cork, Cork, Ireland
e-mail: tim.ohiggins@ucc.ie

J. Zaucha, K. Gee (eds.), *Maritime Spatial Planning*,
https://doi.org/10.1007/978-3-319-98696-8_3

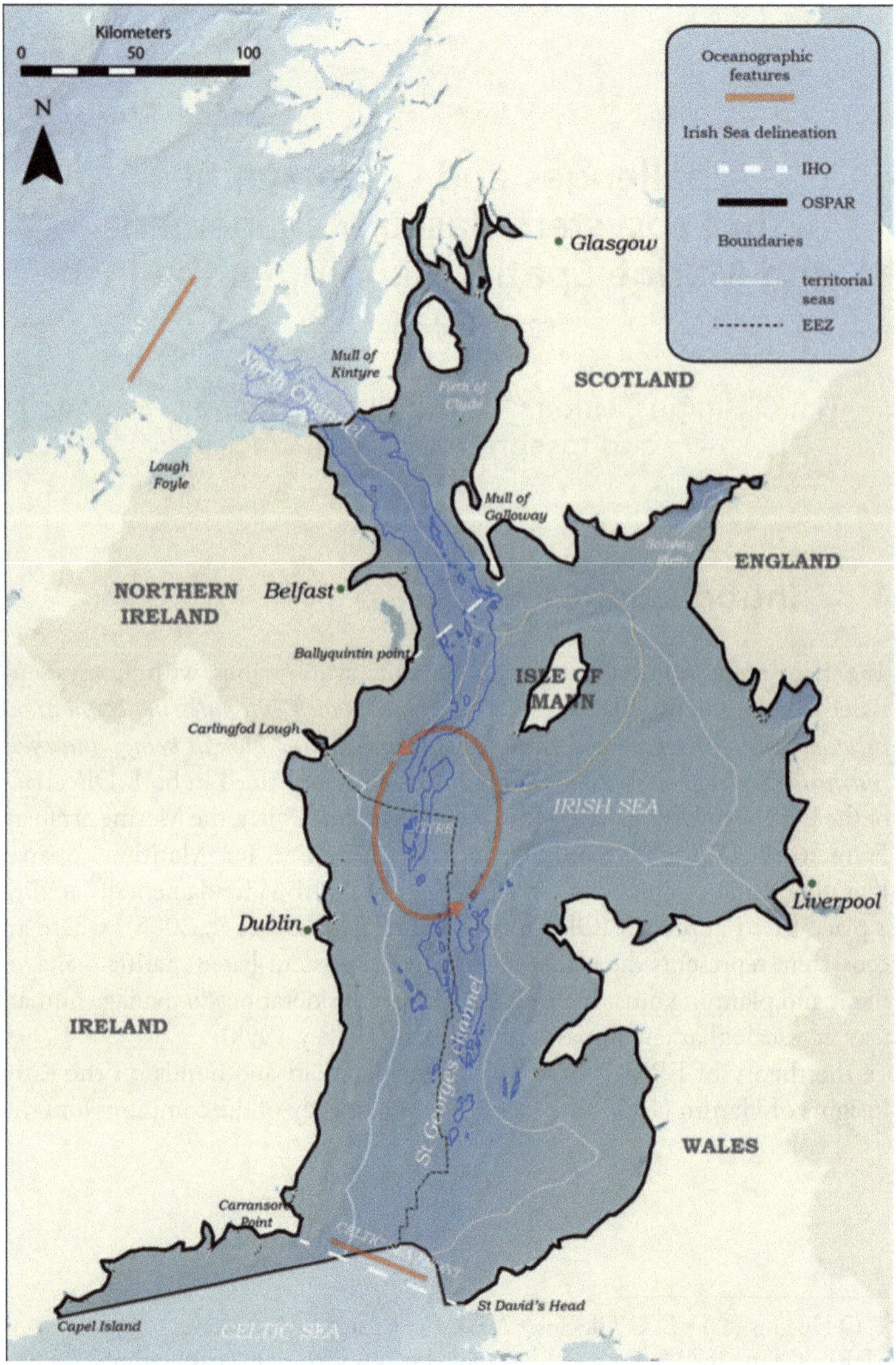

Fig. 3.1 Main geographic and oceanographic features of the Irish Sea. Data sources: Background bathymetry from http://www.emodnet-bathymetry.eu/. EEZ and territorial seas and boundaries from http://marineregions.org/. Location of oceanographic features re-drawn from information contained in Simpson 1974 and Simpson and Hunter 1976

necessity to develop appropriate institutions for the management of common pool resources. Hardin's solution was the assignment of property rights to common pool resources. The problem of institutional fit has been the subject of intense enquiry (Folke et al. 2007), and empirical research has illustrated that there are many different types of management systems that can evolve or be applied to effectively manage common pool resources. While these systems may have some common properties, these tend to be context specific and there are no one-size-fits-all solutions (Dietz et al. 2003; Ostrom 2009). Successful management systems often occur where specific social conditions are present. These include a shared common understanding of the problems generated by poor management as well as shared norms of reciprocity and trust which build social capital (Ostrom 2003).

Another important element in the modern conception of the Ecosystem Approach is the inclusion of ecosystem services (MEA 2003; 2005; Tallis et al. 2010) and the recognition of multiple different types of values not all of which are readily amenable to economic valuation (O'Higgins 2017). The new EU Directive on Maritime Spatial Planning obliges all Member States to establish and implement maritime spatial plans with the aim:

> to contribute to the sustainable development of energy sectors at sea, of maritime transport, and of the fisheries and aquaculture sectors, and to the preservation, protection and improvement of the environment, including resilience to climate change impacts. In addition, Member States may pursue other objectives such as the promotion of sustainable tourism and the sustainable extraction of raw materials. (Article 5(2))

However, the operationalisation of EBM in Marine/Maritime Spatial Planning (MSP) is not simple. Apart from the sea being dynamic and three dimensional, the major challenge is that the marine space remains a public good, remote from, but valued by, the public (Potts et al. 2016) and requires effective public representation in the processes of decision-making and trading off of multiple competing objectives.

Box 3.1 The Bull Lagoon, Dublin Bay: An Historical, Social-Ecological Perspective

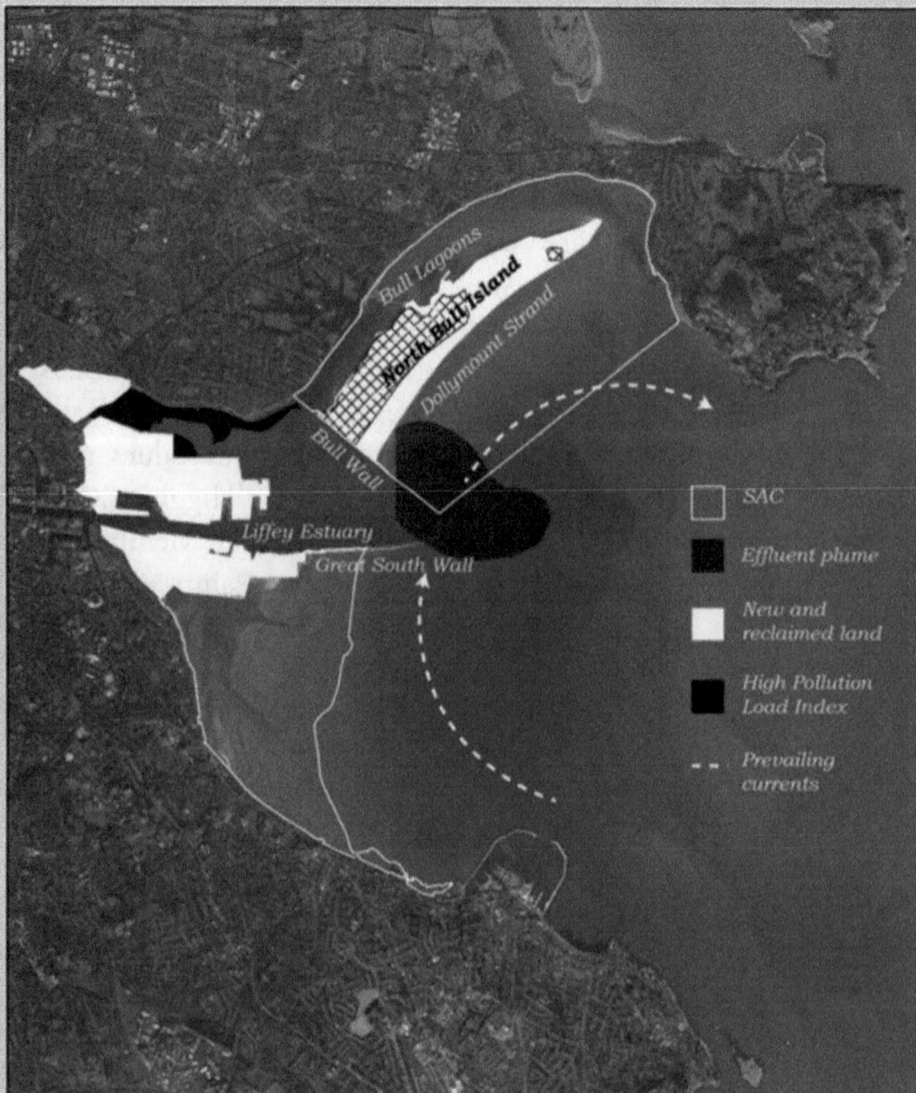

Fig. 3.2 Map of Dublin Bay showing human modification. The hatched area shows the extent of the Bull Island prior to 1913. Effluent data and Pollution Load Index from O'Higgins and Wilson (2005) and O'Higgins (2006)

At the turn of the eighteenth century, Dublin was a major and expanding trading port of the British Empire. Shipping activities in Dublin Port at the time were hampered by the presence of shifting shoals and sandbanks known as the North and South Bulls. The Ballast Board, established in 1786 to improve the Port of Dublin, oversaw the first major modification of Dublin Bay, the finalisation of the Great South Wall. At the same time, studies were initiated for construction of a wall on the northern bank of the River Liffey. The North or Bull Wall (first conceived by Captain William Bligh of "Mutiny on the Bounty" infamy) commenced construction and was completed by the early 1820s. The wall was successful in improving the port, and the actions of the clockwise prevailing currents within Dublin Bay resulted in the gradual formation of a sandy spit, the North Bull Island. Over time the island grew and became a popular recreational location. On the landward side, the sedimentation of fine particulate matter from the cities' effluent (human and other) developed rich muddy sediments (the Bull Lagoons) supporting a diverse intertidal fauna, and avian fauna, annually attracting migrating flocks of Brent Geese. The Bull Island was designated a United Nations Educational, Scientific and Cultural Organization (UNESCO) biosphere reserve in 1981.

Today the North Bull Island is joined to the mainland by two causeways. Dollymount Strand, on the eastern side of the island, is a popular recreational area for walking, a designated bathing area and is popular for kite surfing. The island and lagoons also hold multiple environmental designations and is a Special Area of Conservation (SAC). As the city has grown, the same prevailing currents which resulted in the formation of the Bull Island have also carried (primary treated) sewage effluent from the Liffey mouth onto Dollymount Strand. A major capital investment in sewage treatment in 2003 brought secondary and tertiary treatment, but insufficient capacity has resulted in intermittent bathing water quality on Dollymount Strand, and efforts to maintain the Blue Flag status of the beach have faltered. In 2015, the whole of Dublin Bay was declared a UNESCO biosphere reserve.

The legacy of human modifications has resulted in an ecosystem in north Dublin Bay which is largely anthropogenic, suffers serious and persistent environmental problems, is highly valued for recreational use and is globally recognised in terms of natural heritage.

In the past decade, MSP has been used as a practical tool in applying EBM (Domingues-Tejo et al. 2016; Crowder and Norse 2008). The coupling of MSP and EBM has been seen as necessary and offers an approach for ensuring sustainable development where MSP defines high-level objectives and policies for spatial and temporal ordering of human activities (Ansong et al. 2017; Domingues-Tejo et al. 2016; Douvere 2008) by assessing the cumulative impacts of multiple human activities on the ecosystem at the appropriate scale (Stelzenmüller et al. 2018). Despite the mandate to adopt the Ecosystem Approach to management, this approach does not reflect the historic sectoral management practices around Europe, for example, in the application of the Common Fisheries Policy (CFP). Management options are constrained by legacy effects (O'Higgins et al. 2014), that is, historic and current drivers and pressures set the context in which future activities occur. At the European scale, there has been limited success in measurement or mapping of marine ecosystem services for marine environments, which is hampered by both the lack of data on economic values (Pendleton et al. 2007) and the lack of reliable information on the ecosystem processes associated with specific marine habitats (Maes et al. 2014). Developing the mechanisms and institutions to effectively manage shared marine areas at the regional scale is also a particular challenge (Van Tatenhove et al. 2014).

Some novel approaches are beginning to emerge with the potential for improving understanding of the dynamics of localised ecosystem service supply (Alexander et al. 2012; Potts et al. 2013) and novel approaches to understanding the trade-offs involved in MSP are gaining increasing popularity (Mayer et al. 2013). Here we examine the Irish Sea, taking a social-ecological systems approach to examine MSP. First the geographic and historical context of the Irish Sea is introduced, and the complexity of governance in the region is discussed. Next, the history of cooperation under the EU's environmental Directives is examined in the context of EBM. Major physical (oceanographic) features and the sectors that exploit them are discussed in the context of governance boundaries and institutional fit. Some examples of developing best practice in the Irish Sea which have emerged as part of the Supporting Implementation of Maritime Spatial Planning in the Celtic Seas (SIMCelt) project and other EU projects are identified and some promising avenues for developing a more holistic, ecosystem-based approach are identified. Finally, some potential future challenges for management of the Irish Sea are identified based on the emerging challenge of UK exit from the EU in March 2019.

2 Irish Sea History, Geography and Politics

There are several possible geographical definitions of the Irish Sea. Figure 3.1 illustrates the limits of the Irish Sea according to the International Hydrographic Organization (IHO), as well as those used for operational management purposes under the Oslo–Paris (OSPAR) Convention.

In terms of oceanography, this southern boundary also reflects a physical discontinuity, the Celtic Sea front (shown in red) (Simpson and Hunter 1974) where deeper stratified waters of the Celtic Sea meet the shallow tidally mixed waters of St. George's Channel (Simpson 1976). A second tidal front occurs to the north, between the Isle of Man and the island of Ireland, which defines the southern boundary of a gyre circulation system, which is characterised by stratification and associated with a fishery for *Nephrops norvegicus* (the Dublin Bay Prawn or Langoustine). Northwards, the next major physical discontinuity in water column characteristics occurs at the Islay front outside the formal bounds of the Irish Sea.

Political boundaries within the Irish Sea are complex, to the west of the Irish Sea is the island of Ireland. Ireland is divided into two jurisdictions, the Republic of Ireland, comprised of 26 counties, and Northern Ireland, one of the devolved administrations of the UK, made up of six counties. To the east of the Irish Sea lies Britain, comprised of England (home of the central UK administration), Wales and Scotland (also UK devolved administrations). The Isle of Man, a UK protectorate, sits between Britain and Ireland. Under international law, the UK and Ireland as well as the Isle of Man claim territorial seas to 12 nm from the baseline and individual devolved administrations within the UK have responsibility for specified activities within their territorial sea. In addition, both the UK and Ireland have claimed Exclusive Economic Zones beyond their territorial sea, small portions of which occur within the Irish Sea (Fig. 3.1 dashed line) though formal maritime boundaries in the border bays of Carlingford Lough and Lough Foyle have never been agreed on.

Historically all the administrations bordering the Irish Sea were under British jurisdiction, and the areas' main cities share this common history of development. Several major cities are located on the shores of the Irish Sea including Glasgow (pop. 0.6m) and Belfast (pop 1.2m) to the north, both major historical ship building centres as well as Dublin (pop. 1.3m), which was once considered the second city of the British Empire, and Liverpool (pop. 1.38m), its major trading port. The free movement of people between Ireland and the UK remains a legacy of this shared history. The example of Dublin Bay (Box 3.1) illustrates how the legacy of a large-scale geopolitical process, the expansion of the British Empire, has affected the supply and demand for the production of ecosystem services (recreational and cultural benefits) over long timescales resulting in a distinct and highly valued, nested social-ecological system embedded within the physical and social context of the larger Irish Sea.

3 The EU as a Driving Force for Environmental Efforts

Apart from international law such as the United Nations Convention on the Law of the Sea (UNCLOS), the Convention on Biological Diversity, the Aarhus Convention, and the Espoo Convention, a common basis for cooperation between Ireland and the UK (including its devolved administrations) in addressing environmental conflicts and the management of activities occurring in, or impacting upon, the Irish Sea has occurred as the result of membership of the EU (and its precursors). Both the UK and Ireland joined in 1973 and are subject to EU law, much of which mandates regional cooperation. As EU Member States, both sovereign countries have been subject to the provisions of the CFP for the last 40 years, under its common legal basis.

The Birds Directive (EEC, 1979 as amended 2009) established the basis for international cooperation on the management of wild birds and was subsequently complemented by the Habitats Directive (EEC 1992), which together provide for the protection of rare and threatened species and natural habitat types through the Natura 2000 network. The European Court of Justice confirmed in 2004 that the provisions of the Habitats Directive extend to the limit of the Exclusive Economic Zone (200 nautical miles) and Member States must designate SACs and Special Protection Areas to protect listed habitats and species.

The adoption of the Water Framework Directive (WFD) (EC, 2000) obliges Member States to meet Good Ecological Status (GES) in transitional (estuarine) and coastal waters and provides a mandate for regional cooperation. It was the first piece of EU legislation to introduce management of river basins and adjacent coastal waters at the catchment scale, through the establishment of River Basin Districts (RBDs). The WFD requires transboundary cooperation for international RBDs. The introduction of the MSFD (EC, 2008) considerably expanded the legal basis for regional cooperation with respect to the marine environment. The Directive mandates that Member States use EBM to achieve Good Environmental Status (GEnS) on a regional basis and contains 11 descriptors to assist Member States in interpreting what GEnS should look like in practice. The descriptors include commercial fisheries, biodiversity and eutrophication which are already regulated by the EU's CFP, Habitats Directive and the WFD, respectively and a suite of relatively new descriptors including marine litter and the introduction of energy (including underwater noise). The Maritime Spatial Planning Directive (2014/85/EU) (EC 2014a) complements the transboundary approach of the MSFD by ensuring that there is a sustainable balance between Member State economic ambitions and the achievement of GEnS.

While these Directives have provided a common framework for environmental protection in the marine environment, engagement with, and implementation of, the Directives have varied between the devolved authorities in the UK and also between the UK and Ireland. Generally speaking, the UK has traditionally engaged more proactively with environmental legislation and implemented more stringent measures than strictly necessary, sometimes referred to as "gold plating", while in the Republic of Ireland transposition and implementation have sometimes been more reactive, in response to infraction proceedings or the potential for these. By contrast, the Scottish Government has pursued a very proactive approach towards MSP and in the development of the Spatial Data Infrastructure, and Scotland is seen to be leading the way in the MSP process. The Scottish Marine Plan Interactive, their national digital atlas, contains a shapefile of a Scottish Exclusive Economic Zone, perhaps belying the ambition for independence of the governing Scottish National Party, which may to some extent explain their proactive approach towards EU policy and its implementation.

Under the first implementation cycle of the MSFD, the European Commission recognised a number of serious challenges to implementation at regional scales (EC 2014b). Most recently, the revised Commission Decision on Descriptors (EC 2017) sets out more rigorous definitions of GEnS criteria, meaning that the second cycle of implementation is likely to be more demanding in terms of implementation. The regional cooperation and mandate for more participatory "bottom-up" approaches under the MSFD and MSP Directives have resulted in several efforts to develop regional and sub-regional fora for marine environmental management. There have been a number of European research projects including the Partnerships Involving Stakeholders in the Celtic Sea Ecosystem (PISCES), Celtic Seas Partnership and Transboundary Planning in the European Atlantic (TPEA) projects (with the aim of harmonising regional cooperation and developing Ecosystem Approaches) to management in the region (see Chap. 6 in this volume). However, as with the WFD, the timescales for national implementation of the MSFD have resulted in limited harmonisation of approaches at the regional level. Supported by EU research funding, project-based efforts have each brought together various stakeholders to encourage multisectoral perspectives for incorporation into regional management. However, while these projects have provided a platform for consideration of different perspectives in developing management plans, they have no legal standing, are time-limited, and while it may be politically expedient for national governments to engage with such groups, there is no legal requirement to follow up on any specific recommendations. The same can be said for non-statutory national initiatives, mainly those advanced by the UK government to support regional EBM and MSP at the Irish Sea scale such as the Irish Sea Pilot project (2002–2004) and the Marine Spatial Planning Pilot (2004–2006). While these projects engaged government officials from the Republic of Ireland, the Isle of Man, the devolved administrations of the UK and many Irish Sea stakeholders and supported the statutory institutionalisation of MSP in the UK, they did not result in a more formal or statutory approach to partnership between the UK and Ireland. Transboundary working and partnership at the Irish Sea scale has mainly been at the strategic level of sharing information (Kidd and McGowan 2013) while operational cooperation has been very limited, though cooperation is a legal requirement under Article 11 of the MSP Directive.

Outside of the EU framework the only institution with an established legal basis for cooperation on matters relating to the marine environment of the North-East Atlantic, including the Irish Sea, is the OSPAR Commission created under the OSPAR Convention for the Protection of the Marine Environment of the North-East Atlantic. The Convention objectives are taken forward through the adoption of decisions which are legally binding on the Contracting Parties, and as such it represents a forum for regional cooperation, and beyond which it may become increasingly important as the UK plans to exit the EU. OSPAR objectives and approaches are very much in line with those of the EU and OSPAR structures have been used as a forum to generate a common basis for cooperation for the WFD, the Habitats Directive and, more recently, and with relative success compared to other regional seas (EC 2014b), for the MSFD. Many of the transitional waters on both sides of the Irish Sea are considered to be OSPAR problem areas or OSPAR potential problem areas in terms of eutrophication (OSPAR 2008). OSPAR offers the potential for continued regional cooperation in tackling eutrophication beyond the proposed UK exit in March 2019.

3.1 Managing Multiple Sectors

Much of the maritime activity in the Irish Sea has developed independently from the relatively new concepts of EBM and MSP and has been influenced only marginally by environmental legislation or formal MSP process. Maritime transport is the principal and traditional economic activity making use of marine spaces. In the Republic of Ireland, maritime transport accounts for 85% of the total volume of goods and 56% of the total value of goods traded nationally (Vega and Hynes 2017). In 2015, 27 million tonnes, 55% of the total volume of goods received or forwarded by ship in Ireland passed through the Irish Sea, including 84% of goods traded by sea with the UK. For the UK, this volume makes up a much smaller, but nevertheless significant, proportion of total maritime trade (approximately 10%).[1] Figure 3.3 shows the relative density of shipping and Automatic Identification System (AIS) ship track data for the passenger/roll-on roll-off vessels in the Irish Sea.

[1]Total UK maritime Freight for 2015 was 182,535,000 tonnes (Dept. of Transport Statistics, 2016).

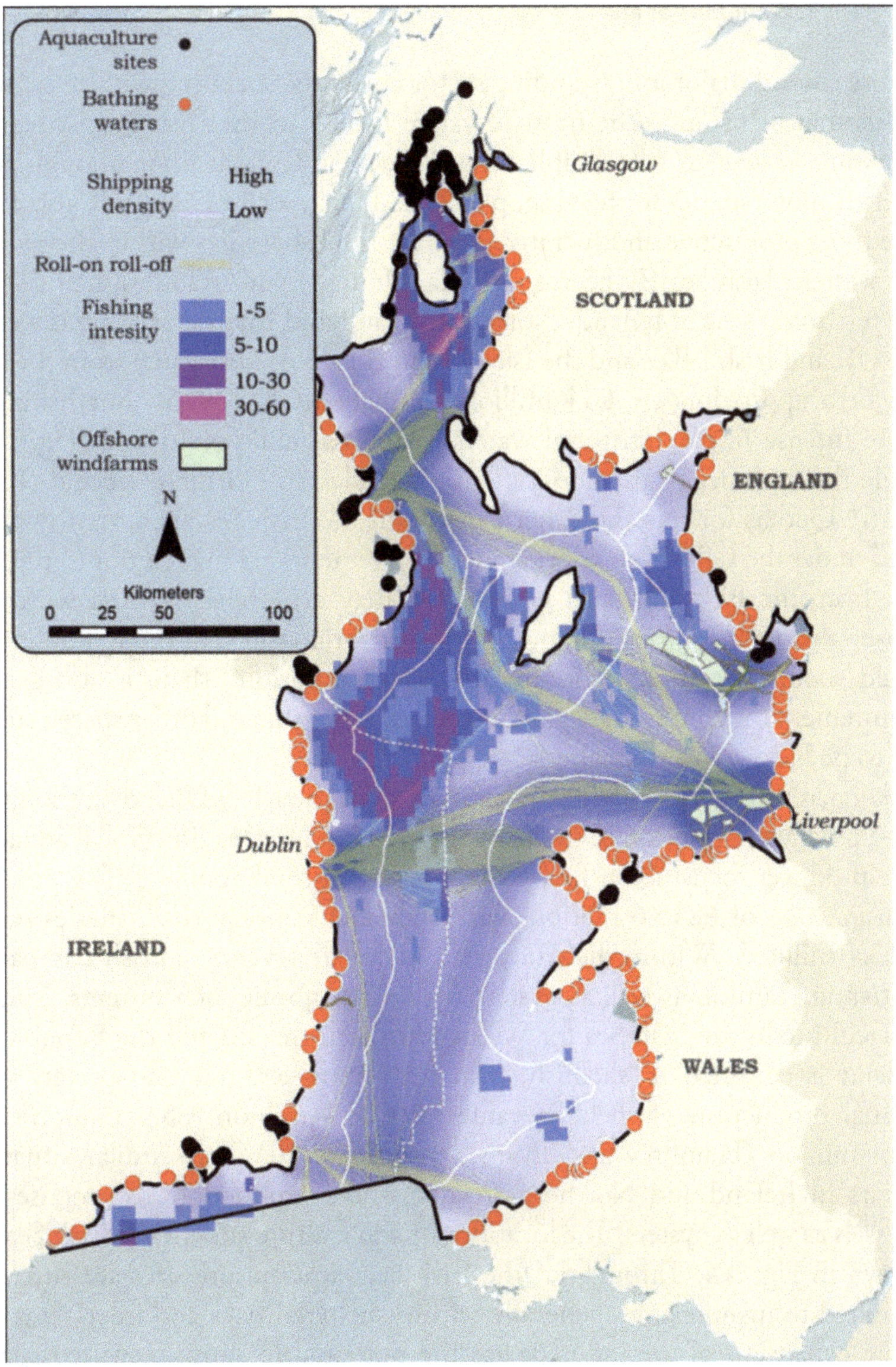

Fig. 3.3 Map of sectors explicitly addressed in the MSP Directive. Fishing pressure is expressed as swept area ratio. Renewable energy is shown in pale green. Main roll-on roll-off shipping routes are shown in yellow. Fishing pressure is shown in blue to purple. Designated bathing waters are shown in red. Aquaculture sites are shown as black circles. Data sources: Background bathymetry from http://www.emodnet-bathymetry.eu/. Aquaculture sites—https://atlas.marine.ie/#?c=53.9043:-15.8972:6. Shipping density and roll-on roll-off are based on data from https://data.gov.uk/. Fishing Intensity: OSPAR, https://odims.ospar.org/. Offshore wind farms: http://www.emodnet-humanactivities.eu/search-results.php?dataname=Wind+Farms+%28Polygons%29

The second major long-standing sector currently operating within the Irish Sea is that of fishing. The most lucrative fishery in the Irish Sea is that for *Nephrops norvegicus* (the Dublin Bay Prawn or Langoustine), though relatively minor fisheries for herring, plaice, haddock, whiting and sole also exist. The most productive and lucrative area for the fishery lies within the gyre of the western Irish Sea (ICES region VIIa, Unit 15) and straddles the limits of the territorial seas of Ireland, Northern Ireland and the Isle of Man as well as the UK and Irish EEZs and the Isle of Man (Fig. 3.4). The catch from the area is worth approximately €54 million annually (ICES 2016), but fishing is more intense in the territorial waters of the Republic of Ireland (Fig. 3.4), while the majority of the quota (75%) is landed in Northern Ireland (ICES 2016). Quotas for this and other fisheries in the Irish Sea are currently managed under the CFP. If no new fishing arrangements are made prior to the UK exit from the EU, Northern Irish fishers may no longer have access to the more valuable Nephrops grounds in the territorial waters of the Republic and could stand to lose out economically in this location. There is also a clear requirement for continued regional cooperation if this and other shared stocks are to be harvested sustainably.

Aquaculture is also specifically referred to in the MSP Directive, though it has a patchy distribution in the Irish Sea. Scotland is a leading global aquaculture producer focusing on farmed salmon (with total annual finfish production in 2014 of €855.6 million) with a smaller national shellfisheries sector (€13.1 million). Within the study area, there are several Scottish companies cultivating salmon as well as oysters within the fjordic loch systems. Marine aquaculture in the Irish Sea for Wales, Northern Ireland and the Republic of Ireland is confined to shellfish, principally mussels, but also oysters with annual production of shellfish values at €18.7 million, €5.9 million[2] and €8.5 million (Hambrey and Evans 2016; BIM 2014). A similar situation occurs in Ireland and Northern Ireland where production also focuses on mussels as well as oysters. The locations of aquaculture sites in the Irish Sea are shown in Fig. 3.4. Throughout the Irish Sea, aquaculture sites are currently confined to inshore sites, generally within sheltered bays and inlets, but offshore expansion of the industry has the potential to cause increased spatial conflict with other activities.

Both the UK and Ireland have ambitious targets for the development of offshore energy in the Irish Sea. Wind farms in the Irish Sea alone have an installed capacity of over 2 GW (ABPmer 2016) and account for about 2.6%

[2] NI production values also include Lough Foyle outside the Irish Sea.

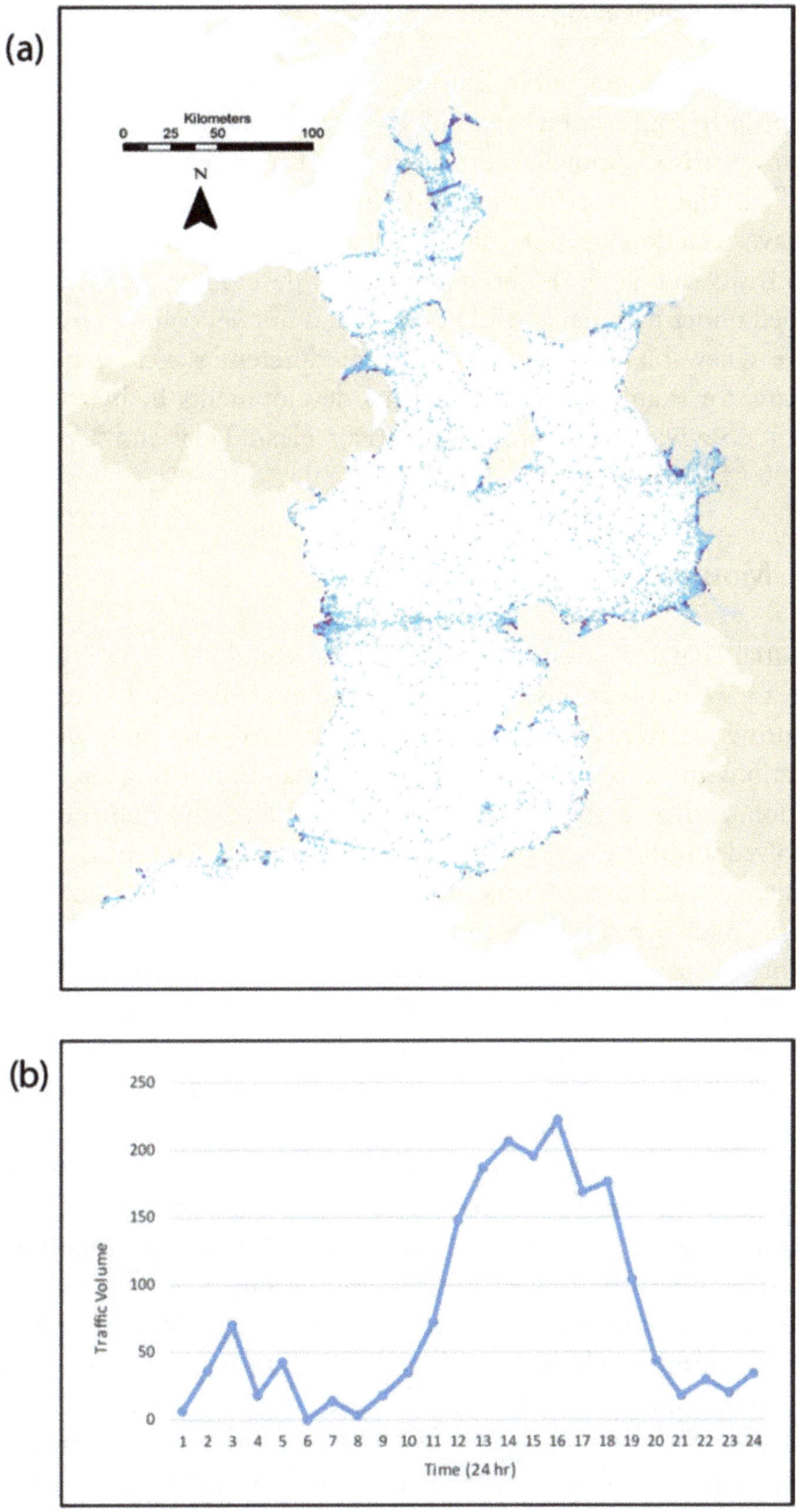

Fig. 3.4 Examples of crowd-sourced and open data. (**a**) Photography User Days based on the InVEST model (Adamowicz et al. 2011). (**b**) Data from Dublin's traffic monitoring system showing temporal patterns in beach use for Dollymount Strand. Data source: https://data.gov.ie/dataset/volume-data-for-dublin-city-from-dublin-city-council-traffic-departments-scats-system

of consented offshore wind in Europe.[3] Development of offshore wind in the Republic of Ireland commenced in 2003 with the construction of the Arklow Bank array (Risø National Laboratory 2004), but subsequent development has stalled. The majority of the wind farms in the Irish Sea (ten wind farms sites) have been developed in the English and Scottish territorial waters of the eastern Irish Sea (Fig. 3.3). Though the major development of offshore energy envisaged under national and EU policies has not yet come to fruition, other offshore renewable energy projects are at different stages of development. There are, for example, two major tidal developments in Northern Ireland waters at an advanced planning stage (Fair Head Tidal and Tidal Ventures, both 100 MW projects).

3.2 Management Challenges

The spatial characteristics of specific sectoral activities operating within the Irish Sea have implications for their management. Both offshore energy and aquaculture are relatively static, occurring at fixed sites and within specific jurisdictions and consequently both operational monitoring and overarching regulation of these activities occurs at national level through the responsible or devolved authority, where applicable. As a result, data on the location of particular activities, for example, are held by different institutions with different data policies, procedures and requirements, and there is no centralised repository of spatial data for all aquaculture sites or renewable energy sites in the Irish Sea. Most data available are based on the static boundaries of national jurisdictions. For non-mobile activities, this situation may be sufficient to enable local management. It may, however, be considered an obstacle from the perspective of more holistic regional EBM. For example, development of offshore energy farms has the potential to reduce visual amenity across international boundaries and potentially resulting in conflicts.

For mobile sectors operating within the area, for example, maritime transport and fisheries, the lack of a centralised resource for the collection and analysis of spatial information is perhaps more of a problem. The AIS system used to monitor vessels over the length of 15 m generate high volumes of almost continuous spatial information as do the Vessel Monitoring Systems (VMS) used for the monitoring of fishing effort. The patterns contained within this data are of vital importance not just for maritime safety but also in

[3] https://windeurope.org/wp-content/uploads/files/about-wind/statistics/WindEurope-Annual-Offshore-Statistics-2016.pdf.

the assessment of the levels of activity within the Irish Sea ecosystem and the impacts on the environment. While there have been centralised efforts (under the auspices of the International Council for the Exploration of the Seas) to make VMS data available for analysis, national approaches to integration and compilation of data have differed, resulting in duplication of efforts and inability to cross-compare nationally analysed data. For AIS, the sheer volume and the distributed nature of the data have meant that there has been little coordinated effort in data synthesis at the scale of the Irish Sea.

While information overload is the problem for some sectors, for other activities, a lack of information hampers local and regional ability to make informed choices. The latter is particularly true for recreational activities. For example, there are 160 designated bathing waters in the Irish Sea, which undergo regular water quality monitoring, but for those waters, there is more accurate information on the number of faecal coliforms in the water column than on the number of users of the bathing waters, and still less is known about how recreational use changes with variations in water quality. The MSP Directive also makes provision for incorporating the objectives of sustainable tourism development in spatial plans. As with bathing water quality, the relationships between tourism and environmental quality are poorly understood. While regional tourism statistics do exist, these are difficult to relate directly to specific environmental features, and recreational activities which utilise ecosystem services are often not part of the market economy, therefore estimating their value and consequently weighting them against other activities with well-constrained spatial scales and known market values remains a major challenge. While the paucity of appropriate ecosystem services data to support decision-making is not confined to the Irish Sea, the complexity of the governance structures including two nations, three devolved authorities and one Crown protectorate, each with their own unique economic and social conditions and national priorities, can result in additional complexity in terms of sourcing, harmonising and centralising data.

4 Good Practices: The Solway Firth Partnership

Despite the complexity of governance and the data challenges identified above, there are a number of emerging initiatives and technologies that offer the potential to assist MSP at the regional scale.

Integrated management and planning of marine resources across borders offers an approach to ensure that shared resources and ecosystem units are effectively managed. However, differences in timelines for the implementation of MSFD and MSP pose challenges to the management and planning of shared local and communal resources. One of the initiatives in Britain fostering formal cross-border working and local input into decision-making is the use of coastal and marine partnerships. One such partnership is the Solway Firth (SWF) Partnership.

The SWF is a unique ecosystem which lies between England and Scotland (Fig. 3.1) and is home to various national and international conservation sites (Ramsar site, Special Protection Area and SAC), historical and archaeological sites. It also hosts the largest offshore wind farm in Scottish waters (Robin Rigg). Although this ecosystem is managed and regulated primarily under two pieces of legislation (the UK Marine and Coastal Access Act (MCAA) 2009 for the English side and the Marine (Scotland) Act 2010 for the Scottish), there is policy convergence under the UK Marine Policy Statement (HM Government 2011). This policy statement derives from Section 44 of the MCAA, whereby a joint policy statement outlines the general policies of the four respective administrations that contribute to the achievement of sustainable development of the UK's marine area. Approaches being applied in the SWF to ensure joint initiatives include:

- coordination of data sharing facilitated by Scotland's National Marine Plan interactive (webGIS) & the UK Marine Science Co-ordination Committee research platform;
- harmonisation of public budget and funding available on each side of the border from local Councils and state agencies. The SWF Partnership has developed a common business plan for the SWF;
- coordination of SWF Regional Plan with Scotland National MSP;
- coordination of SWF Regional Plan with Scottish sectoral marine plans for offshore wind, wave and tidal energy; and
- joined up stakeholder involvement in the MSP process.

Through the SIMCelt project, practical approaches for planning across borders in the SWF were explored by increasing awareness of transboundary issues, highlighting conflicts in cross-border planning and management, enhancing integration and cooperation between the devolved authorities. Such lessons, joint initiatives, policies and funding will be relevant across the whole Irish Sea, especially on the island of Ireland to foster cooperation and integration to ensure effective EBM.

While hard, geographic data are often available for specific sectors, for example, the maps of fishing (a provisioning ecosystem services) shown in Fig. 3.2, finding appropriate data for the incorporation of cultural ecosystem services and recreational values into marine planning at appropriate scales remains a challenge. The potential for new sources of data to inform spatial planning is beginning to emerge. Figure 3.4a shows the levels of recreational photography within the Irish Sea area, based on the number of Photography User Days, and was calculated using InVEST data modelling suite (see Adamowicz et al. 2011 for detail). While patterns of photography are clearly linked to patterns of travel (photographs are clustered around the main ferry routes), some clusters of photographs found offshore do not match up with expected travel patterns and may indicate the existence of features of particular importance in terms of recreational and cultural values. Figure 3.4b shows temporal patterns in recreational beach use in Bull Island Dublin (see Box 3.1) inferred from traffic data on the "Dublin Bay Dashboard" developed as part of Celtic Seas Partnership Project illustrating how existing public data can be used to gather information on recreational use patterns.

With the increasing amount of spatial and temporal information being generated from the bottom-up by members of the public and local government initiatives (Dublin's traffic monitoring system), developing appropriate techniques for gathering and analysing such "big data" provides a promising avenue for incorporating semi-quantitative ecosystem services data into spatial planning, which may help to better represent data on public values into management of the public good that are the seas.

5 Future Management

The EU has provided the legislative framework and common basis for cooperation on maritime affairs and marine environmental management and protection over the last 40 years. Though concerted regional efforts have been sporadic and project-based, nevertheless these efforts have helped to develop an international community of best practice and expertise in marine planning and environmental management in the UK and Ireland. With the UK decision to exit from the EU, the future basis for cooperation is less certain.

One high-profile issue in the referendum campaign was the CFP, and it is highly likely that the UK will now enforce a more restrictive regime on international vessels fishing within its national waters. Such a change has clear implications for the management of the *Nephrops* fishery in the Irish Sea.

There is no clear basis for common future exploitation of this shared resource in this area where boundaries are still contested.

At present, the split of Nephrops fisheries quotas is made centrally at the EU level; in the absence of this, an alternative process will be needed to allocate and enforce quotas. In the absence of a local cross-border management arrangement, it is entirely possible that the fishery might return to an open-access regime with its inherent tendencies towards the tragedy of the commons. Alternatively, restricting access to the fishery for non-national vessels could potentially revive historic tensions between north and south.

The examples of successful local regional cooperation in the Solway Firth, explored as part of the SIMCelt project, may provide a model for local management of the resource on a transboundary (Irish all island) cooperation basis, in keeping with the concepts of an Ecosystem Approach. However, such cooperation would necessitate the development of appropriate local cross-border institutions. The Loughs Agency, as one of the North-South Implementation bodies under the Good Friday Agreement, provides such a role, but has a remit only to manage fisheries and aquaculture. More effective management would require an institution with a wider remit.

While fisheries represent a high-profile and contentious example of potential future conflicts, maritime transport in the Irish Sea is particularly vital to Ireland and not insignificant to the UK. Under any planning framework (whether inside or outside of the EU), efficient transport is likely to be of the highest priorities when considering maritime development. There is a tradition of free trade and transport across the Irish Sea, driven by markets and their inherent efficiencies, which is centuries old and, at least in terms of its spatial patterns, is unlikely to be affected by changes in obligations for environmental protection or for MSP. The legacy of historic shipping and its infrastructure will continue to shape the patterns of transport in the Irish Sea (just as they have shaped the social and ecological development of the Dublin Bay social-ecological system). Nevertheless, depending on the nature of future EU–UK trade, customs and tariffs arrangements, volumes of ship traffic could potentially stand to change, potentially favouring more direct routes between Ireland and continental Europe.

In terms of marine renewable energy development, given the short distances between countries, and across the Irish Sea, in the absence of a harmonised approach to marine planning, unilateral decisions of individual nations (or of devolved authorities on certain matters) within their own territorial waters risk imposing externalities, dis-benefits in terms of cultural and amenity values (cultural ecosystem services) on the coastlines of other countries. The two states bordering the Irish Sea have had an uneasy relationship in the past and, if planning conflicts are to be avoided, some means of operational cross-border cooperation will need to be maintained. OSPAR may continue to provide a mechanism for such cooperation.

Overall, the activities currently occurring within the Irish Sea are strongly influenced by their history of development. As illustrated by the case of Dublin Bay, nested sub-systems of human uses, both commercial and recreational, have evolved over time and are influenced by global trade, transport and economy as well as local patterns of physical and social phenomena. The challenges and the potential for regional MSP in terms of governance, harmonisation of information and joined approaches are beginning to emerge, yet given the unknown nature of the new relationship developing between the UK and the EU, the future of the MSP process within the Irish Sea is highly uncertain. The impact of the MSP Directive on the activities occurring within the Irish Sea and the patterns of resource use and exploitation have yet to be experienced. Whatever the future political context following the UK departure from the EU, effective sustainable management for the Irish Sea will require ecosystem-based approaches, which reflect the complexity of the Irish Sea and its nested social and ecological sub-systems and involve transboundary cooperation. Whether the political and economic conditions will favour such approaches will be critical in determining the outlook for the Irish Sea environment and the ecosystem services it provides to the people on its shores.

Acknowledgements This material is based upon works supported by Science Foundation Ireland (SFI) under Marine and Renewable Energy Ireland (MaREI) Centre (12/RC/2302). The Open Access fee of this chapter was provided from the same source.

References

ABPmer. (2016). *Future Trends in the Celtic Seas: Summary Report*. ABPmer Report No. R.2584a. A Report Produced by ABPmer and ICF International for the Celtic Seas Partnership, August.

Adamowicz, W. L., Naidoo, R., Nelson, E., Polasky, S., & Zhang, J. (2011). Nature-Based Tourism and Recreation. In P. Kareiva, G. Daily, T. Ricketts, H. Tallis, & S. Polasky (Eds.), *Natural Capital: Theory and Practice of Mapping Ecosystem Services*. New York: Oxford University Press.

Alexander, K., Janssen, R., Arciniegas, G., O'Higgins, T., Eikelboom, T., & Wilding, T. (2012). Interactive Marine Spatial Planning: Siting Tidal Energy Arrays Around the Mull of Kintyre. *PloS One*. https://doi.org/10.1371/journal.pone.0030031

Ansong, J., Gissi, E., & Calado, H. (2017). An Approach to Ecosystem-Based Management in Maritime Spatial Planning Process. *Ocean and Coastal Management, 141*, 65–81.

BIM. (2014). *BIM Annual Aquaculture Survey*. 16 pp. http://www.bim.ie/media/bim/content/publications/BIM,Aquaculture,Survey,2014.pdf. accessed 7/9/18

Crowder, L., & Norse, E. (2008). Essential Ecological Insights for Marine Ecosystem-Based Management and Marine Spatial Planning. *Marine Policy, 32*, 772–778.

Dietz, T., Ostrom, E., & Stern, C. (2003). The Struggle to Manage the Commons. *Science, 302*, 1907–1912.

Domingues-Tejo, E., Metternicht, E., Johnston, E., & Hedge, L. (2016). Marine Spatial Planning Advancing the Ecosystem-Based Approach to Coastal Zone Management: A Review. *Marine Policy, 72*, 115–130.

Douvere, F. (2008). The Importance of Marine Spatial Planning in Advancing Ecosystem-Based Sea Use Management. *Marine Policy, 32*, 762–771.

EC. (2014a). Directive 2014/89/EU of the European Parliament and of the Council of 23rd of July 2014 Establishing a Framework for Maritime Spatial Planning. *Official Journal of the European Union*. L257/135.

EC. (2014b). Report from the Commission to the Council and the European Parliament on the First Phase of Implementation of the Marine Strategy Framework Directive (2008/56/EC). COM (2014) 97 final.

EC. (2017). Commission Decision (EU) 2017/848 of 17 May 2017 Laying Down Criteria and Methodological Standards on Good Environmental Status of Marine Waters and Specifications and Standardized Methods for Monitoring and Assessment, and Repealing Decision 2010/477/EU.

EC. (2000). Directive 2000/60/EC of the European Parliament and of the Council of 23 October 2000 Establishing a Framework for Community Actions in the Field of Water Policy. *Official Journal of the European Communities*. 2000; L327, 1.22.12.2000.

EC. (2008). Directive 2008/56/EC of the European Parliament and of the Council of 17 June 2008 Establishing a Framework for Community Action in the Field of Marine Environmental Policy. *Official Journal of the European Union*. 2008.2 L 164/19.

EC Directive 2009/147/EC of the European Parliament and of the Council of 30 November 2009 on the Conservation of Wild Birds (Codified Version). *Official Journal of the European Union*. 2009; L20/7.

EEC. (1979). Council Directive 79/409/EEC of 2 April 1979 on the Conservation of Wild Birds. *Official Journal of the EEC*. 1979 C24.

EEC. (1992). Council Directive 92/43/EEC of 21 May 1992 on the Conservation of Natural Habitats and of Wild Fauna and Flora. *Official Journal of the EEC*. 1992; L206/7.

Folke, C., Pritchard, L., Berkes, F., Colding, J., & Svedin, U. (2007). The Problem of Fit Between Ecosystems and Institutions: Ten Years Later. *Ecology and Society, 12*(1), 30 Retrieved from http://www.ecologyandsociety.org/vol12/iss1/art30/.

Hambrey, J., & Evans, S. (2016). *SR694. Aquaculture in England, Wales and Northern Ireland* (p. 162). Seafish.

Hardin, G. (1968). The Tragedy of the Commons. *Science, 162*, 1243–1248.

HM Government. Marine and Coastal Access Act 2009. Retrieved from http://www.legislation.gov.uk/ukpga/2009/23/contents.

HM Government. (2011). UK Marine Policy Statement. HMSO ISBN 978 0 10 8510434. Retrieved from http://www.official-documents.

ICES. (2016). *ICES Advice on Fishing Opportunities Catch and Effort in the Celtic Seas Ecoregion*. ICES Copenhagen.

Kidd, S., & McGowan, L. (2013). Constructing a Ladder of Transnational Partnership Working in Support of Marine Spatial Planning: Thoughts from the Irish Sea. *Journal of Environmental Management, 126*, 63–71.

Lackey, R. T. (1998). Seven Pillars of Ecosystem Management. *Landscape and Urban Planning, 40*(1–3), 21–30.

Maes, J., Teller, A., Erhard, M., et al. (2014). *Mapping and Assessment of Ecosystem and Their Services. Indicators for Ecosystem Assessments Under Action 5 of the EU Biodiversity Strategy to 2020*. Luxembourg: Publications Office of the European Union. ISBN 978-92-79-36161-6.

Mayer, I., Zhou, Q., Lo, J., Abspoel, L., Keijser, X., Olsen, E., Nixon, E., & Kannen, A. (2013). Integrated, Ecosystem-Based Marine Spatial Planning: Design and Results of a Game-Based, Quasi-Experiment. *Ocean and Coastal Management, 82*, 7–26.

McLeod, K. L., Lubchenco, J., Palumbi, S. R., & Rosenberg, A. A. (2005). *Scientific Consensus Statement on Marine Ecosystem-Based Management*. Prepared by Scientists and Policy Experts to Provide Information About Coasts and Oceans to U.S. Policy-Makers. Retrieved from http://www.compassonline.org/sites/all/files/document_files/EBM_Consensus_Statement_v12.pdf.

MEA. (2003). *Ecosystems and Human Well-Being: A Framework for Assessment*. Washington, DC: Island Press.

MEA. (2005). Ecosystems and Human Well-Being: Current State and Trends: Findings of the Condition and Trends Working Group. Retrieved from http://www.unep.org/.

Mee, L. D., Cooper, P. C., Gilbert, A. J., Kannen, A., & O'Higgins, T. (2015). Sustaining Europe's Seas as Coupled Social-Ecological Systems. *Ecology and Society, 19*(3). Retrieved from https://doi.org/10.5751/ES-07143-200101.

O'Higgins, T. (2006). *Dublin Bay Water Quality Monitoring Programme Biological Monitoring Report* (p. 57). Dublin City Council.

O'Higgins, T. G. (2017). You Can't Eat Biodiversity: Agency and Irrational Norms in European Aquatic Environmental Policy. *Challenges in Sustainability, 5*(1), 43–51.

O'Higgins, T. G., & Wilson, J. G. (2005). Impact of the River Liffey Discharge on Nutrient and Chlorophyll Concentration in the Liffey Estuary and Dublin Bay (Irish Sea). *Estuarine Coastal and Shelf Science, 64*, 323–334.

O'Higgins, T. G., Cooper, P., Roth, E., Newton, A., Farmer, A., Goulding, I., & Tett, P. (2014). Temporal Constraints on Ecosystem Management: Definitions and Examples from Europe's Regional Seas. *Ecology and Society, 19*(3): 12 pages.

Olesen, E., Kleeivn, A. R., Skjoldal, H. R., & von Quillfeldt, C. H. (2011). Place-Based Management at Different Spatial Scales. *Journal of Coastal Conservation, 15*, 257–269.

OSPAR. (2008). Eutrophication Status of the OSPAR Maritime Area-Second OSPAR Integrated Report. ISBN 978-1-906840-13-6.

Ostrom, E. (2003). How Types of Goods and Property Rights Jointly Affect Collective Action. *Journal of Theoretical Politics, 15*, 239–270.

Ostrom, E. (2009). A General Framework for Analysing Sustainability of Social-Ecological Systems. *Science, 325*, 419–422.

Pendleton, L., et al. (2007). Is the Non-market Literature Adequate to Support Coastal and Marine Management? *Ocean and Coastal Management, 50*, 363–378.

Potts, T., Alexander, K., & O'Higgins, T. (2013). *Supporting Marine Spatial Planning with Local Socioeconomic Data* (p. 21). Scotland Centre for Expertise in Water.

Potts, T., Pita, C., O'Higgins, T., & Mee, L. D. (2016). Who Cares? European Attitudes Towards Marine and Coastal Environments. *Marine Policy, 72*, 59–66.

Risø National Laboratory. (2004). *Offshore Wind Energy and Industrial Development in the Republic of Ireland*. Dublin: Sustainable Energy Ireland.

Scottish Government. Marine (Scotland) Act (2010). Retrieved from http://www.legislation.gov.uk/asp/2010/5/pdfs/asp_20100005_en.pdf.

Simpson, J. H. (1976). A Boundary Front in the Summer Regime of the Celtic Sea. *Estuarine Coastal and Shelf Science, 4*(1), 71–81.

Simpson, J., & Hunter, J. (1974). Fronts in Irish Sea. *Nature, 250*(5465), 404.

Stelzenmüller, V., Coll, M., Mazaris, A. D., Giakuomi, S., Katsanevakis, S., Portman, M. E., Degen, R., Mackelworth, P., Gimple, A., Albano, P. G., Almpanidou, V., Claudet, J., Essl, F., Evagelopoulos, T., Heymans, J. J., Genov, T., Kark, S., Micheli, F., Pennino, M. G., Rilov, G., Rumes, B., Steenbeek, J., & Ojaveer, H. (2018). A Risk-Based Approach to Cumulative Effect Assessments for Marine Management. *Science of the Total Environment, 612*, 1132–1140.

Tallis, H., Levin, P. S., Ruckelshaus, M., Lester, S. E., McLeod, K. L., Fluharty, D. L., & Halpern, B. S. (2010). The Many Faces of Ecosystem-Based Management: Making the Process Work Today in Real Places. *Marine Policy, 34*, 340–348.

Van Tatenhove, J., Raakjaer, J., van Leeuwen, J., & van Hoof, L. (2014). Regional Cooperation for European Seas: Governance Models in Support of the Implementation of the MSFD. *Marine Policy, 50*, 364–372. https://doi.org/10.1016/j.marpol.2014.02.020.

Vega, A., & Hynes, S. (2017). Ireland's Ocean Economy. Socio-Economic Marine Research Unit, NUI, Galway. Pp 67 ISSN 2009-6933.

4

Systematic Conservation Planning as a Tool to Advance Ecologically or Biologically Significant Area and Marine Spatial Planning Processes

Linda R. Harris, Stephen Holness, Gunnar Finke, Stephen Kirkman, and Kerry Sink

1 Introduction

It is no coincidence that human population densities are three times higher along coastal margins compared to the global average (Small and Nicholls 2003). People love the sea. It features prominently in many cultures, traditions, myths and legends, with our connection ranging from occasional

L. R. Harris (✉)
Department of Zoology, Institute for Coastal and Marine Research, Nelson Mandela University, Port Elizabeth, South Africa
e-mail: Linda.Harris@mandela.ac.za

S. Holness
Department of Zoology, Nelson Mandela University, Port Elizabeth, South Africa

G. Finke
Deutsche Gesellschaft für Internationale Zusammenarbeit (GIZ) GmbH, Swakopmund, Namibia

S. Kirkman
Oceans and Coasts, Department of Environmental Affairs, Cape Town, South Africa

K. Sink
South African National Biodiversity Institute, Cape Town, South Africa

Institute for Coastal and Marine Research, Nelson Mandela University, Port Elizabeth, South Africa

J. Zaucha, K. Gee (eds.), *Maritime Spatial Planning*,
https://doi.org/10.1007/978-3-319-98696-8_4

holidays through to complete dependence for livelihoods. Unsurprisingly, use of the abundant and rich resources and services provided by the global oceans has escalated rapidly, with increasing and diversifying ocean-based resource extraction, shipping and trade, and recreational activities. Even in just a recent five-year period, nearly 66% of all oceans and 77% of Exclusive Economic Zones (EEZs) showed increases in cumulative impacts from anthropogenic activities (Halpern et al. 2015).

With increasing uses and users of the ocean comes increasing conflict. This conflict exists as both user-user conflicts, where competing sectors require use of the same space, and user-environment conflicts, where an activity negatively impacts the natural environment. Studies that sought to reduce these conflicts have shown the benefits of zoning the ocean in space and time. They demonstrated that a planned use of the marine environment can minimise losses and maximise gains for conflicting sectors whilst still protecting and conserving the underlying ecosystems and their associated biodiversity (e.g., Klein et al. 2009; White et al. 2012). Thus, if all users are willing to compromise and perhaps forego some of their ideals in cases of unavoidable conflicts, the overall outcome is that many more objectives can be achieved and many more benefits won.

The challenge, therefore, is to develop science-based methods that can help resolve as many of these conflicts in an open, fair and robust way, such that social, economic and ecological objectives can be met in a single solution. This chapter considers two existing tools—Ecologically or Biologically Significant Areas (EBSAs) and Marine Spatial Planning (MSP)—and describes how Systematic Conservation Planning (SCP) can advance and link these two processes. The efficacy of this SCP approach is discussed in the context of developing countries currently seeking sustainable ocean-resource use whilst simultaneously aiming to grow their national economies. The broad applicability of the method is also showcased by including countries with contrasting data availability. The International Union for Conservation of Nature (IUCN) definition of a "Protected Area" (PA) is used throughout, with "reserve" referring to the stricter Category 1a and 1b PAs (see Dudley 2008).

2 Spatial Prioritisation and Planning

2.1 Lessons from Land

The discipline of land-use planning has a much longer history than that of sea-use planning, providing opportunity for the latter processes to learn from what has gone before and to build on what is currently considered best practice. Further, our understanding of the relationship between humans and the environment has grown substantially in the last few decades. Prior to the 1960s, humans were considered separate from the environment; conservation was framed as "nature for itself", with areas of wilderness locked away in reserves (Mace 2014) and the remainder seen largely as available for almost any other human use. Placement of these reserves was generally ad hoc, often in areas unsuitable for productive agriculture or human habitation, and mostly ignored fundamental conservation issues such as biodiversity representation (Pressey 1994). In hindsight, this was an inefficient strategy. Despite the very low opportunity cost and limited conflict with other sectors, reserve networks spanned a much larger area than was required to achieve the same conservation benefits, with the additional disadvantage of carrying higher operational costs (Pressey 1994).

Over the turn of the century, we have progressed through periods of framing conservation as "nature despite people", where avoiding extinction and loss was our focus; to "nature for people", as the value of ecosystem services was recognised and explored; to "people and nature", where people are now considered part of ecological systems (Mace 2014). No longer is the focus on those isolated reserve "islands" in a landscape we were otherwise content to modify at will. Rather, we recognise the need to create shared landscapes between people and nature, with strong emphasis on maintaining ecological processes, adaptability and resilience in this social-ecological space (Dudley 2008; Mace 2014).

This modern framing of conservation and management is exemplified in South Africa, where the term "conservation planning" was replaced with "biodiversity planning" among practitioners and in policy. The former term was widely misinterpreted as strictly reserve design and PA expansion, whereas the intent was rather spatial prioritisation for land-use planning and decision-making. In this process, SCP is used to identify priority areas for biodiversity (Critical Biodiversity Areas [CBAs] and Ecological Support Areas [ESAs]), a desired state or management objective is set for these areas, and then activities compatible with achieving or maintaining that state are specified (SANBI 2017). Although only a subset of CBAs are PAs, biodiversity in all CBAs receive some form of protection because of the additional policies and regulations in place to regulate activities within them. ESAs are similar, although the focus in such areas is more on maintaining ecological processes that support ecosystem form and function, particularly for safeguarding biodiversity in CBAs and delivering ecosystem services, for example, corridors along which species can migrate in response to climate change. Some ecosystem modification is permissible in ESAs, provided the ecological condition of the site remains above a specified threshold.

It is only recently that our growing contemporary viewpoint is that people are part of nature and not separate from it—neither are we the sole benefactors of what nature provides. Best practice in conservation and land-use planning is now understood to be managing landscapes as social-ecological systems using multidisciplinary processes that aim to achieve social, economic and ecological objectives in an open, fair and transparent way (Ban et al. 2013). Formal reserves still, and will always, have their place. However, there is much more emphasis today on using the land "beyond the fence" more coherently and sustainably, such that ecosystems retain their resilience and adaptive capacity, especially in the face of accelerating global change.

2.2 Application in the Sea

> Aichi Target 11: By 2020, at least 17 per cent of terrestrial and inland water, and 10 per cent of coastal and marine areas, especially areas of particular importance for biodiversity and ecosystem services, are conserved through effectively and equitably managed, ecologically representative and well connected systems of protected areas and other effective area-based conservation measures, and integrated into the wider landscape and seascapes. (Convention on Biological Diversity 2011)

MSP is a modern solution to a modern problem, so it reflects our contemporary understanding of conservation and management, described earlier. It is considered "a practical way to create and establish a more rational organisation of the use of marine space and the interactions between its uses, to balance demands for development with the need to protect marine ecosystems, and to achieve social and economic objectives in an open and planned way" (DEFRA 2008, cited in Ehler and Douvere 2009). It explicitly aims to analyse and allocate parts of the ocean to the various human uses, in both space and time, in such a way that it reduces conflict and achieves social, economic and environmental objectives (Ehler and Douvere 2009; see also Chap. 1 of this book).

MSP initiatives invariably have a strong political or government-driven process behind them, with the intent of achieving an overarching goal—usually sustainable development (Gilliland and Laffoley 2008). To succeed, MSP must adopt principles of ecosystem-based management. Critically, therefore, a core objective in the plan must be to maintain the underlying environment "in a healthy, productive and resilient condition so that it can provide the services humans want and need" (McLeod et al. 2005). Safeguarding biodiversity is thus the foundation of sustainable development: the demands placed on the ocean space must not exceed its capacity to provide and meet those demands (Gilliland and Laffoley 2008). Consequently, it is imperative that the MSP includes Marine Protected Area (MPA) networks and other effective area-based conservation measures to mitigate user-environment conflict. It cannot focus solely on resolving user-user conflict.

Given the lessons learnt in the terrestrial environment, the currently limited extent of MPAs globally (Sala et al. 2018) is a strong concern, but it could also be viewed as an opportunity. We are now poised to take the tools and principles we have learnt on land, adapt them for the sea and plan efficiently for a sustainable future, with biodiversity appropriately represented in complementary MPAs. In this way, we can avoid two important pitfalls: first, inefficient and insufficient MPA networks do not deliver optimal benefits; second, they may bring an illusion of accomplishment, with no perceived need for well-located MPAs. This provides a clear motivation to fully consider the biodiversity represented in sites and the potential benefits from MPAs rather than rushing to declare MPAs with limited biodiversity and ecosystem-service value simply to meet internationally agreed area-based targets. As on land, no-take reserves are one of several conservation and management tools and serve a critical role in safeguarding biodiversity. However, beyond reserves, it is important that we create shared seascapes with nature, zoning the ocean into areas that support activities compatible with the underlying biodiversity features such that, despite partial ecosystem modification, ecological form and function are maintained.

2.3 The Role of Ecologically or Biologically Significant Areas

At the seventh Convention of Parties (COP 7) to the Convention on Biological Diversity (CBD) in 2004, an Ad Hoc Open-Ended Working Group on Protected Areas (hereafter, Working Group) was established, inter alia, to explore options for establishing MPAs in Areas Beyond National Jurisdiction (ABNJ) in a way that was science based and consistent with international law. After a series of meetings and discussions, the Working Group proposed the concept of Ecologically or Biologically Significant Marine Areas (EBSAs; see Dunn et al. 2014 for a full history), which has since been applied in areas under national jurisdiction as well.

The intent of a global MPA network is "To maintain, protect and conserve global marine biodiversity through conservation and protection of its components in a biogeographically representative network of ecologically coherent sites" (UNEP-CBD 2007, 2009), where EBSAs (with enhanced management) were intended to be a core part of the initial steps towards identifying and creating this network (UNEP-CBD 2007). The Working Group proposed seven criteria by which EBSAs are evaluated. Candidate sites are ranked (as high, medium or low) for their uniqueness or rarity; special importance for life history stages of species; importance for threatened, endangered or declining species and/or habitats; vulnerability, fragility, sensitivity or slow recovery; biological productivity; biological diversity; and naturalness—they must meet at least one of these to qualify as an EBSA (UNEP-CBD 2007, 2009). At COP 9, the criteria were adopted by the CBD, and it was also noted that MPA networks needed to take into account EBSAs; biodiversity representation across a suitable bioregionalisation; connectivity among sites; replication of features; and adequacy and viability of sites (UNEP-CBD 2007, 2009). However, very little guidance was provided on how countries ought to do this, other than "iteratively use qualitative and/or quantitative techniques" (UNEP-CBD 2007).

Following adoption of the criteria, the CBD Secretariat arranged workshops to assist countries and regions in identifying EBSAs, with the first set of EBSAs formally recognised at COP 11 in 2012. Today, 279 EBSAs are recognised worldwide (Johnson et al. 2018). Importantly, EBSA status itself does not require any conservation, protection or management interventions. However, at COP 10, in 2010, governments were encouraged to cooperate to identify and adopt appropriate conservation and sustainable-use measures in EBSAs within their EEZs and territorial waters, including establishing networks of representative MPAs. In this way, countries could potentially use EBSAs to identify areas for formal protection towards achieving Aichi Target 11. Additionally, negotiations are underway towards an instrument under which marine biodiversity (e.g., in EBSAs) could be protected in ABNJ (UN General Assembly 2017), such that these important areas could also contribute towards the global MPA network.

How countries identified EBSAs at the workshops was largely an expert-based approach. Although the seven criteria do make EBSA identification systematic to some degree and the principles for network design are useful, the loose guidance for applying these makes it difficult to assess if networks of EBSAs or MPAs are indeed sufficiently representative, connected, replicated, adequate and viable (see also Bax et al. 2016). These are especially important shortcomings if EBSAs are the mechanism that a country might choose to underpin their national MPA networks towards achieving Aichi Target 11 and perhaps ultimately for similar targets to be met in ABNJ.

To this end, further development of the EBSA process was encouraged at COP 13 in 2016. Some research teams have attempted this, for example, by advocating for a multi-criteria approach with thresholds per criterion conducted for separate habitat types (Clark et al. 2014). Exploring different methods for applying the criteria is needed to advance this aspect of the EBSA process. However, a multi-criteria analysis, particularly on a per-habitat basis, still does not give an indication if the replication and representation is sufficient; does not account for complementarity, which will be essential for conservation efficiency in the MSP context; does not address issues of connectivity, among both ecosystem types or biodiversity features and EBSAs in the network; and is strongly dependent on data to evaluate the criteria against the set thresholds.

Another gap in the EBSA process is that there is currently no system in place to identify and recognise areas that are not EBSAs in their own right but still need special management because they support ecosystem (and EBSA) function and contribute to securing long-term persistence of biodiversity features and processes. These areas are much like the ESAs in South African terrestrial biodiversity planning. Nevertheless, with appropriate conservation and management measures, EBSAs could easily be the tool by which countries can achieve internationally codified conservation targets. They could also form the ecological basis of an ecosystem-based MSP. This provides a key imperative to address the shortcomings in the current EBSA process, notably around biodiversity representation and persistence.

2.4 Systematic Conservation Planning: The Tool for the Job

SCP is a spatial prioritisation tool that supports decision-making about actions (usually with limited resources) that optimise benefits for biodiversity at the least cost to society. It is based on two key objectives: representation and persistence (Margules and Pressey 2000); biodiversity must be adequately represented in comprehensive PA networks such that species, features and processes can persist in perpetuity. This framing requires the conservation problem to be spatially explicit and target driven (recognising that non-target-based approaches have also been developed, e.g., Zonation; Moilanen et al. 2009).

SCP software, for example, Marxan (Ball et al. 2009), generally relies on an optimisation algorithm. Thus, it has strong focus on using the principle of complementarity to achieve the user-defined targets in the most efficient spatial configuration. However, because planning is for real-world implementation, the distribution of other activities within the planning domain must be accounted for: the fewer conflicts proposed PAs create with existing uses and users, the greater the likelihood of implementation. Thus, where options exist to meet biodiversity targets in areas that have no or few competing uses, those sites should be preferably selected. Consequently, SCP algorithms are designed to select sites with biodiversity value in a configuration that meets targets for biodiversity representation in the most spatially efficient way, avoiding competing land or sea uses where possible. Plans also need not focus solely on delineating conservation areas. For example, SCP tools have been very successful in zoning the oceans to optimise both socio-economic and conservation objectives simultaneously (e.g., Klein et al. 2008).

Comprehensive cover of biodiversity (at all organisational levels) in a planning domain is often an unrealistic ideal, forcing planners to use surrogates, typically habitats or ecosystem types, with some additional key biodiversity features, such as threatened species, unique features and important ecological processes. Setting targets for biodiversity features is often a contentious debate, with options ranging from codified targets, such as Aichi Target 11, to species- or ecosystem-specific targets based on minimum viable population sizes or species-area curves. How rigorously targets need to be addressed may depend on the nature of the planning problem. However, experience has shown that pragmatic decisions can circumvent issues like "how much is enough", and that adopting heuristic or codified targets provides an excellent, practical solution in the interim until better information becomes available. This is particularly the case when protection levels are well below any target (i.e., near zero, as is generally the case in the oceans). Finding optimum targets matters more when protection levels might be approaching those values.

SCP deliberately incorporates past conservation efforts and seeks to find complementary solutions to existing PA networks, thereby minimising any inefficiency in past ad hoc delineations. Further, because the spatial prioritisation problem is solved using an algorithm that was designed to be a scenario-planning, decision-support tool, it can generate multiple alternate solutions among which decision-makers can choose and trade-offs analysed and compared (Harris et al. 2014b). This flexibility in finding solutions across the planning domain is also very powerful for negotiations. In cases where some PA sites are acceptable to stakeholders and others rejected, the algorithm can be rerun with the acceptable portions hardwired into the final solution, and alternative areas sought to meet the remainder of the biodiversity targets. Another benefit is that the site-selection frequency can be used to guide delineation of both core areas for conservation and supporting areas. The outputs can also be interrogated to determine reasons for site selection. In turn, this could be used as a robust guide for management and stakeholder negotiation: knowing which biodiversity features are meeting their targets in a particular area will give an indication as to which activities are compatible and thus locally permissible.

The current shortcomings highlighted in the EBSA identification and delineation process thus appear to fit the strengths of SCP. Using this spatially explicit, target-driven tool could assist in selecting sites that are more representative of biodiversity and address replication, connectivity, adequacy and viability. Emphatically, the SCP process does not replace the criteria-based EBSA identification process in any way. Rather, it provides a more robust method of applying the criteria than expert judgement alone (see Table 4.1). The additional benefits of the SCP approach are as follows: first, it explicitly addresses the objective of creating a "biogeographically representative network of ecologically coherent sites" (UNEP-CBD 2007, 2009); second, it inherently seeks conflict avoidance, making implementing the encouraged conservation and sustainable-use measures within EBSAs more likely.

At this point, undertaking data-driven SCP to identify EBSAs may seem ideal but entirely impossible in data-poor areas where no maps exist on which to base the planning. With this in mind, the Benguela Current Large Marine Ecosystem (BCLME) in Southern Africa provides a robust test of using SCP to advance EBSA and MSP processes. On the one hand, South Africa is one of the global leaders in SCP (Balmford 2003) and thus has comprehensive spatial data available for the marine environment (Sink et al. 2011; Majiedt et al. 2012; Sink et al. 2012). On the other hand, Namibia and Angola do not have the required data available on which planning can be based. Further, data issues notwithstanding, there is a very clear and recognised need for MSP to enhance sustainable development in this region, with legislative frameworks currently being developed.

3 Spatial Planning in the Benguela Current Large Marine Ecosystem

3.1 One of the Most Productive Marine Areas in the World

The BCLME spans the West African Coast, including the EEZs of Angola, Namibia and the west coast of South Africa (Fig. 4.1). It is one of the four

Table 4.1 SCP elements that link to each of the seven EBSA criteria

EBSA criterion	SCP element
Uniqueness or rarity	Unique sites or features are considered as "irreplaceable" in an SCP context, and thus will always be selected because they are the only place where targets for that feature can be met.
Special importance of life history stages of species	Usually, these sites of importance are included in the spatial prioritisation as an explicit feature (e.g., turtle nesting beaches) with a representation target.
Importance for threatened, endangered or declining species and/or habitats	All habitat types and species that are included in the spatial prioritisation have their own representation target. Features (habitats, species) that are threatened or declining will have few options where these targets can be met, and thus will have high selection frequency. The ecosystem threat status analysis that follows the condition assessment (see Sect. 3.4) can also contribute to this criterion.
Vulnerability, fragility, sensitivity or slow recovery	This criterion is accounted for in two possible ways. For species, they could be included as a separate feature with a representation target (e.g., vulnerable marine ecosystems). For habitats, the cumulative pressure assessment explicitly scores recovery time as one of the assessment metrics.
Biological productivity	Productivity can be included either as a map of chlorophyll-*a* intensity (or similar), from which the areas with higher values will be preferentially selected to meet targets, or it could be included as a feature map of upwelling cells with a representation target.
Biological diversity	Sites with high biological diversity can be mapped either as a separate feature with a target or if multiple biodiversity layers are included in the spatial prioritisation, then diverse areas will be preferentially selected because they are efficient sites in which biodiversity targets can be met.
Naturalness	This is accounted for in the site condition assessment, where sites in good condition (less degraded) are preferentially selected over sites in fair or poor condition where the option exists.

major eastern boundary upwelling systems and one of the most productive marine areas globally (Heileman and O'Toole 2009). This productivity, coupled with strong contrasts in habitat types (Harris et al. 2013) concomitantly supports a rich diversity and great abundance of fauna and flora, and a high biomass of commercially important species. It includes unique features and species, supports key ecological processes and provides important ecosystem services. In short, the BCLME represents a system of global and regional importance that comprises a wealth of natural resources.

The three states thus rely strongly on the BCLME to sustain their national economies (Hamukuaya et al. 2016). In consequence, the region has high levels of commercial resource extraction, largely fishing and mining, with ocean-based economic development set to increase further through national

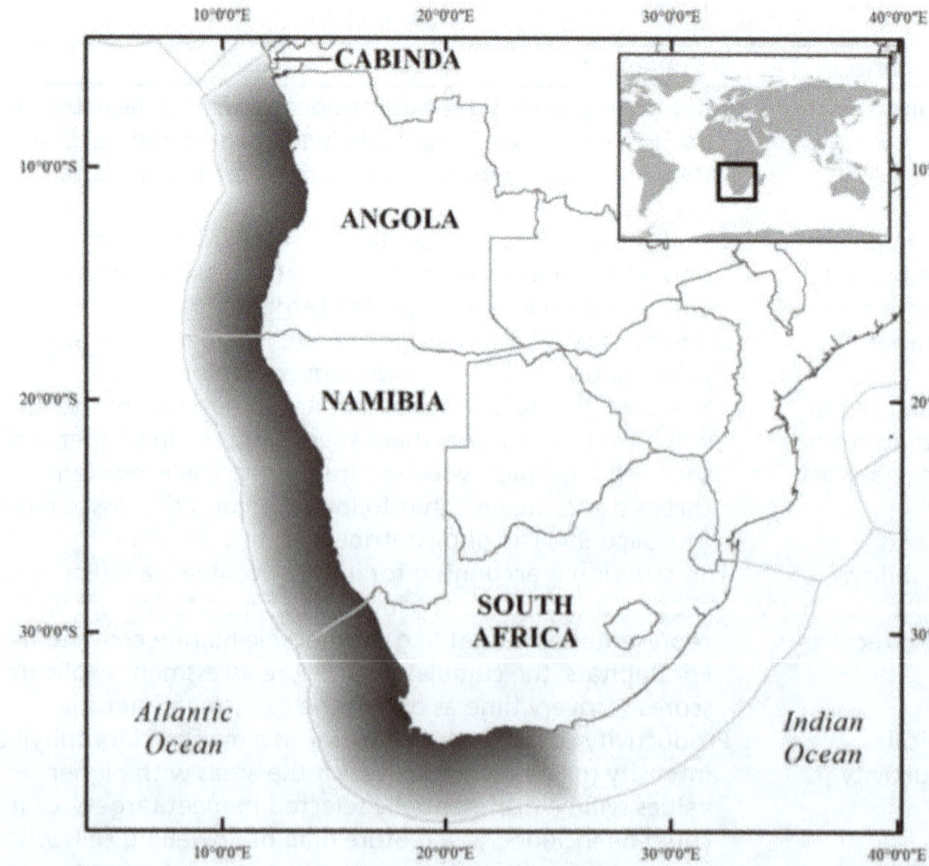

Fig. 4.1 The Benguela Current Large Marine Ecosystem (shaded grey) includes the EEZs of Angola, Namibia and western South Africa in the south-east of the Atlantic Ocean. Note that Cabinda is an exclave of Angola. World EEZ (version 6) boundaries available from http://marineregions.org/. For BCLME region, see BCC (2014)

and regional initiatives. These initiatives include the Benguela Current Convention's (BCC) Strategic Action Programme 2015–2019 (BCC 2014) and Operation Phakisa in South Africa (Republic of South Africa 2014) and are backed by strong government support and political will.

Although there are very clear socio-economic benefits intended through these development strategies, there are also notable ecological concerns of intensifying the current pressure levels on the BCLME. Already this significant system is under threat from existing resource extraction (Boyer and Hampton 2001) and associated pressures, for example, from ports and shipping, coastal development and various forms of pollution (Holness et al. 2014). These are compounded by global-change pressures that, inter alia, are shifting species distributions with knock-on effects through food webs that are stressing further the already threatened top predators (e.g., Pichegru et al. 2010). However, we have never been in a position that is as strong as it is today to take cognisance of the system-level complexity and plan for a sustainable future. Given the key role that the BCLME plays in the three respective countries, and in global ecological processes, safeguarding this natural capital for the generations to come is both imperative and a moral obligation to our children's children.

3.2 Ecosystem-Based Sustainable Development for the Benguela Current Large Marine Ecosystem

Recognising the need for sustainable development and ecosystem-based management, the three countries ratified the BCC (BCC 2013). Building on a strong history of cooperative governance in the BCLME (Hamukuaya et al. 2016), the BCC has taken a proactive role in developing robust conservation and management strategies for the region. One of their first projects was a Spatial Biodiversity Assessment that aimed to design a Spatial Management Plan for the BCLME as a whole, including identification of priority areas for MPAs (Holness et al. 2014). This was followed with a (current) regional co-operation project: The Marine Spatial Management and Governance Project (MARISMA), 2014–2020. The aims of MARISMA are to build capacity in the BCC and its contracting parties and for them to contribute to the sustainable management of the Benguela Current's marine biodiversity and marine natural resources. In so doing, MARISMA intends to directly support countries to achieve their obligations as signatories to the CBD.

The approach taken in the MARISMA Project is to safeguard the natural capital of the BCLME by identifying EBSAs for effective management, including conservation and protection, in a region-wide MSP that allows for socio-economic development in a sustainable manner. Consequently, there are three work areas in the MARISMA Project: the EBSAs' work stream informs the MSP work stream, which is supported by the cross-cutting focus on capacity development, awareness raising and dissemination of results, experiences and products. The case study in this chapter focusses largely on the EBSA work stream of the MARISMA Project and the role that SCP can play in advancing EBSA delineation and integration into MSP processes.

3.3 Ecologically or Biologically Significant Area Identification and Delineation

The first of the CBD's regional EBSA identification workshops was held for the Western South Pacific and the Wider Caribbean and Western Mid-Atlantic Regions in 2011, resulting in 47 EBSA being adopted at COP 11 in 2012. The success of the process led to seven more regional meetings the following year, including the South East Atlantic (SEA) and the Southern Indian Ocean (SIO) Regional Meetings, at which EBSAs in the BCLME and the rest of South Africa were identified, respectively. At COP 12 in 2014, 157 more EBSAs were adopted, including 12 from the BCLME (and an additional seven in the SIO portion of South Africa). Given that EBSA identification was an expert-based approach, delineation of the focus areas was largely coarsely done, with the boundaries poorly linked to the shape of the underlying biodiversity features (Fig. 4.2). Further, South Africa had many more EBSAs in the BCLME region compared to those in Namibia and Angola, simply because spatial prioritisation for marine biodiversity in the former country was already well underway (Sink et al. 2011; Majiedt et al. 2012).

The BCLME case highlights clearly the inherent pitfalls of the expert-based EBSA identification and delineation process. We support the sentiment that progress should not be delayed in the search for refined data and perfect processes (Johnson et al. 2018). However, we also acknowledge that, although excellent for providing a pragmatic first step and guiding larger-scale prioritisation and management, the rough boundaries of the EBSAs are too coarse to be useful for integration into any Spatial Management Plans that also need to

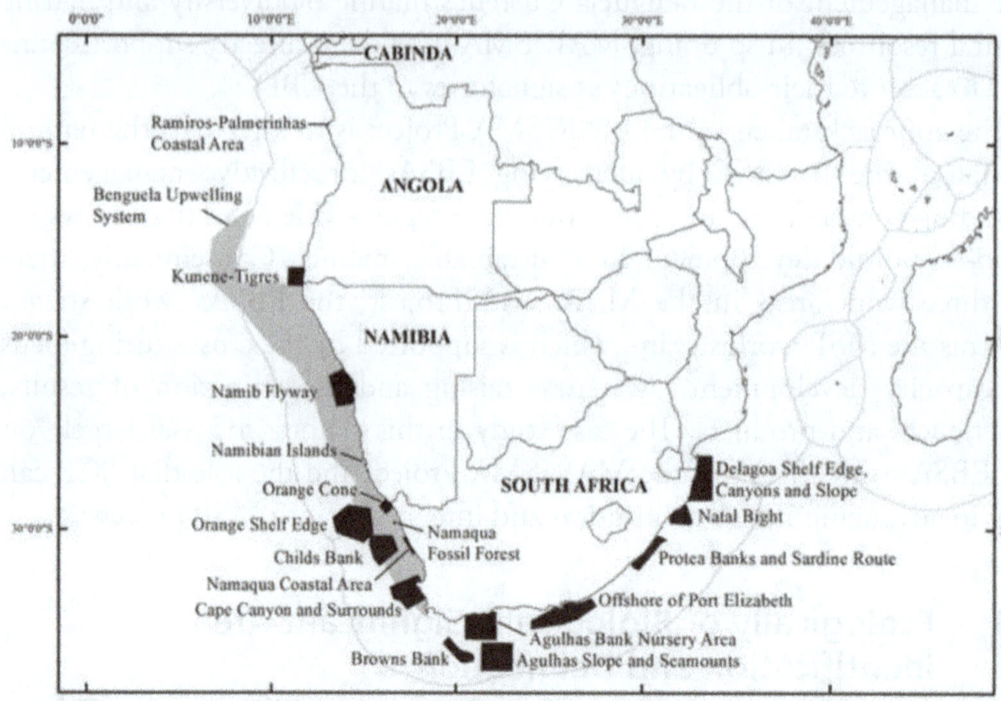

Fig. 4.2 Original set and delineation of EBSAs adopted for the BCLME and SIO portion of mainland South Africa within the respective countries' EEZs. EBSAs in the surrounding high seas are excluded, except for the Benguela Upwelling System EBSA (light grey) that falls mostly within the BCLME. World EEZ (version 6) boundaries available from http://marineregions.org/. For EBSA boundaries, see https://www.cbd.int/ebsa

include other stakeholders. Further, the current EBSA networks are not necessarily representative of the local or regional biodiversity patterns and processes and might, rather, reflect a country's progress in marine conservation initiatives. Ultimately, using the ad hoc, expert-based approach, there is no way to assess with confidence that a proposed EBSA network captures all important sites for a sufficient sample of the region's biodiversity.

As discussed in Sect. 2.4, SCP is proposed to be a particularly useful tool to address these pitfalls. However, the foundation input layers were not available for the three countries. Notably missing was a comprehensive map of ecosystem types across the region that could serve as the primary surrogate of marine biodiversity for the BCLME. Despite this, it is possible to build such datasets with limited information and resources. Coastal habitat types can be mapped from Google Earth imagery (Harris et al. 2013), and offshore habitat types can be delineated by combining bathymetric data (e.g., from the General Bathymetric Chart of the Oceans (GEBCO), available at: www.gebco.net) with a pelagic bioregionalisation based on a cluster analysis of multiple physical variables that can be measured using remote sensing (Roberson et al. 2017).

The benefit of these desktop approaches to mapping ecosystem types is that the bulk of the underlying data are freely available online. Where in situ data from field-based surveys exist, these can easily be incorporated into the ecosystem-type map, either as an independent ground-truthing dataset or to help delineate biotopes or seascapes (Karenyi et al. 2016). Key features, such as seamounts, can also be included from either existing spatial datasets (Yesson et al. 2011) and/or mapped specifically for the project. The additional feature detail in the final output map depends entirely on what is available, but it must be comprehensive in coverage, delineated at an appropriate spatial scale and integrated into one single map product.

The second input dataset that is required is a cumulative pressures map from which ecosystem-type condition can be assessed. If only a limited portion of the sea is allocated for conservation protection, it is preferable for targets to be met in places where the features are in a good ecological condition, meaning that biodiversity and ecological processes are still well intact at the selected sites. The premise is that the more activities there are at a site, and the greater intensity of the respective activities, the more degraded a site becomes (i.e., lower naturalness). It is fully recognised that the complexities of interactions among pressures—positive, neutral and negative—are not accounted for in this approach, but it is a sufficiently robust assumption to make for this assessment where site condition is a relative measure. As for the ecosystem type map, this input layer needed to be custom built for the BCLME.

Raw data on the distribution and intensity of activities and pressures can be sourced from the various industries and sectors through stakeholder engagement: workshops, formal data requests and in-person visits to key data holders. Datasets could include fishing effort and catch, shipping lanes, mining locations and volumes of wastewater discharge out of different pipelines, some of which data are freely available online. Aggregating the suite of pressure data largely follows the cumulative threat assessment methodology developed by Halpern and colleagues (Halpern et al. 2007) that has since been broadly applied (Halpern et al. 2008; Sink et al. 2012; Halpern et al. 2015). These data were sourced from industries in the BCLME (Holness et al. 2014), but if no data exist, a country could use the data from the global assessments (Halpern et al. 2008, 2015) or online databases (e.g., ICCAT 2018).

In this method, each pressure is mapped to a predetermined grid, and pixel values are assigned as the intensity of each activity scaled from 0 to 1, with the upper tail cut at the 80th (or similar) percentile such that all values above that threshold are scored as 1. The size of the grid pixels depends on the resolution of the input data and size of the planning domain, with a 5' grid commonly used in national or regional marine plans. The functional impact and recovery time of a pressure to a particular ecosystem type is scored (by experts, supported by published studies where available), with that score multiplied by the intensity of each pressure in each pixel depending on which pressure-ecosystem combinations are present. Finally, the values are summed per pixel to give an overall cumulative pressure score per pixel across the whole planning domain. Based on these cumulative pressure scores, condition can be ranked as good, fair or poor, where biodiversity pattern and process are, respectively, intact, degraded or lost (Sink et al. 2012).

The third fundamental input layer is a map of existing PAs. As discussed in Sect. 2.4, SCP is definitively efficient and seeks to incorporate existing conservation action and meet outstanding targets in complementary areas. Countries may have these datasets readily available, but if not, the World Database on Protected Areas serves a free global map (available at https://www.protectedplanet.net). Therefore, even if countries start the process with seemingly no data, the three primary maps on which SCP-based EBSA identification is based can be constructed largely from freely available data. In other words, the simplest form of SCP can be used to delineate EBSAs in any country, anywhere in the world.

Of course, the more data available, the more planners can be confident that the EBSA network adequately represents all important sites for a sufficient sample of the region's biodiversity. Although a very good surrogate of biodiversity patterns and processes, ecosystem types may not adequately account for or highlight areas that are important for particular life-history stages, such as breeding and foraging grounds of top predators and migratory species. Any additional biodiversity data such as key species' distributions, internationally recognised sites (e.g., World Heritage Sites, RAMSAR Sites, Important Bird and Biodiversity Areas), areas of high diversity and areas that support key ecological processes can also be included as input datasets. These data were collated for the BCLME (Holness et al. 2014).

The final required input layer is a cost map. There are many ways in which this "cost" can be defined and quantified, but at its core, it represents the penalty to other stakeholders within the planning domain if a site is selected for conservation. This could be measured as opportunity cost, the market value to purchase an area or some other metric that gives a relative indication of potential conflict over a site. In the BCLME context, cost was customised per country to reflect socio-economic priorities from their respective key industries (Holness et al. 2014).

The next step in the process is to compile a list of representation targets for each of the input features. Planners are strongly encouraged to avoid the "target trap" in target-driven SCP (SANBI & UNEP-WCMC 2016). Although it is certainly ideal to have empirical targets derived for each input feature based on their detailed ecological requirements (Desmet and Cowling 2004; Harris et al. 2014a), heuristic or codified values do work especially well, as discussed in Sect. 2.4 earlier. For ecosystem types, planners often set the target at 20% of the historical extent and slightly higher for biodiversity features and ecological processes (Holness et al. 2014); Aichi Target 11 would also work well as a starting point.

With the four key maps and a list of targets, planners can then run SCP software, such as Marxan (Ball et al. 2009), to identify networks of areas that may qualify as meeting the EBSA criteria. The most useful output from Marxan is the selection frequency map, which sums the number of times an area is selected to meet targets out of a user-defined number of repeated runs of the algorithm. Thresholds of selection can be used to identify potential EBSAs (e.g., selection frequency of >80%), with those areas iteratively locked into the solution, along with the existing PAs, until all targets are met. Planners may also wish to include areas of lower selection frequency that serve as ESAs, or "support EBSAs", such as those with a selection frequency of >65%, to ensure persistence of the biodiversity features within the planning domain.

This process can also be repeated iteratively in a stakeholder negotiation process, and/or stakeholders can be presented with a series of candidate EBSA network options and what they might be trading off if one option is selected over another (Harris et al. 2014b).

The SCP approach allows candidate EBSAs to be delineated in a way that matches the underlying biodiversity features much more closely than the current, largely geometric shapes drawn by experts over the broader focus area. This carries three benefits for easier adoption into MSP processes: first, they are not "spatially greedy" areas that unnecessarily exclude other stakeholders from an area; second, they have been designed deliberately to avoid conflict with competing sectors as far as possible through inclusion of those industries in the condition assessment and cost layer; and third, the design is science based and thus easier to defend when challenged by other stakeholders in a negotiation. The latter is an especially important point that has similarly motivated others to improve application of the criteria to strengthen the transparency and robustness of EBSA delineation (Clark et al. 2014).

3.4 Ecologically or Biologically Significant Area Status Assessment and Management Options

With the data inputs described earlier, planners can easily undertake two ecosystem-level assessments that can serve as headline indicators: threat status and protection level (SANBI & UNEP-WCMC 2016). The proportion of each ecosystem type evaluated as good, fair or poor is compared against specific thresholds from which the threat status is assigned. Ecosystem types are said to be Critically Endangered when the proportion in good condition is less than or equal to the biodiversity target (in the BCLME case, 20%). Endangered systems should trigger a warning and are thus recommended to be the biodiversity target +15% (i.e., in the BCLME case, 35%). Vulnerable and Least Threatened ecosystem types have more generous thresholds: in the BCLME case, Vulnerable ecosystem types have <80% of their historical extent in good or fair condition; Least Threatened, >80% (Holness et al. 2014). The second headline indicator is protection levels, where the proportion of each ecosystem type that is protected is determined relative to its target, and the ecosystem type is assigned a rank of well, moderately, poorly or not protected. At this point, the outputs can be interrogated and proposed EBSA descriptions prepared for consideration by the CBD, with a strong scientific basis for the criteria ranks from the SCP process (refer to Table 4.1).

3.5 Integrating Ecologically or Biologically Significant Areas into Marine Spatial Planning Processes

Summary statistics of the earlier indicators can be calculated per EBSA and used to guide conservation and management actions, for example, if an EBSA contains ecosystem types that are poorly or moderately protected, the management action might be to proclaim the EBSA as an MPA and apply relevant management measures. Ecosystem types that need notable intervention are those that are both threatened and not well represented in MPAs. Each individual EBSA could also be assessed to determine the reasons for its selection during the SCP process by identifying the features it contains, site cost and condition, which in turn will help guide sea use in that area (see also Dunstan et al. 2016). For example, if a site is selected because it contains key benthic features, the EBSA could be zoned as a special management area where benthic trawling is prohibited but large pelagic longlining is permitted, depending on the activity-feature compatibility.

Once conservation and management recommendations per EBSA are listed, these can be very easily integrated into an MSP. Recall that to legitimately achieve sustainable use of marine resources, it is critical to first secure the natural capital from which the production services flow. This might mean reserve proclamation for some EBSAs but could also take the form of a restricted-use area (e.g., IUCN PA Categories V or VI) where only activities compatible with the local biodiversity features are allowed. The latter might be especially relevant for "support EBSAs" or ESAs, and in such cases, the suite of compatible activities listed for that area (as extracted from the SCP process) could guide and inform MSP negotiations around user-environment conflicts. Stakeholders and decision-makers need to remain cognisant of the need to secure the nature capital during negotiations, such that short-term socio-economic gains do not come at the expense of long-term losses, for both nature and people (Harris et al. 2018). It has been argued previously that EBSAs could be used to implement MSP through an adaptive hierarchical framework (Dunstan et al. 2016). The process presented in this chapter provides a simpler, spatially explicit variation of the EBSA-MSP integration to achieve ecosystem-based management. This spatialisation of the planning problem (gained through the SCP approach) is proposed here to be one of the most important steps in achieving sustainable development.

3.6 Progress in the Real World

Although the MARISMA Project is still ongoing, progress towards the final outcomes is well underway. EBSA boundaries have been refined and supporting ESAs identified (Fig. 4.3a). Further, Operation Phakisa in South Africa is supporting MPA proclamation, with 22 MPAs gazetted for public comment in 2016 (Republic of South Africa 2016). If proclaimed, they would take the country from <0.4% to 5% marine protection, with a further 5% protection to follow that would then fulfil South Africa's obligation to achieve Aichi Target 11. These MPA boundaries were derived from the ongoing spatial prioritisation (SCP) in the country (Sink et al. 2011; Majiedt et al. 2012) that had also supported the EBSA identification. Consequently, the proposed MPAs will contribute to securing the critically important biodiversity features within the EBSAs (Fig. 4.3b).

The SCP method and outputs have been invaluable during negotiation with industries that have competing interests within these proposed MPAs. As described earlier, it is allowing for iterative boundary refinement throughout the negotiation process. Further, it allows site-specific interrogation of the biodiversity features and key threatening activities within the sites such that stakeholder negotiation and MPA regulations can be targeted, transparent and informed (Fig. 4.3c). Once the ESAs are identified, it is envisaged that these will be integrated in the national emerging MSP, with restrictions on threatening activities in the remaining portions of the EBSAs and in the ESAs such that key marine biodiversity pattern and process is safeguarded for the future.

4 The Value Added by Taking a Systematic Conservation Planning-Based Approach

Inevitably, the success of any MSP will depend on implementation and compliance. The more governments and stakeholders are engaged in the planning process, the greater their sense of plan ownership, and the higher the likelihood that oceans will be developed sustainably. It is important to recognise that political involvement in EBSA delineation and integration into MSP does not mean that the scientific process is compromised if SCP is the decision-support tool. Rather, SCP advances empirical ecosystem-based MSP in the real world through the following seven attributes.

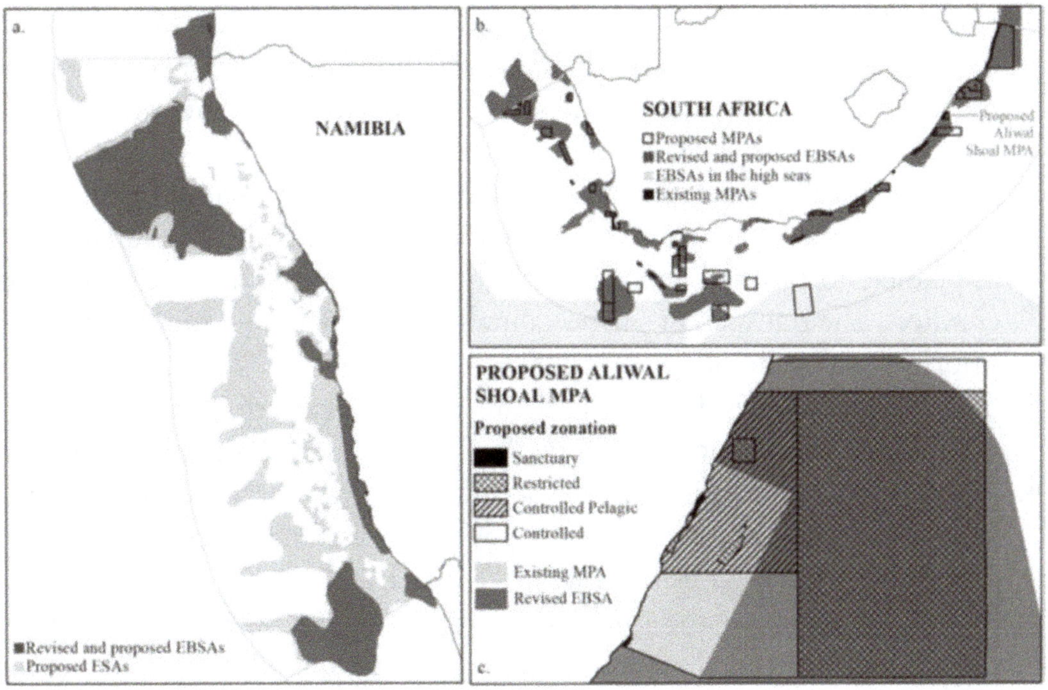

Fig. 4.3 Illustration of the advances in EBSA and MSP processes that can be achieved by SCP. (**a**) Draft revision of existing EBSAs and proposed EBSAs and ESAs in the Namibian EEZ designed using SCP; (**b**) existing and proposed MPAs as gazetted (Republic of South Africa 2016) relative to the revised and proposed EBSAs in South Africa (note that the two EBSAs in the adjacent high seas are also shown in light grey); and (**c**) example of the site-level interrogation of SCP inputs and outputs that can guide both MPA regulations and spatial management of activities in the rest of the EBSA and surrounding areas. In this case, the SCP accounted for existing protection provided by the Aliwal Shoal MPA by including that area in the new delineation of proposed MPAs (and proposed revision of EBSA boundaries). The SCP data supported fine-scale planning of the proposed zonation of the proposed MPA, with the different zones allowing only those activities that are compatible with the underlying biodiversity features. This application could be extended through all EBSAs in an MSP to minimise user-environment conflicts; beyond MPAs and EBSAs, the MSP could focus on resolving only user-user conflicts. Refer to the Government Gazette (Republic of South Africa 2016) for full details on the proposed MPAs, draft regulations and allowed activities per proposed zone. World EEZ (version 6) boundaries available from http://marineregions.org/. Proposed MPAs from Republic of South Africa (2016)

1. SCP supports the key goal of a sustainable ocean system through identification of the most important areas required for securing representation and persistence of ocean features. In so doing, it contributes to securing the natural capital by prioritising sites for conservation action.
2. SCP is underpinned by quantitative targets. This allows easy alignment with CBD, Aichi or any other codified targets, and it helps address the question of sufficiency.

3. SCP is definitively spatially explicit and spatially efficient. This is imperative in an MSP context where planners must balance the many requirements set by multiple stakeholders. The more spatially efficient each sector can be, the lower the chance of unnecessary conflict.
4. SCP specifically considers the potential effects that a conservation area would have on other activities and deliberately avoids spatial overlap as far as possible to facilitate reduced MSP negotiations over user-environment conflicts. The transparent process allows stakeholders to understand sector priorities and trade-offs.
5. SCP can rapidly develop and evaluate alternate scenarios or spatial management options, which is essential in a stakeholder negotiation process.
6. SCP allows identification of the specific pressures acting on specific high-value biodiversity features, which helps to move management action from generic approaches to being truly place based.
7. SCP helps to assess the qualitative EBSA criteria, which are currently ranked for a site as high, medium or low, with no quantitative guidance for what these relative measures mean. This, in turn, makes applying the criteria more consistent among EBSAs in different regions. Ultimately, it makes the EBSA identification and delineation process more science based.

From these attributes, SCP clearly has much to add to both EBSA and MSP processes. Data availability (or lack thereof) should not be seen as a hindrance to its application. As demonstrated through the BCLME case, it is possible to build the required datasets with relatively few resources, largely from existing spatial information that is freely available online. Planning can be as elegant or as simple as the data allow and still achieve robust outputs.

The complexity of governing modern society within a dynamic ocean space that has inherent large-scale connectivity necessitates innovative and creative solutions to conservation and management. These solutions need to allow socio-economic development in a three-dimensional environment, whilst still maintaining ecosystem health and function, all in the face of accelerating global change. Importantly, these solutions must follow good governance practices and thus must be transparent, fair and founded in robust, defendable science to the equitable benefit of all. At all times, we must retain cognisance of the consequences that the industrial revolution had on the environment, notably the acceleration in global climate change that it triggered. As we embark on a similar industrial revolution in the oceans, we have the opportunity to take what we have learnt and leave a sustainable legacy for future generations.

Acknowledgements The MARISMA Project is funded by the German Federal Ministry for the Environment, Nature Conservation and Nuclear Safety (BMU) through its International Climate Initiative (ICI) and further made possible through considerable in-kind contributions by the BCC and its contracting parties. It is implemented by GIZ (Deutsche Gesellschaft für Internationale Zusammenarbeit GmbH; German Development Cooperation) in partnership with the BCC and its contracting parties. The authors thank the above consortium and the people behind it for enabling implementation of the MARISMA Project. They also thank the many scientists and regional workshop participants who have contributed to the SCP, EBSA and MPA processes in South Africa outside of the MARISMA Project. The Open Access fee of this chapter was provided by the project "Economy of Maritime Space" funded by the Polish National Science Centre.

References

Ball, I., Possingham, H., & Watts, M. (2009). Marxan and Relatives: Software for Spatial Conservation Prioritization. In A. Moilanen, K. Wilson, & H. Possingham (Eds.), *Spatial Conservation Prioritization*. Oxford: Oxford University Press.

Balmford, A. (2003). Conservation Planning in the Real World: South Africa Shows the Way. *Trends in Ecology & Evolution, 18*, 435–438.

Ban, N. C., Mills, M., Tam, J., Hicks, C. C., Klain, S., Stoeckl, N., Bottrill, M. C., Levine, J., Pressey, R. L., Satterfield, T., & Chan, K. M. (2013). A Social–Ecological Approach to Conservation Planning: Embedding Social Considerations. *Frontiers in Ecology and the Environment, 11*, 194–202.

Bax, N. J., Cleary, J., Donnelly, B., Dunn, D. C., Dunstan, P. K., Fuller, M., & Halpin, P. N. (2016). Results of Efforts by the Convention on Biological Diversity to Describe Ecologically or Biologically Significant Marine Areas. *Conservation Biology, 30*, 571–581.

BCC. (2013). Benguela Current Convention. The Secretariat, Benguela Current Commission, Swakopmund, Namibia. Retrieved from http://www.benguelacc.org/index.php/en/publications.

BCC. (2014). Strategic Action Programme 2015–2019. The Secretariat, Benguela Current Commission, Swakopmund, Namibia. Retrieved from http://www.benguelacc.org/index.php/en/publications.

Boyer, D. C., & Hampton, I. (2001). An Overview of the Living Marine Resources of Namibia. *South African Journal of Marine Science, 23*, 5–35.

Clark, M. R., Rowden, A. A., Schlacher, T. A., Guinotte, J., Dunstan, P. K., Williams, A., O'Hara, T. D., Watling, L., Niklitschek, E., & Tsuchida, S. (2014). Identifying Ecologically or Biologically Significant Areas (EBSA): A Systematic Method and Its Application to Seamounts in the South Pacific Ocean. *Ocean & Coastal Management, 91*, 65–79.

Convention on Biological Diversity. (2011). *Strategic Plan for Biodiversity 2011–2020 and the Aichi Targets: "Living in Harmony with Nature"*. Montreal, Canada: UNEP.

Desmet, P., & Cowling, R. (2004). Using the Species-Area Relationship to Set Baseline Targets for Conservation. *Ecology and Society, 9*, 11.

Dudley, N. (Ed.). (2008). *Guidelines for Applying Protected Area Management Categories*. Gland, Switzerland: IUCN.

Dunn, D. C., Ardron, J., Bax, N., Bernal, P., Cleary, J., Cresswell, I., Donnelly, B., Dunstan, P., Gjerde, K., Johnson, D., Kaschner, K., Lascelles, B., Rice, J., von Nordheim, H., Wood, L., & Halpin, P. N. (2014). The Convention on Biological Diversity's Ecologically or Biologically Significant Areas: Origins, Development, and Current Status. *Marine Policy, 49*, 137–145.

Dunstan, P. K., Bax, N. J., Dambacher, J. M., Hayes, K. R., Hedge, P. T., Smith, D. C., & Smith, A. D. M. (2016). Using Ecologically or Biologically Significant Marine Areas (EBSAs) to Implement Marine Spatial Planning. *Ocean & Coastal Management, 121*, 116–127.

Ehler, C., & Douvere, F. (2009). *Marine Spatial Planning: A Step-by Step Approach Towards Ecosystem-based Management*. Manual and Guides No 153 ICAM Dossier No 6. Paris: Intergovernmental Oceanographic Commission UNESCO IOC, 99 pp.

Gilliland, P. M., & Laffoley, D. (2008). Key Elements and Steps in the Process of Developing Ecosystem-Based Marine Spatial Planning. *Marine Policy, 32*, 787–796.

Halpern, B. S., Selkoe, K. A., Micheli, F., & Kappel, C. V. (2007). Evaluating and Ranking the Vulnerability of Global Marine Ecosystems to Anthropogenic Threats. *Conservation Biology, 21*, 1301–1315.

Halpern, B. S., Walbridge, S., Selkoe, K. A., Kappel, C. V., Micheli, F., D'Agrosa, C., Bruno, J. F., Casey, K. S., Ebert, C., Fox, H. E., Fujita, R., Heinemann, D., Lenihan, H. S., Madin, E. M. P., Perry, M. T., Selig, E. R., Spalding, M., Steneck, R., & Watson, R. (2008). A Global Map of Human Impact on Marine Ecosystems. *Science, 319*, 948–952.

Halpern, B. S., Frazier, M., Potapenko, J., Casey, K. S., Koenig, K., Longo, C., Lowndes, J. S., Rockwood, R. C., Selig, E. R., Selkoe, K. A., & Walbridge, S. (2015). Spatial and Temporal Changes in Cumulative Human Impacts on the World's Ocean. *Nature Communications, 6*, 7615.

Hamukuaya, H., Attwood, C., & Willemse, N. (2016). Transition to Ecosystem-Based Governance of the Benguela Current Large Marine Ecosystem. *Environmental Development, 17*, 310–321.

Harris, L., Holness, S., Nel, R., Lombard, A. T., & Schoeman, D. (2013). Intertidal Habitat Composition and Regional-Scale Shoreline Morphology Along the Benguela Coast. *Journal of Coastal Conservation, 17*, 143–154.

Harris, L., Nel, R., Holness, S., Sink, K., & Schoeman, D. (2014a). Setting Conservation Targets for Sandy Beach Ecosystems. *Estuarine, Coastal and Shelf Science, 150*(Part A), 45–57.

Harris, L. R., Watts, M. E., Nel, R., Schoeman, D. S., & Possingham, H. P. (2014b). Using Multivariate Statistics to Explore Trade-Offs Among Spatial Planning Scenarios. *Journal of Applied Ecology, 51*, 1504–1514.

Harris, L. R., Nel, R., Oosthuizen, H., Meÿer, M., Kotze, D., Anders, D., McCue, S., & Bachoo, S. (2018). Managing Conflicts Between Economic Activities and Threatened Migratory Marine Species Toward Creating a Multiobjective Blue Economy. *Conservation Biology, 32*, 411–423.

Heileman, S., & O'Toole, M. J. (2009). I West and Central Africa: I-1 Benguela Current LME. In K. Sherman & G. Hempel (Eds.), *The UNEP Large Marine Ecosystems Report: A Perspective on Changing Conditions in LMEs of the World's Regional Seas*. Nairobi, Kenya: United Nations Environment Programme.

Holness, S., Kirkman, S., Samaai, T., Wolf, T., Sink, K., Majiedt, P., Nsiangango, S., Kainge, P., Kilongo, K., Kathena, J., Harris, L. R., Lagabrielle, E., Kirchner, C., Chalmers, R., & Lombard, A. (2014). *Spatial Biodiversity Assessment and Spatial Management, Including Marine Protected Areas*. Final Report for the Benguela Current Commission Project BEH 09-01.

ICCAT. (2018). *International Commission for the Conservation of Atlantic Tunas Task II: Catch-Effort*. Madrid: ICCAT Retrieved from https://www.iccat.int/en/t2ce.asp.

Johnson, D. E., Barrio Froján, C., Turner, P. J., Weaver, P., Gunn, V., Dunn, D. C., Halpin, P., Bax, N. J., & Dunstan, P. K. (2018). Reviewing the EBSA Process: Improving on Success. *Marine Policy, 88*, 75–85.

Karenyi, N., Sink, K., & Nel, R. (2016). Defining Seascapes for Marine Unconsolidated Shelf Sediments in an Eastern Boundary Upwelling Region: The Southern Benguela as a Case Study. *Estuarine, Coastal and Shelf Science, 169*, 195–206.

Klein, C. J., Chan, A., Kircher, L., Cundiff, A. J., Gardner, N., Hrovat, Y., Scholz, A., Kendall, B. E., & Airame, S. (2008). Striking a Balance Between Biodiversity Conservation and Socioeconomic Viability in the Design of Marine Protected Areas. *Conservation Biology, 22*, 691–700.

Klein, C. J., Steinback, C., Watts, M., Scholz, A. J., & Possingham, H. P. (2009). Spatial Marine Zoning for Fisheries and Conservation. *Frontiers in Ecology and the Environment, 8*, 349–353.

Mace, G. M. (2014). Whose Conservation? *Science, 345*, 1558–1560.

Majiedt, P., Holness, S., Sink, K., Oosthuizen, A., & Chadwick, P. (2012). *Systematic Marine Biodiversity Plan for the West Coast of South Africa*. Cape Town, South Africa: South African National Biodiversity Institute.

Margules, C. R., & Pressey, R. L. (2000). Systematic Conservation Planning. *Nature, 405*, 243–253.

McLeod, K. L., Lubchenco, J., Palumbi, S. R., & Rosenberg, A. A. (2005). *Scientific Consensus Statement on Marine Ecosystem-Based Management*. Signed by 221 Academic Scientists and Policy Experts with Relevant Expertise and Published by the Communication Partnership for Science and the Sea. Retrieved from http://compassonline.org/?q=EBM.

Moilanen, A., Kujala, H., & Leathwick, J. R. (2009). The Zonation Framework and Software for Conservation Prioritization. In A. Moilanen, K. A. Wilson, & H. P. Possingham (Eds.), *Spatial Conservation Prioritization*. New York: Oxford University Press.

Pichegru, L., Ryan, P. G., Crawford, R. J. M., van der Lingen, C. D., & Grémillet, D. (2010). Behavioural Inertia Places a Top Marine Predator at Risk from Environmental Change in the Benguela Upwelling System. *Marine Biology, 157*, 537–544.

Pressey, R. L. (1994). Ad Hoc Reservations: Forward or Backward Steps in Developing Representative Reserve Systems? *Conservation Biology, 8*, 662–668.

Republic of South Africa. (2014). *Operation Phakisa*. Pretoria: RSA Retrieved November 2015, from http://www.operationphakisa.gov.za.

Republic of South Africa. (2016). National Environmental Management: Protected Areas Act, 2003 (Act No. 57 of 2003): Draft Notices and Regulations to Declare a Network of 22 New Proposed Marine Protected Areas. Government Gazette No. 10553. Volume 608, No. 39646, February 3.

Roberson, L. A., Lagabrielle, E., Lombard, A. T., Sink, K., Livingstone, T., Grantham, H., & Harris, J. M. (2017). Pelagic Bioregionalisation Using Open-Access Data for Better Planning of Marine Protected Area Networks. *Ocean & Coastal Management, 148*, 214–230.

Sala, E., Lubchenco, J., Grorud-Colvert, K., Novelli, C., Roberts, C., & Sumaila, U. R. (2018). Assessing Real Progress Towards Effective Ocean Protection. *Marine Policy, 91*, 11–13.

SANBI. (2017). *Technical Guidelines for CBD Maps: Guidelines for Developing a Map of Critical Biodiversity Areas and Ecological Support Areas Using Systematic Biodiversity Planning*. First Edition (Beta Version). Compiled by Driver, A., Holness, S. & Daniels, F. South African National Biodiversity Institute, Pretoria.

SANBI & UNEP-WCMC. (2016). *Mapping Biodiversity Priorities: A Practical, Science-Based Approach to National Biodiversity Assessment and Prioritisation to Inform Strategy and Action Planning*. Cambridge, UK: UNEP-WCMC.

Sink, K. J., Attwood, C. G., Lombard, A. T., Grantham, H., Leslie, R., Samaai, T., Kerwath, S., Majiedt, P., Fairweather, T., Hutchings, L., van der Lingen, C., Atkinson, L. J., Wilkinson, S., Holness, S., & Wolf, T. (2011). *Spatial Planning to Identify Focus Areas for Offshore Biodiversity Protection in South Africa. Final Report for the Offshore Marine Protected Area Project*. Cape Town: South African National Biodiversity Institute.

Sink, K., Holness, S., Harris, L., Majiedt, P., Atkinson, L., Robinson, T., Kirkman, S., Hutchings, L., Leslie, R., Lamberth, S., Kerwath, S., von der Heyden, S., Lombard, A., Attwood, C., Branch, G., Fairweather, T., Taljaard, S., Weerts, S., Cowley, P., Awad, A., Halpern, B., Grantham, H., & Wolf, T. (2012). *National Biodiversity Assessment 2011: Technical Report. Volume 4: Marine and Coastal Component*. Pretoria: South African National Biodiversity Institute.

Small, C., & Nicholls, R. J. (2003). A Global Analysis of Human Settlement in Coastal Zones. *Journal of Coastal Research, 19*, 584–599.

UN General Assembly. (2017). A/RES/72/249: International Legally Binding Instrument Under the United Nations Convention on the Law of the Sea on the Conservation and Sustainable Use of Marine Biological Diversity of Areas Beyond National Jurisdiction. Retrieved from http://www.un.org/depts/los/general_assembly/general_assembly_resolutions.htm.

UNEP-CBD. (2007). Report of the Expert Workshop on Ecological Criteria and Biogeographic Classification Systems for Marine Areas in Need of Protection. UNEP/CBD/EWS.MPA/1/2. Azores, Portugal, October 2–4.

UNEP-CBD. (2009). *Azores Scientific Criteria and Guidance for Identifying Ecologically or Biologically Significant Marine Areas and Designing Representative Networks of Marine Protected Areas in Open Ocean Waters and Deep Sea Habitats.* Montreal, Canada: Secretariat of the Convention on Biological Diversity.

White, C., Halpern, B. S., & Kappel, C. V. (2012). Ecosystem Service Tradeoff Analysis Reveals the Value of Marine Spatial Planning for Multiple Ocean Uses. *Proceedings of the National Academy of Sciences, 109*, 4696–4701.

Yesson, C., Clark, M. R., Taylor, M. L., & Rogers, A. D. (2011). The Global Distribution of Seamounts Based on 30 Arc Seconds Bathymetry Data. *Deep Sea Research Part I: Oceanographic Research Papers, 58*, 442–453.

5

Can Classical Location Theory Apply to Sea Space?

Jacek Zaucha

1 Introduction

Economists assume that consumer preferences are usually revealed through the market mechanism in which demand and supply are confronted. A competitive market is believed to ensure efficient allocation of resources providing the highest level of consumer utility and producer profit. Both are assumed to behave in an entirely rational way, and furthermore, it is also assumed that their choice is immediately mirrored in the changes of the prices of goods and factors of production. However, these beneficial market outcomes might become suboptimal due to time-inconsistent preferences, information asymmetries, unequal market power of some producers or consumers, the occurrence of externalities or the non-excludable and non-rivalrous character of consumption of some goods or services (i.e. public goods). Such a situation is known as market failure. For this reason, in some cases, the market is supplemented with the public choice mechanism under which democratically elected public bodies aggregate consumer preferences and reveal them (impose them) in the form of various administrative decisions and economic incentives (e.g. subsidies, taxes, auctions for location permits). Marine spatial planning (MSP) is part of public choice.

This chapter compares the two mechanisms outlined earlier (market and public choice) for the allocation of marine space to various uses. Both of them

J. Zaucha (✉)
Institute for Development and Maritime Institute, University of Gdańsk, Gdańsk, Poland
e-mail: j.zaucha@instytut-rozwoju.org

J. Zaucha, K. Gee (eds.), *Maritime Spatial Planning*,
https://doi.org/10.1007/978-3-319-98696-8_5

are assumed to shape spatial patterns in the sea. The chapter aims at answering the question whether MSP can neglect market outcomes and to what extent the market can neglect MSP. The market process is analysed within the framework of location theory, a well-established branch of economics. First, different location models based on market principles are analysed in order to find the most suitable one for maritime space. Second, Thünen's[1] way of thinking is applied in order to predict hypothetical spatial development patterns at sea that might emerge under market rule. Third, these results are confronted with existing patterns of offshore spatial development in Poland. Fourth, the outcomes of administrative decisions are added to this picture. The final part of the chapter is devoted to the discussion of the interplay of market and public choice (MSP) as mechanisms shaping spatial development offshore.

2 Location Theory

As noted by Fujita (2010), classical location theory emphasizes market-driven mechanisms that shape spatial patterns, that is, spatial development. According to Blaug (1985, 614), the theory of spatial economics[2] focuses on area and distance. Contemporary research in this field adds new characteristics to the understanding of area, such as density (intensity of economic activity per square km) or institutional tissue (World Bank 2009). Spatial economics, using an economic approach, explains how space (e.g. distance, economies of agglomeration) affects the decisions of economic agents. It contains two groups of models explaining spatial development: those assuming an a priori existence of certain nodal points in a space of higher economic density (markets, production or extraction locations) and those models treating space as fully homogeneous and isotropic, where only an interplay between economic factors diversifies its economic density. In both models, economic agents act in their own self-interest. Firms choose locations to maximize their profits, and consumers choose locations maximizing their utility level. Those choices manifest themselves in the so-called spatial/ground/location rent (or bid rent), that is, the amount of money the users of a given part or land are willing to pay for earning the right to its usage. The amount corresponds to the profits or utility provided by a given piece of land.

[1] Johann Heinrich v. Thünen is a founder of spatial economics. In the nineteenth century, he developed the first rigorous approach to explain the formation of spatial patterns (concentric rings of various types of agriculture crops) around a pre-set market for agricultural products.

[2] Spatial economics covers also location theory.

In the first class of models, rent refers to the distance and net profits derived from the usage of a given piece of land, that is, the revenue from a given piece of land minus the costs, including transport costs, the latter of which depends on the distance to suppliers and points of sale. In some models, instead of revenue, there are only costs. In these models, the a priori assumptions about the organization of space allow for the use of the concept of perfect competition, ignoring the fact that the cost of covering distance might result in monopolistic competition or even an oligopoly market. The second class of models, in its reasoning, has to add an economic mechanism that leads to the distortion of spatial homogeneity. As a rule, it is assumed that there are interactions between two forces shaping the socio-economic space: the centripetal force that favours concentration (e.g. economies of agglomeration) and the centrifugal force causing dispersion (e.g. costs of covering the distance). Therefore, the second type of reasoning is quasi-dynamic or anticipatory—choice changes the underlying parameters. The first class of models (in line with their assumptions) has a quasi-static or adaptive character because choice is based on known parameters, that is, existing patterns of nodal points (although in these models spatial reallocation can occur as a result of changes in the productivity of some areas, e.g. new discoveries of natural resources or changes in the transport techniques). Thünen's and Weber's models and Launhardt's sales areas, as well as Palander's market area theory (Blaug 1985, 618–626), belong to the first class of models. Christaller's central place theory (1933) and Lösch's economic region (1940), however, have elements of the second approach even though they are situated in the first class of models. However, both in Lösch and Christaller, the a priori assumption is that population is distributed evenly in space. In the second class, we have Isard's regionalism (1956) and new economic geography (Krugman 1991a, b; Fujita et al. 2000). Most of the aforementioned models are microeconomic and emphasize business decisions, although regionalism and even Christaller's theory are characterized by a macroeconomic approach. Only the models of the new economic geography, submerged in the realities of monopolistic competition that results from the very nature of space, belong to the class of formalized equilibrium models. These issues are described in detail in the literature consulted (Blaug 1985; Ponsard 1988; Zaucha 2007).

In the models based on monopolistic competition, the aforesaid economies of agglomeration (agglomeration externalities) play a decisive role. One of the most comprehensive attempts to examine them is to be found in Fujita and Thisse (2002). The authors state that the "fundamental trade-off of a Spatial Economy" are economies of scale and transport costs (Fujita and Thisse 2002, 93). Increasing returns, along with an increase in production scale, are the result of externalities stemming from proximity to other businesses, supply chain efficiency and customer perceptions, as well as more efficient (better specialized) labour resources (Fujita and Thisse 2002, 98). All this contributes to the economies of agglomeration that, according to McCann (2013, 54–56), include internal returns to scale (which require the concentration in a single place of significant capital and labour inputs), economies of localization (physical proximity of enterprises in the same sector) and economies of urbanization (proximity of enterprises of various sectors). Their appearance was recognized by Marshall (1920, 225) at the beginning of the previous century. According to the new economic geography (Krugman 1991b, 101–113), such economies are reinforced by the influx of workers, encouraged by the relatively higher wage levels in places where such externalities emerge. This in turn allows for an increase in the number of services and goods produced in a given location, which is important in the situation of consumer preferences for variety. As a result, there are processes of catastrophic[3] agglomeration followed by a spatial bifurcation of the economy. However, they are countered by the costs of overcoming the resistance of space (e.g. transport costs). When they are high, the local market does not allow for the emergence of large business entities as they would not have the sufficient market to be served by them. The economies of scale would reduce costs and prices, but this effect would be offset by the high costs of supplying consumers. Production has to take place close to the consumer, and consumers do not look for employment outside of the their place of residence, because of the low concentration of production in space. Only the falling costs of trade allow the economies of agglomeration to become visible and concentration to become irreversible. However, in a situation of zero or very low costs of this type (Internet, relatively low transportation costs, telework), the economies of agglomeration spill out. They are no longer limited to certain places of high density of economic activity since, with no distance, they work everywhere (skilled workers can work through Internet regardless their location, ideas and know-how easily spread out in the space). Hence, dispersion tendencies may emerge. Here the favourable factor is the lower wages outside the existing production centres and the non-mobile local assets, that is, territorial capital (Zaucha 2007, 64–66). As Przygodzki points out (2016, 84), “Functional and relational elements are the most recent and most interesting development factors” of space. Hence, he points to the important role of “social absorption, diffusion and processing of knowledge and experience, common learning, establishing and maintaining

[3] Catastrophic means that firm or consumer location changes in a discontinuous way.

territorial co-operation." In a similar vein, Johansson and Quigley (2003) emphasize that networks of assets dispersed in space can be a good substitute for agglomeration processes. Paradoxically, Christaller's elaboration of the theory of central places, more than 80 years ago, arose from very similar assumptions to those described earlier. In his reasoning, there are also centripetal forces in the form of minimum sales thresholds (minimum number of consumers ensuring profitability of production) and centrifugal in terms of reach (the maximum distance a consumer is willing to travel to buy a given good or service).

Unfortunately, contemporary economics' inspiring approach to the economic mechanisms of spatial development (economies of agglomeration and distance) cannot, to date, be applied to maritime space (at least to sea areas). From the entire array of elements pertaining to the new economic geography mainly local assets (territorial capital) and the cost of transport also appear at sea. The economies of agglomeration, even if they do occur in marine space, have very high transaction costs[4] or barriers of nonconformity and temporal friction. This is due to the specificity of this space, characterized, for example, by the lack of inhabitants, positive externalities (related mainly to costs of shared use of resources) and the differences in market power among the users of marine space.

3 The Location Theory Applied to Sea Space

Referring to marine space, one may think that it would be appropriate to consider the return to models which accept the already existing (a priori) set-up of human activities. This is plausible because development of the terrestrial structures (e.g. port cities, transhipping terminals) related to sea exploitation predates maritime spatial development. The "nodal" elements appeared ashore while the seas constituted economic space functionally linked to them. The most promising model in this situation appears to be a relatively old agricultural one developed by Thünen which assumes the existence of a pre-existing marketplace. This model is still used in the analysis of the spatial development of cities (McCann 2013, 107–153).

The legitimacy of such a choice is based on the fact that there are already established "sea gateways" on land aiding the economic activity of people at sea. They are characterized by a specific hierarchy similar to Christaller's

[4] The possibility of lowering them is discussed in the concluding part of the chapter.

pattern since certain gates serve many functions (e.g. ports are bases for sailing, fishery, wind power stations, marine-mining or marine tourism), while others, for example, beach resorts or piers, serve a limited range of functions. The difference in relation to Christaller's concept of central places is that multifunction centres (gates) need not support the sea activity that is typical for their monofunctional equivalents (e.g. port cities do not always have beaches). All of the above calls for a more thorough consideration of models proposed by Thünen and Weber[5] since other models seem to have lesser explanatory potential.[6]

The essence of Thünen's concept is an exogenously given sales market and two parameters shaping spatial patterns around it: net benefits per unit area of the cultivation of different agricultural products and costs of their transportation. Near the market area, there are cultivated goods that yield high profits and have high transportation costs. Further away, there appears to be a place for less profitable and expensive farming, while at an even greater distance, those goods appear that are least effective at using the soil but also cheapest in transport per unit. The result is the appearance of Thünen's famous location circles (Blaug 1985, 619). A number of assumptions were made in the model, the most important of which deal with constant economies of scale, homogeneous soil fertility, lack of restrictions on the side of productive resources that are available everywhere in the same proportions and so on.

It seems that location rent in maritime space is shaped in a similar manner. The Thünen model foundations are generally fulfilled. A certain problem in this respect is the heterogeneous productivity of the space resulting from natural conditions, such as the existence of deposits, fishing grounds or areas particularly predestined for offshore wind energy. However, a similar dilemma appears on land, which was analysed with reference to urban space and its ecological values by McCann (2013, 127). This may bring about a concavity in certain fragments of the rent function, which means that functions of the rents can intersect at several points (different economic activities are not located any longer in the same distance from the city centre). Consequently, similar manners of reaping benefits from the sea may appear in several zones at various distances from the land gateways. However, if maritime space lacks

[5] At the beginning of the twentieth century, Alfred Weber formulated the "least-cost model" of location of industrial plant, allowing him to explain industrial location decisions at a macro-scale.

[6] Lösch is concerned with the economic region and, above all, the effects of spatial competition of producers while assuming a uniform distribution of population. This condition is not fulfilled at sea. There are no Launhardt markets or Palander's market areas at sea. These theories would be able to explain the location of certain land-based marine management entities.

suitable characteristics, it may not be used for economic purposes at all. This is similar to the situation on land, where under certain circumstances, poor soil quality may result in land lying idle within a certain distance from the centre (negative location rent). This results from the fact that the rent curve in this situation becomes a discontinuous, non-monotonic[7] function (Ponsard 1988, 39).

Taking all this into account, it is plausible that market forces at sea could lead to the formation of Thünen's semicircles (assuming that the coastline is straight) around the sea gateways (ports, bathing beaches, etc.). The first circle includes functions typical for their proximity to the port (anchorages, dumping sites), while others will be farther away, for example, wind energy at sea, and even farther—fishing. However, the denser spreading of bathing areas will result in a narrower strip for bathing along the coast determined by the overlapping half-circles of traditional coastal tourism (Fig. 5.1). If necessary, however, they will have to allow for other ways of using the sea, ones that have a higher degree of location rent (such as port complexes).

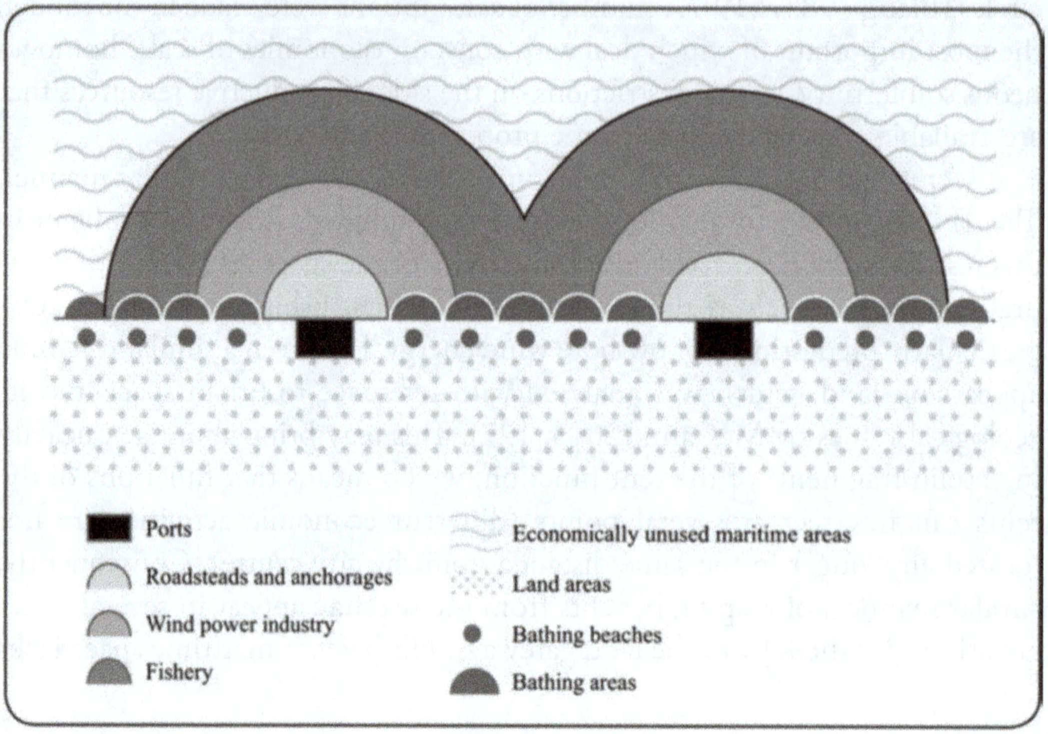

Fig. 5.1 Location semicircles (in sea areas) around the sea gateways of different importance (Zaucha 2018)

[7] The function is non-monotonic when it is growing at certain intervals and decreasing at other intervals.

Taking into account differences of marine areas' suitability for various purposes (i.e. its non-homogeneity), location circles will not be regular in shape due to dissimilarities in the productivity of different marine areas. In certain circumstances, economic activity might be spatially limited, for example, where a possibility exists of obtaining specific ecosystem services or abiotic benefits (gravel extraction is only possible in areas with gravel deposits). Location rent will nonetheless play a significant role. For example, coastal defence, if applied to densely inhabited or economically developed pieces of coast, usually offers larger benefits in comparison to gravel extraction so that coastal defence can even stop commercial gravel extraction due to rent differences. If the rent associated with renewable energy sources or mining is higher than what can be derived from fishery, the latter activity must operate at a further distance (Fig. 5.2).

Some of the patterns of maritime spatial development are dependent on several land gates at the same time. For example, facilities of the offshore wind power industry have to be located at a proper distance to service ports and, especially, to shore power connections (connecting wind farms to the power grid) which may not be at the service ports. This placement can be linked to Weber's theory which is based on his analysis of the best potential location for a production facility aiming to minimize transportation costs (access to

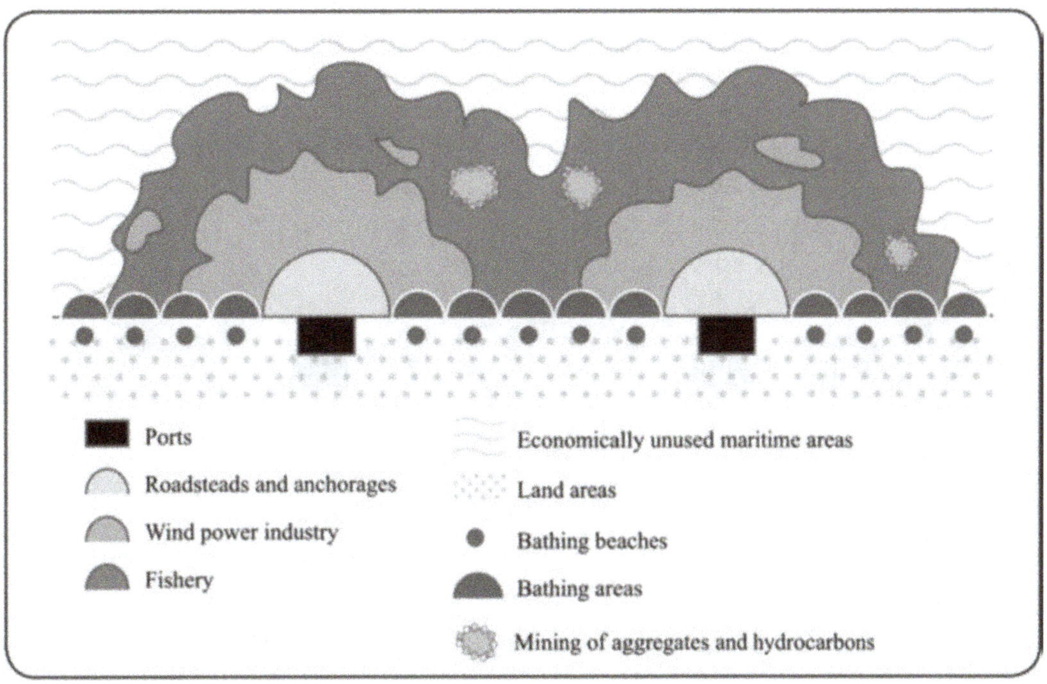

Fig. 5.2 Location zones in sea areas taking into account the phenomenon of the non-homogeneity of marine space and locations of ecosystem services and services exploiting sea abiotic assets (Zaucha 2018)

markets and suppliers). However, it should be kept in mind that the final location of this particular facility at sea depends not only on low transportation costs but also on the differences in marine areas' levels of productivity, which is important when it comes to energy production (for this purpose, some areas are more suitable and some less so).

What is more, additional linear structures (connecting two or more gates on both sides of the sea) have to be added in order to make the model more realistic. Such structures include cables, pipelines or sea lanes. They do not follow the logic of Thünen's model. Minimizing costs between two points is a crucial factor in these situations. Of key importance is the shortest distance, and very rarely do specific features (e.g. depth, bottom habitats) of particular parts of maritime space influence their location. However, practical computation of location rent for shipping might pose a challenge since in some cases, such as navigation, it would be very difficult to attribute costs and revenues to the part of marine space that is used for that purpose (it would require detailed information of each voyage, i.e. its length and net profits).

Nevertheless, it should be assumed that zones based on Thünen's model might exist collectively with linear areas (Fig. 5.3). Due to the multidimensional character of maritime space, in some situations, the summing of various location rents might occur since many users can use the same sea space simultaneously (under certain conditions, navigation does not impact

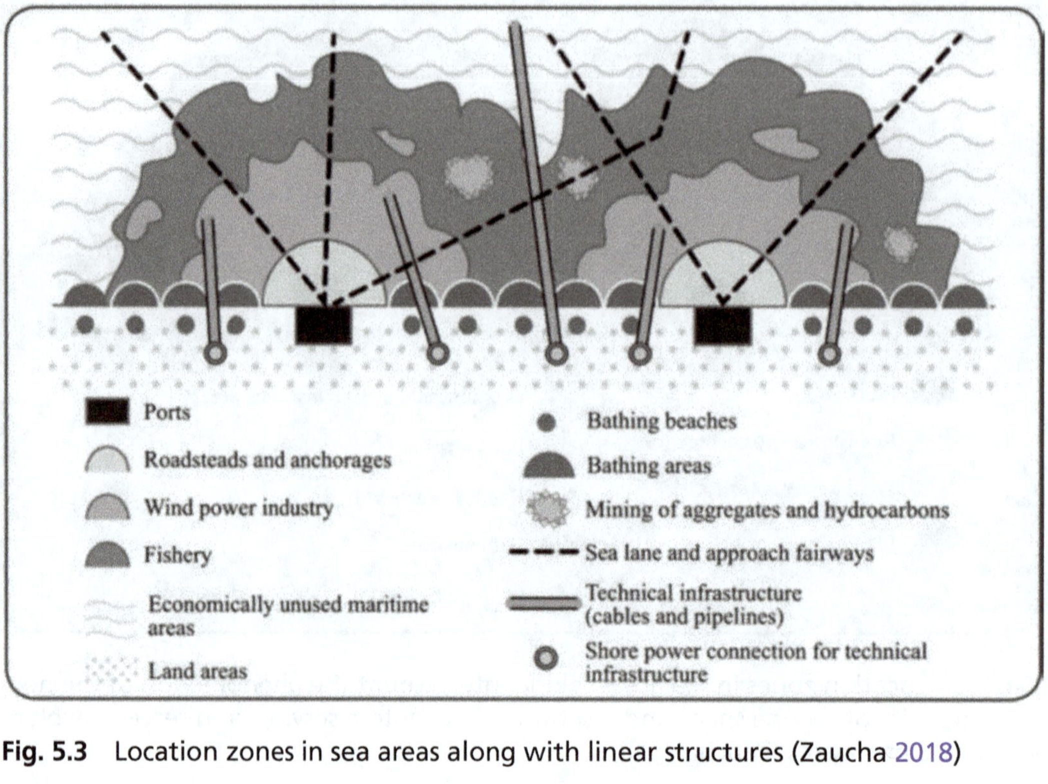

Fig. 5.3 Location zones in sea areas along with linear structures (Zaucha 2018)

pipelines negatively, and cables are outright necessary for the development of the wind power industry, representing synergy). In other situations, changes in the market might allow for the highest rent through the selection of appropriate forms of economic activity. For example, building wind farms may force some changes in sea transport lanes.

The picture presented seems pretty static. However, along with the intensification of the blue economy and blue growth, new functional regions will appear at sea holding the possibility of redefining such regions in the future. Finally, it may happen that the sea-land influence occurs in the opposite direction (from sea to land), thus becoming an economic incentive to create new maritime gateways on land. That could lower the costs of transportation.[8] For example, it is very likely that an increase in the popularity of yachting may result in establishing new marinas. Demand related to ferries can cause more ferry terminals in new locations. The critical mass of particular forms of using marine assets should be the next important factor in the context of presenting models of market processes inducing changes in maritime spatial development. Potentially, new gateways servicing sea space might appear offshore and perhaps some of them could offer economies of agglomeration.

The deductive approach presented earlier finds support in the empirical material gathered in the course of the preparation of a maritime spatial plan for Poland. Prior to the emergence of MSP activities in Poland, marine space was developed according to the demand of investors and sea users. The outcome is visualized in the form of the map of existing and planned sea uses of Polish maritime areas in the report entitled Study of Conditions of Spatial Development of Polish Sea Area (Zaucha et al. 2016).[9] The map confirms a picture that the above theory would also predict. Port-related activities, traditional tourism and recreation and offshore wind farms are sorted according to their distance from the shore. For example, for offshore wind farms, depth seems to play an important role. For other activities like fishing or oil extraction, natural conditions and oceanographic characteristics seem to be decisive. Surprisingly, environmental protection tends to cluster near the shore as well, possibly due to the photic conditions there and a larger amount of easily available information on birds and habitats. Thus in economic terms, the

[8] New investments of this type are limited by their costs which should be lower than the discounted (on the day of the opening of such gateways) amount of benefit presented in the form of lower transportation expenses or an increase of existing revenues or even the creation of new ones obtained from new forms of exploiting sea areas.

[9] http://www.umgdy.gov.pl/wp-content/uploads/2015/04/INZ-UM_Map1_Present_and_future_use.png.

photic zone produces more benefits (ecological values) in comparison to other types of marine space. The ultimate monetary value of those benefits, however, depends on the value system of a given society.

In general, all this roughly confirms the importance of location rent as a guiding location force and its dependence on the distance to the land gateways servicing the sea areas at least for some sea uses.

4 Maritime Spatial Planning as a Public Choice Mechanism for Marine Governance

Maritime space as a precious development asset (natural capital) is considered to be a perfect example of market failure. The main reason is the lack of private ownership restricting the proper functioning of prices in their function of balancing demand with supply. Maritime space is considered to be a common-pool resource (Ostrom et al. 1994, 7), which is characterized by competitiveness of consumption and the inability to exclude anyone from it, that is, non-excludability (Daly and Farley 2011, 169). There is a lack of clearly defined property rights or those rights are acting in a limited way. According to game theory, it is profitable to maximize individual payouts here and now at the expense of the resource itself. In addition to the above, market-driven allocation of marine space suffers from information asymmetries, importance of externalities provided by the marine ecosystems and the unequal market power of some sea users. Moreover, status differences of particular users of maritime space may be perceived as a problem. The offshore wind industry, for example, has to bear the cost of functioning at sea, which has a substantial impact on its services' prices. Other users, such as sailors, are allowed to access the sea for free. This disrupts the effective development of maritime space by using market mechanisms. One should also keep in mind that maritime space should be maintained for future generations that are not able to reveal their preference at the market. Due to all these reasons, marine space requires collective governance mechanisms. In economics, these are called public choice mechanisms and are associated with administrative decisions. MSP forms its core.

Public choice (Stiglitz 1999, 157–188) is a form of aggregating individual preferences into collective preferences in cases of market failure. It entails joint decision-making in a democratic manner; that is, it involves voting. As a consequence, selected people are entitled to make specific decisions concerning public goods (including key components of social life, such as social justice, biodiversity). They are also entitled to choose the methods used in supporting or neutralizing externalities in some special situations of market failure. The act of voting provides the necessary social legitimization. Democratic decision-making is not the only means of ensuring public choice, but it is the dominant model in Europe. Without such legitimization, decisions made by the public administration would take a voluntaristic form. The decision-making body, in a democratic process, is able to assign some elements of aggregation preferences to the executive body. One should keep in mind that this is a slightly simplified picture of public choice, as in reality, public choice governance is composed of myriad interactions between various decision-making and executive bodies, including stakeholders, with and without jurisdiction both in vertical and horizontal dimensions (e.g. Hassler et al. 2018). The outcome is agreement on the key societal goals and their execution within a framework of various policies.

The first stage of public choice is, most frequently, related to axiological matters, that is, determining goals that should be achieved in compliance with the social welfare function (Stiglitz 1999, 98), provided there are no market mechanisms responsible for achieving those specific goals (such as inherited altruism). The social welfare function contains every significant value, not only public justice but also, for example, the beauty of specific landscapes. Agreeing on the catalogue of key societal values allows for the establishment of methods of how the administration engages in the economic process, which means the implementation of strategies, policies and specific programmes. This impacts resource management, for example, by deciding which resources should be spent on activities outside and inside the market. There are multiple options:

- Modification of the market processes by tax systems (fees associated with the usage of maritime space or subsidies (renewable energy sources) or other activities (e.g. restoration of information symmetry by the publication of research results))
- Allocation of public funds on goals that are not included in market mechanisms (e.g. navigational signs that guarantee safety of navigation, social capital, social justice by supporting fisheries organizations)
- Changes in producer and consumer behaviour by legal regulations and various forms of rationing or administrative regulations and by supporting the formation of proper institutions (e.g. binding maritime spatial plans, licensing, arrangement, rules and conditions of using common resources)
- Changes in producer and consumer behaviour by educational activities, capacity building and awareness raising

In the literature, it is generally assumed that the purpose of MSP is sustainable development, although the importance given to individual dimensions of sustainable maritime development varies (see: Saunders et al. 2016). The European Commission is also explicitly in favour of this type of development, as expressed in various documents including the Maritime Spatial Planning Directive (EC 2014a, Article 5). Many documents issued by the European Commission refer to sustainability in their titles, for example, the Sustainable Blue Growth Agenda for the Baltic Sea Region (EC 2014b). It is generally accepted that sustainable development encompasses ecological, social and natural spheres in their specific spatial dimensions (Dühr 2011). The environmental dimension can be related to the Marine Strategy Framework Directive (MSFD; EC 2008), the economic dimension to Blue Growth strategies (EC 2014b; Varjopuro et al. 2015, Schultz-Zehhden et al. in this book) and the social dimension to stakeholder participation and knowledge building (Zaucha et al. 2017). Despite some strong critique, sustainability still seems to be politically attractive, as evidenced by the adoption in 2015 (23 years after the first Earth Summit) of the document "Transforming Our World: the 2030 Agenda for Sustainable Development" (UN 2015). Goal 14 has a clear reference to sustainability at sea.

The ecological dimension of sustainable development can also be expressed as resilience, understood as an ability of ecosystems to absorb shocks, various pressures and disturbances by renewing, reorganizing and developing while maintaining their essence and preceding functions (Walker et al. 2004). Davoudi et al. (2016) propose an expanded understanding of the notion: evolutionary resilience, focusing on the management aspect. In this theory, in addition to persistence, which is concerned only with ecosystems, flexibility plays a key part (understood as an ability of an ecosystem to choose alternative paths of development), as do resourcefulness, transformability and, above all, readiness (preparedness) to meet challenges. Weig (2016), dwelling on evolutionary economic geography and complexity theory, also explains how path dependency can avoid lock-ins through learning processes, building up resilience as an emergent pattern. Thus understood, resilience is both a paradigm and a development pattern of broader socioecological systems, and thus a complement to sustainable development, since the transformational element of the resilience concept brings with it dynamism and adaptivity.

Another alternative to sustainable development or evolutionary resilience as a public choice key objective may be the goal of minimizing spatial conflicts at sea. This makes sense in the case of so-called win-win solutions. A spatial order emerges, but this can encourage more intensive use of maritime space, which in turn makes this kind of approach no longer robust.

Marine governance as a function of conflict minimization has been prevalent in most coastal countries until recently. This era, however, has ended with the emergence of new forms of sea use and more intensive use of marine space that would require trade-offs.

Other goals frequently applied in relation to terrestrial spatial development such as quality of life, territorial cohesion or spatial integration (Costanza et al. 2008; Zaucha and Szlachta 2017, 19–22; Doucet 2013) appear to be of lesser importance at sea due to the limited presence of human beings there or the lack of clarity regarding the goals' substance and content, particularly quality of life (cf. Bok 2010). These goals should be treated as part of broader development paradigms, that is, sustainable development or resilience. Such an approach to quality of life is, for example, seen in OECD (Organisation of Economic Co-operation and Development) research (2013, 29).

All the axiological issues mentioned earlier are only the tip of the iceberg in terms of problems arising from public choice in the context of maritime spatial development. Even assuming that risk aversion, typical for public authorities, has been overcome and that authorities have managed to successfully aggregate private preferences into the public ones regarding maritime space, the public choice process immediately encounters several other challenges.

The first dilemma concerns temporal aspects of the aggregation of individual preferences. For instance, the desirable proportions of elements constituting sustainable development might evolve in time. These proportions depend on social prosperity and social awareness. During processes related to public choice, organized groups of stakeholders, who may convince authorities that they are speaking for the entire society and all those concerned, can emerge as a threat. The ease of this operation is proportional to the magnitude of transaction cost associated with the participation in the public choice processes and the magnitude of the individual loss perceived by non-organized individuals as a result of non-resisting the vested interests.

The second issue is associated with the multilevel character of public choice. Preferences aggregated locally may differ from those at regional, national or EU level. Externalities of energy production can serve as an example. Offshore energy can be treated as desirable at national level but can be opposed at the local level due to landscape pollution. This applies not only to EU shared policies but also to the exclusive ones. The Common Fisheries Policy puts emphasis on the sustainable use of resources of marine biota, while on the local level there are frequent demands for additional maritime space for fishers in order to protect their cultural role as part of a specific landscape and touristic values (externalities). Sometimes, legislators might even consider local and regional preferences more important than national ones; however, as described earlier, this might easily serve the needs of vested interests.

The third potentially faulty element is associated with the agency dilemma (principal-agent problem) (Mitnick 2006). Legislature aggregates public preferences with regard to marine space, but its ability to control the executive bodies is limited. This is because the latter group often has more information in certain fields of knowledge, and therefore, decision-makers have various problems evaluating the agents' level of involvement and reasons for failure in this context. In such cases, the phenomenon of subjective risk-taking—also known as moral hazard—can occur. Another problem is adverse selection, that is, an agent's choice to act in a negative way (from the principal's perspective) on the basis of information that is being held back by them. The result is insufficient effort of the executive authority focusing on activities based on self-interest.

The fourth problem relates to frequent changes in terms of goals, preferences and directions taken during spatial development. Public choice is characterized by its dynamic nature. There is no denying that voters' preferences change in response to stimuli, such as the available information or the state of the economy, and that this causes changes in policies and programmes. Nonetheless, private investors require a predictable economic horizon for their decisions, especially when the long-time rate of return is concerned. Distrust towards the policy stability of a given country or region may discourage investments. As a result, the most desirable patterns of maritime spatial development might not be realized even though they may have been previously declared (in the course of the public choice) and investment from the public sector may be necessary to fill in the gap. Economic praxis shows that investors start to act if, at the time of bearing the cost, the discounted future profits are higher than the discounted costs themselves. Risk provokes a more conservative assessment of profits, while uncertainty disrupts this process.

The fifth problem is related to deficiencies in putting forward aggregated public choice preferences towards maritime space. The above-mentioned instruments might appear insufficient and the various governance processes might not sufficiently reinforce each other. For instance, MSP can reserve areas for offshore renewable energy, but the absence of adequate feed-in tariffs or limits in transmission capacities of a public grid might make this effort futile.

All these situations are related to *governance failure*. This does not mean that public choice has no real influence on the way maritime space is developed. On the contrary. It is only thanks to better or worse public choice decisions that some functions can be assigned to sea space, such as:

- Preservation of environment (externalities, public good)
- Landscape protection (public good)
- National defence (public good)
- Underwater cultural heritage (externalities)
- Living organisms' well-being (common resources)
- Basic scientific research (public good)
- And those left unused for the use by future generations (inter-generational justice)

This situation is presented in Fig. 5.4.

By doing this, MSP adds a social-spatial rent to the private rent perceived by the business sector. Such rent is related to important social values (e.g. sustainable development), positive and negative externalities and can be described as an expression of their importance revealed through MSP in comparison to a pure market approach. From a purely economic perspective, MSP is a public choice process. Its essence is in the aggregation of preferences of individuals towards maritime spatial development and the shaping of

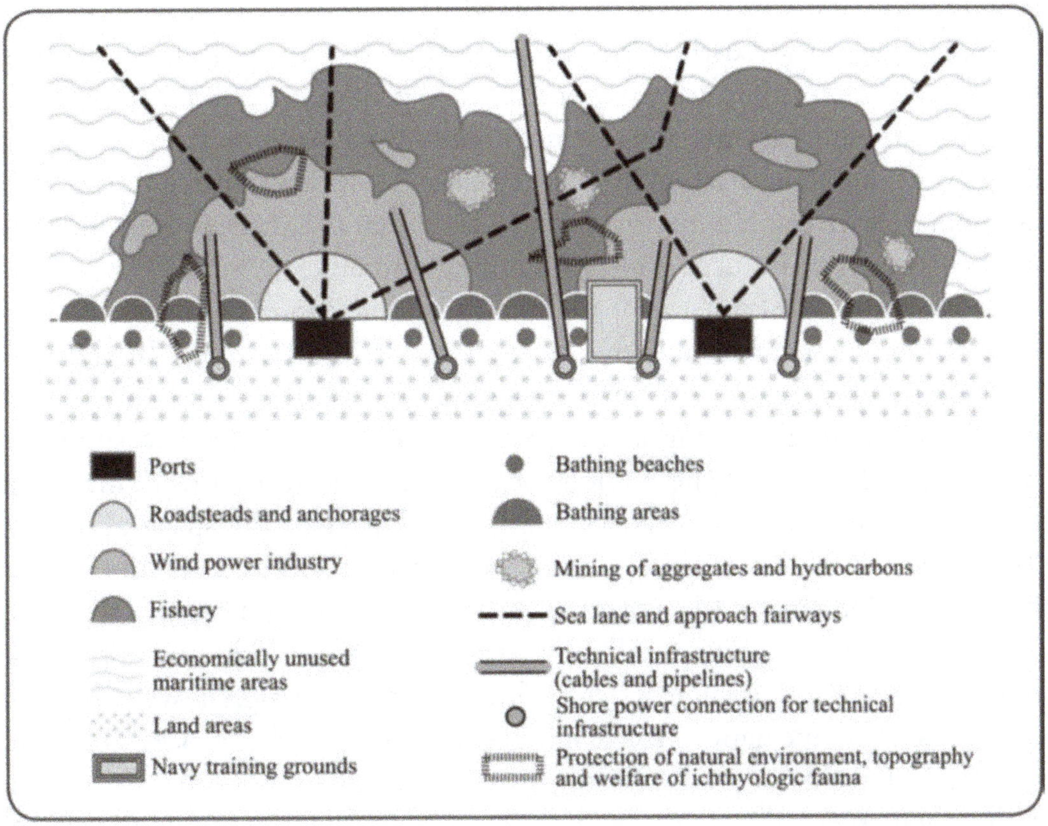

Fig. 5.4 Influence of public choice on the market processes that together shape maritime spatial development (Zaucha 2018)

public decisions on the allocation of sea space to these preferences in a situation of market failure. However, as described earlier, it can fulfil its role provided that other marine governance regimes are properly integrated with or within MSP (see also Chap. 6 in this book).

5 Interplay Between Maritime Spatial Planning and Market Forces

In reality, maritime spatial development is shaped by both public choice (MSP in particular) and market mechanisms. This fact has been widely recognized by maritime stakeholders, for example, in Poland. Out of 70 Polish MSP stakeholders examined by Ciołek et al. (2018), the vast majority view maritime spatial development as an outcome of such a combination. Only a few (eight) declared that MSP should be solely driven by the concerns and ideas of maritime administration and even fewer (five) declared that the market should have a final say in shaping solutions of MSP. These preferences were independent of the level of knowledge on MSP. Here one can see a kind of broad societal consensus, at least in Poland.

According to the existing regulations (EU 2014) in the EU, MSP, along with some other administrative processes (e.g. Natura 2000 management plans, some international conventions like United Nations Convention on the Law of the Sea—UNCLOS), was assigned, at least in formal terms, with a leading role in shaping maritime space. However, this does not preclude the market forces' real influence on the planning outcomes.

By observing MSP in several EU countries, one can easily notice the significance of such forces and the importance of location rent. In the UK, Germany and Belgium, offshore energy has received a prominent role in spatial plans. In all existing plans, particular attention is paid to shipping. Both users offer one of the highest location rents according to estimates of their Gross Added Value (Ecorys 2012) in relation to the space occupied. The exception to this rule is environmental protection as a genuine public choice decision under EU governance, the latter being a regime that also plays an essential role despite low private location rent. However, its social rent (private rent plus value of externalities) seems very high as well. Thus, as on land, on the one hand, MSP acknowledges some market processes (due to important benefits of key MSP stakeholders), while on the other, it corrects some key market failures (e.g. by internalizing externalities within the allocation process, as is the case with environmental protection). Another example of the importance of social rent is the decision in the Polish MSP to pay special attention to the spatial needs of artisanal fishers. Despite its limited profits, this sector has been considered as important and deserving of access to marine space due to symbolic and cultural reasons (i.e. due to its high social rent).

The key problem is, however, that MSP, in many cases, acts under uncertainty. The monetary value of externalities is unknown, and the democratic decision-making in many cases fails to reveal clear preferences to some uses (as it has been done with regard to environmental protection via MSFD). Thus, learning by doing and provoking public debate are the only feasible methods for planners who, as a rule, have no authority to decide on values and societal goals under public choice (see also Chap. 9 in this book).

The question posed at the beginning regarding the relevance of classical location theory in understanding maritime spatial development should therefore garner a positive answer. Nowadays in the EU, both market and MSP shape maritime spatial development. Classical location theory, despite all its shortcomings, might offer an interesting starting point for considering how it plays out in practice. Its strength is in encompassing economic considerations with regard to the usage of maritime space combined with a pre-set structure of existing sea gateways on land. Thünen's model can help in predicting good candidate areas for certain economic development zones in a plan. Such an approach should allow the designation of e.g. investment zones in order, perhaps, also to promote economies of agglomeration.

However, in contrast to Thünen's specific time period, nowadays the situation seems much more complex and dynamic. Game theory and strategic behaviour of developers should also be considered. Also, a key difference is that MSP might become a proactive agent in influencing all cardinal features of Thünen's interplay between costs and revenues. For example, it can influence market mechanisms by the wise management of distance and the development of terrestrial gateways servicing the sea areas. Thus, spatial planning can influence behaviour and prompt the decisions of private businesses. For instance, in Poland the completion of a new motorway network will create economic incentives that might result in the construction of a new large port in the central Polish coast (Komornicki 2015).

Moreover, other more sophisticated mechanisms of influencing market processes are also available for MSP. As pointed out by Zaucha (2007), spatial planning can influence investor decisions:

- by creating expectations of a given course of development in the future (thus, the planning process might cause the location decisions of the private sector even without public investments [in transport or in sea gateways], provided that the perceived benefits [i.e. those resulting from planning] are sufficiently high).
- by revealing important information about the space and/or diminishing risk of conflicts since all of these lower investment costs.

More on this topic can be found in Schultz-Zehden et al. in this book.

The question remains as to what extent MSP can foster economies of agglomeration. As pointed out by the MUSES (The Multi-Use in European Seas - a Horizon 2020 funded project) project,[10] MSP can help in overcoming high transaction costs of multi-uses that are considered to be new and more efficient ways of exploiting marine space. Multi-use in the long run leads to an increase in the productivity of labour and capital (e.g. higher revenues from usage of ships both for servicing offshore wind farms and mariculture co-located with them). This, in turn, might result in a clustering of economic activities in marine space. Additionally, other features might lead to a similar outcome (e.g. bathymetry, availability of light). However, it is not clear whether such islands of higher productivity in the sea would underpin a cumulative causation, that is, forward and backward linkages. On the one hand, a combination of offshore energy and mariculture can attract or even foster entirely new uses, such as tourism related to offshore industries or the construction of electricity filling stations for autonomous ships but, on the other hand, this might increase the cumulative pressure on the sea ecosystem that is essential for the provision of numerous marine ecosystem services. Thus, the environment can pose some limits to the concentration of Blue Growth. Moreover, sea industries share the same value chain only to a limited extent. Many up- and downstream industries connected to the marine sectors are much more productive on land than on sea. The most intriguing question is, therefore, the possibility of the appearance of network agglomeration economies at sea and land. For instance, multi-uses may foster, in an indirect way, learning processes and, thus, agglomeration economies, because people from different sectors need to talk to one another in order to understand the needs of others and to cooperate. All these questions deserve more systematic answers and further research. They will pave the way to a research agenda of marine spatial economists in the years to come. Thus, the future of marine spatial economics resides not only in understanding the

[10] For the project, please consult https://muses-project.eu/.

patterns of private and social location rent and the role of MSP as a vehicle of their integration but also in better exploring interdependencies between marine sectors and their economic results as well as the ways in which MSP can contribute to their hindrance or stimulation.

6 Conclusion

The ultimate conclusion is that allocation of marine space requires both market and public choice mechanisms. MSP neglecting market outcomes would be hardly enforceable due to resistance of many stakeholders. However, MSP seems a key vehicle for delivering important non-market values of society, such as good environmental status, integrity of habitats or safety and security. Therefore, MSP should both support and restrict market forces simultaneously. A key challenge is to achieve a proper mix between market and non-market outcomes and approaches. The final mix depends on the values of a given society, and is dynamic and changes with time and prosperity level. Therefore, MSP should also be seen as a dynamic process deeply rooted in and constantly revealing key societal values, searching for an acceptable proportion between efficiency and other societal values that together constitute the social welfare function.

Acknowledgements Analyses presented in this chapter were financed by the Polish National Science Centre under the project "Economy of Maritime Space," Decision No. 2015/17/B/HS4/00918. The results are presented in Polish in Zaucha (2018). The Open Access fee of this chapter was provided by the same project. The author expresses his gratitude to the colleagues from Maritime Institute in Gdańsk, Institute for Development in Sopot and Faculty of Economics of the University of Gdańsk for support and inspiration in the course of preparation of this chapter.

References

Blaug, M. (1985). *Economic Theory in Retrospect* (4th ed.). Cambridge: Cam Bridge University Press, 737p.

Bok, D. (2010). *The Politics of Happiness: What Governments Can Learn from the New Research on Well-Being*. Princeton, NJ: Princeton University Press, 272p.

Christaller, W. (1933). *Die Zentralen Orte in Suddedeutschlan*. Jena: Gustav Fischer Verlag.

Ciołek, D., Matczak, M., Piwowarczyk, J., Rakowski, M., Szefler, K., & Zaucha, J. (2018). The Perspective of Polish Fishermen on Maritime Spatial Planning. *Ocean and Coastal Management*. https://doi.org/10.1016/j.ocecoaman.2018.07.001.

Costanza, R., Fisher, B., Ali, S., Beer, C., Bond, L., Boumans, R., Danigelis, N. L., Dickinson, J., Elliott, C., Farley, J., Elliott Gayer, D., MacDonald Glenn, L., Hudspeth, T. R., Mahoney, D. F., McCahill, L., McIntosh, B., Reed, B., Turab, R. A., Rizzo, D. M., Simpatico, T., & Snapp, R. (2008). An Integrative Approach to Quality of Life Measurement, Research, and Policy. *S.A.P.I.E.N.S, 1*(1), 17–21.

Daly, H. E., & Farley, J. (2011). *Ecological Economics: Principles and Applications.* Washington; Covelo; London: Island Press, 509p.

Davoudi, S., Zaucha, J., & Brooks, E. (2016). Evolutionary Resilience and Complex Lagoon Systems. *Integrated Environmental Assessment and Management, 12*(4), 711–718.

Doucet, P. (2013). Territorial Integration—Food for Thought. In G. Gorzelak & K. Zawalińska (Eds.), *European Territories: From Cooperation To Integration?* (pp. 27–41). Warsaw: Wydawnictwo Naukowe Scholar.

Dühr, S. (2011). Territorial Cohesion and Its Impact on Sustainable Development. *Baltic 21 Series, 2/2011*, 14–15.

EC. (2008). Directive 2008/56/EC of the European Parliament and of the Council of 17 June 2008 Establishing a Framework for Community Action in the Field of Marine Environmental Policy (Marine Strategy Framework Directive). *Official Journal of the European Union*, L 164/19.

EC. (2014a). Directive 2014/89/EU of the European Parliament and of the Council of 23 July 2014 Establishing a Framework for Maritime Spatial Planning. *Official Journal of the European Union*, L 257/135.

EC. (2014b). Commission Staff Working Document a Sustainable Blue Growth Agenda for the Baltic Sea Region. Brussels, 16.5.2014 SWD(2014) 167 final.

Ecorys. (2012). *Blue Growth Scenarios and Drivers for Sustainable Growth from the Oceans, Seas and Coasts.* Final Report. Rotterdam and Brussels: European Commission, DG MARE.

Fujita, M. (2010). The Evolution of Spatial Economics: From Thünen to the New Economic Geography. *The Japanese Economic Review, 61*(1), 1–32.

Fujita, M., & Thisse, J. F. (2002). *Economics of Agglomeration, Cities, Industrial Location, and Regional Growth.* Cambridge: Cambridge University Press, 465p.

Fujita, M., Krugman, P., & Venables, A. J. (2000). *The Spatial Economy: Cities, Regions, and International Trade.* Cambridge, MA and London: The MIT Press, 367p.

Hassler, B., Gee, K., Gilek, M., Luttmann, A., Morf, A., Saunders, F., Stalmokaite, I., Strand, H., & Zaucha, J. (2018). Collective Action and Agency in Baltic Sea Marine Spatial Planning: Transnational Policy Coordination in the Promotion of Regional Coherence. *Marine Policy, 92*, 138–147.

Isard, W. (1956). *Location and Space-Economy*. New York: John Willey & Sons, 350p.

Johansson, B., & Quigley, J. M. (2003). Agglomeration and Networks in Spatial Economies. *Papers in Regional Science, 83*(1), 165–176.

Komornicki, T. (2015). Present and Future Spatial *Accessibility* of the Polish Sea *Ports. Bulletin of Maritime Institute, 30*(1), 59–71.

Krugman, P. (1991a). Increasing Returns and Economic Geography. *Journal of Political Economy, 99*(3), 483–499.

Krugman, P. (1991b). *Geography and Trade*. Leuven; London; Cambridge, MA: MIT Press, Leuven University Press, 142p.

Lösch, A. (1940). *Die räumliche Ordnung der Wirtschaft*. Jena: Gustav Fischer Verlag.

Marshall, A. (1920). *Principles of Economics* (8th ed.). London: Macmillan, 754p.

McCann, P. (2013). *Modern Urban and Regional Economics*. Oxford: Oxford University Press, 408p.

Mitnick. (2006). The Origins of Agency Theory. Retrieved March 23, 2018, from http://www.pitt.edu/~mitnick/agencytheory/agencytheoryoriginrev11806r.htm.

OECD. (2013). *OECD Guidelines on Measuring Subjective Well-Being*. Paris: OECD Publishing.

Ostrom, E., Gardner, R., Walker, J., Agrawal, A., Bloomquist, W., Schlager, E., & Yan, T. S. (1994). *Rules, Games and Common-Pool Resources*. Ann Arbor: University of Michigan Press, 392p.

Ponsard, C. (Ed.). (1988). *Analyse économique spatial*. Paris: PUF 452p.

Przygodzki, Z. (2016). Kapitał terytorialny w rozwoju regionów. *Acta Universitatis Lodziensis Folia Oeconomica, 2*(319), 83–97.

Saunders, F., Gilek, M., Gee, K., Göke, C., Hassler, B., Lenninger, P., Luttmann, A., Morf, A., Piwowarczyk, J., Schiele, K., Stalmokaite, I., Strand, H., Tafon, R., & Zaucha, J. (2016). *Exploring Possibilities and Challenges for MSP Integration in the Baltic Sea*. Stockholm: Bonus Baltspace. Retrieved March 25, 2018, from https://www.baltspace.eu/published-reports.

Stiglitz, J. (1999). *Economics of the Public Sector* (3rd ed.). New York and London: W.W. Norton & Company.

UN. (2015). Transforming Our World: The 2030 Agenda for Sustainable Development. A/RES/70/1. United Nations. Retrieved September 7, 2018, from https://sustainabledevelopment.un.org/post2015/transformingourworld/publication.

Varjopuro, R., Soininen, N., Kuokkanen, T., Aps, R., Matczak, M., & Danilova, L. (2015). Communiqué on the Results of the Research on Blue Growth in the Selected International Projects Aimed at Enhancement of Maritime Spatial Planning in the Baltic Sea Region (BSR). *Bulletin of the Maritime Institute in Gdańsk, 30*(1), 72–77.

Walker, B., Holling, C. S., Carpenter, S. R., & Kinzig, A. (2004). Resilience, Adaptability and Transformability in Social-Ecological Systems. *Ecology and Society, 9*(2), 5.

Weig, B. (2016). *Resilienz komplexer Regionalsysteme*. Berlin: Springer Publishing.

World Bank. (2009). *Reshaping Economic Geography*. Washington, DC: World Bank, 383p.

Zaucha, J. (2007). *Rola przestrzeni w kształtowaniu relacji gospodarczych. Ekonomiczne fundamenty planowania przestrzennego w Europie Bałtyckiej*. Gdańsk: Uniwersytet Gdański, 371p.

Zaucha, J. (2018). *Gospodarowanie przestrzenią morską*. Warszawa: Instytut Rozwoju i Sedno.

Zaucha, J., & Szlachta, J. (2017). Territorial Cohesion: Origin, Content and Operationalization. In J. Bradley & J. Zaucha (Eds.), *Territorial Cohesion: A Missing Link Between Economic Growth and Welfare: Lessons from the Baltic Tiger* (pp. 23–47). Gdańsk: Uniwersytet Gdański.

Zaucha, J., Matczak, M., & Pardus, J. (with 44 coauthors). (2016). Study of Conditions of Spatial Development of Polish Sea Area. Retrieved March 29, 2018, from http://www.umgdy.gov.pl/?cat=96.

Zaucha, J., Gilek, M., Hassler, B., Luttmann, A., Morf, A., Saunders, F., Piwowarczyk, J., Gee, K., & Tusrki, J. (2017). *Bonus Policy Brief: Challenges and Possibilities for MSP Integration in the Baltic Sea*. Stockholm: Bonus Baltspace. Retrieved March 25, 2018, from https://www.baltspace.eu/published-reports.

6

Maritime Spatial Planning and the EU's Blue Growth Policy: Past, Present and Future Perspectives

Angela Schultz-Zehden, Barbara Weig, and Ivana Lukic

1 Introduction

Blue Growth, as a cross-cutting policy tool in Europe, has continuously evolved with the development of Maritime Spatial Planning (MSP) and several related maritime policies. The EU's overarching Integrated Maritime Policy (IMP), set in place in 2007 (EC 2012a), seeks to provide a more coherent approach to maritime issues, with increased coordination between different policy areas. In particular, it pursues three main targets: (1) sustainable development of the European maritime economy, (2) protection of the environment and (3) cooperation of all maritime players across sectors and borders.

To reach these goals, IMP suggests several tools and cross-cutting policies including Blue Growth, marine data and knowledge, integrated maritime surveillance, MSP, maritime security as well as sea-basin strategies. While the Marine Strategy Framework Directive (MSFD) stands for the environmental pillar, MSP is presented as the economic pillar (EC 2012b).

Outside Europe, MSP is often seen as a tool to ensure the needs of marine nature conservation, while also serving Blue Growth desires. MSP is therefore understood in a much broader sense as being almost more connected to the overarching IMP policy rather than the narrow understanding provided by

A. Schultz-Zehden (✉) • B. Weig • I. Lukic
s.Pro | sustainable projects GmbH, Berlin, Germany
e-mail: asz@sustainable-projects.eu

J. Zaucha, K. Gee (eds.), *Maritime Spatial Planning*,
https://doi.org/10.1007/978-3-319-98696-8_6

the EU Marine Spatial Planning Directive (MSPD; EC 2014c), which, however, should always be understood in context of the parallel provisions under the MSFD. As discussed in other chapters and elsewhere, under IMP, both Directives should be integrated in principle. But better coordination between all IMP cross-cutting policies has yet to be achieved in Europe (Fritz and Hanus 2015). Integration mechanisms across the EU occur at very strategic and high government levels. In many EU member states, the designated authorities in charge of the implementation of the MSPD and MSFD differ and do not want to be held responsible for each other. Thus, integration efforts may not have yet trickled down to the more practical implementation level (Ansong et al. 2018).

Moreover, some of the discussions about the role of MSP and its relationship to Blue Growth emerge from different understandings of MSP, Blue Growth or IMP—not only between European and non-European processes but also within Europe.

Therefore, in this chapter, we first shed light on the original rationale behind the Blue Growth policy in Europe and discuss its evolution. We then show how this is supported by economic figures of maritime sectors across European sea-basins and countries, including potential variations. We further explore original expectations of how MSP should contribute to promoting Blue Growth and how these have evolved as a result of projects, studies and planning processes undertaken in the meantime. We conclude with related practical implications for the work of maritime spatial planners in Europe.

Since all three policies (IMP, Blue Growth and MSP) are targeted towards all European member states, the chapter also always takes a European-wide perspective. It shows not only differences in what Blue Growth may mean across all European sea-basins but also the possible consequences for MSP processes across Europe. This is important due to the MSPD, which requires all EU member states to develop MSP plans by 2021. As a result, MSP is no longer pursued only by countries which are 'pushed' to find suitable space for offshore wind but also by those where offshore wind does not play a role.

2 The Evolution of the Blue Growth Policy in Europe

2.1 The Origin of the EU Blue Growth Policy as a Way to Address the 2010 Economic Crisis

Building upon parallel efforts on Blue Growth from the Agenda 2010 process (Barbesgaard 2016) and OECD, FAO (2014) and UNEP (2012) initiatives, among others, further development of the Blue Growth concept and its consequent policy in Europe can be traced back to the year 2010. The concept was highly influenced by the economic crises at the time and the need to find adequate policy responses. While acknowledging the role of global Blue Growth initiatives around the time of the global financial crisis, the focus of this chapter is on Blue Growth evolution in Europe.

The Europe 2020 Strategy suggested a way out of the economic crises by fostering smart, sustainable and inclusive growth (EC 2010). The strategy offered a vision of Europe's social market economy for the twenty-first century, by focusing on knowledge and innovation, based on the concepts of smart specialization (Foray 2015), a resource-efficient, greener and more competitive economy, and a high-employment economy-enabling economic, social and territorial cohesion (EC 2010).

Blue Growth—defined as '*smart, sustainable and inclusive economic and employment growth from the oceans, seas and coasts*' (ECORYS et al. 2012: 26)—stands for the maritime pillar of the Europe 2020 strategy. In fact, the sea and the coasts have always been drivers of the European economy and centres for new ideas and innovation. In contrast to earlier times, additional new reasons, such as rapid progress in development of offshore technologies, potential for further exploration of marine resources and the relatively low emission of greenhouse gases in seaborne transport, led to the conclusion that the maritime economy could become one of the main drivers for fostering smart, sustainable and inclusive growth (EC 2012a). Whereas it was suggested that innovative approaches, new technologies and synergies would be important factors leading to a growing maritime economy, it was also argued that supporting policy measures were necessary for Blue Growth to develop to its full potential (ECORYS et al. 2012).

As discussed in the following section, the focus of the resulting EU blue growth policy has slightly changed during the last eight years, as evidenced by several development stages.

2.2 The Development of Blue Growth and Support Approaches in the EU

The *first phase (2010–2013)* of the EU Blue Growth policy was characterized by a general discussion on what is Blue Growth, how to support it and if support is needed at all. This stage was highly influenced by the initial study on Blue Growth scenarios and drivers, which analysed six maritime functions and 27 subfunctions (ECORYS et al. 2012). The study concludes that all Blue Growth activities highly depend on suitable framework conditions. Adequate infrastructure, high-skilled staff as well as access to low-skilled workers are as important as public acceptance, a solid international legal framework and good governance at local and regional levels (ECORYS et al. 2012). Moreover, the study concludes that blue activities differ across Europe—therefore, sea-basin-specific studies and strategies should be elaborated. Due to possible heterogeneity within sea basins, it was additionally recommended to focus on specific blue clusters and develop tailor-made policy measures (ECORYS et al. 2012).

The study also analysed synergies between the 27 blue subfunctions, with emphasis on fields with a relatively high probability of cross-innovation. This was based on the claim that synergies result from shared suppliers, activities, input factors or common use of infrastructure. A focus of Blue Growth should thus be on promoting synergies, which enable the whole to be more than the sum of its parts. However, tensions between different blue activities were also identified, arising from mutually exclusive activities claiming limited space. To enable Blue Growth, it was seen as essential to avoid tensions and support the use of synergies, which could be accomplished through MSP (ECORYS et al. 2012).

Subsequently, the European Commission endorsed its Blue Growth Strategy 'Opportunities for marine and maritime sustainable growth' in September 2012 (EC 2012a). By that time, the Commission stressed that Blue Growth was sufficiently covered and supported by already existing initiatives related to MSP and Integrated Coastal Zone Management (ICZM), 'Marine Knowledge 2020', the MSFD, FP7 Ocean of Tomorrow calls and many others. The Commission actually suggested only five focus areas for policy action: (1) blue energy; (2) aquaculture; (3) maritime, coastal and cruise tourism; (4) marine mineral resources and (5) blue biotechnology. Maritime transport is left out, with reference to specific ongoing EU initiatives already in place. The Commission also emphasized that this list is not exhaustive, as new fields might emerge (EC 2012a).

On 8 October 2012, the Informal Minister Conference on Integrated Maritime Policy in Nicosia (Cyprus) approved the Limassol declaration on 'A Marine and Maritime Agenda for Growth and Jobs'. In contrast to the Commission, the ministers broadened the scope of Blue Growth actions to six priorities by also including shipping and shipbuilding, while leaving out blue biotechnology. The suggested policy actions are, however, in line with the Commission's suggestions with the main focus being laid on reducing administrative and regulatory burdens and removing bottlenecks for innovation and investment (Limassol Declaration 2012).

The first phase of the blue growth policy ends with a resolution of the European Parliament (2013) on 'Blue Growth: Enhancing sustainable growth in the EU's marine, maritime transport and tourism sectors', which highlights in particular the role of maritime transport and tourism. The resolution addresses other aspects such as the significance of Blue Growth as part of the Europe 2020 Strategy, the importance of regional sea-basin strategies and the central role of MSP as enabler of Blue Growth. In addition, the Parliament points to the need for harmonizing planning processes at the interface between maritime and land-based planning as well as closing knowledge gaps on maritime activities (European Parliament 2013).

2.3 Fostering Blue Growth via Stimulating Innovation in the Blue Sectors

During the *second phase (2014–2016)*, the discussion on Blue Growth was directed towards the topic of innovation. According to the European Commission, innovation is a prerequisite for growth and job creation, also in blue sectors. Moreover, innovation is considered to be important for improving environmental conditions. However, several studies had unveiled severe bottlenecks for innovation in Europe in general, with three barriers specific to Blue Growth: (1) gaps in marine/maritime knowledge and data; (2) diffuse research efforts hindering interdisciplinary learning; and (3) lack of scientists, engineers and skilled workers (EC 2014a).

For closing those gaps and to push aside barriers of innovation, the Commission worked out a detailed roadmap (EC 2014b). In this roadmap, the European Marine Observation and Data network (EMODnet) plays a major role for harmonizing data, standardizing access and reducing bureaucracy. EMODnet is intended to include data from diverse sources, including EU research projects, environmental studies conducted in the context of offshore wind farms (OWFs), monitoring instruments such as satellites or floating robots, as well as existing data from fisheries. The aim is to optimize observation networks by collecting data once and use them for many purposes instead of collecting data for specific purposes only. This new paradigm aims at avoiding gaps and duplications by saving costs and improving marine knowledge at the same time (EC 2014b).

The roadmap presented by the European Commission has been criticized in several aspects: a definition of blue economy is missing; lack of attention to the decline of traditional sectors such as small-scale fisheries, shipping and tourism; and consequences of the reduction of EU funds are not taken into account (EESC 2015). Others criticize (CoR 2015) that some of the most important blue sectors such as shipyards, shipping and blue energy are not covered in an appropriate way. A more effective matching between different EU strategies and programmes is requested as well as a specific knowledge and innovation society for blue economy to develop competences and enable better knowledge transfer from science to business. The development of entrepreneurship in blue economy should get more attention. Moreover, it should be considered that Blue Growth does not only take place on sea, but that support is also needed for blue sectors based on land, such as fish processing companies (CoR 2015).

Taking those different opinions and recommendations into account, the European Parliament endorsed its resolution on 'Untapping the potential of research and innovation in the blue economy to create jobs and growth' (European Parliament 2014). In this resolution, the Parliament emphasizes its dissatisfaction with the strict reduction of Blue Growth to five priorities and calls for a more integrative approach, including traditional and young sectors (European Parliament 2014).

2.4 Achievements in Blue Growth Policy

An evaluation of the Blue Growth policy in 2017 (EC 2017) comes to the conclusion that progress can so far primarily be observed in the collection of marine data and investments in research. Initiatives on skills development such as Leadership 2020 or the Commission's Blue Careers Initiative were introduced to close the gap on the labour market. Stakeholder events such as the European Maritime Day, the Blue Business and Science Forum or the Ocean Energy Forum have been established with the aim to bring together industry, finance, academia and public authorities to identify solutions and make investment more attractive. Finally, the adoption of the MSPD and the resulting need of EU member states to develop MSP as an integrative tool to improve maritime governance is noted positively, stressing the relation between MSP and Blue Growth. Weaknesses are still seen in a lack of private risk funding for innovative maritime technologies, which is still hampering maritime innovation to get to the market. Other challenges are rather sector specific (EC 2017).

In April 2017, the responsible ministers of EU member states expressed their continuous support to the Blue Growth policy (Valletta Declaration 2017). However, as already highlighted in the previous Limassol Declaration, the ministers stress again that the future direction of the Blue Growth Strategy should acknowledge the potential and importance of all relevant sectors of the blue economy, crucial for growth in value and jobs, and not only the five priority fields presented by the European Commission in the initial Blue Growth Strategy.

2.5 Future Steps: The Sea Not Only as an Economic Space but Also a Political One

In May 2017, the Committee of the Regions (CoR) offered policy recommendations, which initiated a new, third phase in the European policy on Blue Growth (CoR 2017). They request the Blue Growth Strategy to address the sea as a political topic and not only as a subject for projects. According to the CoR, the new integrated European maritime policy should provide solutions for a substantial broader set of not only economic but also socio-political and environmental challenges: (1) security of Europe's borders; (2) management of migration; (3) development of a maritime policy for EU's neighbourhood, regulation of maritime trade and governance of the oceans; (4) protection of biodiversity, combating climate change and successful energy transition, including transition to renewable fuels for ships; (5) development of the blue economy in traditional sectors such as fisheries, aquaculture, tourism, the maritime industries as well as emerging sectors like marine energy and marine biotechnology; (6) the reconciliation of activities and uses; (7) a coastal and maritime policy based on regions and local authorities; and (8) addressing the specific challenges of Europe's islands and overseas territories.

A coherent maritime territory is seen as the foundation of the blue economy and better interlinkage of land- and sea-based actions are fundamental to achieve this. The Committee emphasizes the importance of regional and sea-basin approaches and calls for cooperation between different levels, regions and sectors. To foster investment in blue economy, regional innovation strategies (RIS3) are suggested as appropriate means. MSP as an integrative tool is expected to play a central role in implementing those ambitious ideas of this new European Blue Growth policy (CoR 2017).

On 27 June 2018, the European Commission published its first annual economic report on the EU's blue economy. This report includes a detailed definition of blue economy but avoids the use of the term 'Blue Growth'. Instead, a distinction is drawn between established and emerging sectors (EC 2018) with aquaculture included under 'established' sectors and offshore wind still included under emerging sectors.

3 Blue Growth: Differences Among European Sea-Basins and Countries

3.1 Results from the Series of Sea-Basin Blue Growth Studies

Following the results of the initial European Blue Growth study (ECORYS et al. 2012) and variations in Blue Growth activities, as well as framework conditions between European sea-basins, a series of studies were commissioned to look into the specifics of Blue Growth sectors of EU member states around the Baltic Sea region (s.Pro 2013), the Mediterranean, Adriatic and Ionian, and Black Sea (EUNETMAR 2014), the North Sea region and English Channel (ECORYS, s.Pro, MRAG 2013), as well as Europe's Atlantic Arc (ECORYS, s.Pro, MRAG 2014).

All four studies followed the same methodological approach, identifying the largest maritime economic activities (MEAs) in terms of gross value added (GVA) and employment, as well as the fastest growing and most promising MEAs in each of the countries. While this harmonized analysis allows for comparison, it also meant that most of the data used in those studies were mainly from European-wide statistics—at that time only available for the years 2008–2010 and thus reflected an outdated picture set in midst of economic crisis. However, qualitative assessments provided in the studies—especially for identifying the most promising MEA—took into account more current expert knowledge. Nevertheless, results should mainly be understood to provide a broad picture of what may constitute Blue Growth in the various countries across Europe.

The synthesis of the sea-basin studies reveals that, not surprisingly, traditional sectors such as fishery, shipping and coastal tourism are the most important MEAs in terms of size (see Fig. 6.1) throughout all European countries. In the North Sea region also oil and gas extraction is relevant. In contract, the list of fastest-growing MEAs identified in the different countries is much longer and more heterogeneous (see Fig. 6.2). This indicates that maritime activities tend to become more diverse. At the same time, the most frequently named fastest-growing MEAs are also among the MEAs which are already the largest in terms of size. Short sea shipping, passenger ferry services and fisheries are traditional sectors, which are large and still growing. Cruise tourism is by far the most important growing sector all over Europe. In addition, at the time of the studies, a significant number of still small but rapidly growing activities emerged, such as offshore wind energy, marine mineral mining,

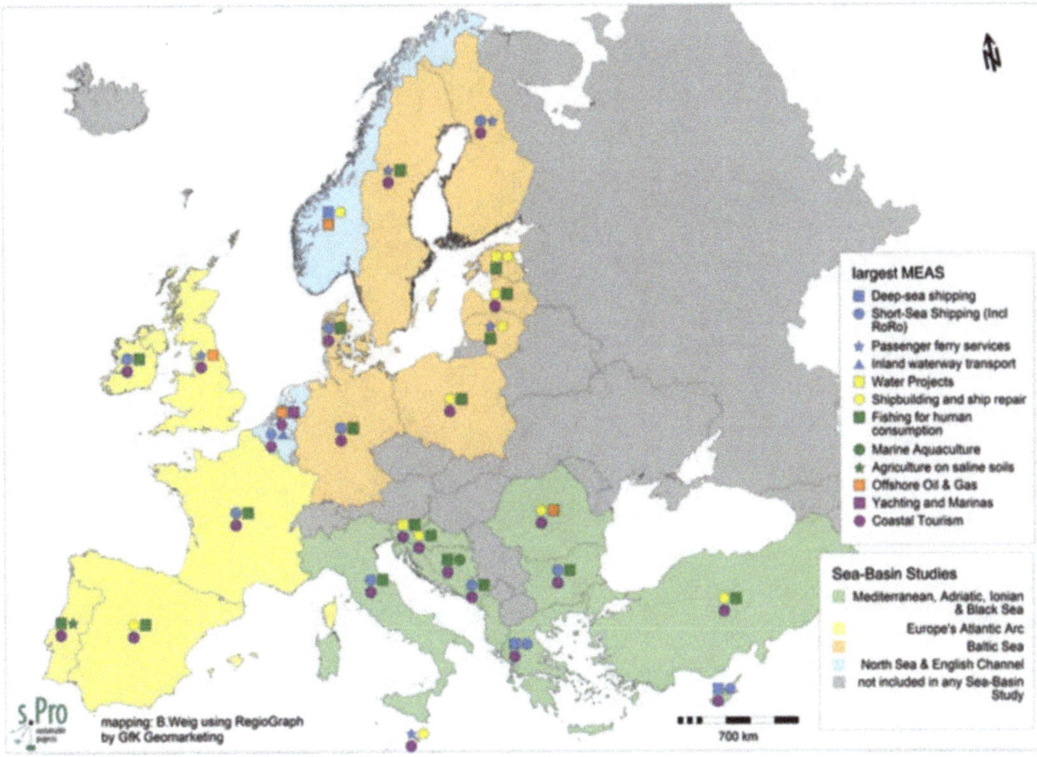

Fig. 6.1 Largest MEAs in terms of GVP and employment

fishing for animal feed, fresh water supply, protection of habitats, protection against flooding and erosion, traceability and security of goods supply chains and environmental monitoring.

The list of most promising MEAs identified for each country (see Fig. 6.3) shows even more different activities. Tourism and shipping are indicated in several countries in the Adriatic, Baltic and North Sea areas. In Portugal, Spain and France focus was placed on energy, monitoring, blue biotechnology and shipbuilding, while all forms of shipping as well as fishing were not seen as promising. In general, coastal tourism is identified in almost all countries (21 out of 28), followed by short sea shipping (17 countries), aquaculture (14 countries), shipbuilding (13) and offshore wind and cruise tourism (11 countries each). Whereas at the time of the studies offshore wind was only seen as fast growing in Germany and Finland, ocean energy was, however, seen as an important emerging topic in all sea basins with the exception of the Mediterranean. In contrast, there is a remarkable concentration of different growing tourism activities in the Mediterranean, especially the Adriatic Sea area.

The synthesis of the four studies demonstrates a large variety of what constitutes Blue Growth between European countries. Moreover, in most

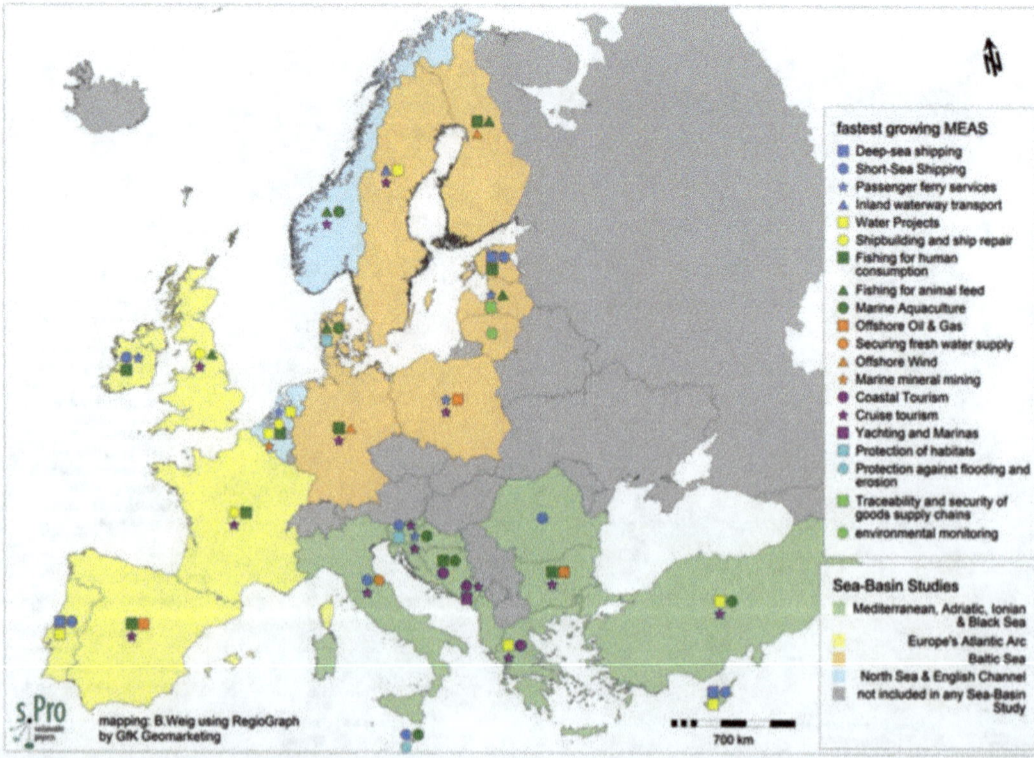

Fig. 6.2 The fastest-growing MEAs

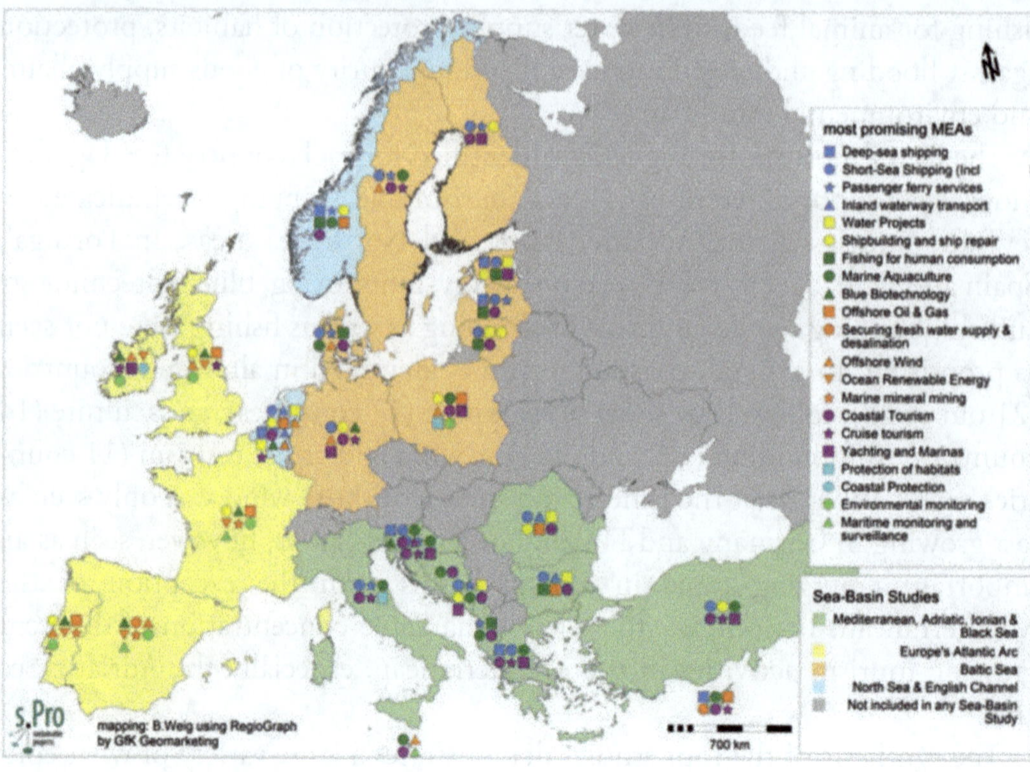

Fig. 6.3 The most promising MEAs

countries the specific Blue Growth fields belong to different maritime functions. While the number of important sectors in terms of size was still modest at that time, the great variety of MEAs in terms of fastest-growing and most promising activities already indicated that the variety of maritime activities will probably rise. This would result in an increased need for MSP to organize a fair, secure and sustainable use of the seas.

An update of those studies has just been published by the European Commission (EC 2018), which confirms most of the assumptions made in the earlier studies. For reasons of data availability, the report focuses on monitoring the development of the established blue sectors (including fisheries, shipping, tourism as well as aquaculture). It comes to the conclusion that these traditional sectors developed positively from 2009 to 2016. Employment rose by 2%, while average wages increased by as much as 14.2% (EC 2018). In 2016, the GVA of those established blue sectors was 9.7% higher than in 2009, while net investment in tangible goods increased by 71.7%. However, the sectors developed differently during and after the years of economic crises. While sectors in the field of living resources (fisheries and aquaculture and processing and retail) and coastal tourism increased in terms of GVA and employment, other sectors such as shipbuilding and repair as well as oil and gas extraction suffered losses (EC 2018). Much of this was foreseen in the earlier studies. Most emerging sectors still lack sufficient data for detailed monitoring. The offshore wind sector is an exception: this sector has grown most rapidly with the number of jobs rising from 20,000 in 2009 to 160,000 jobs in 2016. A total of 91% of the global capacity in terms of gigawatt is located within the EU, with potential for further growth (EC 2018).

In addition, the MUSES (Multi-Use in European Seas) project (Przedrzymirska et al. 2018) has provided a comparison of possible maritime multi-uses across the different European sea-basins. Even though this study is not based on economic data per sector in each country, it sheds an interesting light on the substantial differences between sea basins. This difference is especially prominent between Northern and Southern European countries, where different blue sectors appear to constitute a main driver for the respective national economies. Offshore wind and other renewable ocean energy sources are still the main driver for Blue Growth policy throughout Northern Europe. In contrast, most initiatives in Southern European sea-basins (Atlantic, Mediterranean and Black Sea) focus on smart combinations of new and old maritime economic sectors with tourism, which constitutes the primary expanding income source for those countries. The absence of a strong push for offshore wind development in the Southern European countries has substantial implications on what type of Blue Growth Policy should be emphasized and what may be key objectives and strategic goals for MSPs in those countries.

3.2 Blue Growth Policy: Supporting All Sectors or Only a Few?

As shown from the analysis earlier, a clear definition of what is exactly covered and understood under the European Blue Growth concept is currently still missing. So far, two different understandings compete in official documents. The comprehensive understanding of Blue Growth includes all maritime activities, as well as cross-sector activities and their respective land-based activities. However, the narrow approach so far supported by the European Commission understands Blue Growth as a support tool for young but high growth potential sectors, thus reducing Blue Growth to five sectors. Even though the European Parliament, the European Economic and Social Committee, the Committee of the Regions and others repeatedly propose using the comprehensive approach, the European Commission has so far always come back to the five focus areas, whenever dealing with Blue Growth (EC 2012a, 2014a, 2017).

Taking into account the significant variation among European countries concerning important blue growth activities, and also acknowledging that in most sea basins the five focus areas stressed by the European Commission only play a minor role, a suggested best way forward is to embrace a broader approach of Blue Growth and an accompanying strategy. These should be flexible enough to take into account regional and sectoral characteristics and needs—combined with a profound IMP and readiness for future challenges. Such an inclusive definition would also help to raise awareness of the importance of our seas among a broader range of stakeholders, which in turn is needed to make full use of its potential.

This is also confirmed by the MUSES 'Action Plan' process, which is based on the analysis of possible and promising multi-use concepts throughout European sea-basins (Lukic et al. 2018). The action plan shows that, apart from the need to push for technology breakthroughs for suitable multi-use combinations with offshore energy installations in Northern European countries, embracing cross-sectoral synergies and employing new technologies (also in traditional, developed or even declining sectors) provide totally new opportunities for sustainable Blue Growth in areas where this was previously not expected. The report emphasizes the importance of such multi-use concepts not only for Southern European countries but also for enabling Blue Growth in rural areas and remote island communities throughout the whole of Europe.

For example, pescatourism, an activity that combines fishing and tourism, allows artisanal fishers to diversify their activities and ensure an alternative source of income (Castellani et al. 2017; Vergílio et al. 2017). Also, the identification of multi-use opportunities between underwater cultural heritage (UCH) sites and tourism activities allows for new activities such as diving trails in UCH sites or virtual UCH tours on land. These may provide important sources of funding for UCH or nature protection sites while diversifying tourism offers (Przedrzymirska et al. 2018). These Blue Growth opportunities require policy actions focusing on capacity building of local stakeholders involved. This is quite different from the actions targeted towards the energy sector, which depend on an initial high level and advanced capacity (Lukic et al. 2018).

4 The Role of MSP in the EU Blue Growth Policy

4.1 Blue Growth Policy: More than MSP

As shown earlier, from the very beginning of the emergence of the Blue Growth concept, MSP has been mentioned as an important enabling tool which provides the precondition for maritime activities to thrive. However, as the discussion also demonstrates, MSP is by no means the only and most important policy tool to promote Blue Growth in Europe.

Other tools and support actions, which are not directly under the remit of MSP, are equally—if not even more—important and necessary to address the most urgent current challenges and thus foster the development of the various maritime sectors: (1) regulations, which are important for legal security of blue activities; (2) promotion programmes which foster knowledge transfer from research to business and thus accelerate technology innovations; (3) maritime skills development and training programmes to provide for the skilled labour forces necessary to apply new technologies; (4) efforts to gain better and relevant data and information; (5) initiatives which facilitate and streamline investments including risk funding for innovative maritime technologies; (6) economic support policies, programmes, incentives; and (7) facilitation of events and forums, which bring together industry, finance, academia and public authorities to come to joint solutions. These are just a few of the main actions necessary.

Although not completely representative, this more 'balanced' view on MSP was also confirmed during the interviews undertaken within the framework of the 2013 Baltic Sea Blue Growth study (s.Pro 2013). Of all four IMP areas, MSP was seen as least important by EUSBSR (EU Strategy for the Baltic Sea Region) stakeholders at that time. Additionally, the resulting European Commission working paper towards 'A sustainable Blue Growth Agenda for the Baltic Sea Region' (s.Pro 2017) puts more emphasis on (1) a consistent approach to innovation; (2) skills and qualifications; (3) cluster development; and (4) access to finance for maritime sectors. The follow-up stakeholder process undertaken four years later to promote the implementation of the Baltic Blue Growth Agenda (s.Pro 2017) confirms this by stressing the need to 'remove regulatory barriers', 'foster issue-driven collaboration' as well as increased efforts to 'raise awareness for blue products'.

4.2 Assumed Benefits of MSP to Blue Growth

Nevertheless, the role of MSP as an important facilitator and enabler for fostering the development of maritime sectors should also not be underestimated. A pre-impact study commissioned by the European Commission in the wake of the introduction of the MSPD (EC 2014c) pointed to the following three main economic effects of MSP:

1. MSP was anticipated to result in higher efficiency and therefore cost reductions for governments due to enhanced coordination, integrated decision-making and simplified decision processes. Whereas the initial set-up of an integrated and aligned MSP process would involve additional costs, these should ultimately lead to overall cost reductions in the long run due to lower administrative, employment and overhead costs per procedure or activity of governmental bodies working in the maritime field.
2. Proper MSP was also seen to lead to reduced transaction costs for industry across the following four dimensions:

   - In view of the common knowledge base created through MSP processes, individual businesses would likely have less search costs in relation to finding the right location, where their maritime activity can take place.
   - MSP was also foreseen to create substantial savings in legal costs, that is, those costs which occur due to determining that a business action is legitimate and in compliance with agreements.

- Also with regard to businesses, administrative costs deriving from application and approval processes for permits, licenses and certification would be reduced, assuming that MSP would lead to more efficient and aligned decision procedures of the various government authorities involved.
- MSP should also reduce business costs in order to reduce conflicts with existing or emerging maritime activities in the given space or impacted by the use. MSP should enable governments to incorporate interests of stakeholders and thus prevent such conflicts to happen in the first place.

3. Lastly, due to the enhanced legal certainty and security provided through maritime spatial plans, in particular, spatial allocations for each maritime sector, the general investment climate for the blue economy was foreseen to improve—meaning that investors would increase and accelerate investments into established as well as new blue sectors.

This basic set of positive effects of MSP on stimulating Blue Growth was confirmed by a later study undertaken by the World Ocean Council (WOC 2016). Even though described in different terms, the paper presents the same potential benefits of an integrated MSP approach to ocean industries and thus the overall economy. This is remarkable as the information basis of that paper differed substantially from the earlier study undertaken on behalf of the European Commission. Instead of predominantly relying on government authorities from across Europe, the WOC included also input from the ocean industry itself and non-European sources (mainly from US and Australia) in addition to European cases (e.g. Norway, Germany and the UK East Inshore Plan).

4.3 Assumptions and Concerns on MSP Processes

The WOC paper, however, also pointed to some limits and concerns, which were voiced by the private sector regarding MSP. These mainly referred to the fact that MSP can only act as a Blue Growth facilitator if the process leading to an agreed plan is carried out properly:

- Whereas MSP may ultimately lead to streamlined processes, industry was concerned that its introduction would initially lead to uncertainty, delay and thus negative economic effects for business and communities.
- Many economic benefits associated were only associated with statutory, legally binding MSP processes, as opposed to non-legally binding or pilot MSP processes.

- Some industry sectors claimed that conflict resolution is only required for coastal, shallow areas, which is in high demand by many diverse users. They therefore questioned the need for MSP processes for offshore areas, where little conflicts occur and thus initial costs associated with an MSP process may not be outweighed by potential benefits.
- Furthermore, industry voiced concern on whether MSP processes can really be designed in such a way to allow them to adequately participate in the development and design of the MSP, which is in turn, however, a precondition to generate the economic benefits of an MSP process. Most stakeholders understood that MSP may not always end in a 'win-win' situation with all user needs accommodated in every location. It was, however, seen as necessary that all uses and resources are considered simultaneously and that this needs to remain a dynamic process in view of emerging and future uses.
- Moreover, there was concern whether current MSP processes have sufficient tools available to quantify and evaluate trade-offs among competing uses, users and finite resources, and thus accurately determine positive and negative consequences of a plan.
- Benefits associated with an increased knowledge base were only seen as possible if planning processes were to disclose the source and methodology used.

Within Europe, most of these process principles were already defined in 2008 by the European Commission in the 'Roadmap on Maritime Spatial Planning'; most of which are reflected in the 'Minimum Requirements' as stipulated by the EU MSPD:

EU Roadmap: key principles (2008)	Requirements: EU MSP Directive (2014)
• Use MSP according to area and type of activity • Define objectives to guide MSP • Develop MSP in a transparent manner • Encourage stakeholder participation • Coordinate within member states—simplify decision processes • Ensure legal effect of national MSP • Implement cross-border cooperation and consultation • Incorporate monitoring and evaluation in planning process • Achieve coherence between terrestrial and MSP—relation with ICZM • Create a strong data and knowledge base	Member states shall • take into account land-sea interactions; • take into account environmental, economic and social aspects, as well as safety aspects; • aim to promote coherence between MSP and the resulting plan or plans and other processes, such as ICZM or equivalent formal or informal practices; • ensure the involvement of stakeholders • organize the use of the best available data • ensure trans-boundary cooperation between member states • promote cooperation with third countries in accordance

5 How to Implement MSP as to Lead to Blue Growth

5.1 MSP Projects and the MSP Platform Supporting EU MSP Implementation

As shown earlier (EC 2011), the adoption of the EU MSPD in 2014 marks a milestone also for Blue Growth as it obliges all coastal EU member states to have MSPs in place by 2021. The question on whether MSP has a positive effect on the blue economy is, however, highly dependent on how member states actually implement MSP and thus bring the key principles of the MSP Roadmap into life.

In order to assist EU member states in the implementation of the EU MSPD, the European Commission is continuously providing funding for MSP-related projects. Moreover, since early 2016, the so-called European MSP Platform facilitates European-wide knowledge exchange and generation of MSP implementation practices.

As part of the service, the study 'MSP for Blue Growth' (s.Pro, ECORYS 2018) examined different projects, practices, approaches and lessons learnt that may help EU member states render their MSP processes more effective in developing sustainable Blue Growth. The study is less focused on providing evidence on whether or not MSP can be seen as a key tool for achieving sustainable Blue Growth, but is rather designed to provide practical guidance to maritime spatial planners as well as related stakeholders on how to realize this. Specifically, it covers the following related aspects: (1) How to develop visions that can be effectively used in MSP? (2) What kind of future trends impact sector development and how do they influence the MSP process? (3) How can MSP authorities monitor whether they are on the right track with the Blue Growth objectives of their MSPs?

5.2 Guidance on How to Take Sector Considerations on Board

MSP projects especially in the Baltic Sea such as the BaltSeaPlan Vision 2030, PartiSEApate, BaltSpace, Baltic SCOPE and Baltic LINes (Varjopuro et al. 2015) as well as some EU-wide research projects (esp. with focus on aquaculture; i.e. AquaSpace, Co-Exist) (s.Pro, msp-platform, 2018) have started to look into sector-specific aspects that maritime spatial planners should take into account to enable sustainable Blue Growth. Most notably, the PartiSEApate project was the first ever MSP project, which systematically reached out to sector stakeholders to initiate a pan-Baltic dialogue on how best to integrate them into MSP processes both at national and at transnational level (Schultz-Zehden and Gee 2015). Moreover, the project provided key recommendations on the future Baltic Sea-wide MSP governance system suggesting among others to build much closer links with the existing transnational sector-specific organizations (Schultz-Zehden and Gee 2016).

The 'MSP for Blue Growth' study is, however, the first initiative which has comprehensively assessed this for all nine key maritime sectors across the whole of Europe. The resulting sector fiches not only provide information on the current nature of spatial requirements of the various sectors but also consider the implications of future developments and their consequences for sector requirements in a given maritime space, as well as offering concrete recommendations on how both planners and sectors may inform each other to create suitable MSP solutions which unlock the respective Blue Growth potentials in a sustainable manner.

The following paragraphs provide for a snapshot of the main factors planners have to consider in relation to the various sectors at stake.

The Traditional, Big Sectors

Shipping: Maritime Spatial Planning Important Role

While shipping is guided by freedom of navigation and thus allowed everywhere, MSP is highly important for ensuring that important routes are kept free as shipping is in conflict with all fixed installations. Nevertheless, most MSP initiatives have found it difficult to involve the sector.

Most MSP processes start off with existing IMO shipping routes. Even though these can be changed in principle and in some instances have also been earmarked as useful to be shifted, for example, due to environmental considerations, such changes are a lengthy process. Thus, first-generation MSPs normally have to take existing IMO routes as a 'given'.

The actual size of the free shipping lane depends on traffic volume and size of ships: The higher the traffic volume and the bigger the ships, the wider and deeper the free shipping lane needs to be. In addition, increased emergence of weather extremes requires availability of space to which ships can deviate to avoid bad weather.

Data and information on current ship traffic patterns are continuously improving but are not always easily accessible and need a lot of processing capacity. Moreover, MSP processes need to anticipate future shipping routes. In doing so, planners have to have information on future port developments: Which of them will accommodate the larger ships? What are the land connections and who do they service? Are there new ports upcoming, which may influence shipping routes?

Also other new ship traffic routes are expected to emerge in view of ship maintenance, short sea shipping and recreational as well as tourism-oriented shipping. These routes follow completely different patterns. The spatial implications of autonomous vessels are not yet clear, but in the near future, there is a need to allocate separate, exclusive test beds free of other uses.

Planning approaches currently differ between countries: some take a maximum approach taking into account also future port developments. Others rather take a minimum approach by focusing on the most important routes and those areas, which ships need to avoid.

Fishery: Maritime Spatial Planning to Be Integrated into Overall Fisheries Policy

Similar to shipping, fishing has a long tradition of claiming space and has not been easy to integrate into MSP processes, as those are often understood to take space away, while not being able to provide the necessary spatial security. In fact, currently most plans only consider fishing, when allocating space to other uses, but do not allocate specific areas to fishery.

Planners are in need of much better information and tools, which enable them to better consider relevant areas for fishing and fish species according to life stages (incl. spawning areas). Similar to shipping, continuous improvements are, however, made, for example, in Vessel Monitoring System (VMS) and Automatic Identification System (AIS) data systems and models, which will facilitate planning in the future.

But at the same time, MSP is by no means the only instrument for fisheries management. Maritime spatial planners have no influence, for example, on fish quota, meaning that in case of closure of some areas more fishing is taking place in less space. Thus, interaction between MSP and the sector should take place at a much earlier stage, in order to provide for better linkage and integration of MSP into the overall evolvement of fisheries policy, including cross-border considerations.

Coastal and Maritime Tourism: Maritime Spatial Planning Indirect, but Important Role

Continuous growth of coastal and cruise tourism and the related infrastructure developments and subsequent impacts have implications on MSP decisions and vice versa. These land-sea interactions are highly important and a good interlinkage between maritime and terrestrial planning including efficient multi-level governance is crucial. Moreover, MSP processes and authorities may play an important role in fostering synergies between maritime uses, which are beneficial for tourism.

Appropriate assessment tools are only slowly evolving, but more and more research efforts are undertaken to foster a better understanding also on concepts such as cultural landscapes; recreational values and attractive living areas, which are closely connected.

Ocean Energy Sectors

Offshore Wind: Maritime Spatial Planning Important Direct Role (but Sector not Relevant for Some Sea-Basins)

As evidenced before, the sector shows continued growth and thus growing demand for space in many Northern EU countries and is expected to emerge also in numerous countries where currently no offshore wind is in place. However, this is less likely in Southern Europe.

The sector is also important in view of long-time impacts of spatial decisions. Once installed, the infrastructure remains in place for a long time with considerable implications for other maritime uses.

MSP considerations differ depending on the method for designation of OWF zones. Some governments allocate specific sites for OWF development and thus also cover for the Environmental Impact Assessment (EIA) and related grid connections. This method is a valuable tool for large-scale deployment in short term. In case of an open door policy, large zones are designated as search areas for industry, with more responsibility on their side to conduct the EIA and organize the grid connections. Such an approach is more prone to foster innovative, market-based blue energy solutions.

Moreover, MSP authorities play an important role in decisions on whether OWF areas are open or closed to other uses such as fishery, aquaculture, recreation/tourism purposes as well as conservation needs. They are also in charge of the Strategic Environmental Assessments (SEAs), which in turn should facilitate EIAs. Better SEAs and more multi-use options may decrease resistance to new developments.

At the same time, technological advances may have substantial spatial implications. The emergence of floating wind farms, possible connections of OWF to interconnectors and new energy storage systems open up new potential sites especially further offshore and substantially influence the related environmental impacts.

Cables and Pipelines: Maritime Spatial Planning Important Direct Role (Long Time Horizon)

The installation of new submarine cables and pipelines has to be taken into account by MSP in view of the potential for more efficient use of space by bundling corridors for electricity and telecom cables and pipelines, while also considering the related increased risk factor in case of damage.

Grids and interconnectors are important in facilitating more ambitious energy system scenarios and improved routing and installation criteria can lead to avoidance of conflicts, for example, with fishing. There is potential to facilitate better siting due to improvements in submarine 3D bathymetric mapping. At the same time, it should be noted that often general data is missing and that, in some cases, MSP authorities are not always in charge of the detailed planning of related routes.

Tidal and Wave: Maritime Spatial Planning May Facilitate Development

The sector is only relevant in the North Sea and the Atlantic Ocean. It is important to consider synergies with offshore wind energy infrastructure including vessels, grids, cables as well as onshore transmission. MSP requires accurate resource mapping of tidal and wave power potential to be able to locate areas of interest.

Oil and Gas: Maritime Spatial Planning Should Only 'Consider' the Sector

The sector is only relevant in a few EU countries, and new sites are only foreseen in very limited cases. Thus for MSP and its role in Blue Growth, it is mainly important to integrate the current sites (also those to be decommissioned) and the related maritime activities in light of creating synergies with other uses.

Other Place-Based Maritime Sectors

Marine Aquaculture: Maritime Spatial Planning Can Foster Sector, but Depends on Sector Input and Actions

MSP has a potentially important role in fostering the development of marine aquaculture, but this has to be done in strong cooperation with other Blue Growth policy areas as the sector itself is not strong enough to provide the necessary push.

Whereas visual and physical impacts of marine aquaculture may be similar regardless of which type and species cultured, feasible environmental condition requirements as well as vice versa environmental impacts of the given aquaculture vary substantially between the different forms of aquaculture (e.g. shellfish, seaweed, fish) as well as between the different species.

Planners and the sector should work together to identify new and better areas for aquaculture potential. MSP may support the sector by providing better and open access to relevant data, which is are otherwise not available to small individual aquaculture companies.

Moreover, MSP can stimulate the creation of clusters of farms by allocating aquaculture zones. So far, the small-sized aquaculture companies tend to prioritizes coastal space, even though offshore areas may substantially increase social acceptance and reduce conflict fields with other uses.

Marine Aggregates and Mining: Maritime Spatial Planning Important Direct Role

There is increasing demand for dredging sites for sand and gravel with spatial allocation depending on the resource. Dredging is necessary due to coastal defence and protection but may, at the same time, itself have substantial environmental impacts. It is important to follow technology developments which may improve sustainability.

Actual dredging only requires limited areas, but it is important that the seabed at these locations is not negatively impacted by other sectors (known as mineral safeguarding). Planning has to consider not only the actual locations but their surroundings. Moreover, industry investments have a much longer time horizon (30 years) than MSP; thus also future revisions of MSPs need to safeguard these time horizons.

Contrary to general belief, there is, however, potential for combinations with numerous other sectors, especially in light of the temporal aspect of when the actual dredging is carried out, but better evidence is required.

Summary

Taken together, the sector fiches show the large diversity of the spatial characteristics and time horizons of the various sectors. Whereas it has been commonly assumed that MSP is specifically important for supporting suitable siting of emerging place-based industries in ocean energy (especially offshore wind and increasingly the related cables and pipelines), the fiches also show the role of well-informed MSP for ensuring the ongoing development and evolution of traditional activities such as shipping, fishery and tourism.

The sector analysis highlights the importance of aligning MSP processes with other related Blue Growth policies and the need for much closer cooperation with sectors at an early stage of planning. Rather than seeing MSP as an isolated task, the study indicates that MSP processes and the work of individual sector organizations, such as Regional Fisheries Advisory Councils, should be better aligned. For other sectors, the study describes the limited impact of MSP, such as on aquaculture development, unless there is a much stronger connection created at an early planning stage to allow for better input by the sector itself to indicate optimal sites. Moreover, it underlines the importance of a much stronger merging, not only of maritime and terrestrial spatial planning but the overarching regional development programmes especially in view of tourism and port development (EC 2018).

These are just a few of the useful study insights, in light of the increasing number of MSPs being developed in the coming years in countries with no offshore wind or other renewable energy industries acting as main drivers. Therefore, Blue Growth is foreseen to be generated by other maritime sectors in those countries (EC 2018).

5.3 MSP for Blue Growth Is About Strategic Planning for the Future

The study also underlines the importance of interlinking MSP processes with the development of maritime visions and strategies (EC 2018). The earlier MSP economic impact studies mainly emphasize the ability of MSP to reduce or avoid conflicts among sectors, which occur most often in coastal, shallow areas (World Ocean Council 2016). This fact however neglects the more strategic planning function of MSP processes and resulting plans in terms of promoting Blue Growth.

In that sense, MSPs should no longer only be developed in reaction to pressure from already existing strong industry demands, nor should MSP be limited to the function of minimizing current conflicts or preventing such conflicts to happen in the future. It is actually more of a tool to put the maritime space on the 'economic' agenda by showing the sustainable development potential of the sea to inspire new stakeholders to take advantage of that Blue Growth potential and to find the right place to do so. With that understanding, MSP can also be of high Blue Growth relevance for less crowded sea areas, as it may show optimal, new areas for certain maritime activities. By focusing in parallel on (terrestrial) areas in need of economic development, MSP can thus inspire relevant initiatives in other policy fields (i.e. food security, cohesion policy). It was in this spirit that the first Lithuanian MSP was developed (Schultz-Zehden and Gee 2013).

The EU MSPD does not oblige EU member states to develop a maritime vision or strategy as part of the MSP process. Nevertheless, numerous countries (e.g. Belgium, the Netherlands, Sweden) have opted for a broader understanding of their MSP process by combining it with a vision or strategy development processes.

As shown in the MSP for Blue Growth study (s.Pro and ECORYS 2018), vision processes are an important tool to promote collaboration between sectors and stakeholders—who may currently experience tensions—by instigating a dialogue on a positive, joint future to work towards. Vision processes are also useful, in that they draw attention to uses, which are not present so far, as well as other emerging issues (such as climate change or other broader demographic, political or economic developments). Moreover, the future development of some sectors, such as ocean energy and marine aggregates, depends on a long-term framework providing stable locations. Planning periods of these sectors go well beyond the typical six-year horizon of the MSP, and the resulting structures can remain fixed for decades. Therefore, the development of a long-term maritime vision or a strategy has an important role to provide certainty for these sectors, which exceeds political cycles and may even provide the basis to derive smart objectives for the given MSP process. The task of the MSP is consequently to link this desired future to present conditions and related spatial planning needs.

Currently, however, hardly any MSP process has a systematic monitoring and evaluation framework in place, which not only requires to set objectives but also translate them into measurable targets against current baselines (MSP for Blue Growth study: Indicator Handbook, s.Pro and ECORYS 2018). Moreover, there is the urgent need to develop and introduce tools to be able to carry out a more systematic cost-benefit analysis of the provisions of a draft maritime spatial plan, for example, assessing costs of a rerouting of shipping set against the benefits of avoiding a sensitive area (Jay 2017).

6 Conclusions

6.1 MSP as Part of the Overarching Framework of an Integrated Maritime Policy

MSP is a powerful tool for Blue Growth, but it can only realize its full potential by being strongly interconnected not only with the whole set of other Blue Growth measures but also as part of the overarching framework of an IMP. For example, development of maritime visions and strategies as part of an MSP process may function not only as a preparatory step for MSP but also provide a long-term overarching framework for an IMP. Such vision can also serve to address wider national priorities and link MSP to other planning frameworks, including integrated coastal zone management, territorial development planning, and other relevant policies including food security, research and innovation or cohesion policy.

6.2 MSP Is About Planning for the Future

MSP is not only about reducing current conflicts but also about providing a vision for the sea as a source of sustainable national development. It has an important Blue Growth function not only for coastal areas but also by putting the open maritime space onto the economic agenda. MSP, if orchestrated and aligned with other policy tools, has the power to initiate Blue Growth also in currently still unused sea areas and showing the whole maritime space as a development field. On this premise, MSP processes should consider rural development areas while at the same time providing indication of potential strategic resource areas where traditional and new offshore technologies and uses can be developed.

6.3 Blue Growth Potential Is Reaching Beyond Five Key Sectors

While development in key Blue Growth sectors can satisfy high-level policy goals and bring prosperity over the long run, local coastal communities very much depend on traditional uses for their day-to-day livelihoods. Therefore, the scope of Blue Growth policies should also take into consideration potential for growth in traditional sectors through innovation and implementation of multi-use concepts and sector combinations. MSP in countries where there is no 'push' from Blue Growth sectors such as ocean energy may orient towards Blue Growth through diversification of traditional sectors such as fisheries and tourism. Moreover, cooperation with sectors such as fishing, aquaculture as well as tourism has to start at a much earlier pre-planning stage to integrate MSP with sector policies.

In countries with present key Blue Growth sectors, taking this wider approach to Blue Growth and considering cross-sectoral synergies can ensure a more sustainable integration of new and emerging sectors into existing contexts and more local socio-economic benefits.

6.4 Integration Through MSP and of MSP

MSP is not only an important tool to support emerging sectors but also key to secure the traditional, more mature sectors such as shipping, fishing and tourism. Realizing that Blue Growth potential lies not only in the given five key sectors originally associated with Blue Growth, and that MSP can integrate a wider set of maritime policies, allows for addressing a much wider set of challenges and unlocking a larger future development potential. For this to happen, it should, however, be understood that MSP may not only act as the integrative tool, but that MSP should also integrate itself much earlier into the overarching as well as sector-specific Blue Growth policies both at EU and at national level.

Acknowledgements We are grateful to the project 'Economy of maritime space' funded by the Polish National Science Centre for contributing the Open Access fee for this chapter and facilitating our discussions and preparation of the book.

References

Ansong, J., O'Hagan, A. M., & MacMahon, E. (2018). Existing Mechanisms for Cooperation on MSP in the Celtic Seas (Deliverable 14). EU Project Grant No.: EASME/EMFF/2014/1.2.1.5/3/SI2.719473 MSP Lot 3. *Supporting Implementation of Maritime Spatial Planning in the Celtic Seas (SIMCelt)* (pp. 74). University College Cork.

Barbesgaard, M. (2016). *Blue Growth: Saviour or Ocean Grabbing?. An International Colloquium Global Governance/Politics, Climate Justice & Agrarian/Social Justice: Linkages and Challenges.* Colloquium Paper No. 5. Retrieved from https://www.tni.org/files/publication-downloads/18-icas_cp_doerr.pdf.

Castellani, C., Carrer, S., Bocci, M., Ramieri, E., Depellegrin, D., Venier, C., Sarretta, A., & Barbanti, A. (2017). Case Study 6: Coastal and Maritime Tourism and O&G Decommissioning as Drivers for Multi-Use in the Northern Adriatic Sea. MUSES Deliverable: D3.3—Case Study Implementation—Annex 9. MUSES Project.

CoR. (2015). Opinion of the Committee of the Regions—Innovation in the Blue Economy: Realising the Potential of Our Seas and Oceans for Jobs and Growth (2015/C 019/05).

CoR. (2017). Opinion of the European Committee of the Regions on 'A New Stage in the European Policy on Blue Growth (2017/C 306/11).

ECORYS, et al. (2012). *Blue Growth—Scenarios and Drivers for Sustainable Growth from the Oceans, Seas and Coasts*. Final Report. Commissioned by DG. MARE. Retrieved from https://webgate.ec.europa.eu/maritimeforum/system/files/Blue%20Growth%20Final%20Report%2013092012.pdf.

ECORYS, s.Pro, MRAG. (2013). *Study on Blue Growth and Maritime Policy Within the EU North Sea Region and the English Channel*. Final Report FWC MARE/2012/06—SC E1/2012/01. (Conducted for DG Maritime Affairs and Fisheries).

ECORYS, s.Pro, MRAG. (2014). *Study on Deepening Understanding of Potential Blue Growth in the EU Member States on Europe's Atlantic Arc*. Final Report FWC MARE/2012/06—SC C1/2013/02. (Conducted for DG Maritime Affairs and Fisheries).

EESC. (2015). Opinion of the European Economic and Social Committee on the 'Communication from the Commission to the European Parliament, the Council, the European Economic and Social Committee and the Committee of the Regions—Innovation in the Blue Economy: Realising the Potential of Our Seas and Oceans for Jobs and Growth' (2015/C 012/15).

EUNETMAR. (2014). *Studies to Support the Development of Sea Basin Cooperation in the Mediterranean, Adriatic and Ionian, and Black Sea*. Final Report FWC MARE/2012/07—Ref. No 2. (Conducted for DG Maritime Affairs and Fisheries).

EC. (2010). Europe 2020—A Strategy for Smart, Sustainable and Inclusive Growth (COM(2010) 2020).

EC. (2011). *Study on the Economic Effects of Maritime Spatial Planning*. Final Report. Retrieved from https://ec.europa.eu/maritimeaffairs/sites/maritimeaffairs/files/docs/body/economic_effects_maritime_spatial_planning_en.pdf.

EC. (2012a). Blue Growth Opportunities for Marine and Maritime Sustainable Growth (COM(2012) 494 final).

EC. (2012b). Progress of the EU's Integrated Maritime Policy (COM(2012) 491 final).

EC. (2014a). Innovation in the Blue Economy: Realising the Potential of Our Seas and Oceans for Jobs and Growth (COM(2014) 254 final/2).

EC. (2014b). Marine Knowledge 2020: Roadmap (SWD(2014) 149 final).

EC. (2014c). Directive 2014/89/EU of the European Parliament and of the Council of 23 July 2014 Establishing a Framework for Maritime Spatial Planning.

EC. (2017). Report on the Blue Growth Strategy Towards More Sustainable Growth and Jobs in the Blue Economy (SWD (2017) 128 final).

EC. (2018). The 2018 Annual Economic Report on EU Blue Economy. Retrieved July 2, 2018, from https://ec.europa.eu/maritimeaffairs/sites/maritimeaffairs/files/2018-annual-economic-report-on-blue-economy_en.pdf.

European Parliament. (2013). Blue Growth—Enhancing Sustainable Growth in the Marine, Maritime Transport and Tourism Sectors (2012/2297(INI)).

European Parliament. (2014). Research and Innovation in the Blue Economy to Create Jobs and Growth European Parliament Resolution of 8 September 2015 on Untapping the Potential of Research and Innovation in the Blue Economy to Create Jobs and Growth (2014/2240(INI).

FAO. (2014). *The State of World Fisheries and Aquaculture 2014*. Rome: Food and Agriculture Organization of the United Nations.

Foray, D. (2015). *Smart Specialisation. Opportunities and Challenges for Regional Innovation Policy*. Abingdon and New York: Routledge.

Fritz, J. S., & Hanus, J. (2015). The European Integrated Maritime Policy: The Next Five Years. *Marine Policy, 53*, 1–4.

Jay, St. (2017). Issue Paper: Marine Spatial Planning, Assessing Net Benefits and Improving Effectiveness. Edited by OECD. Retrieved from https://www.oecd.org/greengrowth/GGSD_2017_Issue%20Paper_Marine%20Spatial%20Planning.pdf.

Limassol Declaration. (2012). Declaration of the European Ministers Responsible for the Integrated Maritime Policy and the European Commission, on a Marine and Maritime Agenda for Growth and Jobs, the "Limassol Declaration". Retrieved from https://ec.europa.eu/maritimeaffairs/sites/maritimeaffairs/files/docs/body/limassol_en.pdf.

Lukic, I., Schultz-Zehden, A., Ansong, J. O., et al. (2018). *Multi-Use Analysis*. Edinburgh: MUSES Project.

Przedrzymirska, J., Zaucha, J., et al. (2018). *Multi-use Concept in European Sea-basins*. Edinburgh: MUSES Project.

Schultz-Zehden, A., & Gee, K. (2013). BaltSeaPlan Findings—Experiences and Lessons from BaltSeaPlan.

Schultz-Zehden, A., & Gee, K. (2015). Toward Sectoral Stakeholder Involvement in a pan-Baltic MSP Dialogue. *Bulletin of the Maritime Institute in Gdańsk, 30*(1), 139–149.

Schultz-Zehden, A., & Gee, K. (2016). Towards a Multi-Level Governance Framework for MSP in the Baltic. *Bulletin of the Maritime Institute in Gdańsk, 31*(1), 34–44.

s.Pro. (2013). *Study on Blue Growth, Maritime Policy and the EU Strategy for the Baltic Sea Region*. Final Report FWC MARE/2012/07—Ref. No 1. (Conducted for DG Maritime Affairs and Fisheries).

s.Pro. (2017). Towards an Implementation Strategy for the Baltic Blue Growth Agenda. Commissioned by the European Commission.

s.Pro. (2018). European MSP Platform. Retrieved June 14, 2018, from https://www.msp-platform.eu/.

s.Pro, ECORYS. (2018). *Maritime Spatial Planning (MSP) for Blue Growth*. Final Technical Study. Written by the European MSP Platform Under the Assistance Mechanism for the Implementation of Maritime Spatial Planning (Conducted for the Executive Agency for Small and Medium-Sized Enterprises/DG Maritime Affairs and Fisheries).

UNEP, FAO, IMO, UNDP, IUCN, World Fish Center, GRIDArendal. (2012). Green Economy in a Blue World. Retrieved from www.unep.org/greeneconomy and www.unep.org/.

Valletta Declaration. (2017). Declaration of the European Ministers Responsible for the Integrated Maritime Policy on Blue Growth, the "Valletta Declaration". Retrieved from http://data.consilium.europa.eu/doc/document/ST-8037-2017-INIT/en/pdf.

Varjopuro, R., Soininen, N., Kuokkanen, T., Aps, R., Matczak, M., & Danilova, L. (2015). Communiqué on the Results of the Research on Blue Growth in the Selected International Projects Aimed at Enhancement of Maritime Spatial Planning in the Baltic Sea Region (BSR). *Bulletin of the Maritime Institute in Gdańsk, 30*(1), 72–77.

Vergílio, M., Calado, H., & Varona, M. C. (2017). Case Study 3B: Development of Tourism and Fishing in the Southern Atlantic Sea (Azores Archipelago—Eastern Atlantic Sea). MUSES Deliverable: D3.3—Case Study Implementation—Annex 6. MUSES Project.

World Ocean Council. (2016). Ocean Industries and Marine Planning. Retrieved from http://oceancouncil.org/wp-content/uploads/2016/05/Ocean-Industries-and-Marine-Planning_22-Mar-2016.pdf.

7

Socio-cultural Dimensions of Marine Spatial Planning

Emma McKinley, Tim Acott, and Tim Stojanovic

1 Introduction

Within wider marine governance and management, there is an increasing call for greater levels of effective public involvement in marine and coastal issues (McKinley and Fletcher 2010, 2012). Related to this is a need to develop improved understandings and conceptualisations of societal relationships and interactions with the sea. The intricacies, interdependencies and factors influencing these relationships are increasingly being viewed through a socio-cultural lens (Bryce et al. 2016). Relations between society and the sea can be underpinned by a broad array of religious, aesthetic, economic and place-based values. Socio-cultural is a broad term that incorporates these many different facets of human society, including attitudes, values, behaviours as well as the structures that frame social organisations and actions. Although hailed as a mechanism through which sustainable management of global marine and coastal resources can be achieved, to date, MSP has given limited treatment to the socio-cultural components of marine use within the planning process,

E. McKinley (✉) • T. Acott • T. Stojanovic
School of Earth and Ocean Sciences, Cardiff University, Cardiff, UK

Greenwich Maritime Centre, University of Greenwich, London, UK

School of Geography and Sustainable Development, University of St. Andrews, St Andrews, Scotland
e-mail: McKinleyE1@cardiff.ac.uk

J. Zaucha, K. Gee (eds.), *Maritime Spatial Planning*,
https://doi.org/10.1007/978-3-319-98696-8_7

instead being dominated by economic, ecological and administrative considerations.

This chapter examines a selection of key concepts currently underpinning 'socio-cultural' thinking and draws on examples of social and cultural assessments in marine and other planning contexts. We provide an overview of the breadth of concepts and ideas that fall within this umbrella term of 'socio-cultural', before focussing on three key aspects: Cultural Ecosystem Services (CES), Societal Connection to the Sea, and Well-being. Reviewing the use of these concepts in marine planning, the chapter then discusses evidence gaps, key challenges and a series of recommendations as to how contemporary marine planning can better include socio-cultural dimensions.

This chapter:

- describes the range of theoretical perspectives which underpin research on social and cultural dimensions of the oceans, and related debates;
- illustrates examples of how CES, marine citizenship and well-being are applied in MSP;
- discusses the challenges involved in developing a socio-cultural evidence base, particularly in light of the political ecology of coastal space and development; and
- presents evidence as to why a deeper consideration of socio-cultural aspects could be of value to marine and coastal planning.

Recent global initiatives, including, for example, the Aichi Targets and the UN Sustainable Development Goals (SDGs) (UN 2015), place the relationship between society and the natural world at the forefront of international policy development. The connections between society and the sea are dynamic and complex, influenced across spatial and temporal scales by an evolving social, cultural, economic, political and environmental landscape. In order to realise the potential of effective MSP, there is a real need for the sociocultural components of our relationship with the global seas to be better understood and more appropriately embedded within MSP. Evidence from marine planning documents suggests that global marine and coastal governance is developing towards more participatory, integrated and increasingly holistic approaches; there has been a proliferation of MSP efforts worldwide in the 2010s. Cultural components have begun to be considered in these planning efforts, but the efforts have arguably been basic and at a low baseline. In nations with indigenous communities such as Australia and Canada, marine plans have begun to acknowledge cultural perspectives of native societies, particularly where this relates to notions of tenure and rights in the sea. In another example, the UK Marine Policy Statement (2011) considers topics such as 'Seascape', 'Cultural heritage' and 'CES', whilst the draft and published marine plans in the devolved nations England (East, 2014) Scotland (2015), Wales (2017) and Northern Ireland (2018) consider those concepts plus other themes such as sustainable 'coastal communities', 'social values' and 'well-being' within their remit. However, how this is realised within the MSP process and operationalised within on the ground, marine and coastal management remains to be seen.

As we continue to understand the role of socio-cultural dimensions within MSP, it is first necessary to consider what this term is actually referring to. Evidence of the terms 'social' and 'cultural' being used interchangeably within environmental governance discourse is commonplace. Within this chapter, we first examine what is meant by 'socio-cultural', investigating the diverse and wide-ranging interpretations of these terms and how they are currently, and could potentially be, used within MSP. Through the chapter, we then explore a sample of emerging concepts from within the socio-cultural arena and consider how these concepts can be more effectively embedded within marine planning.

2 What Do We Mean by Socio-cultural?

The social and cultural dimensions of the marine environment are numerous and multifaceted. Cultural interactions between people and the environment are pivotal in the context of broader attitudes and behaviours (Bryce et al. 2016). Relationships between people and the ocean can shape sense of place, personal identity and a broad array of leisure, recreation and work opportunities. Relations can be underpinned by a broad array of religious, aesthetic, economic and place-based values. Socio-cultural is a broad term that incorporates these many different facets of human society, including attitudes, values, behaviours as well as the structures that frame social organisations and actions. Table 7.1 presents an overview of key terms encompassed within the terminology and language of 'sociocultural', including CES, ocean literacy, notions of 'value', place attachment/sense of place and well-being, among others. We acknowledge this diversity of terms, and in the following sections explore a selection of these different approaches to socio-cultural research from the perspective of ecosystem services (ES); ocean literacy, marine citizenship and behaviour change; and well-being.

Table 7.1 Key socio-cultural concepts and their potential application in marine spatial planning

Concept	Definition and potential applications in MSP	Key references and cross referencing to chapters and chapter sections.
Cultural ecosystem services (CES)	Defined as "*the nonmaterial benefits people obtain from ecosystems through spiritual enrichment, cognitive development, reflection, recreation, and aesthetic experiences*" (MA, 2003). This definition is widely contested. It has been explicitly used as a framing for MSP in a few examples.	See Sect. 3.1
Ocean literacy	Understanding of the impact of the sea on human life, and of people on the sea—a relatively recent term that has the potential to engender greater levels of public awareness, knowledge and capacity to support MSP implementation.	See Sect. 3.2
Marine citizenship	Understanding of the individual rights and responsibilities towards the marine environment, having an awareness and concern for the marine environment and the impacts of individual and collective behaviour, and supporting public capacity to have a role in ensuring ongoing sustainable management of the marine environment.	See Sect. 3.2
Attitudes and perceptions	Public perceptions of marine issues that explore broadscale and regionally distinct social perspectives of marine environments.	See Sect. 3.2
Well-being	Measures of the quality of life. Reflected in marine plan policies which are related to blue space and its increasingly recognised impact on human health and well-being, and potential criteria for evaluating the outcomes of marine planning.	See Sect. 3.3
Cultural heritage	Sets of buildings, monuments or sites, and also intangible heritage such as cultural knowledge or practice, which relate to the marine environment and resources. Built heritage is often highlighted in conservation and tourism aspects of marine plans.	Alegret and Carbonell (2016)

(continued)

Table 7.1 (continued)

Concept	Definition and potential applications in MSP	Key references and cross referencing to chapters and chapter sections.
Seascape	"An area of sea, coastline and land, as perceived by people, whose character results from the actions and interactions of land with sea, by natural and/or human factors." Occasionally developed as supporting evidence for marine planning through Seascape characterisation, Seascape assessments or Visual impact assessments.	Natural England (2012, p. 1); Falconer et al. (2013)
Human activities	Overviews of sectoral activities in space and time. Cultural importance of these human activities to society. Often quantified and mapped in marine planning, challenging to assess cultural significance.	Stojanovic and Farmer (2014); Smith et al. (2012)
Social values (monetary and non-monetary)	Recognition and consideration of a diverse range of social values, including drawing on environmental economic valuation techniques but also broader social values.	See Chap. 8 in this book on social sustainability.
Socio-demographics	Includes the traditional metrics considered within socio-demographics (e.g. gender, age, employment, income, education level) but also encompasses other more recent concepts including coastal typologies and population projections. Phenomena including mobility, migration, social justice and equity.	Links to the work on social values and how background and demographics can influence individual value systems.

3 Sociocultural Concepts and Their Place Within MSP

3.1 Cultural Ecosystem Services

Adopting an ecosystem-based approach is becoming increasingly important within marine management and decision-making. The European Commission's EU Directive on MSP 2014/89 states that "applying an ecosystem-based approach" (Art 5, sec 1) is key to marine planning. Consideration of ES is deemed to be central to the ecosystem approach. The concept of CES is particularly relevant to MSP, in that its focus is the socio-cultural benefits people derive from nature. However, of the four ecosystem service categories, CES provides the most difficult challenges for identification and assessment. It is not the purpose of this chapter to review contemporary debates in ES; however, in thinking about the application of CES to MSP, it is important to recognise that many similar challenges apply.

With an intellectual tradition dating back to the late 1970s (Gómez-Baggethun et al. 2010), the concept of ES was placed firmly on the policy agenda through the publication of the seminal Millennium Ecosystem Assessment (MA 2003). The aim of the MA was '*to provide an integrated assessment of the consequences of ecosystem change for human well-being and to analyze options available to enhance the conservation of ecosystems and their contributions to meeting human needs*' (MA 2003, p. 2). A framework was provided that distinguished four types of ES—supporting, regulating, provisioning and cultural. CES are defined as '*the nonmaterial benefits people obtain from ecosystems through spiritual enrichment, cognitive development, reflection, recreation, and aesthetic experiences*' (MA 2003, p. 58) and may refer to cultural diversity, spiritual and religious values, knowledge systems, educational values, inspiration, aesthetic values, social relations, sense of place, cultural heritage values and recreation and ecotourism.

Since its publication, the issue of CES has been the most problematic category (Satz et al. 2013), with many attempts to articulate relationships between culture and other services (Fish et al. 2016). Although rooted in economics and natural sciences, ES work over recent years has seen closer engagement with the social sciences, with particular emphasis on values and value deliberation (Kenter et al. 2015, 2016a; Cooper et al. 2016). While social science has arguably been admitted to the ES 'club', there are still concerns that the ecosystem framework provides extensive epistemological challenges when thinking about culture and the idea of CES (Leyshon 2014). There is increasing interest to understand CES not just from the social sciences but also from an arts perspective (Edwards et al. 2016). In a recent overview of ES, Costanza et al. (2017) state that cultural services was the least developed category when the MA was published. They point towards the large numbers of papers on CES that have since been published, indicating that there has been some development in this area in the last decade or so. However, there remain major concerns regarding the ability of ES to adequately represent sociocultural perspectives across different world views. For instance, the Intergovernmental Science-Policy Platform on Biodiversity and Ecosystem Services (IPBES) is proposing an approach called nature's contribution to people (NCP) that builds on the ES concept but more strongly recognises the 'central and pervasive role that culture plays in defining links between people and nature' (Diaz et al. 2018, p. 270) and elevates the role of indigenous and local knowledge. This approach is strongly criticised by the editor of the journal *Ecosystem Services* (Braat 2018). At the very least, this exchange demonstrates the evolving nature of the concept and deep divisions that exist.

Although research into ES has grown considerably over the past decade, most studies have had a terrestrial focus, and there is a knowledge gap relating to marine and coastal ES (Liquete et al. 2013). In their systematic review, Liquete et al. (2013) identified 145 papers that specifically assessed marine and coastal ES. They conclude that social sciences are under-represented in the studies, and one of the main gaps are indicators related to cultural services. Beaumont et al. (2007) provided an overview of the ES provided by marine biodiversity. Börger et al. (2014) highlight various valuation techniques which could be applied at different stages of marine planning and regulation, and Arkema et al. (2015) apply an ES framework to the coastal and marine planning process in Belize, highlighting the difference in resultant evaluations of future scenarios when these benefits are included in the planning process. Despite these studies, there are relatively few examples of published work that explicitly connect CES and MSP, although there is a fast-growing body of work that deals with the CES of marine and coastal spaces that might be of interest to marine spatial planners. Examples of the former include Ruiz-Frau et al. (2013), who state that MSP should account for all aspects of value associated with marine biodiversity, meaning that a holistic approach is needed that includes ecological, social and economic aspects. Using a questionnaire approach, their study focussed on providing an economic assessment of non-extractive uses of marine biodiversity.

Guerry et al. (2012) make the case for using the broader ES concept in MSP, '*The framework of ecosystem services enables the explicit examination of trade-offs in services and it provides a quantitative approach for assessing the value of MSP versus sectoral or uncoordinated planning*' (p. 108). To achieve this, they developed an approach called the marine Integrated Valuation of Ecosystem Services and Tradeoffs (InVEST) which was designed to assess the multiple services provided by marine ecosystems. Cultural services are included as part of their framework and recognition that understanding and accounting for cultural values (such as existence, subsistence and aesthetic values) are fundamentally important for coastal communities. InVEST was designed to provide results grounded in both local ecological knowledge and, also, reflect diverse values, conflicts and aspirations.

Where InVEST takes a whole ES view, Gee et al. (2017) focus specifically on understanding the importance of culture and suggest that cultural values associated with the sea tend to be a neglected aspect of MSP. They discuss the sociocultural evidence gap is a result of the difficulties in defining and eliciting cultural values but also in attributing values to particular places that can then be used in the context of area-based approaches to management. For the authors, a CES approach is just a starting point for thinking about how communities are connected to the sea, and they propose a method for developing 'spatialised' community-based narratives that can be used to identify 'culturally significant areas'.

The importance of developing participatory mapping of ES as a way to navigate coastal values is explored by Klain and Chan (2012), who suggest that monetary and biophysical dimensions tend to dominate spatial planning. They use social value mapping methods to explore how tangible and intangible values are associated with particular locations in the hope of highlighting the underappreciated ways in which ecosystems are important to people. Their study concludes that '*many people attach strong and diverse values to nature, but that spatially identifying and quantifying the importance of particular places is only possible for some people and values. This suggests that planning and decision-making will be most effective and appropriate when they include a deliberative component*' (p. 112). In a general review of the priorities for coastal and marine spatial planning (Halpern et al. 2012), the authors suggest that an ecosystem-based process should be preserving critical ES; a key hurdle is how to measure and compare very different ES such as cultural values versus a more easily marketable service or benefit, such as seafood. MSP is seen as an important step in the implementation of comprehensive ecosystem-based management.

Moving beyond the literature that explicitly frames a socio-cultural approach in ES in the context of MSP, there is a growing body of work dealing with the ES of marine and coastal spaces which might find application in MSP. CES feature to a lesser or greater extent. The intention here is to selectively draw from this literature to examine the inclusion of CES in broader policy-relevant studies. Turner and Schaafsma (2015) provide a broad overview of coastal ES in their edited book in which Saunders et al. (2015) suggest that social information is often lacking in the context of coastal ES data. Luisetti et al. (2014) suggest that coastal zone ES that can be valued in economic terms with CES considered as meaningful places supplying a range of goods and benefits. Barbier et al. (2011) give a broad review of the value of estuarine and coastal ES, and Hattam et al. (2015) examine ES broadly in the marine environment and then examine a case example of the Dogger Bank in more detail. Both studies make reference to CES as one type of service.

Fletcher et al. (2011), Jobstvogt et al. (2014) and Potts et al. (2014) all consider CES in the context of marine protected areas (MPAs) and marine habitats. They point out the links with human activities such as sport, recreation and nature watching, but all highlight the paucity of data available for making assessments. Fletcher et al. (2014) examined marine CES in the Black Sea. The importance of understanding people's 'experiences' of the sea (beyond recreation) and the deep sense of connectivity that goes beyond the physical properties of objects is stressed. The research illustrates the broad range of sociocultural considerations that are relevant to MSP beyond leisure and recreational opportunities. Where broader ecosystem service studies are carried out, recreation is often the focus of the study (see, e.g. Hynes et al. 2018). However, Baulcomb et al. (2015) suggest their work is the first non-market valuation study to formally consider culture as a generator of ES in a marine environment. They propose an approach to CES valuation that pairs ecological and cultural insight within an ES typology. Pushing the methodological boundaries, Kenter et al. (2016b) integrate deliberative monetary valuation, storytelling, subjective well-being and psychometric approaches to elicit the CES values in proposed UK MPAs. Their study explicitly considers the role of shared values in decision-making. Bryce et al. (2016) recognise the difficulties of assessing CES and suggest a novel framework developed by the UK National Ecosystem Assessment (UKNEA) to evaluate the CES benefits of 151 UK marine sites to recreational sea anglers and divers. Ranger et al. (2016) describe an approach for exploring deeply held cultural values using the Community Voice Method (CVM) set within a deliberative-democratic framework for decision-making with regard to MPAs.

Murray et al. (2016) consider the importance of finding better ways to incorporate social data into decision-making processes and uses the idea of marine socio-ecological systems and integrated ecosystem assessment. Their findings highlight the tension between the need to reduce complexity into measurable indicators and the danger of valuing only what can be quantified (often in an economic sense). Mixed methods using quantitative and qualitative approaches are suggested as a middle ground.

In summary, there is a growing research interest in cultural values and ES (including well-being, which is discussed in Sect. 3.3) allied to management and decision-making around the marine environment. Within the CES literature, there are significant debates on how to appropriately measure or assess the cultural values of ecosystems with inputs from economics, social science and, to a lesser extent, the arts and humanities. There has been some direct attempt to reflect on CES in the context of MSP, but by far the largest number of reports are those that study marine and coastal environments through the lens of CES in the broader context of management, rather than specifically for MSP. Incorporating sociocultural values into decision-making encounters challenges of using assessment methods that are acceptable within a decision-making framework. The area of shared values and value deliberation seem to offer potential ways forward for capturing a range of deeply held cultural values alongside other ES assessment procedures for MSP.

3.2 Societal Connection to the Sea: Values, Perceptions and Citizenship

Our global seas and coasts provide a rich diversity of goods and services that communities of all shapes and sizes derive a range of benefits from. These interactions and exchanges have a direct influence on the relationships and sense of connection garnered between the marine environment and individual people, as well as society as a whole. Increasingly, social science disciplines and techniques are being employed as a mechanism to delve more deeply into these interactions, resulting in what has been a recent exponential growth in research around the emerging themes (within a marine and coastal context) of public perceptions-based research (Jefferson et al. 2014, 2015; Potts et al. 2016), social values (Ives and Kendal 2014; Schwartz 1992), marine citizenship (McKinley and Fletcher 2010, 2012; Fletcher et al. 2012) and ocean literacy (Costa and Calderia 2018; Uyarra and Borja 2016; Steel et al. 2005).

Cumulatively, these concepts (as described and defined in Table 7.1) allow us to develop a comprehensive understanding of the various ways in which society interacts with, and uses, the marine environment. Importantly, with a common grounding in the social sciences, these concepts provide a 'social lens' through which use of the sea can be viewed. Using techniques, theories and concepts from sociology, psychology and environmental economics, there is a growing recognition that society comprises multiple audiences and that this heterogeneous group possesses a highly mixed set of values, attitudes, perceptions, beliefs and experiences about the marine environment, all of which must in some way be taken account of within MSP.

'Social values', as a concept, has been considered within the disciplines of psychology, philosophy, economics, human geography, anthropology and, now increasingly, within the world of environmental management and conservation (Ives and Kendal 2014). The varying interpretations of the word 'values' itself is challenging; however, the concept can be broadly defined as the underlying values held by individuals and society as a collective, and the value attributed to items, things and places (Ives and Kendal 2014). For the purposes of this discussion, 'social values' is also taken to encompass the parallel notions of public perceptions, attitudes and beliefs. Increasingly, the concept of social values and their role is becoming more and more commonplace within environmental decision-making (Tallis et al. 2008). Within a marine context, this can be seen throughout the conversation surrounding the recent European Maritime Spatial Planning Directive, for example. There is evidence of this move towards a more inclusive way of thinking in, for example, the UK's recently published 25 Year Environment Plan (UK Government 2018). This shift in the debate surrounding MSP, while welcome, poses a number of critical questions—what are the social values that should be taken account of within MSP? How do coastal communities (and those not directly living or working at the coast) attribute value (both monetary and non-monetary) to different aspects of their marine environment? How might these values change as a result of MSP-related decision-making, such as licensing of certain activities? What impact might this have on the social and cultural character of the region/town/coastline, and how will this influence its capacity to deliver on sustainable development and the blue growth agenda underpinning MSP?

Traditionally, emphasis has been placed on ascertaining the monetary value of marine and coastal ecosystems, with economic-based metrics used as leverage within decision-making and governance (Tallis et al. 2008). More recently, however, there is a recognition of the importance of other value types, acknowledging that many marine and coastal resources are not easily marketable, making economic valuation complex, partial and even inappropriate in some cases (Kallis et al. 2013; Dempsey and Robertson 2012). By understanding the diverse set of values (both monetary and non), views, perceptions and attitudes held by society about both the local and global marine environment, there is an opportunity for MSP to more effectively recognise the intrinsic complexity of societal interactions with the global seas. Marine and coastal governance, and by association MSP, is at the cusp of a wave, with public interest in marine issues recently reaching a peak through the 'Blue Planet Effect' (The Guardian 2018) and other studies that explore public awareness of marine issues (see for example, Potts et al. 2016). This is therefore an important potential juncture to link public attitudes and MSP processes. Recognising the role that this evidence could play in MSP is just the first step; there is now a real need to develop standardised, effective pathways to incorporate these data in MSP.

Following on from the dialogue around social values, the interconnected concepts of marine citizenship and ocean literacy are an ideal framework to be embedded more effectively within the broader MSP processes as a way of ensuring sustainable use of marine resources, with guidance and management accepted and implemented more easily as a result of a more marine aware society (McKinley and Fletcher 2010, 2012; Fletcher et al. 2012). Marine citizenship, derived from the traditional concepts of environmental citizenship (e.g. see Hawthorne and Alabaster 1999), builds on early concepts of ocean citizenship (Fletcher and Potts 2007), which is defined as 'an awareness of the rights and responsibilities towards the marine environment, and an awareness and capacity to engage in management' (McKinley and Fletcher 2010). Marine citizenship sets out a framework that takes the influence of socio-demographic factors (e.g. gender, age, employment history, education level and ethnicity) on individual and collective values into account (see Table 7.1). Complementary to the notion of marine citizenship is the parallel agenda of ocean literacy (Uyarra and Borja 2016; Steel et al. 2005)—each of these seeks to understand and inculcate greater levels of public engagement with marine issues, ultimately leading to positive behaviour change. A relatively new term, ocean literacy was initially coined in 2004 and defined as 'the understanding of the ocean's influence on humans and of our influence on the ocean' (Uyarra and Borja 2016). Fundamental to the concept of ocean literacy, and indeed that of marine citizenship, are the interdependencies that characterise the society-sea relationship, and that society has both rights and responsibilities towards the marine environment, its resources and their use and experience of it. By leveraging the models of ocean literacy, or that of highly engaged marine citizens, and entrenching these processes as fundamental components of the marine planning process, MSP has the potential to not only deliver sustainable development and management of blue space but also engender more marine-aware communities.

Despite garnering increasing attention as a recognised evidence gap, there appears to be limited emphasis on these sociocultural aspects of human interactions with the global marine environment. A review of the UK's Marine Policy Statement (DEFRA 2011), a high-level policy document signed by the four devolved administrations of the UK, *and* the current versions of the devolved marine plans highlighted a significant lack of consideration of the social values, perceptions- and attitude-based data commonly associated with the concepts discussed in this section, with no explicit mention of these concepts in any of the documents. Despite the use of a somewhat narrow lens to view societal relationships with the sea (i.e. through resource use, and the blue growth agenda), global conversations are making increasing moves to take account of the less tangible aspects of 'value'. Wales, for example, has recently introduced new legislation centred on achieving social, cultural, economic and environmental well-being, through the aspirational Well-being of Future Generations (Wales) Act (2015) (see Sect. 3.3 for more on well-being within MSP). On an international stage, the objectives set out by the Aichi Targets and the UN SDGs (specifically SDG 14) echo this attempt to consider society as part of the environmental system. The concepts and frameworks discussed in this section are a valuable lens through which the complex web of societal connections with the sea can be explored. To date, they have not been utilised to their maximum potential to realise change (see, e.g. Costa and Calderia 2018)—MSP is an opportunity to move this forward.

3.3 Well-being

Well-being is a multidimensional concept in the social sciences which refers to happy, healthy, prosperous or flourishing people or communities. It is an idea which can be traced back to ancient philosophy and notions of 'the good life'. More recently, the term has come to prominence globally as nations and international organisations have begun to promote novel measures of the success of policies (OECD 2011). Thus, marine plans and policies will be increasingly scrutinised to consider their potential to contribute to greater well-being for people. Some have argued that current governance overemphasises economic prosperity, and well-being is a counterbalance for planning systems to consider the broader contributions to people's quality of life (Stiglitz et al. 2009). In response to this growing interest, a range of disciplines with varying interpretations of well-being have begun to research well-being in relation to the oceans.

Firstly, in environmental psychology and medicine,[1] there has been a particular focus on 'blue space'—coined in contradistinction to 'green space' to reflect the importance of water-based environments and the health benefits which people get from engaging with coastal or ocean outdoor spaces. For example, Wyles et al. (2017) show that visits to coastal locations produce higher psychological restoration benefits than urban greenspaces, based on an extensive survey. Furthermore, White et al. (2017) report improved emotional and cognitive restoration due to engagement with wildlife, especially where marine wildlife exhibits fascinating behaviours, based upon a study using an experimental approach. A range of recent studies consider how factors such as

[1] Studies in medicine have also considered how degraded marine environments can have negative consequences for physical health through disease or injury. Examples include increased risk of drowning due to floods, exposure to pathogens or harmful algal blooms.

type of activity, frequency of exposure and proximity of habitation to the coast affect psychological benefits or reported life satisfaction. Overall reviews conclude that there is moderately strong empirical evidence for the positive effect of blue space on mental health, with less consistent evidence on the relationship to general health, obesity or cardiovascular outcomes (Zufferey 2015; Gascon et al. 2017). Research in environmental economics also considers the premium that people are willing to put on interactions with blue space, such as a home with a sea view. However, the economic concept of welfare is narrower than well-being, which extends beyond financial or material benefits or the health of an economy (McGregor and Pouw Nicky 2017).

Secondly, in cultural geography, anthropology and sociology, there have been investigations of the relationship between affect, emotion and place in marine and coastal settings. In contrast to psychological and health studies, these are often based upon qualitative studies which undertake detailed exploration of phenomena, in order to understand the nature of human experience in or by the oceans. For example, Foley (2015) uses oral histories of outdoor swimming in coastal locations to document numerous responses to immersion in the oceans as 'a place where the body can let go'. Kearns and Collins (2012) report feelings of anxiety and anger when coastal developments are seen to conflict with notions of sacredness or place attachment, while Urquhart and Acott (2013) consider how the collective identity of coastal communities draws upon fishers' engagement with oceanic spaces, particularly emotional attachment to places in the sea, and a sense of freedom. The importance of place as a dimension of well-being is explored in Acott and Urquhart (2017) and White (2017) as part of an edited volume on social well-being and the values of small-scale fisheries (Johnson et al. 2017).

Thirdly, in political economy, development studies and sustainability science, there is considerable research exploring the links between the quality of the environment, access to resources and well-being in terms of overall quality of life (Breslow et al. 2016; Biedenweg et al. 2016; Daw et al. 2015). Well-being is a key term in a number of frameworks—for example, it is postulated as the ultimate 'good' to which the benefits of ES contribute (Russell et al. 2013). One reason that people have argued for the use of well-being in planning is its ability to connect political narratives with people's everyday lives, in contrast with more monodimensional measures such as gross domestic product (GDP) or other sustainability indicators which seem removed from their existence (McGregor 2014). Well-being presents a framing for stakeholder engagement in marine planning. Thus, an identified problem such as deprivation in coastal communities, can be engaged via community debate about which qualities of the marine area or future scenarios for the oceans will give rise to improved well-being.

Debate continues about how well-being should be measured and operationalised. This is perhaps unsurprising, given the multidimensional character of the concept. One way of categorising different measures of well-being is to consider broadly material/objective, subjective and relational measures (McGregor et al. 2015). Material/objective measures include tangible assets or physiological characteristics. Subjective measures include people's reported perceptions such as their emotional state or level of satisfaction. Relational measures consider life lived in the context of social interactions. Coulthard (2012) presents a 3D conceptualisation of well-being which draws upon each of these dimensions. Despite this rich picture of well-being, there still remains a challenge of how to aggregate these different measures to inform MSP.

4 MSP and Socio-cultural Dimensions: A Critique

The studies discussed in the sections earlier indicate a growing socio-cultural evidence base that could be used within MSP and have showcased examples of the opportunities and potential associated with this. However, it should be recognised that it would not be simple to just draw upon these studies to support plan/policy development; challenges remain. The following section provides a short commentary on our views as to the key challenges associated with embedding sociocultural factors within MSP.

- Qualitative studies, which are the most common approach within sociocultural studies, although insightful, are sometimes difficult to generalise beyond the context in which they were developed. There is, therefore, a need for greater standardisation of methods and approaches, and development of effective pathways to utilise data of a qualitative nature within MSP. Furthermore, there is a need to identify effective pathways that support the incorporation of diverse epistemologies that offer rich insights into cultural beliefs, values and practice into MSP.
- Quantitative large-scale studies tend to be favoured as an evidence base in planning processes, but it should be recognised that they also have some weaknesses. For example, because of the need for large data sets to support multivariate analyses, they tend to depend upon existing data sets which were not designed for the purpose of MSP and make broad assumptions

for many of their measures—for example, assuming protected area status as a proxy for high-quality biodiversity or pristine site, when measuring the effects of environmental quality on well-being.

- Challenges remain as how the ideas of space and place across marine and terrestrial borders can be reconciled in ways that satisfy the needs of both groups of planners and decision-makers.
- From a well-being perspective, the literature discussed in Sect. 3.3 presents a range of findings about the connections between well-being and blue space. However, it is not yet sufficiently advanced to operationalise this in a therapeutic framework that can be effectively embedded within MSP.

  - A marine plan truly based around achieving well-being outcomes might lead to a different set of priorities.
  - Evidence from studies such as those discussed may support policies such as improved coastal access, but there are trade-offs between this and conservation objectives. Considering how well-being arises in an offshore space and accrues to different land-based populations or interest groups is complicated for marine planning to consider.
  - Marine plans themselves are not always the appropriate regulatory framework for well-being policy—these policy interventions might be developed in other fields such as public healthcare. Nevertheless, plan policies could encourage developments which support this kind of outcome.
  - Well-being, therefore, represents a measurable outcome for marine planning systems and plans. However, to date, few national marine planning systems have engaged with this topic extensively or set up an established metric to evaluate this outcome.

- Within current MSP processes, there is a widespread lack of understanding and, therefore, reliable and credible evidence associated with these more sociocultural components of marine management.

  - Socio-cultural aspects of societal relationships with the sea are subject to spatial and temporal variation, as well as having the unique issue of being a landscape/environment that is often quite removed from everyday public experience. This lack of public awareness and connection with the marine environment poses a real challenge to MSP.
  - A layer of complexity is added through a domination of studies that explore and/or measure socio-cultural metrics at a local or regional level. Scaling this up to a national, or even a regional MPA level, is challenging, and there has been limited success to date.

– MSP has an opportunity to lead the way in developing a standardised approach to this evidence need, supporting the realisation of global goals of a broader, more holistic approach to marine planning and wider marine governance.

MSP is still a relatively new mechanism within broader marine and coastal management. The emerging and evolutionary nature of this process suggests there to be scope for the MSP process to evolve, and to establish mechanisms for the inclusion of these sociocultural aspects of societal relationships with the global seas. In summary, it is clear that, despite international goals and an ever-growing emphasis on the importance of considering the 'human' element of interactions within environmental governance, a lack of understanding about the flows and pathways to impact between these socio-cultural dimensions and MSP remains. The reason for the exclusion of these forms of evidence may range from resource constraints, to the complexities of knowledge generation, to whether the overall framing of the marine planning initiative is sympathetic to this kind of knowledge. Yet it is the socio-cultural dimension and the key concepts explored in this chapter that often provide the basis for engaging the public within the planning process and demonstrating the societal relevance of MSP. We, therefore, contend that there is much benefit to the future development of this knowledge base to support MSP.

Acknowledgements We are grateful to the project "Economy of maritime space" funded by the Polish National Science Centre for contributing the Open Access fee for this chapter and facilitating our discussions and preparation of the book.

References

Acott, T. G., & Urquhart, J. (2017). Exploring CES and Wellbeing Through a Place Based Approach: The Case of SSF Along the English Channel. In D. Johnson, T. G. Acott, et al. (Eds.), *Social Wellbeing and the Values of Small-Scale Fisheries.* Springer.

Alegret, J.-L., & Carbonell, E. (2016). Maritime Heritage Conservation. In H. D. Smith, J. L. S. De Vivero, & T. S. Agardy (Eds.), *Routledge Handbook of Ocean Resources and Management.* Routledge.

Arkema, K. K., Verutes, G. M., Wood, S. A., Clarke-Samuels, C., Rosado, S., Canto, M., Rosenthal, A., Ruckelshaus, M., Guannel, G., Toft, J., Faries, J., Silver, J. M., Griffin, R., & Guerry, A. D. (2015). Embedding Ecosystem Services in Coastal Planning Leads to Better Outcomes for People and Nature. *Proceedings of the National Academy of Sciences of the United States of America, 112*, 7390–7395.

Barbier, E. B., Hacker, S. D., Kennedy, C., Koch, E. W., Stier, A. C., & Silliman, B. R. (2011). The Value of Estuarine and Coastal Ecosystem Services. *Ecological Monographs, 81*, 169–193.

Baulcomb, C., Fletcher, R., Lewis, A., Akoglu, E., Robinson, L., Von Almen, A., Hussain, S., & Glenk, K. (2015). A Pathway to Identifying and Valuing Cultural Ecosystem Services: An Application to Marine Food Webs. *Ecosystem Services, 11*, 128–139.

Beaumont, N. J., Austen, M. C., Atkins, J. P., Burdon, D., Degraer, S., Dentinho, T. P., Derous, S., Holm, P., Horton, T., Van Ierland, E., Marboe, A. H., Starkey, D. J., Townsend, M., & Zarzycki, T. (2007). Identification, Definition and Quantification of Goods and Services Provided by Marine Biodiversity: Implications for the Ecosystem Approach. *Marine Pollution Bulletin, 54*, 253–265.

Biedenweg, K., Stiles, K., & Wellman, K. (2016). A Holistic Framework for Identifying Human Well-Being Indicators for Marine Policy. *Marine Policy, 64*, 31–37. https://doi.org/10.1016/j.marpol.2015.11.002.

Börger, T., Beaumont, N. J., Pendleton, L., Boyle, K. J., Cooper, P., Fletcher, S., Haab, T., Hanemann, M., Hooper, T. L., Hussain, S. S., Portela, R., Stithou, M., Stockill, J., Taylor, T., & Austen, M. C. (2014). Incorporating Ecosystem Services in Marine Planning: The Role of Valuation. *Marine Policy, 46*, 161–170. https://doi.org/10.1016/j.marpol.2014.01.019.

Braat, L. C. (2018). Five Reasons Why the Science Publication "Assessing Nature's Contributions to People" (Diaz et al. 2018) Would Not Have Been Accepted in Ecosystem Services. *Ecosystem Services, 30*, A1–A2.

Breslow, S. J., Sojka, B., Barnea, R., Basurto, X., Carothers, C., Charnley, S., Coulthard, S., Dolšak, N., Donatuto, J., García-Quijano, C., Hicks, C. C., Levine, A., Mascia, M. B., Norman, K., Poe, M., Satterfield, T., Martin, K. S., & Levin, P. S. (2016). Conceptualizing and Operationalizing Human Well-Being for Ecosystem Assessment and Management. *Environmental Science and Policy, 66*, 250–259. https://doi.org/10.1016/j.envsci.2016.06.023.

Bryce, R., Irvine, K. N., Church, A., Fish, R., Ranger, S., & Kenter, J. O. (2016). Subjective Well-Being Indicators for Large-Scale Assessment of Cultural Ecosystem Services. *Ecosystem Services, 21*, 258–269.

Cooper, N., Brady, E., Steen, H., & Bryce, R. (2016). Aesthetic and Spiritual Values of Ecosystems: Recognising the Ontological and Axiological Plurality of Cultural Ecosystem 'Services'. *Ecosystem Services.*

Costa, S., & Calderia, R. (2018). Bibliometric Analysis of Ocean Literacy: An Underrated Term in the Scientific Literature. *Marine Policy, 87*, 149–157.

Costanza, R., De Groot, R., Braat, L., Kubiszewski, I., Fioramonti, L., Sutton, P., Farber, S., & Grasso, M. (2017). Twenty Years of Ecosystem Services: How Far Have We Come and How Far Do We Still Need to Go? *Ecosystem Services, 28*, 1–16.

Coulthard, S. (2012). What Does the Debate Around Social Well-Being Have to Offer Sustainable Fisheries? *Current Opinion in Environmental Sustainability, 4*(3), 358–363. https://doi.org/10.1016/j.cosust.2012.06.001.

Daw, T. M., Coulthard, S., Cheung, W. W. L., Brown, K., Abunge, C., Galafassi, D., Peterson, G. D., McClanahan, T. R., Omukoto, J. O., & Munyi, L. (2015). Evaluating Taboo Trade-Offs in Ecosystems Services and Human Well-Being. *Proceedings of the National Academy of Sciences of the United States of America, 112*(22), 6949–6954. https://doi.org/10.1073/pnas.1414900112.

DEFRA. (2011). UK Marine Policy Statement. Retrieved June 20, 2018, from https://www.gov.uk/government/publications/uk-marine-policy-statement.

Dempsey, J., & Robertson, M. M. (2012). Ecosystem Services: Tensions, Impurities and Points of Engagement Within Neoliberalism. *Progress in Human Geography, 36*(6), 758–779.

Edwards, D. M., Collins, T. M., & Goto, R. (2016). An Arts-Led Dialogue to Elicit Shared, Plural and Cultural Values of Ecosystems. *Ecosystem Services, 21*, 319–328.

Falconer, L., Hunter, D. C., Telfer, T. C., & Ross, L. G. (2013). Visual, Seascape and Landscape Analysis to Support Coastal Aquaculture Site Selection. *Land Use Policy, 34*, 1–10. https://doi.org/10.1016/j.landusepol.2013.02.002.

Fish, R., Church, A., & Winter, M. (2016). Conceptualising Cultural Ecosystem Services: A Novel Framework for Research and Critical Engagement. *Ecosystem Services.*

Fletcher, S., & Potts, J. (2007). Ocean Citizenship: An Emergent Geographical Concept. *Coastal Management, 35*(4), 511–524. https://doi.org/10.1080/08920750701525818.

Fletcher, S., Saunders, J., & Herbert, R. J. H. (2011). A Review of the Ecosystem Services Provided by Broad-Scale Marine Habitats.pdf. *Journal of Coastal Research, 64*, 378–383.

Fletcher, S., Jefferson, R. L., & McKinley, E. (2012). Saving the Shallows: Focusing Marine Conservation Where People Might Care. *Aquatic Conservation: Marine and Freshwater Ecosystems.*

Fletcher, R., Baulcomb, C., Hall, C., & Hussain, S. (2014). Revealing Marine Cultural Ecosystem Services in the Black Sea. *Marine Policy, 50*, 151–161.

Foley, R. (2015). Swimming in Ireland: Immersions in Therapeutic Blue Space. *Health and Place, 35*, 218–225. https://doi.org/10.1016/j.healthplace.2014.09.015.

Gascon, M., Zijlema, W., Vert, C., White, M. P., & Nieuwenhuijsen, M. J. (2017). Outdoor Blue Spaces, Human Health and Well-Being: A Systematic Review of Quantitative Studies. *International Journal of Hygiene and Environmental Health, 220*(8), 1207–1221. https://doi.org/10.1016/j.ijheh.2017.08.004.

Gee, K., Kannen, A., Adlam, R., Brooks, C., Chapman, M., Cormier, R., Fischer, C., Fletcher, S., Gubbins, M., Shucksmith, R., & Shellock, R. (2017). Identifying Culturally Significant Areas for Marine Spatial Planning. *Ocean & Coastal Management, 136*, 139–147.

Gómez-Baggethun, E., De Groot, R., Lomas, P. L., & Montes, C. (2010). The History of Ecosystem Services in Economic Theory and Practice: From Early Notions to Markets and Payment Schemes. *Ecological Economics, 69*, 1209–1218.

Guerry, A. D., Ruckelshaus, M. H., Arkema, K. K., Bernhardt, J. R., Guannel, G., Kim, C.-K., Marsik, M., Papenfus, M., Toft, J. E., Verutes, G., Wood, S. A., Beck, M., Chan, F., Chan, K. M. A., Gelfenbaum, G., Gold, B. D., Halpern, B. S., Labiosa, W. B., Lester, S. E., Levin, P. S., Mcfield, M., Pinsky, M. L., Plummer, M., Polasky, S., Ruggiero, P., Sutherland, D. A., Tallis, H., Day, A., & Spencer, J. (2012). Modeling Benefits from Nature: Using Ecosystem Services to Inform Coastal and Marine Spatial Planning. *International Journal of Biodiversity Science, Ecosystem Services & Management, 8*, 107–121.

Halpern, B. S., Diamond, J., Gaines, S., Gelcich, S., Gleason, M., Jennings, S., Lester, S., Mace, A., Mccook, L., Mcleod, K., Napoli, N., Rawson, K., Rice, J., Rosenberg, A., Ruckelshaus, M., Saier, B., Sandifer, P., Scholz, A., & Zivian, A. (2012). Near-Term Priorities for the Science, Policy and Practice of Coastal and Marine Spatial Planning (CMSP). *Marine Policy, 36*, 198–205.

Hattam, C., Atkins, J. P., Beaumont, N., Borger, T., Bohnke-Henrichs, A., Burdon, D., De Groot, R., Hoefnagel, E., Nunes, P. A. L. D., Piwowarczyk, J., Sastre, S., & Austen, M. C. (2015). Marine Ecosystem Services: Linking Indicators to Their Classification. *Ecological Indicators, 49*, 61–75.

Hawthorne, M., & Alabaster, T. (1999). Citizen 2000 Development of a Model of Environmental Citizenship. *Global Environmental Change, 9*, 25–43.

Hynes, S., Ghermandi, A., Norton, D., & Williams, H. (2018). Marine Recreational Ecosystem Service Value Estimation: A Meta-Analysis with Cultural Considerations. *Ecosystem Services*.

Ives, C. D., & Kendal, D. (2014). The Role of Social Values in the Management of Ecological Systems. *Journal of Environmental Management, 144*, 67–72.

Jefferson, R. L., Bailey, I., Laffoley, D. D., Richards, J. P., & Attrill, M. (2014). Public Perceptions of the UK Marine Environment. *Marine Policy, 43*, 327–337.

Jefferson, R., McKinley, E., Capstick, S., Fletcher, S., Griffin, H., & Milanese, M. (2015). Understanding Audiences: Making Public Perceptions Research Matter to Marine Conservation. *Ocean and Coastal Management*. https://doi.org/10.1016/j.ocecoaman.2015.06.014.

Jobstvogt, N., Watson, V., & Kenter, J. O. (2014). Looking Below the Surface: The Cultural Ecosystem Service Values of UK Marine Protected Areas (MPAs). *Ecosystem Services, 10*, 97–110.

Johnson, D., Acott, T. G., Stacey, N., & Urquhart, J. (Eds.). (2017). *Social Wellbeing and the Values of Small-Scale Fisheries*. Springer.

Kallis, G., Gomez-Baggethun, E., & Zograios, C. (2013). To Value or Not to Value? That Is not the Question. *Ecological Economics, 94*, 97–105.

Kearns, R., & Collins, D. (2012). Feeling for the Coast: The Place of Emotion in Resistance to Residential Development. *Social & Cultural Geography, 13*(8), 937–955. https://doi.org/10.1080/14649365.2012.730150.

Kenter, J. O., O'brien, L., Hockley, N., Ravenscroft, N., Fazey, I., Irvine, K. N., Reed, M. S., Christie, M., Brady, E., Bryce, R., Church, A., Cooper, N., Davies, A., Evely, A., Everard, M., Fish, R., Fisher, J. A., Jobstvogt, N., Molloy, C., Orchard-Webb, J., Ranger, S., Ryan, M., Watson, V., & Williams, S. (2015). What Are Shared and Social Values of Ecosystems? *Ecological Economics, 111*, 86–99.

Kenter, J. O., Bryce, R., Christie, M., Cooper, N., Hockley, N., Irvine, K. N., Fazey, I., O'brien, L., Orchard-Webb, J., Ravenscroft, N., Raymond, C. M., Reed, M. S., Tett, P., & Watson, V. (2016a). Shared Values and Deliberative Valuation: Future Directions. *Ecosystem Services, 21*, 358–371.

Kenter, J. O., Jobstvogt, N., Watson, V., Irvine, K. N., Christie, M., & Bryce, R. (2016b). The Impact of Information, Value-Deliberation and Group-Based Decision-Making on Values for Ecosystem Services: Integrating Deliberative Monetary Valuation and Storytelling. *Ecosystem Services, 21*, 270–290.

Klain, S. C., & Chan, K. M. A. (2012). Navigating Coastal Values: Participatory Mapping of Ecosystem Services for Spatial Planning. *Ecological Economics, 82*, 104–113.

Leyshon, C. (2014). Cultural Ecosystem Services and the Challenge for Cultural Geography. *Geography Compass, 8*, 710–725.

Liquete, C., Piroddi, C., Drakou, E. G., Gurney, L., Katsanevakis, S., Charef, A., & Egoh, B. (2013). Current Status and Future Prospects for the Assessment of Marine and Coastal Ecosystem Services: A Systematic Review. *PLoS One, 8*, e67737.

Luisetti, T., Turner, R. K., Jickells, T., Andrews, J., Elliott, M., Schaafsma, M., Beaumont, N., Malcolm, S., Burdon, D., Adams, C., & Watts, W. (2014). Coastal Zone Ecosystem Services: From Science to Values and Decision Making; a Case Study. *Science of the Total Environment, 493*, 682–693.

McGregor, J. A. (2014). Human Well-Being and Sustainability: Interdependent and Intertwined. In *Handbook of Sustainable Development* (2nd ed., pp. 217–233).

McGregor, J., & Pouw Nicky, A. (2017). Towards an Economics of Wellbeing. *Cambridge Journal of Economics, 41*(4), 1123–1142. https://doi.org/10.1093/cje/bew044.

McGregor, J. A., Camfield, L., & Coulthard, S. (2015). Competing Interpretations: Human Well-Being and the Use of Quantitative and Qualitative Methods. In K. Roelen & L. Camfield (Eds.), *Mixed Methods Research in Poverty and Vulnerability: Sharing Ideas and Learning Lessons* (pp. 231–260). London: Palgrave Macmillan.

McKinley, E., & Fletcher, S. (2010). Individual Responsibility for the Oceans? An Evaluation of Marine Citizenship by UK Marine Practitioners. *Ocean and Coastal Management, 53*(7), 379–384.

McKinley, E., & Fletcher, S. (2012). Improving Marine Environmental Health Through Marine Citizenship: A Call for Debate. *Marine Policy, 36*, 839–843. https://doi.org/10.1016/j.marpol.2011.11.001.

Millennium Ecosystem Assessment. (2003). Chapter 1: MA Conceptual Framework. In *Millennium Ecosystem Assessment, Ecosystems and Human Well-being: A Framework for Assessment*. Island Press.

Murray, G., D'anna, L., & Macdonald, P. (2016). Measuring What We Value: The Utility of Mixed Methods Approaches for Incorporating Values into Marine Social-Ecological System Management. *Marine Policy, 73*, 61–68.

Natural England. (2012). *An Approach to Seascape Character Assessment*. Report NECR105. Peterborough: Natural England.

OECD. (2011). *How's Life? Measuring Well-Being*. Paris: OECD.

Potts, T., Burdon, D., Jackson, E., Atkins, J., Saunders, J., Hastings, E., & Langmead, O. (2014). Do Marine Protected Areas Deliver Flows of Ecosystem Services to Support Human Welfare? *Marine Policy, 44*, 139–148.

Potts, T., Pita, C., O'Higgins, T., & Mee, L. (2016). Who Cares? European Attitudes Towards Marine and Coastal Environments. *Marine Policy, 72*, 59–66.

Ranger, S., Kenter, J. O., Bryce, R., Cumming, G., Dapling, T., Lawes, E., & Richardson, P. B. (2016). Forming Shared Values in Conservation Management: An Interpretive-Deliberative-Democratic Approach to Including Community Voices. *Ecosystem Services*.

Ruiz-Frau, A., Hinz, H., Edwards-Jones, G., & Kaiser, M. J. (2013). Spatially Explicit Economic Assessment of Cultural Ecosystem Services: Non-extractive Recreational Uses of the Coastal Environment Related to Marine Biodiversity. *Marine Policy, 38*, 90–98.

Russell, R., Guerry, A. D., Balvanera, P., Gould, R. K., Basurto, X., Chan, K. M. A., Klain, S., Levine, J., & Tam, J. (2013). Humans and Nature: How Knowing and Experiencing Nature Affect Well-Being. *Annual Review of Environment and Resources, 38*, 473–502. https://doi.org/10.1146/annurev-environ-012312-110838.

Satz, D., Gould, R. K., Chan, K. M., Guerry, A., Norton, B., Satterfield, T., Halpern, B. S., Levine, J., Woodside, U., Hannahs, N., Basurto, X., & Klain, S. (2013). The Challenges of Incorporating Cultural Ecosystem Services into Environmental Assessment. *Ambio, 42*, 675–684.

Saunders, J., Beaumont, N., Atkins, J. P., Lannin, A., Lear, D., Ozdemiroglu, E., & Potts, T. (2015). A Review of Marine and Coastal Ecosystem Services Data and Tools to Incorporate This into Decision-Making. In K. R. Turner & M. Schaafsma (Eds.), *Coastal Zones Ecosystem Services: From Science to Values and Decision Making*. Springer International Publishing.

Schwartz, S. H. (1992). Universals in the Content and Structure of Values: Theoretical Advances and Empirical Tests in 20 Countries. *Advances in Experimental Social Psychology, 25*, 1–65.

Smith, H. D., Ballinger, R. C., & Stojanovic, T. A. (2012). The Spatial Development Basis of Marine Spatial Planning in the United Kingdom. *Journal of Environmental Policy and Planning, 14*(1), 29–47. https://doi.org/10.1080/1523908x.2012.663192.

Steel, B. S., Smith, C., Opsommer, L., Curiel, S., & Warner-Steel, R. (2005). Public Ocean Literacy in the United States. *Ocean and Coastal Management, 48*(2), 97–114.

Stiglitz, J. E., Sen, A. K., & Fitoussi, J.-P. (2009). Measuring Economic Performance and Social Progress, Paris, Commission on the Measurement of Economic Performance and Social Progress.

Stojanovic, T. A., & Farmer, C. J. Q. (2014). The Development of World Oceans & Coasts and Concepts of Sustainability. *Marine Policy, 42*, 157–165.

Tallis, H., Kareiva, P., Marvier, M., & Chang, A. (2008). An Ecosystem Services Framework to Support Both Practical Conservation and Economic Development. *PNAS, 105*(28), 9457–9464.

The Guardian. (2018). The Blue Planet Effect: Why Marine Biology Courses Are Booming. Retrieved June 20, 2018, from https://www.theguardian.com/education/2018/jan/12/blue-planet-effect-why-marine-biology-courses-booming.

Turner, K. R., & Schaafsma, M. (Eds.). (2015). *Coastal Zones Ecosystem Services: From Science to Values and Decision Making*. Springer International Publishing.

UK Government. (2018). A Green Future—Our 25 Year Plan to Improve the Environment. Retrieved June 20, 2018, from https://www.gov.uk/government/publications/25-year-environment-plan.

United Nations. (2015). Sustainable Development Goals. Retrieved June 15, 2018, from https://sustainabledevelopment.un.org/?menu=1300.

Urquhart, J., & Acott, T. G. (2013). Re-connecting and Embedding Food in Place: Rural Development and Inshore Fisheries in Cornwall, UK. *Journal of Rural Studies, 32*, 357–364.

Uyarra, M. C., & Borja, A. (2016). Ocean Literacy: A 'New' Socio-Ecological Concept for a Sustainable Use of the Seas. *Marine Pollution Bulletin, 104*, 1–2.

White, C. (2017). Symbols of Resilience and Contested Place Identity in the Coastal Fishing Towns of Cromer and Sheringham, Norfolk, UK: Implications for Social Wellbeing. In D. Johnson, T. G. Acott, N. Stacey, & J. Urquhart (Eds.), *Social Wellbeing and the Values of Small-Scale Fisheries* (pp. 45–74). Springer.

White, M. P., Weeks, A., Hooper, T., Bleakley, L., Cracknell, D., Lovell, R., & Jefferson, R. L. (2017). Marine Wildlife as an Important Component of Coastal Visits: The Role of Perceived Biodiversity and Species Behaviour. *Marine Policy, 78*, 80–89. https://doi.org/10.1016/j.marpol.2017.01.005.

Wyles, K. J., White, M. P., Hattam, C., Pahl, S., King, H., & Austen, M. (2017). Are Some Natural Environments More Psychologically Beneficial Than Others? The Importance of Type and Quality on Connectedness to Nature and Psychological Restoration. *Environment and Behavior*. https://doi.org/10.1177/0013916517738312.

Zufferey, J. (2015). Relationships Between Health and Green and Blue Spaces: A Synthesis of Empirical Research, 2003–2014. *Natures Sciences Societes, 23*(4), 343–355. https://doi.org/10.1051/nss/2015057.

8

Adding People to the Sea: Conceptualizing Social Sustainability in Maritime Spatial Planning

Fred P. Saunders, Michael Gilek, and Ralph Tafon

1 Introduction

Over the last 30 years, the Brundtland's conception of sustainable development (SD) has gained a firm place on global policy agendas. Here SD seeks to achieve multidimensional goals by linking ecological, social and economic well-being. These three pillars are seen as compatible and mutually supportive rather than completely separated. Furthermore, SD has been the 'go to' concept to address multidimensional problems in an iterated and holistic way in natural resource planning and management. However, during this time, academic and political attention has largely centred on environmental and economic sustainability, thereby leaving social sustainability relatively under-theorized and under-elaborated in policy practice (Boström 2012; Murphy 2012). This is exemplified in the case of marine spatial planning (MSP). In this chapter, we aim to contribute to filling this gap by exploration of what the social pillar of SD in MSP could or should mean and suggestions on how it could be furthered in practice. This is not to say that economic and environmental sustainability are not vitally important, but with the intention to broaden out the sustainability ambition in MSP.

MSP aims to achieve SD by balancing a range of economic, social and environmental goals in decision-making over use of marine space. The EU Maritime Spatial Planning Directive (2014/89/EU) can be seen as a recent

F. P. Saunders (✉) • M. Gilek • R. Tafon
Södertörn University, Huddinge, Sweden

J. Zaucha, K. Gee (eds.), *Maritime Spatial Planning*,
https://doi.org/10.1007/978-3-319-98696-8_8

attempt to legislate to achieve this ambition across European seas by harmonizing environmental protection with economic development opportunities. MSP is a worldwide marine governance phenomenon with marine plans in place or under development in Australia, Canada, China, Ecuador, Mozambique, Namibia, New Zealand, Philippines, Seychelles, USA and Vietnam, among others (UNESCO n.d.).

MSP policymaking, at least in Europe, commonly explicitly incorporates environmental (protection) and economic (Blue Growth) components and goals, but very rarely, if at all, are social aspects elaborated or addressed.[1] As others have pointed out (cf. Boström 2012), this oversight is not atypical in natural resource management, where it is commonly (and we argue wrongly) assumed that social benefits will flow through (or trickle down) by realizing a 'balance' between economic growth and environmental protection without paying explicit attention to the social pillar of SD (e.g. Gilek et al. forthcoming). Arguably social concerns (democratic decision-making, welfare of different groups, etc.) is captured partially within both concepts in different ways—in economics because of its concern for society-wide material development[2] (with the built-in assumptions that this benefits everyone) and in environmental protection because ensuring the continuance of (environmental) conditions (as the underpinning resource base for economic ambitions) remains sustainable (enough) so as not to (overly) disrupt market potentiality/capital accumulation. In addition to undermining the orthodox view of SD which posits that the ecological must be somehow interwoven with the economic and the social, the MSP approach, characterized above, accentuates a rather inconclusive understanding of the relationship between the multiple dimensions of SD and is evasive in regard to social sustainability.

Concerns about disharmony between different dimensions of sustainability also extend to acknowledging 'conflict urgencies' between social justice and environmentalism (i.e., what should get priority in orchestrating planning 'balance'), which has often been a point of debate in the broader SD discourse between intra- and intergenerational equity (cf. Dobson 2003; Campbell 2013). In addition, assumptions of harmony between different goals of

[1] There are exceptions of course. For example, in the draft Welsh National MSP there is a section devoted to 'Ensuring a strong, healthy and just society', where normative ambitions concerning the role of MSP in providing societal benefits are described. Still these ambitions are not embedded in the role of Blue Growth; rather the references to economic uses/interests are dominated by sectoral planning.

[2] As pointed out by Campbell (2016), it is well worth noting that there is 'no singular, homogenous "economic" interest' (p. 389). Referring to economic priorities purely as Blue Growth or even 'sustainable growth' (with environmental protection in mind) as an MSP goal fails to consider other economic-related factors, such as uneven distribution of wealth and access to resources.

sustainability have been implicated in creating a 'post-political' MSP (Flannery et al. 2018; Flannery et al. 2016; Tafon 2017; Tafon et al. 2018; Jones et al. 2016; Ritchie 2014; Kidd and Ellis 2012). Post-political processes are those run by government, with *a priori* or fixed goals (sometimes not explicitly stated), that give the illusion of creating authentic spaces of public engagement, while limiting possibilities for meaningful debate and consequential action. This critique reflects growing concerns (most vocally among critical planning and social science scholars) that the stated ambitions of MSP are not being realized in practice. Among serious criticisms pointed at MSP are that it is largely devoid of social context (Flannery et al. 2018), eschews meaningful inclusion of dissenting stakeholders (Ritchie 2014), draws on limited (mostly) technical knowledge input (Ritchie and Ellis 2010; Kidd and Ellis 2012), does little to address uneven power relations among stakeholders (Ritchie and Ellis 2010; Kidd and Ellis 2012; Tafon 2017), is mostly concerned to give effect to a state agenda that privileges elite or powerful groups (Jones et al. 2016; Tafon et al. 2018), and lacks meaningful consideration of the distribution of the cost and benefits of marine use (Jentoft 2017; Flannery et al. 2016). This omission of the social pillar across a range of issue areas has come under increasing scrutiny recently as commentators from different fields have urged for public governance to pay more attention (and give increasing priority) to redressing growing forms of inequality, rather than solely focusing on economic growth as the de facto socio-economic goal and measure of progress.[3]

While MSP is often boosted as a promising means of pluralistic marine governance able to mediate tensions between competing values and interests to reach a 'common public interest' on how we are to use the sea, this recent burst of critical evaluation of MSP practice indicates that it is far from living up to these expectations. While the critical literature mentioned above provides us with insights into the shortfalls of MSP, little effort has been invested in how to meaningfully elaborate and incorporate social sustainable dimensions into MSP. Integral to understanding what social sustainability could/should mean in MSP are questions over: what should the goals of MSP be, who should decide over access to marine resources, how should these decisions be made and who should benefit from them. This underlines a need to reconsider MSP in terms of social sustainability constitutively (what is the purpose

[3] See Milanovic (2013) and Piketty (2014) on general problems of growing inequality (between and within countries); Muraca (2012) on degrowth and Tafon (2017) and Flannery et al. (2016) for accounts of problems of exclusive pursuit of economic growth in MSP.

of MSP), procedurally (how should it be done) and substantively (what should be distributive outcomes of MSP). These insights suggest that recasting MSP to take greater account of social sustainability would necessitate greater social inclusion, redressing power inequalities and making trade-offs with substantive consideration to the equity of outcomes. This reflects a need to develop joined-up thinking on sustainability to consider the possibilities of addressing environmental concerns, while centring prospects for justice and equity through economic development (Agyeman et al. 2003).

The chapter is structured in the following way. First, we identify and discuss different features of social sustainability in MSP. Then we synthesize the social sustainability features discussed in the previous section into a conceptual approach that we propose to examine social sustainability in MSP. The chapter finishes by conjecturing on how the framework could be utilized to further lift the importance of engaging in social sustainability in MSP.

2 Developing Social Sustainability in MSP

What could social sustainability look like in MSP? First, we must acknowledge that in conceptualizing social sustainability, normative, analytical and political aspects will be inevitability difficult to differentiate. Whereas the normative strives to set standards on how society ought to develop or be considered in public governance initiatives like MSP (expressed e.g. in policy or political programmes such as SD goals), the analytical looks at how social sustainability values, such as participation, equity, social cohesion and so on, are adopted and given effect. Within a sustainability approach, the social ambitions of sustainability are indelibly intertwined with the fate of ecological and economic dimensions in the sense that economic and ecological dimensions can influence the ability to achieve social sustainability and vice versa. Nevertheless, all three dimensions should not be merely seen as the means to achieve another but as legitimate in themselves.

The broad question confronted here is how to examine and understand social values in MSP, and how these, in turn, relate to the marine environment, marine uses and activities, and marine planning both as an institution and practice. This means that first we must try to delineate what values and associated concepts should be included in the social sustainability pillar of MSP. This should respond to what the critics of MSP see as current shortfalls but also the need for a conceptual approach that encompasses both procedural and substantive aspects and what might be practically achievable in different MSP practices across different contexts.

2.1 Deeping Democratic Decision-Making in MSP

First, there have been numerous scholarly interventions arguing for the deepening of democracy in MSP decision-making processes (cf. Jentoft 2017; Ritchie and Ellis 2010; Jones et al. 2016; Tafon 2017). These writings variously call for a wholesale reconstitution of MSP and/or reform to create more socially cooperative approaches to MSP. There are two key themes to these calls: (1) more voices need to be included and (2) these voices need to be heard in a way that makes a meaningful difference to MSP outcomes (i.e., equity).

Jones et al. (2016) drawing on a diverse range of MSP cases in Europe found that MSP practice exhibits an exclusive approach to participation and when a wider range of stakeholders were included, commonly their involvement was not meaningful. 'Meaningful' implies here, the inclusion of actors, so that they have a capacity to shape and influence marine planning decisions that they have an interest in (material and non-material) and that affect them. This critique, similarly to debates over SD more generally, laments what some see as the unquestioned underlying, but overwhelming purpose of MSP—to deliver Blue Growth (or economic growth through sea use). This driver is seen as steering the dominant strategic sectoral stakeholder engagement approach that Jones et al. (2016) and others see as (a problem) characterizing MSP practice. Presenting almost fully formed plans as one-way flow of information—or consultation—is also seen as a lower order and perhaps even undermining form of participation that can work to undermine the legitimacy of public governance initiatives like MSP by exacerbating power differences and elitism. As Metzger et al. (2017) and others have noted, who frames what a stakeholder is in planning or 'stakeholderness' (conferring in MSP 'the property of being considered legitimately concerned', p. 2) will affect how and what stakeholders are represented and included. If the preponderance of MSP's strategic focus is on Blue Growth and/or environmental protection, there is little likelihood that social sustainability (and associated 'stakes') will be seen as a legitimate concern without some form of explicit representation and/or constitutive reform of MSP.

When thinking of sectors in MSP (those seen to have a stake), it may be important to keep in mind that, as discussed here, they are made up of public, private and voluntary entities (Kidd and Shaw 2014). The extent of inclusion of these multiple spheres of society (i.e., where boundaries of a 'sector' are drawn for inclusion—government, civil society, business, general public, vulnerable social groups, unions) is important in thinking about democracy in marine governance and will affect other aspects of planning—such as what stakeholders are included, when and where they are included, what influence they have and what type of knowledge is valued in decision-making. So as Kidd and Ellis (2012) point out in advocating a deliberative approach, realizing mutual benefits across spheres of interest will also likely require institutional steering where differences in power are managed[4] and where antagonistic differences can be put on the table, weighted 'equitably' and addressed. In this way stakeholder engagement in MSP would work towards giving opportunities for different values, interests and types of knowledge to be expressed, exchanged and considered in a transparent way (McCann et al. 2014; Jentoft 2017).

The inherently political character of MSP raises thorny questions about how to develop proactive integrative planning processes to support this engagement of interested and affected stakeholders across multiple sectors, scales and administrative boundaries in MSP decision-making over time (Olsen et al. 2014). Of course in pluralistic governance approaches such as MSP (in ambition at least), weak or marginalized groups may not 'automatically' become represented, so this suggests that MSP needs to be cognizant of ways to give voice to these groups: particularly ways to redress the lack of social resources which undermine possibilities to realize what Pansardi (2016) calls substantive political equality (which arks back to the importance of considering inequalities in economic and symbolic power resources). Of course, this may be easier said than done, especially in settings characterized by asymmetrical power relations or where complex inter- and intra-sector dynamics are at play, such as in Poland, where there are several sub-sectors of fisheries with different histories and variable claims over resources (Saunders et al. 2016; Tafon forthcoming). Even if these difficulties could begin to be overcome (or more likely, put aside), the deliberative/agonism debate in planning theory tells us that there are unresolved questions of power, as well as ambiguity of the magnitude of (democratic or constitutive—how radical) shift desirable and/or possible (Bond 2011). Planners will also often cite resource and capacity limitations as constraints to conducting 'more democratic engagement' in MSP, but as has been noted, meaningful 'participation may be time consuming, but may also reduce [both] transaction costs at some later stage in the process' (Jentoft 2017, p. 34), and the intensity of conflict during the implementation stage of plans (Tafon et al. 2018). In this way, it may help to establish long-term buy-in and planning continuity, plus also of course, deliver on stated sustainability commitments.

[4] At least to the point where there are not dominant stakeholders.

Agonistic theorists would see this an insufficient response—as this perspective is not just concerned with extending democratic processes but opening them up to currently excluded or non-present discourses (e.g., the material and non-material distributional implications of MSP decision-making; more contestatory voices) (Tambakaki 2017). While there are more critical voices who advocate agnostic planning approaches (cf. Flannery et al. 2016; Tafon et al. 2018) in MSP, more common is a call for more 'genuinely' deliberative approaches to address competing marine use options, and choices among different knowledge claims and their relation to interests (Ritchie and Ellis 2010; Jones et al. 2016). This would involve knowledge exchange and joint formulation of goals and outcomes among a wide range of stakeholders. While calls for the democratic deliberative reform in MSP adopt a pluralist perception of power, they do not tend to naively assume that all stakeholders have equal power to assert their stake regardless of their resource levels. These calls do, however, tend to adopt a normative view of the value of including stakeholders in MSP—not just to achieve desired MSP decision-making ends, but as an end in itself. That is, in this conception, planning process and outcome are not dichotomous, as a search for the ends is always present within the planning process or means (Scholsberg 2004). This more sanguine critical perspective sees that democratic deliberative reform would be beneficial on a range of fronts related to social sustainability and quality of MSP decision-making. Even if social sustainability were to attain more legitimacy as a concern in MSP (i.e., to the extent that its specific interests could be represented by stakeholders), as the post-political (and Marxist political economy critique) of MSP tells us, there is still no certainty that deliberation *in practice* would substantively affect or generate equitable outcomes. What is in question here is whether *formal* political equality, in terms of entitlements and rights to participate, can sufficiently do away with the problem of elite capture in deliberation (which derives from differences in social resources of power) and ensure *substantive* political equality for marginalized stakeholders or their power resources to see their interests considered in decision-making processes (Pansardi 2016, p. 98).

Dryzek and Pickering (2017) describes the foundations of deliberative democracy as legitimacy, representation, communication, pluralism and consensus. While the argued benefits—learning through knowledge/interest exchange, generation of trust, jointly developed objectives and so on—of deliberative democratic practice are many and hard to disregard, adopting deliberative planning in MSP presents many challenges. These include problems involving stakeholderness (or recognition of who is to be represented), how participatory processes are designed, the inevitable unevenness of power relations in interaction and the assumed neutral stance of the planner (Forester 2006). These challenges also relate to the multidimensional character of MSP, in cases where there may be little scope or will for delegation of influence because a decision has already been made (Reed et al. 2017).

Often MSP is strategic in orientation and is being undertaken at national or regional level rather than local scale, so in addition to the problems of centralized decision-making, the seemingly lack of 'tangible effects of planning consequences' may make it more difficult to motivate widespread involvement. As inferred above, this is because there is commonly an explicit or implicit commitment to a pre-existing outcome (at least in broad terms) (Jones et al. 2016). This may be so in some cases of MSP, but Tafon et al. (2018) in a recent study involving MSP and wind farms in Estonia found that the degree and quality of involvement is also likely to depend on how controversial the issue of concern is (or the tangibility of the stake) at a local or broader scale, how formal guidelines are interpreted over who can be a stakeholder and more generally how planners design and conduct engagement processes. Enacting deliberation does not, as Forester (2006) points out, mean gullibly accepting at face value claims of preferences and interests but subjecting them to examination and scrutiny in exchange (negotiation) with others. Such interaction, Forester (2009) argues, poses opportunities to shift conflicts (a preoccupation of MSP) by moving beyond confrontation and stalemate, in visualizing and putting on the table the specific needs of the stakeholders. Deliberation processes search for consensus—to underpin and legitimate planning decisions.[5] Critics of deliberation see consensus as a cover for power and as we have noted in MSP, there is most commonly a high degree of what we can call power stratification among actors (Tafon forthcoming). Putting aside the likely significant problems of negating backstage lobbying (shadow planning) used by actors to further vested interests, how to effectively acknowledge and set aside stakes in formalized deliberative planning has proved difficult. We must also keep in mind the messiness of planning practice, including the complexity of mediating negotiations involving actors with divergent values, traditions, epistemologies and ontologies. More so, in identifying and analysing constraints to, and/or possibilities for human well-being, attention should encompass areas of visible

[5] Several examples of deliberative practices are described by Dryzek and Niemeyer (2012) in Foundations and Frontiers of Deliberative Governance. Erik Ohlin Wright (2010, 2013), within his more radical 'Envisioning Utopias' framework, presents numerous examples of deliberative democracy (what Fung and Wright (2003) refer as 'Empowered Participatory Governance') through his myriad publications and a comprehensive personal website (https://www.ssc.wisc.edu/~wright/).

conflict (i.e. actual deliberation processes) and include seemingly 'non-conflict' situations in which hidden exclusions may have grave consequences for the substantive outcomes of the decision-making process.

2.2 Meaningful Inclusion of Socio-cultural Values and Benefits

Another strand of discernible inquiry emerging in the MSP literature, related to social sustainability, is how to include socio-cultural values and benefits in MSP. Much of the work in MSP (and on natural resource more generally) on socio-cultural aspects has been in the cultural ecosystem services research vein, deriving and spawning from approaches and issues mapped out in the Millennium Ecosystem Assessment (MEA) (Guerry et al. 2012; Ansong et al. 2017; Blake et al. 2017). These writings tend to conceive cultural values and related benefits as non-material values (delivering intangible benefits) placed on marine environments that tend to generate a sense of place and identity (perhaps realized as interests) (Gee et al. 2017). It is argued in this literature that MSP has largely overlooked culturally linked intangible values, which is problematic because they contribute to human well-being and are thought to have a strong bearing on how we conceptualize sustainability (Soini and Birkeland 2014). Most work undertaken either presents cultural service typologies and/or propose ways to better incorporate cultural values in MSP (adopting MEA typologies[6]), so as to take them into account in MSP decision-making (Gee et al. 2017). Excising such cultural values and benefits for consideration in MSP is problematic because they are never stand-alone but depend on practices and cultural frameworks to be reproduced, recognized and valued. As Kenter et al. (2011) comment, the range of benefits produced by cultural ecosystem services (as they are defined) are elusive to capture, particularly in natural resource management and planning approaches, such as MSP dominated by quantitative approaches as they are. Others such as Flannery et al. (2016), in an MSP context, discuss the undesirability of incorporating the social/cultural knowledge of fishers into what they characterize as 'formalized rational planning processes'. The argument here is that it is neither viable nor desirable to classify discrete cultural values through a spatialized zoning process, as MSP attempts to do. Furthermore, that such an approach

[6] According to Small et al. (2017), the MEA was central in advancing the cultural ecosystem service concept, by purporting to show how ecosystem degradation jeopardized human well-being and by developing a nomenclature that categorized and described the 'diverse services' that ecosystems provide to people.

would contradict indigenous epistemology, which, it is argued, sees the 'marine estate' as a 'contiguous area of land and sea (an unarticulated whole), that defines their cultural identity, and of which they are part' (Flannery et al. 2016, p. 130). What is more, it is suggested that the representation of socio-cultural values in predominantly spatial and economic terms misses not only the infinite spatialities and the intrinsic affective character of these values, as well as their incommensurability with material (economic) benefits, but may also contribute to the further marginalization of stakeholders who hold these values (Tafon 2017). The concerns reflected in Flannery et al.'s (2016) critique can be seen as a response to an observable tendency in MSP to privilege quantitative data (seen as objective knowledge of an external reality), without critically considering its limitations or biases (Saunders et al. 2017). This includes the natural sciences and some forms of socio-economic knowledge, which share a similar epistemology and are presented in a similar format—making them more amenable to policy decision-making and perhaps also to spatialized planning. Subjective knowledge such as that underpinning the construction and valuation of cultural ecosystem services are likely to be derived from (acknowledged) feelings and/or experiences through either individual or collective processes. This knowledge and related values is less amenable to quantification and inevitably considered suspect (i.e., irretrievably imbued with what are unrevealed interests) in governance process such as MSP.

A Foucauldian perspective sees that knowledge is power and power is knowledge (i.e., the two concepts are not only inseparable, but parasitical on each other), thereby questioning the objectivity of knowledge regardless of how it is produced (i.e., by scientific methods or not), what form it is in (quantitative, qualitative etc.) or whom it appears to serve (Clegg et al. 2014). The corollary of this view is that actors who have claims to truth through the legitimacy of their knowledge can exercise power in circumstances where these truth claims apply; likewise actors with more power resources can easily impose the 'objectivity' and thus, legitimacy of their knowledge (Foucault 1980). In MSP settings in our case, where quantitative, expert knowledge commonly prevails, incorporating socio-cultural values and interests into MSP would not necessarily mean privileging the associated knowledge or falling into endless relativism, but it would mean that a wider gamut of values associated with human well-being and social groups' welfare (both material and non-material) could be subject to the planning process. So, the key aim of MSP 'would not be to admit and consider 'unequivocal knowledge', but to generate dialogue and exchange in decision-making that leads to more equitable outcomes (Kidd and Ellis 2012: 50). Here, 'socio-cultural' would go beyond listing a menu of values but also distinguish between notions of history, attachment, social justice, productive practices, voice and how these social phenomena dialectically interact with non-human nature in marine space and in turn affect human, community and societal well-being. Additionally, such an approach would distinguish between objects of value, held values and the valuing process. This would require broadening the scope of what is valued in MSP, how it is valued as well as what is considered 'valid knowledge'.

The concerns in the literature mostly seek to answer how to effectively give expression (and consideration) to these mostly identity and place-based values in MSP. Problems that have been confronted in MSP in including cultural values and benefits have to do with the ambiguously broad range of elements captured by the category (ranging from education benefits, seascape aesthetics to spiritual benefits), some of which would seem to contradict the non-material (interpreted as non-economic) category that all socio-cultural benefits tend to be lumped into. How to give spatial expression, and relatedly value, to these widely differentiated experiences of human-nature interaction has proved difficult and to some degree contentious and are confronted with a myriad of conceptual and methdological problems, that is, the values are abstract, intangible and difficult to quantify (Blake et al. 2017). While commentators such as Small et al. (2017) propose that changing the referent from cultural ecosystem services to the more descriptive, 'non-material ecosystem services' would provide a more accurate label about what is meant, it does not resolve the deeper critiques and problems that have plagued the cultural ecosystems services approach. Key sticking points are: how to denote value in those society-nature experiences (captured in the concept of socio-cultural values) in a nuanced and sensitive way that can meaningfully reflect their broad contribution (benefits) to the amorphous notion of human well-being (in ways that are seen to be legitimate by those who hold such values), and who are/should be the beneficiaries of such benefits?

The methodological (and more substantive) problems referred to above in most instances might be possible to be at least partially redressed by giving (more) equal consideration to monetary and non-monetary representations of value, as well as the development of approaches that allow for the recognition and elicitation of shared, plural and cultural value—some of which may 'never' be 'on the table' for negotiation. One approach that may have merit in a spatialized planning context, such as MSP (particularly in coastal contexts) is participatory mapping (using deliberative interaction) which would support spatial consideration of often specific and localized knowledge and values not suited to the more abstract monetary evaluation approaches (cf. Kenter 2016;

Blake et al. 2017). As an added value, such knowledge will ultimately increase our understanding of ecological processes, which is a significant problem in MSP.

2.3 Planning for Equity

The problems of integrating equity into MSP not only lie with methodology (how can we do it?) but also interpretations of what social justice is in deciding how benefits should be distributed. Equity here implies a need for fairness in the distribution of resources and the entitlement of everyone to an acceptable quality and standard of living (Beder 2001). According to Fainstein (2010) an equity perspective not only insists that stakeholders should be treated fairly in both MSP process and outcomes, but also that already socially and economically disadvantaged groups are not further disadvantaged. Most of the critique of MSP has been on calls to reform the role of social agency (participatory influence), whilst underplaying the importance of political economy (as a capacity to act). Arguably this emphasis may work to inflate claims for the possibilities of social action and change, so here we contribute to correcting this bias.

MSP has been accused of marginalizing weak and/or excluding stakeholders in decision-making processes (Flannery et al. 2018). This variously refers to low socio-economic groups, immigrants, traditional users or the less well educated, indigenous groups as well as to professional groups without a strong voice. A prominent example is small-scale artisanal fishers who have found it difficult in many MSP settings to represent their interests against other, more powerful groups (Jentoft 2017). So, while we focused on fairness of process in the Deeping Democratic Decision-making in MSP section above, this in insufficient, as the fairness of the distribution of resources or the relative deprivation compared with others must also be part of a social sustainability agenda (Halpern et al. 2013). Thinking about equity as outcomes in MSP helps us to put into focus how to interrogate the role of MSP in distributing benefits and costs across the differences axes of society. Views about what may constitute an equitable planning in MSP will differ. Equity acknowledges that individuals and social groups start from different places, histories, inheritances, positions of discrimination, marginalization, advantage and so on. So, equity in MSP could be seen as not doing more harm to already disadvantaged or vulnerable social groups and making decisions about the sea towards equality (acknowledging that people/groups flourish in different ways; relying on different values/benefits/conditions).

Critical planning theorists argue that, even if democratic deliberative planning recognizes the unequal power of different stakeholders to engage in planning, it is still unlikely to result in more equitable MSP outcomes.[7] The kernel of argument is that the huge power disparity between stakeholders in terms of their economic resources, education, status and organizational skills is likely to undermine the possibility of reaching equitable outcomes through deliberative processes. In other words, more democratic planning is likely to fail to challenge embedded power imbalances and therefore sustain inequality. While we see this as a plausible position to hold, we do not see these two efforts at social sustainability reform to be mutually exclusive and therefore argue that efforts should be made for MSP to be both more democratic in process and equitable in distribution of outcomes. Of course, as argued earlier, it is important to adjust the political economy of participation, in terms of levelling stakeholders' power resources (economic and symbolic) to see their interests considered in participation.

While here we mostly refer to equity in terms of 'current' distributional concerns, we should also touch on intergenerational equity as a key SD principle. A key role of MSP is to provide a basis for marine use that takes account of current uses, while being future oriented. This ambition, to 'balance' between the consideration of current imperatives and desirable future states, is similar to the intergenerational aims and orientation of SD. In an ideal sense, MSP can be seen as both facilitating and giving certainty to desirable future marine activities, while ensuring that such activities do not impinge on achieving 'good environmental status' and/or undermine the conditions of social sustainability. In discussing MSP and sustainability, Qiu and Jones (2013) argue that the environment can be depicted either as a competing sectoral interest ('soft sustainability') or as a special concern with recognition of ecological limits that frame development possibilities ('hard sustainability'). This hard demarcation separating the two sides of the debate focuses on the degree of permissible substitutability between the economy and the environment or between 'natural capital' and 'manufactured capital', which has for a long time been a feature of the broader SD discussion. Where MSP lands in the 'hard/soft' debate in different settings is seen as a reflection of the relative importance accorded to different values and interests in marine planning. The hard sustainability approach then would, it is assumed, reflect a position that is concerned for future generations. That is, even though future generations may gain from maritime

[7] This point is reflected in Pansardi's concept of substantive political equality.

development and by association economic progress, such gains might be outweighed by environmental deterioration (Beder 2001). This dichotomy highlights the 'forgottenness' of the social pillar—that it is not even on the agenda of what is important to 'choose' between. One could ask, taking inspiration from Raworth's (2017) 'doughnut economics', where are 'fairness limits relating to the current generation in this debate? Incorporating the social more explicitly in the hard/soft debate and in MSP more generally requires not only thinking about future environmental impacts but a repurposing of MSP to place greater emphasis on engaged governance and more active consideration of the distributional effects of planning processes and decisions, particularly towards vulnerable social groups. Without adequate attention to where the costs and benefits of sea use flow (now and in the future), for example, it is likely that unequal socio-spatial distribution of social and environmental costs (Temper et al. 2015), may work to further disadvantage already vulnerable groups.

A key question for social sustainability therefore is how the costs and benefits of MSP decisions are distributed within society and where (and how) different elements of quality of life are affected (e.g. work, recreational access, aesthetics, money etc.) (Flannery et al. 2016). This reflects a broader concept of equity in MSP concerned with taking greater account of both material and non-material values and benefits linked to well-being and quality of life, rather than the current utilitarian notions embedded in MSP as a driver of economic growth. Here a multilevel analytical framework is likely to be important as distribution effects of MSP can vary considerably between individuals within a community, between communities and between regions. That is, it would require that explicit attention be paid to distributional fairness in MSP, including development of (deliberative) techniques able to indicate the likely distributional impact on different social groups of different MSP scenarios. That is, we would need to identify pertinent social groupings relevant to the MSP situation and be able to determine whether 'everybody' is getting a fair share of whatever there is to get. Embracing equity in MSP therefore throws up myriad complexities, including the challenge of 'visualising and mapping fairness' with a particular eye on the already disadvantaged, rather than assuming distribution will flow evenly across society. For example, a recent study in Germany, while not focused on disadvantaged groups, showed that the economic and employment benefits of offshore wind farming are not concentrated on the coast (as might be assumed) but distributed throughout the country (Weig 2017).

Another possible response to the equity problems in MSP, albeit one that would not deal with the manifest structural inequalities (as they variably exist in different settings), would be to engage 'MSP equity planners'[8] with the specific mandate to advocate equitable outcomes (or social sustainability) (Fainstein 2010; Davidoff 1982; Tafon forthcoming). An equitable planning capacity would bring an equity lens to bear on all MSP processes, outcomes (including implementation). Otherwise, it is likely that the politically backed strategic imperatives of Blue Growth and the hard science-backed environmental imperatives are still likely to hold sway despite any claims of a rational planning capacity (or the role of a neutral planner) to objectively balance competing arguments in support of objectives across the triad of sustainability dimensions.

2.4 Social Cohesion

MSP has been much concerned with achieving planning coherence between different MSP administrative areas (cross-boundary/border integration) but far less with social cohesion within MSP jurisdictions. Indeed, what contribution MSP can make to social cohesion is a vexing but nevertheless important question if we are to further thinking on social sustainability in MSP. In an overarching sense, social cohesion concerns the processes (i.e., shared views, values, norms perceptions and behaviours) underpinning social relations (individuals, social groups, communities etc.) (Prell et al. 2009). It should be observed that 'too tight a lock in' when striving for broader social cohesion as a nation building project could result in dire conservatism. To avoid this, we argue that social cohesion at a policy operational level must embrace diversity and equity; otherwise there is a risk that such a programme could manifest as a hegemonic programme of assimilation towards dominant interests and cultural identities, rather than fostering the contribution of social innovation, including citizens' practices that incorporate counter-hegemonic values and ideals.

Social cohesion as a sustainability concern is interested in improving the structure and quality of societal relations. More particularly, it is concerned to accommodate diversity while promoting equality. Conceptually it connects to the diversity alluded to in the socio-cultural dimension and is concerned with how

[8] Derived from Davidoff's (1965) idea of the advocate planner, where the notion of a neutral planner (in urban planning settings) was not seen as a feasible. In advocacy planning, planners explicitly advocate particular issues (Davidoff placed particular focus on the poor and disadvantaged groups in articulating this position). In refining his view, Davidoff (1982) later saw the key role of the planner to plan and organize for social equity.

diverse social groups can interact to support the flourishing of new (equitable) social relations. Berger-Schmitt (2002) and Fainstein (2010) distinguish two core dimensions of social coherence: (1) an equality dimension involving reducing disparities and combating social exclusion; and (2) a social capital dimension aimed at strengthening social relations, interactions and ties. As Fainstein (2010) describes, (social) cohesion alludes to social bonds and trust. It implies inclusion, which infers that inequality is likely to further fragmentation and/or decrease trust in society. Social inclusion seeks to embrace differences between diverse social groups along a range of axes, such as place-based identities, ethnicity and gender among others. Inasmuch as it relates to participatory processes, it is also concerned with how such processes result in just outcomes for various social groups. In an MSP context, a concern for social cohesion would mean concern for how to foster collaborative planning, to ensure different social groups can exchange views with the possibility of engaging in social learning to aid mutual understanding and connections. In cases of intractable conflict, it might involve re-representing what problems are, so that the MSP process does not exacerbate existing schisms in society by exclusionary processes or by reinforcing existing privileges (intentionally or unintentionally). For example, in Poland, reframing 'the fisheries problem' in the Polish National MSP as protecting a fragile ecosystem (the Baltic Sea) where small-scale fisheries can flourish (and contribute to coast sustainability by reproducing cultural heritage and related knowledge and practices) may generate the conditions conducive to forming less agonistic relations in MSP in general and in relation to other MSP stakeholders, such as conservation NGOs (Saunders et al. 2016). Even if problem representations are not reconstituted in such a dramatic way as suggested in the Polish example, this point highlights the importance of meaningful recognition and inclusion of stakeholders in thinking about what the problem is that MSP is trying to solve.

3 A Proposal for a Social Sustainably Conceptual Framework

While the social sustainability features described in Table 8.1 clearly interrelate and overlap in practice, the conceptual thinking underpinning each of them is distinctive and when taken together they contribute towards conceiving social sustainability as a pillar of sustainability—covering both substantive and procedural aspects of MSP. The conceptual framework synthesizes the discussion of each social sustainability dimension discussed above and poses

Table 8.1 A conceptual approach to examine social sustainability in MSP

Social sustainability feature	Description	Related questions	Analytical insight
Deepening democracy in decision-making	Who or what is included and how they are included.	• Are the rationales for participation articulated? • What are the specific conditions and arrangements for participation (in relation to a specific MSP process or event)? • Who is included or excluded? • What issues are organized into, or out of, debate? • What emerges from the enactment of participatory practice? • How do informal and formal processes interact? • How broad are the participatory spaces? • What provisions are there for including capacities/ knowledge from different parts of civil society? • Are there processes to weigh up different types of knowledge input in terms of credibility and salience? • Are the opportunities for ongoing exchange of views and experiences between experts and non-experts?	Whose interests (legitimately) matter in specific MSP contexts (inclusion/exclusion), including the range of actors' values, experience that is considered in MSP, including non-expert knowledge
Meaningful inclusion of socio-cultural values, knowledge and benefits	Consideration of diverse, usually placed-based values, knowledge and benefits (material and non-material)	• Has particular consideration been given on how to include different socio-cultural knowledge? • How are material and non-material benefits spatially valued and measured? • How broad is the range of socio-cultural values and benefits included? • How is socio-cultural (including experiential) knowledge weighted and evaluated in MSP? • How are social groups with particular socio-cultural relationships to place included, specified and considered in MSP decision-making?	Whether (and how) particular social groups, placed-based knowledge, values and benefits have been effectively considered and reflected in MSP outcomes?

(continued)

Table 8.1 (continued)

Social sustainability feature	Description	Related questions	Analytical insight
Planning for equity	The distributional effects (now and in the future)	• How are vulnerable social groups (potentially) affected and dealt with, and were they empowered to participate? • Is there explicit/specific acknowledgement/consideration of distributional implications (who wins and who loses)? • How are the interests of future generations reflected/represented in MSP processes? • Is there commitment to protect future generations by consideration of the implications of current planning scenarios on future generations (socio-environmental scenario planning)?	Indicates the extent that a diversity of social concerns are mapped and engaged within MSP, including how fairly they are considered in the planning processes.
Social cohesion	Creating opportunities that promote the harmonious societal coexistence or, at least, minimize potential for explicit and socially harmful conflict	• Has there been commitment and effort to deal with conflict/tension through transparent and deliberative means? • Has there been an exercise to map and make tangible connections between different social groups' values and interests in marine planning processes? • Does MSP explicitly consider involvement of and implications for different socio-cultural groups? • Have there been specific actions to redress or minimize negative social consequences of Blue Growth or environmental protection (win-win)? • Is there a commitment to minimize the potential of socially harmful tension/conflict in MSP decision-making?	The extent to which consideration has been given to reducing existing fault lines of societies, fostering social ties and building of trust between different social groups.

questions that could begin to direct an empirical examination in specific MSP contexts. This chapter does not present the only possible interpretation of social sustainability as it relates to MSP (e.g., it does not explicitly address social inclusion, social capital, environmental justice, health, education, empowerment, well-being). For instance more radical conceptions of social sustainability in MSP deriving theoretical inspiration from, for instance, Marxist political economy (radical material redistribution) or post-structuralist theory (radical democratic transformation), among others. The framework (as presented here or in an adapted form) could be used as a platform for theory building to analyse social sustainability in MSP, underpinned by alternative theorizing adapted to suit different analytical or practice-based objectives. Granting greater visibility and tangibility to social sustainability as an issue of concern could be achieved by generating more detailed analysis of social sustainability in MSP practice. A reiterative approach that expands and enriches conceptual thinking through learning from wider discussions about aspects of social sustainability in conversation with ongoing examination of MSP practice is needed.

Economic, environmental and political crises at a local or broader scale may also influence social activity at the local scale. Focusing on the contributory factors of urban social sustainability highlights scale as an important issue. A number of factors can relate to multiple scales: social cohesion is often discussed at a national scale, employment at city or district scale, while others such as social interaction and local environmental quality relate to activity and places on a local and spatial scale. Economic, environmental and political crises at a local or broader scale may also influence social activity at the local scale. Focusing on the contributory factors of urban social sustainability highlights scale as an important issue. A number of factors can relate to multiple scales: social cohesion is often discussed at a national scale, employment at city or district scale, while others such as social interaction and local environmental quality relate to activity and places on a local and spatial scale.

4 Concluding Remarks

In the short term at least, it is highly unlikely that MSP will shift from a focus on supporting economic growth coupled with a concern for environmental protection. A programme to give greater attention and increased legitimacy to social sustainability in MSP is confronted by several challenges, some of which have been touched on in this chapter. Amongst these are the lack of agreement on what social sustainability means. Bebbington and Dillard (2009) argue for instance that social sustainability poses greater problems in terms of specification, understanding and communication (and perhaps quantification) than the economic (as growth) and environmental (as protection) pillars of sustainability. Given that MSP is largely a state endeavour, this problem manifests itself in how to build common norms of social sustainability within borders. Building such norms across borders is likely to be even more difficult with countries at different stages of economic development and with different traditions, capacities and views about participatory democracy and welfare (redistributive) economics. Also, as we have noted, there is also the risk of internal contradictions within the social sustainability pillar itself, as calls for present generation equity (maritime development and redistribution) clash with intergenerational equity (environmental protection).

This does not mean that increased consideration cannot be given to fulfilling social sustainability ambitions within such a framework. Achieving enhanced focus on equity of outcomes from MSP is likely to be more difficult than deepening stakeholder engagement given that some (limited) strides have already been made in this direction—although creating genuine spaces where democratic struggle (agonistic decision-making incorporating contestability, openness, and/or strides towards less inequality in decision-making) is still likely to prove difficult to achieve in many settings. Incumbent interests in science, government and industry are unlikely to willingly cede power to focus more on equity; so if social innovation in MSP is to come about, an ensemble of strategies will need to be drawn on to reconfigure existing power relations at various levels. Flyvbjerg et al. (2012), drawing on Foucault, suggest that we should look for 'tension points' or 'lines of fragility in the present' that could be exploited to open possibilities for transformation.

In the concluding part of a chapter like this, a list of strategies for transformation is always going to be inadequate (demanding more serious and extended treatment), but here we proffer some actions for change, including through advocacy in existing MSP forums, awareness and conscious raising campaigns (e.g., of uneven power relations across scales, inequitable treatment, conflicts), giving greater visibility to initiatives that are seeking more democratic and equitable change (Gaventa 2006). Scholars as activists and analysts could *humbly* play a key role in highlighting who is excluded or disadvantaged through uneven distribution, as well as, supporting grass-roots mobilizations to magnify their voices with technical/institutional knowledge. While it is contested, arguably such a shift in thinking in resource planning has occurred elsewhere, for example, in Bolivia and Ecuador through the 'neo-extractivist movement', where governance transitions in these countries expressly look to pursue a greater focus on social recognition, inclusion and equity, while pursuing growth through sustainable resource use (Acosta 2013). The notion of an 'equity planner' may also promise a 'rebalancing' of emphasis in MSP. However, unless this is undergirded by a more constitutive commitment to addressing features of social sustainability discussed in the chapter, it is unlikely to result in a significant reorientation of the existing MSP preponderancy of balancing environmental protection with economic growth.

Acknowledgements We would like to thank the editors of this book and reviewers of our chapter for their helpful and constructive comments to improve the text presented here. This work was the result of (1) the BONUS BALTSPACE project and was supported by BONUS (Art 185), funded jointly by the EU, Swedish Research Council FORMAS and other Baltic Sea national funding institutions and (2) research funding from The Foundation for Baltic and East European Studies.

References

Acosta, A. (2013). Extractivism and Neoextractivism: Two Sides of the Same Curse. In M. Lang & D. Mokrani (Eds.), *Beyond Development. Alternative Visions from Latin America* (pp. 61–86). Amsterdam/Quito: Transnational Institute/Rosa Luxemburg Foundation.

Agyeman, J., Bullard, R., & Evans, B. (2003). *Just Sustainabilities: Development in an Unequal World*. London: Earthscan/MIT Press.

Ansong, J., Gissi, E., & Calado, H. (2017). An Approach to Ecosystem-Based Management in Maritime Spatial Planning Process. *Ocean and Coastal Management, 141*, 65–81.

Bebbington, J., & Dillard, J. (2009). Social Sustainability: An Organization Level Analysis. In J. Dillard, M. King, & V. Dujon (Eds.), *Understanding the Social Dimension of Sustainability* (pp. 157–173). Taylor and Francis.

Beder, S. (2001). Equity or Efficiency. *Engineers Australia*, May, p. 39.

Berger-Schmitt, R. (2002). Considering Social Cohesion in Quality of Life Assessments: Concept and Measurement. *Social Indicators Research, 58*, 403–428.

Blake, D., Augé, A. A., & Sherren, K. (2017). Participatory Mapping to Elicit Cultural Coastal Values for Marine Spatial Planning in a Remote Archipelago. *Ocean and Coastal Management, 148*, 195–203.

Bond, S. (2011). Negotiating a "Democratic Ethos": Moving Beyond the Agonistic: Communicative Divide. *Planning Theory, 10*(2), 161–186.

Boström, M. (2012). A Missing Pillar? Challenges in Theorizing and Practicing Social Sustainability. *Sustainability: Science, Practice, & Policy, 8*, 3–14.

Campbell, S. (2013). Sustainable Development and Social Justice: Conflicting Urgencies and the Search for Common Ground in Urban and Regional Planning. *Michigan Journal of Sustainability, 1*(1), 75–91.

Campbell, S. (2016). The Planner's Triangle Revisited: Sustainability and the Evolution of a Planning Ideal That Can't Stand Still. *Journal of the American Planning Association, 82*(4), 388–397.

Clegg, S., Flyvbjerg, B., & Haugaard, M. (2014). Reflections on Phronetic Social Science: A Dialogue Between Stewart Clegg, Bent Flyvbjerg and Mark Haugaard. *Journal of Political Power, 7*(2), 275–306.

Davidoff, P. (1965). Advocacy and Pluralism in Planning. *Journal of the American Institute of Planners, 31*, 331–338.

Davidoff, P. (1982). Comment. *Journal of the American Planning Association, 48*(2), 179–180.

Dobson, A. (2003). Social Justice and Environmental Sustainability: Ne'er the Twain Shall Meet? In J. Agyeman, R. Bullard, & B. Evans (Eds.), *Just Sustainabilities: Development in an Unequal World*. London: Earthscan/MIT Press.

Dryzek, J. S., & Niemeyer, S. (2012). *Foundations and Frontiers of Deliberative Governance*. Oxford: University Press.

Dryzek, J. S., & Pickering, J. (2017). Deliberation as a Catalyst for Reflexive Environmental Governance. *Ecological Economics, 131*(C), 353–360.

Fainstein, S. (2010). *The Just City*. Ithaca, NY: Cornell University Press.

Flannery, W., Ellis, G., Nursey-Bray, M., et al. (2016). Exploring the Winners and Losers of Marine Environmental Governance/Marine Spatial Planning: Cui bono? Etc. *Planning Theory & Practice, 17*(1), 121–151.

Flannery, W., Nealy, N., & Luna, L. (2018). Exclusion and Non-participation in Marine Spatial Planning. *Marine Policy, 88*, 32–40.

Flyvbjerg, B., Landman, T., & Schram, S. (Eds.). (2012). *Real Social Science: Applied Phronesis* (pp. 285–297). Cambridge: Cambridge University Press.

Forester, J. (2006). Exploring Urban Practice in a Democratizing Society: Opportunities, Techniques, and Challenges. *Development South Africa, 23*(5), 569–586.

Forester, J. (2009). *Dealing with Differences: Dramas of Mediating Public Disputes*. New York: Oxford University Press.

Foucault, M. (1980). *Power/Knowledge: Selected Interviews and Other Writings 1972–1977* (C. Gordon, Ed.). London: Harvester.

Fung, A., & Wright, E. O. (Eds.). (2003). *Deepening Democracy Institutional Innovations in Empowered Participatory Governance, The Real Utopias Project IV*. London: Verso.

Gaventa, J. (2006). Finding the Spaces for Change: A Power Analysis. *IDS Bulletin, 37*(6), 23–33.

Gee, K., Kannen, A., Adlam, R., Brooks, C., Chapman, M., Cormier, R., Fischer, C., Fletcher, S., Gubbins, M., Shucksmith, R., & Shellock, R. (2017). Identifying

Culturally Significant Areas for Marine Spatial Planning. *Ocean & Coastal Management, 136*, 139–147.

Gilek, M., Saunders, F., & Stalmokaitė, I. (forthcoming). The Ecosystem Approach and Sustainable Development in Baltic Sea Marine Spatial Planning—The Social Pillar, a Slow Train Coming. In D. Langlet & R. Rayfuse (Eds.), *The Ecosystem Approach in Ocean Planning and Governance*. Netherlands: Brill Open.

Guerry, A., Ruckelshaus, M., Arkema, K., et al. (2012). Modelling Benefits from Nature: Using Ecosystem Services to Inform Coastal and Marine Spatial Planning. *International Journal of Biodiversity Science, Ecosystem Services & Management, 8*(1–2), 107–121.

Halpern, B. S., Klein, C. J., Brown, C. J., Beger, M., Grantham, H. S., Mangubhai, S., Ruckelshaus, M., Tulloch, V. J., Watts, M., White, C., & Possingham, H. P. (2013). Achieving the Triple Bottom Line in the Face of Inherent Trade-Offs Among Social Equity, Economic Return, and Conservation. *Proceedings of the National Academy of Sciences of the United States of America, 110*, 6229–6234.

Jentoft, S. (2017). Small-Scale Fisheries Within Maritime Spatial Planning: Knowledge Integration and Power. *Journal of Environmental Policy & Planning, 19*(3), 266–278.

Jones, P. J. S., Lieberknecht, L. M., & Qiu, W. (2016). Marine Spatial Planning in Reality: Introduction to Case Studies and Discussion of Findings. *Marine Policy, 71*, 256–264.

Kenter, J. O. (2016). Integrating Deliberative Choice Experiments, Systems Modelling and Participatory Mapping to Assess Shared Values of Ecosystem Services. *Ecosystem Services, 21*, 291–307.

Kenter, J. O., Hyde, T., Christie, M., & Fazey, I. (2011). The Importance of Deliberation in Valuing Ecosystem Services in Developing Countries—Evidence from the Solomon Islands. *Global Environmental Change, 21*, 505–521.

Kidd, S., & Ellis. (2012). From the Land to Sea and Back Again? Using Terrestrial Planning to Understand the Process of Marine Spatial Planning. *Journal of Environmental Policy and Planning, 14*(1), 49–66.

Kidd, S., & Shaw, D. (2014). The Social and Political Realities of Marine Spatial Planning: Some Land-Based Reflections. *ICES Journal of Marine Science., 71*(7), 1535–1541.

McCann, J., Smythe, T., Fugate, G., Mulvaney, K., & Turek, D. (2014). *Identifying Marine Spatial Planning Gaps, Opportunities, and Partners: An Assessment.* Narragansett, RI: Coastal Resources Center and Rhode Island Sea Grant College Program. 60 pp.

Metzger, J., Soneryd, L., & Linke, S. (2017). The Legitimization of Concern: A Flexible Framework for Investigating the Enactment of Stakeholders in Environmental Planning and Governance Processes. *Environment & Planning A: Economy and Space, 49*(11), 2517–2535.

Milanovic, B. (2013). Global Income Inequality by the Numbers: in History and Now. *Global Policy, 4*(2), 198–208.

Muraca, B. (2012). Towards a Fair Degrowth-Society: Justice and the Right to a 'Good Life' Beyond Growth. *Futures, 44*, 535–545.

Murphy, K. (2012). The Social Pillar of Sustainable Development: A Literature Review and Framework for Policy Analysis. *Sustainability: Science, Practice, and Policy, 8*(1), 15–29.

Olsen, E., Fluharty, D., Hoel, A. H., Hostens, K., Maes, F., & Pecceu, E. (2014). Integration at the Round Table: Marine Spatial Planning in Multi-stakeholder Settings. *PLoS ONE, 9*(10), e109964.

Pansardi, P. (2016). Democracy, Domination, and the Distribution of Power: Substantive Political Equality as a Procedural Requirement. *Revue Internationale de Philosophie, 70*(275), 91–108.

Piketty, T. (2014). *Capital in the Twenty-First Century*. Cambridge, MA: Harvard University Press.

Prell, C., Hubacek, K., & Reed, M. (2009). Stakeholder Analysis and Social Network Analysis in Natural Resource Management. *Society and Natural Resources, 22*, 501–518.

Qiu, W., & Jones, P. J. S. (2013). The Emerging Policy Landscape for Marine Spatial Planning in Europe. *Marine Policy, 39*, 182–190.

Raworth, K. (2017). *Doughnut Economics: Seven Ways to Think Like a 21st-Century Economist*. Chelsea Green Publishing Company.

Reed, M. S., Vella, S., Sidoli del Ceno, J., Neumann, R. K., de Vente, J., Challies, E., Frewer, L., & van Delden, H. (2017). A Theory of Participation: What Makes Stakeholder and Public Participation in Environmental Management Work? *Restoration Ecology*. https://doi.org/10.1111/rec.12541.

Ritchie, H. (2014). Understanding Emerging Discourses of Marine Spatial Planning in the UK. *Land Use Policy, 38*, 666–675.

Ritchie, H., & Ellis, G. (2010). A System that Works for the Sea? Exploring Stakeholder Engagement in Marine Spatial Planning. *Journal of Environmental Planning and Management, 53*(6), 701–723.

Saunders, F., Gallardo, G., Tuyen, T. V., Raemaekers, S., Marciniak, B., & Díaz, R. (2016). Transformation of Small-Scale Fisheries—Critical Transdisciplinary Challenges. *Current Opinion in Environmental Sustainability, 20*, 26–31.

Saunders, F., Gilek, M., Gee, K., Dahl, K., Hassler, B., Luttmann, A., Morf, A., Piwowarczyk, J., Stalmokaite, I., Strand, H., Tafon, R., & Zaucha, J. (2017). BONUS BALTSPACE Deliverable D2.4: MSP as a Governance Approach? Knowledge Integration Challenges in MSP in the Baltic Sea. Retrieved from https://www.baltspace.eu/files/BONUS_BALTSPACE_D2-4.pdf.

Scholsberg, D. (2004). Reconceiving Environmental Justice: Global Movements and Political Theories. *Environmental Politics, 13*(3), 517–540.

Small, N., Munday, M., & Durance, I. (2017). The Challenge of Valuing Ecosystem Services that Have No Material Benefits. *Global Environmental Change, 44*, 57–67.

Soini, K., & Birkeland, I. (2014). Exploring the Scientific Discourse on Cultural Sustainability. *Geoforum, 51*, 213–223.

Tafon, R. (2017). Taking Power to Sea: Towards a Post-structuralist Discourse Theoretical Critique of Marine Spatial Planning. *Environment and Planning C: Politics and Space*. https://doi.org/10.1177/2399654417707527.

Tafon, R. (forthcoming). Marine Spatial Planning in Poland: Clarifying the (not so) Uncooperative Fisher and the Role of the Advocate "Social" Planner.

Tafon, R., Howarth, D. R., & Griggs, S. (2018). The Politics of Estonia's Offshore Wind Energy Programme: Discourse, Power and Marine Spatial Planning. *Environment and Planning C: Politics and Space*. https://doi.org/10.1177/2399654418778037.

Tambakaki, P. (2017). Agonism Reloaded: Potentia, Renewal and Radical Democracy. *Political Studies Review, 15*(4), 577–588.

Temper, L., Del Bene, D., & Martinez-Alier, J. (2015). Mapping the Frontiers and Front Lines of Global Environmental Justice: The EJAtlas. *Journal of Political Ecology, 22*(1), 255–227.

UNESCO. (n.d.). Marine Spatial Planning Programme. MSP Around the Globe. Retrieved from http://msp.ioc-unesco.org/world-applications/overview/.

Weig, B. (2017). *Spatial Economic Benefit Analysis*. BONUS BALTSPACE Project Report.

Wright, E. O. (2010). *Envisioning Real Utopias*. London and New York: Verso.

Wright, E. O. (2013). Transforming Capitalism Through Real Utopias. *American Sociology Review, 78*(1), 1–25.

9

Politics and Power in Marine Spatial Planning

Wesley Flannery, Jane Clarke, and Benedict McAteer

1 Introduction

Over the past two decades, increasing industrialisation of the marine environment has intensified competition for marine space. This competition has largely been driven by the rapid growth in number and size of spatially fixed marine industries. For example, in the last decade the average size of European offshore wind farms has increased substantially, from 79.6 MW in 2007 to 493 MW for offshore wind farms under construction in 2017 (Wind Europe 2018). In 2015, the global production of aquaculture products was 106 million tonnes, which has been growing at an average annual rate of 6.6% since 1995 (FAO 2017). The rapid growth of both these industries has obvious socio-spatial consequences. For example, there is increasing concern that the growth of offshore wind farms will displace other activities such as fishing (Kafas et al. 2018). Current governance regimes are, however, sectoral and disconnected and, therefore, ill-suited for managing the rapid industrialisation of the marine environment and related issues of stakeholder conflict.

Until relatively recently, marine governance was highly fractured. Marine governance was sectorally divided, with different marine activities managed on an individual basis, and spatially fragmented, with the governance of contiguous marine areas (e.g. territorial sea and Exclusive Economic Zone) being divided across a number of agencies. MSP has developed as a place-based,

W. Flannery (✉) • J. Clarke • B. McAteer
School of Natural and Built Environment, Queen's University Belfast, Belfast, UK
e-mail: w.flannery@qub.ac.uk

J. Zaucha, K. Gee (eds.), *Maritime Spatial Planning*,
https://doi.org/10.1007/978-3-319-98696-8_9

integrated marine governance approach to address the issues that have arisen from sectoral and fragmented management, including increasing user conflicts (Ehler and Douvere 2009). MSP is also promoted as a means of addressing the democratic deficit within marine governance by providing a mechanism through which all those with a stake in marine management can participate in related decision-making processes (Pomeroy and Douvere 2008). In this way, MSP, as a concept, provides the opportunity to imagine a radically different form of marine governance—one that focuses on understanding the complex nature of stakeholder interactions in the marine environment and implements transparent, democratic decision-making.

While MSP, as a concept, promises to overhaul the existing management regime and introduces a new era of democratic, integrated marine governance, its implementation indicates that MSP, as a practice, fails to address issues of politics and power, blunting its radical potential. MSP has partly failed to achieve its radical potential due to the manner in which it has been promoted by the international community (e.g. UNESCO and EU) and implemented by national governments. Although MSP as a concept holds vast transformative potential, the asocial and apolitical framing of MSP all but nullifies its radical utility. While there undoubtedly remains great potential within MSP, "successful implementation can only come by way of acknowledging and addressing unequal power relations and social injustices" (Tafon 2017, p. 3). A number of marine governance scholars have thus appealed for an increased contribution from the social sciences to MSP research (Ritchie and Ellis 2010; Jay et al. 2012; Smith and Jentoft 2017; Tafon 2017; Kelly et al. 2018). We respond to the call for theoretically informed MSP research by arguing that such research must reconceptualise the role of politics and power within MSP processes, move beyond its asocial and apolitical framings and seek to develop ways through which the radical potential of MSP can be realised.

The next section provides an account of MSP and politics arguing that while MSP should be a deeply political process, it has been depoliticised through the adoption of post-political planning processes. This is followed by a deconstruction of 'rationality' within MSP. We argue that rationality is often a product of power and that MSP must acknowledge how the rationalities that underpin MSP are constructed within existing power relations. The chapter concludes with an overview of areas for further research which we think can contribute to realising a more democratised, progressive form of MSP, including how conceptualising MSP as a boundary object may reveal depoliticisation processes and how the use of citizen science may empower stakeholders to counter hegemonic MSP rationalities.

2 Politics and MSP

Marine governance is a political act through which actors negotiate their understanding of a particular problem. As Hajer (1995) demonstrates in his seminal work on environmental discourse, it is no longer a question of if there is an environmental problem, but more a question of how we frame its consequences and champion particular responses. In attempting to legitimise their understanding of problems, and rationalise particular solutions, actors utilise processes of discursive construction, persuasion and even coercion (Hajer 1995; Oreskes and Conway 2010; Metze 2014). Actors seek to ensure that their interpretation of an environmental problem becomes the dominant one, because the manner in which a problem is discursively constructed favours certain ways of acting, while preventing others (Fischer 2003). The same is true for marine planning. Marine problems and their solutions are constructed through discursive practices in which actors attempt to frame marine problems in their favour and limit the potential for rival discourses to take hold. Therefore, MSP is, first and foremost, a political process constituted by numerous discursive struggles to frame marine issues.

While problematising issues within MSP is a profoundly political act, recent MSP processes appear to be devoid of politics, with the logic of Blue Growth seemingly going unchallenged. This raises some fundamental questions for marine social researchers: How is MSP being depoliticised? Which processes are employed to ensure that MSP preserves the status quo? And how do we recapture the radical potentiality of MSP?

Efforts to depoliticise decision-making have been conceptualised as post-political processes (Žižek 1999; Rancière 1999; Mouffe 2005). Post-political processes refer to a situation in which debate and dissensus are increasingly sanitised or co-opted through consensual procedures (Wilson and Swyngedouw 2014). In essence, post-political processes describe a society in which the space of contest or struggle (the political) is increasingly overrun by the promotion of free-market economics and the uncritical adoption of consensual procedures. Post-political practices disempower stakeholders by replacing debate and dissensus with practices of governing concerned with "consensus, agreement, accountancy metrics and technocratic environmental management" (Swyngedouw 2009, p. 604). Post-political practices subsequently frame 'issues' as being beyond politics or as being no longer contestable. Instead, problems are grounded in an all-consuming model of free-market neoliberal capitalism (Wilson and Swyngedouw 2014), the continuation of which becomes the solution to all issues.

Planning afflicted by the post-political condition has been characterised as suffering from a number of highly interrelated symptoms, which function to remove conflicting alternatives from planning processes. These include the advancement of neoliberal policies, choreographed participation, technocratic managerialism, path dependency and the illusion of progressive change (Swyngedouw 2009, 2010, 2011a, b; Allmendinger and Haughton 2011). While this list is not intended to be exhaustive, it comprises the core symptoms of the post-political condition and can be found in emerging MSP practices (Ritchie 2014; Flannery et al. 2016, 2018; Tafon 2017). In the following sections, we discuss each of these symptoms and illustrate how they are apparent in MSP.

The continuation of a society based on free-market neoliberalism is both an aim and an outcome of post-political planning (Swyngedouw 2007; Purcell 2014; Wilson and Swyngedouw 2014; Beveridge and Koch 2017). Post-political processes frame contemporary problems *and* their solutions within the realm of neoliberal logic. Continued economic progression is positioned as the motivating factor behind all planning decisions (Raco 2014). This logic maintains that the neoliberal model cannot be altered, and the solution to any problems of this organisational structure is to be found within itself. The post-political condition is thus a neoliberally motivated art of government, attempting to suppress social orders other than free-market economics (Swyngedouw 2009). In practice, an uncritical neoliberal logic is being developed around MSP, with the dominant discourse framing it as a mechanism for facilitating Blue Growth, reducing the bureaucratic burden on developers and allocating the 'correct' space to industry. Rather than industry learning how to engage with a planning process that should act in the public interest and facilitate sustainable, just use of marine resources, marine planners are told that they "should go out to the sectors and learn to speak their language" (European MSP Platform 2017, p. 30). MSP then becomes a process driven by the logic, language and needs of elite stakeholders rather than by concerns about the public good. The pervasiveness of neoliberal logic allows little room for meaningful discussion about alternative, progressive MSP functions (e.g. environmental justice and coastal poverty alleviation).

Participation processes are carefully choreographed within post-political systems (Allmendinger and Haughton 2012; Raco 2014). In order to maintain a society centred around the promotion of neoliberal logic, post-political planning limits and manages the capacity of stakeholders to participate in decision-making processes. Tokenistic participatory planning is emblematic of the post-political condition (Mouffe 2005; Purcell 2008; Swyngedouw 2009; Ward et al. 2017). Elites choreograph tokenistic participation around a restricted vision of a society based on free-market neoliberalism (Swyngedouw 2011b; Raco 2014). To do this, conflicting alternatives are neutralised within participatory processes characterised by asymmetrical power structures (Mouffe 2005). Within these processes, debate is perceived as an unnecessary complication, and the hegemony is consequently legitimised through 'collaborative' procedures (Wilson and Swyngedouw 2014). These critiques of participatory procedures as being choreographed and tokenistic rhyme with analyses of emerging MSP practice. MSP initiatives have been described in evaluations of various initiatives as being a 'top-down' process (Jones et al. 2016), or characterised by centralised decision-making (Scarff et al. 2015), or the repackaging of historic power dynamics (Flannery et al. 2018).

Managerial-technological apparatuses are positioned within post-political planning as being capable of negotiating complex socio-environmental conflicts (Oosterlynck and Swyngedouw 2010). With the purpose of legitimising dominant agendas, social problems within post-political processes are reduced to technical issues to be overcome by experts (Wilson and Swyngedouw 2014). For example, the complexity of social-ecological relations in the marine environment are increasingly simplified through the use of mapping technologies (Smith and Brennan 2012) and captured in geospatial databases (Boucquey et al. 2016), creating problematic conceptualisations of relationships as being fixed and two-dimensional (Steinberg and Peters 2015). These Geographic Information System (GIS) databases are analysed by technical experts to make 'rational' decisions about marine issues that have been disembodied from their social contexts (Vonk et al. 2005). In this manner, MSP has been reduced to a mere technocratic exercise of allocating space in an efficient manner, dulling its potential for envisaging alternative marine futures.

Decisions made within post-political planning are often path dependent. Path dependency is the process of making decisions so that they fit with past decisions (Haughton et al. 2013). The capacity for planning afflicted by the post-political condition to provide a shift from the shortcomings of historic practices is limited, as it too often reapplies previous management paradigms that suit powerful stakeholders. Such restrictive decision-making is apparent within MSP implementation. For example, the fragmented licensing and management regimes, which gave rise to MSP, will remain in place even as EU member states begin to implement MSP.

Post-political planning provides an illusion of progressive change while maintaining the status quo (Allmendinger and Haughton 2011). While narratives of participatory and egalitarian practices stimulate visions of a new era of sustainable and radical planning, the reality of post-political planning is less remarkable (Swyngedouw 2009). Under decisions restricted by neoliberal logic, post-political planning fails to provide regime shifts towards sustainable development. By entrenching historic practices, post-political planning simply hides existing power dynamics and their affects behind the rhetoric of progressive change. MSP demonstrates this illusion of progressive change as some nations have merely propagated existing asymmetrical social structures under the pretence of participatory governance and do little more than implement the status quo (Flannery et al. 2016).

Emerging MSP practice contains many of the post-political symptoms outlined earlier. If we are to recover the radical potentials of MSP, research needs to explore the processes used to depoliticise MSP. To achieve this, MSP research needs to examine how and why a wide range of stakeholders' support processes which, according to evaluations, contain very little for them. To do so, researchers must explore how 'rationality' and power have been mobilised to shape particular forms of MSP.

3 Rationality, Power and MSP

Rationality and power are central forces within planning practice. These forces are intertwined and, ultimately, shape planning practice in favour of powerful actors. Rationality is offered as an appeal to reason (Flyvbjerg 1998), to accept some form of neutral logic within planning processes. Rationality, however, is context dependent and the context of rationality is power (Flyvbjerg 1998). As such, rationality, in and of itself, can never be viewed as a neutral or unbiased determination, as it will fundamentally be tied to some form of power (Flyvbjerg 1998). The rationalities constructed within planning processes are never impartial, they are "framed on specific, and often unarticulated assumptions and values" that reflect the hegemony or the interests of powerful actors (Flannery et al. 2016, p. 123). Rationality should, therefore, be read as "the legitimizing of power, rather than as a challenge to it" (Jones and Porter 1994, p. 2). As opposed to being a logic that can be deduced from reason alone, rationality is socially constructed within particular contexts which reflect prevailing power relations.

MSP has been presented as a logical idea whose time has come (Ehler 2018) and has been advanced on the basis of at least two main rationalities: (1) the adoption of space as a core component of governance will address issues arising from historic marine management practices; and (2) the adoption of participatory planning will address the democratic deficit in marine governance. However, as outlined earlier, MSP has been bent to suit particular agendas. The logic of these rationalities has been appropriated by powerful actors to shape MSP to their needs. To exemplify the interrelationship between rationality and power, we focus on the connection between MSP (as the embodiment of these two rationalities) and Blue Growth (as the embodiment of power).

The adoption of space as a core component of marine governance is viewed as a mechanism that can ensure the sustainability of marine environments by reducing user conflict and cumulative impacts on ecosystems through the "rational organization of the use of marine space" (Douvere 2008, p. 766). The phrase 'rational organisation' advances the idea that there is an unproblematic spatial logic that can be deployed to organise the many actors who compete for locations. This asocial and apolitical conceptualisation of MSP views it as a logical process, sitting above power, which will produce a rational use of marine areas. This logic is based on an uncritical understanding of the complex social processes that produce space, particularly how space is produced by power.

Power, conceptualised here as Blue Growth, deploys this rationality to further its agenda. Blue Growth is "a complex governmental project that opens up new governable spaces and rationalizes particular ways of governing" (Choi 2017, p. 37). Despite claims that it is a sustainable development paradigm, Blue Growth is increasingly grounded in the logics of capitalist growth with little or no attention being given to issues related to social inequalities (Silver et al. 2015). Blue growth problematises marine governance in terms of its capacity to create ocean and marine areas *for accumulation* (Silver et al. 2015) and structures marine governance around issues related to utility, efficiency and prosperity (Choi 2017). Responding to the Blue Growth agenda, MSP has become a technical issue, focused on the allocation of spaces for accumulation rather than on good governance. Due to the dominance of the Blue Growth discourse within the EU, the problems to be addressed by MSP no longer relate to good environmental governance, but, rather, are concerned with creating the appropriate conditions for the rapid expansion of particular industries. Specifically, these include ocean energy, seabed mining, blue biotechnology, coastal tourism and aquaculture (EC 2014). The marine problem is reduced to ensuring that there is no spatial conflict amongst marine sectors and that the most valuable sectors have access to the spaces they desire. Therefore, rather than being inherently logical, MSP champions a particularly narrow, neoliberal rationality that views space as merely a site of production.

MSP is rationalised as a mechanism for democratising marine governance, which can incorporate the values of all those with a stake in marine ecosystems and, simultaneously, produce consensus and win-win outcomes for conflicting stakeholders (Pomeroy and Douvere 2008; Carneiro 2012; White et al. 2012). This is an oversimplified conceptualisation of participatory planning, one that is removed from the *realpolitik* and exertion of power and influence that permeate natural resource management. The adoption of the logic of participatory planning ignores the way in which powerful actors can use the 'illusion of inclusion' to secure legitimacy for fundamentally undemocratic processes (Purcell 2009). Rather than being a device to overcome democratic deficits, critics of participatory processes argue that "the true purpose of public participation has again become legitimisation rather than involvement in decision-making" (Blowers et al. 2009, p. 312). For critics, these 'legitimatisation' processes are not concerned with strengthening the democratic nature of decision-making, but, rather, are used to co-opt the public into advancing the agendas of elite actors (Flyvbjerg 1998; McGuirk 2001).

Broad-scale participatory processes within MSP processes, wherein stakeholders have little influence, are used to gain legitimacy for Blue Growth objectives. Recent academic evaluations of participation in MSP portray the process as being implemented in a top-down, tokenistic manner, wherein local actors struggle to be valued within decision-making processes (Flannery and Ó Cinnéide 2012; Jones et al. 2016; Jentoft 2017; Smith and Jentoft 2017). These negative evaluations are leading to a growing academic concern "that MSP is not facilitating a paradigm shift towards publicly engaged marine management, and that it may simply repackage power dynamics in the rhetoric of participation to legitimise the agendas of dominant actors" (Flannery et al. 2018, p. 32). Within MSP, relations of power purposefully marginalise particular groups of marine actors and "herd their participation and ways of knowing toward achieving limited policy outcomes" (Tafon 2017, p. 1). Thus, MSP may turn into a 'zero-sum game' (Jones et al. 2016), failing to accomplish some of the democratic goals of 'good governance' which it reportedly aspires to (Jentoft 2017). Therefore, despite all the positivity associated with a shift to a new form of governance, there are growing doubts about MSP's capacity to progress truly democratic processes, particularly as it bends towards serving a narrow Blue Growth agenda.

As Smith and Jentoft (2017, p. 34) assert, "as the theoretical foundation of Marine Spatial Planning was being laid, the issue of power was arguably not sufficiently problematized". MSP, as it is currently operationalised, is neither a neutral nor an objective instrument to decide about conflicting claims. In reality, MSP, like many other systems that measure and organise sociopolitical spaces, may facilitate a model of governance that benefits some to the detriment of others (Jentoft 2017). It would, therefore, be more appropriate to discuss MSP as sites of politics and power (Tafon 2017) that focuses on the production of space. Accordingly, MSP scholarship and practice need to develop a relational understanding of marine space (Jay et al. 2012) and must understand how it is socially produced and how this is related to power.

4 Recentring Politics and Power in MSP Research

MSP offers the potential to reformulate marine governance regimes. There is a fundamental need, therefore, for MSP research to employ theoretical lens which can expose the post-political nature of these planning processes. MSP research needs to explore the processes used to frame MSP so that it favours elite stakeholders. Research must explore why a wide range of stakeholders support MSP processes which, according to evaluations, contain very little for them. To achieve this, issues of politics and power must be brought to the fore in MSP research. Here, we offer two approaches that may help recentre politics and power in MSP research and practice: (1) the adoption of a boundary object lens and (2) citizen science.

4.1 MSP as a Boundary Object

A more power conscious assessment of MSP would take account of how issues are framed within MSP processes, asking how do stakeholders and decision-makers arrive at particular framings, if these are widely accepted and who, if anyone, dominates this process? We suggest that framing MSP negotiations as a boundary object may offer valuable insights into these questions. Star and Griesemer (1989, p. 393) conceptualised 'boundary objects' as objects/things/concepts "which are both plastic enough to adapt to local needs and the constraints of the several parties employing them, yet robust enough to maintain a common identity". Their state of being does not derive from their materiality or tangibility but, instead, derives from the action that they afford (Star 2010). This facilitative character is exemplified by the role a map of California played within the process to establish the Berkley Museum of Vertebrate Zoology in 1907 (Star and Griesemer 1989). The map, as a boundary object, facilitated collaboration between professional biologists and amateur conservationists. While the map maintained its geo-political boundaries between uses, actors interpreted the internal meaning of the map differently; where professional biologists saw 'life zones', amateur conservationists emphasised trails, campsites and places to collect samples (Star and Griesemer 1989). The map's internal ambiguity, but common boundaries, enabled different types of knowledge from distinct actors to be brought together to provide a more complete understanding of the California area. The literature has broadly viewed boundary objects as exclusively facilitating collaboration. Yet by recentralising the role of power, a boundary object must also be understood as providing an illusion of collaboration (Carlile 2002; Oswick and Robertson 2006; Thomas et al. 2007). Such a deceptive role is apparent in Oswick and Robertson's (2006) re-examination of the analysis of public enquiry into the Piper Alpha disaster.[1] The resulting report acted as a boundary object utilising sense-making narratives similar to other reports to gain legitimacy, whilst reinforcing and legitimising structures of historic governance. Boundary objects must therefore be conceptualised as having the capacity to both enable and inhibit interactions (Hawkins et al. 2017).

We argue that MSP must be viewed as a boundary-spanning object that crosses multiple communities and disciplines and provides a platform through which marine problems are framed. By spanning multiple boundaries, MSP enables a 'common' understanding of these framings to be produced and, ultimately, accepted by a diverse range of communities. However, MSP will not span these communities in a neutral manner, and powerful actors can bend it to suit their needs. Adopting a boundary object lens will provide insights into processes that seek to define MSP so that it both achieves the goals of powerful actors yet remains sufficiently elastic so as to prevent debate and dissensus. Conceptualising MSP as a boundary object—something which brings diverse stakeholders together, which each view from their own perspective, yet negotiate a common understanding of—provides a theoretically driven analysis of the processes through which actors collaborate or act so as to deny the actions of others. Examining MSP in this way will facilitate a greater understanding and explanation of the processes of negotiating, co-option and domination that occur within MSP initiatives.

4.2 Citizen Science and MSP

The production and use of knowledge and rationalities within MSP initiatives also need to be examined. This must go beyond identifying power and should explore alternative approaches to knowledge production that could enhance MSP. While knowledge is only one of many resources in the power field, Gaventa and Cornwall (2001) highlight how it, more than any other, determines

[1] Piper Alpha disaster is the world's deadliest oil platform disaster, which resulted in the loss of 167 lives in 1988.

what is conceived as important. In the simplest of terms, it is knowledge which gives weight and legitimacy to particular rationalities. For instance, discourses of 'technical' or 'expert' knowledge are often utilised as a means of legitimising planning decisions. It is within these discourses that the exercise of power operates (Rose and Miller 1992). What is of most importance here is the manner in which knowledge is used by power. The power-knowledge nexus highlights how we come to understand things as being rational. "Power, quite simply, produces that knowledge and that rationality which is conducive to the reality it wants. Conversely, power suppresses that knowledge and rationality for which it has no use" (Flyvbjerg 1998, p. 36). Effectively, then, power has the ability to pick and choose which knowledge is needed for the particular context in question and, subsequently, produces the necessary rationality to create the desired 'reality'. "In modern societies the ability to facilitate or suppress knowledge is in large part what makes one party more powerful than another" (Flyvbjerg 1998, p. 36). MSP research must analyse how certain knowledges are produced and rationalised by powerful actors in MSP processes. Conversely, it should also explore avenues for the production of alternative knowledge and how it may be used to counter hegemonic thinking.

There appears, therefore, to be scope to examine the potential for an increased focus on stakeholder-driven knowledge production. One such example of this is citizen science. Generally, citizen science is seen as a means of opening up knowledge production. Citizen science in a marine context is not a new means of producing knowledge, yet key social aspects of the approach remain relatively under-examined and may provide avenues for instigating a more radical implementation of MSP. While it is important to examine how citizen science projects function, the types of knowledge they produce and where this knowledge goes, emphasis should also be placed on exploring the potential to view citizen science as a means of changing power balances within structures of marine governance. Changing power relations are exemplified when the knowledge produced by a citizen science project leads to a paradigm shift in marine governance. This may be most evident in a project that leads to an alteration of legislation or policy, but can also be achieved by challenging rationales and dominant discourse.

5 Conclusions

The potential of MSP has been lost due to the manner in which it has been translated into practice. MSP is increasingly implemented through post-political processes or used by powerful actors to further the Blue Growth agenda. There is a fundamental need to understand how MSP has been depoliticised and how power shapes MSP practice. We argue that research needs to go beyond describing MSP as a post-political process or as a site of politics and power, and that there is a need to develop critical MSP social science research that can offer avenues to recapture MSP radical potential. We believe this can be achieved by in-depth research into how objectives for MSP initiatives are negotiated in a seemingly non-political manner. Framing MSP as a boundary object around and through which stakeholders negotiate may offer some insight into this process. We also argue that there needs to be a greater understanding about the use of power and knowledge within MSP processes. Here we argue that there is a need to go beyond identifying how power operates and that there needs to be an exploration of more democratic forms of knowledge production within MSP processes.

Acknowledgements We are grateful to the project "Economy of maritime space" funded by the Polish National Science Centre for contributing the Open Access fee for this chapter and facilitating our discussions and preparation of the book. The contributions from Jane Clarke and Ben McAteer were derived from research funded by the Northern Ireland Department for Economy.

References

Allmendinger, P., & Haughton, G. (2012). Post-political Spatial Planning in England: A Crisis of Consensus? *Transactions of the Institute of British Geographers, 37*(1), 89–103.

Beveridge, R., & Koch, P. (2017). The Post-political Trap? Reflections on Politics, Agency and the City. *Urban Studies, 54*(1), 31–43.

Blowers, A., Boersema, J., & Martin, A. (2009). Whatever Happened to Environmental Politics? *Journal of Integrative Environmental Sciences, 6*(2), 97–101. https://doi.org/10.1080/19438150902981633.

Boucquey, N., Fairbanks, L., Martin, K. S., Campbell, L. M., & McCay, B. (2016). The Ontological Politics of Marine Spatial Planning: Assembling the Ocean and Shaping the Capacities of 'Community' and 'Environment'. *Geoforum, 75*, 1–11.

Carlile, P. R. (2002). A Pragmatic View of Knowledge and Boundaries: Boundary Objects in New Product Development. *Institute for Operations Research and the Management Sciences, 13*(4), 442–455.

Carneiro, G. (2012). Evaluation of Marine Spatial Planning. *Marine Policy, 37*, 214–229. https://doi.org/10.1016/j.marpol.2012.05.003.

Choi, Y. R. (2017). The Blue Economy as Governmentality and the Making of New Spatial Rationalities. *Dialogues in Human Geography, 7*(1), 37–41. https://doi.org/10.1177/2043820617691649.

Douvere, F. (2008). The Importance of Marine Spatial Planning in Advancing Ecosystem-Based Sea Use Management. *Marine Policy, 32*, 762–771.

EC. (2014). Infographics: Blue Growth. Retrieved May 2, 2018, from http://ec.europa.eu/maritimeaffairs/policy/blue_growth/infographics/.

Ehler, C. N. (2018). Marine Spatial Planning: An Idea Whose Time Has Come. In Offshore Energy and Marine Spatial Planning. In K. Yates & C. Bradshaw (Eds.), *Offshore Energy and Marine Spatial Planning* (pp. 6–17). London: Routledge.

Ehler, C., & Douvere, F. (2009). *Marine Spatial Planning: A Step-by Step Approach Towards Ecosystem-based Management*. Manual and Guides No 153 ICAM Dossier No 6. Paris: Intergovernmental Oceanographic Commission UNESCO IOC, 99 pp.

European MSP Platform. (2017). *Maritime Spatial Planning for Blue Growth How to Plan for a Sustainable Blue Economy? Final Conference Report*. Retrieved May 2, 2018, from http://msp-platform.eu/sites/default/files/20171123_msp4bg_conferencereport_0.pdf.

FAO. (2017). *FAO Aquaculture Newsletter, No. 56*. Rome: FAO.

Fischer, F. (2003). *Reframing Public Policy: Discursive Politics and Deliberative Practices*. Oxford University Press.

Flannery, W., & Ó Cinnéide, M. (2012). Stakeholder Participation in Marine Spatial Planning: Lessons from the Channel Islands National Marine Sanctuary. *Society and Natural Resources, 25*(8), 727–742.

Flannery, W., Ellis, G., Ellis, G., Flannery, W., Nursey-Bray, M., van Tatenhove, J. P., Kelly, C., Coffen-Smout, S., Fairgrieve, R., Knol, M., & Jentoft, S. (2016). Exploring the Winners and Losers of Marine Environmental Governance/Marine Spatial Planning: Cui bono?/"More than Fishy Business": Epistemology, Integration and Conflict in Marine Spatial Planning/Marine Spatial Planning: Power and Scaping/Surely Not All Planning Is Evil?/Marine Spatial Planning: A Canadian Perspective/Maritime Spatial Planning–"ad utilitatem omnium"/Marine Spatial Planning: "It Is Better to Be on the Train than Being Hit by It"/Reflections from the Perspective of Recreational Anglers…. *Planning Theory & Practice, 17*(1), 121–151. https://doi.org/10.1080/14649357.2015.1131482.

Flannery, W., Healy, N., & Luna, M. (2018). Exclusion and Non-participation in Marine Spatial Planning. *Marine Policy, 88*, 32–40. https://doi.org/10.1016/j.marpol.2017.11.001.

Flyvbjerg, B. (1998). *Rationality and Power: Democracy in Practice*. Chicago: University of Chicago Press.

Gaventa, J., & Cornwall, A. (2001). Power and Knowledge. In P. Reason & H. Bradbury (Eds.), *Handbook of Action Research: Participative Inquiry & Practice* (pp. 70–80). London: Sage Publications.

Hajer, M. A. (1995). *The Politics of Environmental Discourse: Ecological Modernization and the Policy Process*. New York: Oxford University Press.

Haughton, G., Allmendinger, P., & Oosterlynck, S. (2013). Spaces of Neoliberal Experimentation: Soft Spaces, Postpolitics, and Neoliberal Governmentality. *Environment and Planning A, 45*(1), 217–234.

Hawkins, B., Pye, A., & Correia, F. (2017). Boundary Objects, Power, and Learning: The Matter of Developing Sustainable Practice in Organizations. *Management Learning, 48*(3), 292–310.

Jay, S., Klenke, T., Ahlhorn, F., & Ritchie, H. (2012). Early European Experience in Marine Spatial Planning: Planning the German Exclusive Economic Zone. *European Planning Studies, 20*(12), 2013–2031.

Jentoft, S. (2017). Small-Scale Fisheries Within Maritime Spatial Planning: Knowledge Integration and Power. *Journal of Environmental Policy and Planning, 7200*, 1–13.

Jones, C., & Porter, R. (1994). *Reassessing Foucault: Power, Medicine & the Body*. New York: Routledge.

Jones, P. J., Lieberknecht, L., & Qiu, W. (2016). Marine Spatial Planning in Reality: Introduction to Case Studies and Discussion of Findings. *Marine Policy, 71*, 256–264. https://doi.org/10.1016/j.marpol.2016.04.026.

Kafas, A., Donohue, P., Davies, I., & Scott, B. E. (2018). Displacement of Existing Activities. In K. Yates & C. Bradshaw (Eds.), *Offshore Energy and Marine Spatial Planning* (pp. 88–112). London: Routledge.

Kelly, C., Ellis, G., & Flannery, W. (2018). Conceptualizing Change in Marine Governance: Learning from Transition Management. *Marine Policy, 95*, 24–25.

McGuirk, P. M. (2001). Situating Communicative Planning Theory: Context, Power and Knowledge. *Environment and Planning A, 3392*, 195–217.

Metze, T. (2014). Fracking the Debate: Frame Shifts and Boundary Work in Dutch Decision Making on Shale Gas. *Journal of Environmental Policy and Planning, 19*(1), 35–52.

Mouffe, C. (2005). *On the Political*. Routledge.

Oosterlynck, S., & Swyngedouw, E. (2010). Noise Reduction: The Postpolitical Quandary of Night Flights at Brussels Airport. *Environment and Planning A, 42*(7), 1577–1594.

Oreskes, N., & Conway, E. M. (2010). Defeating the Merchants of Doubt. *Nature, 465*(7299), 686.

Oswick, C., & Robertson, M. (2006). Boundary Objects Reconsidered: From Bridges and Anchors to Barricades and Mazes. *Journal of Change Management, 9*(2), 179–193.

Pomeroy, R., & Douvere, F. (2008). The Engagement of Stakeholders in the Marine Spatial Planning Process. *Marine Policy, 32*, 816–822. https://doi.org/10.1016/j.marpol.2008.03.017.

Purcell, M. (2008). *Recapturing Democracy: Neoliberalization and the Struggle for Alternative Urban Futures*. Routledge.

Purcell, M. (2009). Resisting Neoliberalization: Communicative Planning or Counter-Hegemonic Movements? *Planning Theory, 8*(2), 140–165. https://doi.org/10.1177/1473095209102232.

Purcell, M. (2014). Rancière and Revolution. *Space and Polity, 18*(2), 168–181.

Raco, M. (2014). The Post-politics of Sustainability Planning: Privatisation and the Demise of Democratic Government. In J. Wilson & E. Swyngedouw (Eds.), *The*

Post-political and Its Discontents (pp. 25–47). Edinburgh: Edinburgh University Press.

Rancière, J. (1999). *Disagreement: Politics and Philosophy*. Minneapolis: University of Minnesota Press.

Ritchie, H. (2014). Understanding Emerging Discourses of Marine Spatial Planning in the UK. *Land Use Policy, 38*, 666–675.

Ritchie, H., & Ellis, G. (2010). A System that Works for the Sea? Exploring Stakeholder Engagement in Marine Spatial Planning. *Journal of Environmental Planning and Management, 53*(6), 701–723. https://doi.org/10.1080/09640568.2010.488100.

Rose, N., & Miller, P. (1992). Political Power Beyond the State: Problematics of Government. *The British Journal of Sociology, 43*, 173–205.

Scarff, G., Fitzsimmons, C., & Gray, T. (2015). The New Mode of Marine Planning in the UK: Aspirations and Challenges. *Marine Policy, 51*, 96–102.

Silver, J. J., Gray, N. J., Campbell, L. M., Fairbanks, L. W., & Gruby, R. L. (2015). Blue Economy and Competing Discourses in International Oceans Governance. *The Journal of Environment & Development, 24*(2), 135–160. https://doi.org/10.1177/1070496515580797.

Smith, G., & Brennan, R. E. (2012). Losing Our Way with Mapping: Thinking Critically About Marine Spatial Planning in Scotland. *Ocean and Coastal Management, 69*, 210–216.

Smith, G., & Jentoft, S. (2017). Marine Spatial Planning in Scotland. Levelling the Playing Field? *Marine Policy, 84*, 33–41. https://doi.org/10.1016/j.marpol.2017.06.024.

Star, S. (2010). This Is Not a Boundary Object: Reflections on the Origin of a Concept. *Science, Technology, & Human Values, 35*(5), 601–617. Retrieved from http://journals.sagepub.com/doi/10.1177/0162243910377624.

Star, S. L., & Griesemer, J. R. (1989). Institutional Ecology, 'Translations' and Boundary Objects: Amateurs and Professionals in Berkeley's Museum of Vertebrate Zoology, 1907–39. *Social Studies of Science, 19*(3), 387–420.

Steinberg, P., & Peters, K. (2015). Wet Ontologies, Fluid Spaces: Giving Depth to Volume Through Oceanic Thinking. *Environment and Planning D: Society and Space, 33*(2), 247–264.

Swyngedouw, E. (2007). Impossible "Sustainability" and the Post Political Condition. In R. Krueger & D. Gibbs (Eds.), *The Sustainable Development Paradox: Urban Political Economic in the United States and Europe* (pp. 13–40). London: Guilford Press.

Swyngedouw, E. (2009). The Antinomies of the Postpolitical City: In Search of a Democratic Politics of Environmental Production. *International Journal of Urban and Regional Research, 33*(3), 601–620.

Swyngedouw, E. (2010). Apocalypse Forever? Post-political Populism and the Spectre of Climate Change. *Theory, Culture & Society, 27*(2–3), 213–232.

Swyngedouw, E. (2011a). Depoliticized Environments: The End of Nature, Climate Change and the Post-Political Condition. *Royal Institute of Philosophy Supplement, 69*(October), 253–274.

Swyngedouw, E. (2011b). Interrogating Post-democratization: Reclaiming Egalitarian Political Spaces. *Political Geography, 30*(7), 370–380.

Tafon, R. V. (2017). Taking Power to Sea: Towards a Post-structuralist Discourse Theoretical Critique of Marine Spatial Planning. *Environment and Planning C: Politics and Space, 36*(2), 1–16. https://doi.org/10.1177/2399654417707527.

Thomas, R., Hardy, C., & Sargent, L. D. (2007). *Artifacts in Interaction: The Production and Politics of Boundary Objects*. Advanced Institute of Management Research Paper No. 052 (p. 50). Advanced Institute of Management Research, London.

Vonk, G., Geertman, S., & Schot, P. (2005). Bottlenecks Blocking Widespread Usage of Planning Support Systems. *Environment and Planning A, 37*(5), 909–924.

Ward, L., Anderson, M. B., Gilbertz, S. J., McEvoy, J., & Hall, D. M. (2017). Public Stealth and Boundary Objects: Coping with Integrated Water Resource Management and the Post-political Condition in Montana's Portion of the Yellowstone River Watershed. *Geoforum, 83*, 1–13.

White, C., Halpern, B. S., & Kappel, C. V. (2012). Ecosystem Service Trade-Off Analysis Reveals the Value of Marine Spatial Planning for Multiple Ocean Uses. *Proceedings of the National Academy of Sciences, 109*(12), 4696–4701. https://doi.org/10.1073/pnas.1114215109.

Wilson, J., & Swyngedouw, E. (2014). Seeds of Dystopia: Post-politics and the Return of the Political. In J. Wilson & E. Swyngedouw (Eds.), *The Post-political and Its Discontents* (pp. 1–22). Edinburgh: Edinburgh University Press.

Wind Europe. (2018). *Offshore Wind in Europe: Key Trends and Statistics 2017*. Brussels: Wind Europe.

Žižek, S. (1999). *The Ticklish Subject: The Absent Centre of Political Ontology*. London: Verso.

10

Towards a Ladder of Marine/Maritime Spatial Planning Participation

Andrea Morf, Michael Kull, Joanna Piwowarczyk, and Kira Gee

1 Introduction: Why Participation and How Much

Marine/maritime spatial planning (MSP) is increasingly popular as an approach to address conflicts of use in the marine environment and to promote environmental sustainability and blue growth. How and why to involve marine users and society at large is a topic of debate among researchers (see also Chaps. 8, 9 and 13 in this book), but also a concern for practitioners and decision-makers tasked with implementing MSP.

A. Morf (✉)
Swedish Institute for the Marine Environment, University of Gothenburg, Gothenburg, Sweden

Nordregio, Stockholm, Sweden
e-mail: andrea.morf@gu.se

M. Kull
Nordregio, Stockholm, Sweden

J. Piwowarczyk
Institute of Oceanology, Polish Academy of Sciences, Sopot, Poland

K. Gee
Human Dimensions of Coastal Areas, Helmholtz Zentrum Geesthacht, Geesthacht, Schleswig-Holstein, Germany

J. Zaucha, K. Gee (eds.), *Maritime Spatial Planning*,
https://doi.org/10.1007/978-3-319-98696-8_10

Policy documents, such as e.g. global guidelines by the Intergovernmental Oceanographic Commission (IOC)[1], by the European Union (EU),[2] and national legislation, require the involvement of stakeholders and civil society in MSP. This resonates with a larger scale paradigm shift of the last decades from government to governance, understood here as the involvement of societal actors, in various related fields including planning (e.g. Stoker 1998; Selle 1996; Fainstein and Fainstein 1996; Sandercock 1998). Nature conservation and natural resource and environmental management have witnessed a similar shift, expressed for example in the discussion on co-management, Agenda 21, Integrated Coastal Zone Management or the Ecosystem Approach (e.g. Borrini-Feyerabend et al. 2004; NRC 2008). In recent years, a wide range of guidelines on stakeholder involvement in MSP has become available, often resulting from European projects on MSP,[3] which emphasise stakeholder involvement and technical advice as a key to success. However, very few of these project-based documents provide advice that is directly applicable in the more statutory contexts in which MSP in implemented.

Despite the generally greater routine in public participation (Quick and Bryson 2016; Bryson et al. 2012; Bingham et al. 2005), the principles and methods employed across countries and settings vary considerably. Participation can be seen as a discrete act or a set of practices with different elements, tools and methods used (Quick and Bryson 2016) and can be understood as simple provision of information, or deliberation or collaborative decision-making. Expected outcomes and benefits of participation are also diverse, ranging from more principled benefits (e.g. improving the legitimacy of decisions) to more practical benefits (e.g. improving the knowledge base for decisions), as well as efficiency gains (e.g. participation in conflict prevention). At a more theoretical level, the growing practice of participation has given rise to questions regarding the legitimacy and usefulness of participation, representation and inclusion and the nature and role of different kinds of knowledge and expertise. There are recognised challenges in designing

[1] See: http://msp.ioc-unesco.org/ and http://msp.ioc-unesco.org/msp-guides/msp-guides-overview/.

[2] EU Directive 2014/89/EU: "Member States shall establish means of public participation by *informing all interested parties and by consulting the relevant stakeholders and authorities, and the public concerned,* at an *early stage* in the development of maritime spatial plans, in accordance with relevant provisions established in Union legislation." (Article 9, Directive 2014/89/EU, *our emphasis*).

[3] See, for example, EU policy initiatives on integrated coastal management and MSP (EC 2002; EC 2014), EU financed MSP projects including a focus on stakeholder involvement (e.g. PartiSEApate, Baltic SCOPE, Pan-Baltic SCOPE, SIMCelt, Transboundary Planning in the European Atlantic (TPEA)), and the 2018 MSP workshop in Brussels and for the Baltic the HELCOM-VASAB MSP principles and guidelines (2010 and 2016).

participatory processes "well adapted to their context" (Quick and Bryson 2016, see also Chap. 13 in this book).

Arguably, MSP is a particularly complex context for participation both conceptually and practically. MSP is dealing with what is known as "wicked" problems (Rittel and Webber 1973). These are characterised by multiple dimensions, cross-boundary issues and ongoing change both in the natural environment and in the social sphere. This results in the need to take legitimate decisions within a context of high uncertainty and value conflicts (e.g. Jentoft and Chuenpagdee 2009). But although meaningful participation is a core requirement for MSP, it also needs to be effective, as it is costly in time and an effort for both planners and participants.

Presently, institutional systems and practice for MSP are evolving rapidly but unevenly (Kull et al. 2017; Janßen et al. 2018). Given the need to develop more participatory forms of MSP, it is therefore necessary to outline different degrees of participation to see what could, and possibly is being achieved in practice. This particularly applies to the transboundary context within which MSP is mostly taking place.

This chapter develops a structure for practitioners and researchers to systematically reflect on participation in MSP in transboundary contexts. It is based on the metaphor of "ladders of participation", which is broadly used to conceptualise and evaluate different degrees of participation. First, we analyse relevant ladders with the aim to extract key dimensions of participation for MSP. We then present current challenges of stakeholder involvement in MSP, using recent research from the Baltic Sea area from the projects BaltSpace (Morf et al. forthcoming) and Baltic SCOPE (Kull et al. 2017). Section 4 develops a ladder-based conceptual framework that could help assess to what degree various ambitions of participation are being achieved. We then conclude how the ladder could be developed further to assist a more systematic evaluation and learning for developing participation in cross-border MSP.

2 Ladders of Participation and Their Relevance for MSP

2.1 Examples of Ladder Metaphors

During almost five decades, scholars and practitioners in various fields have used ladders, stairways and other metaphors to describe and assess the degree of power sharing, interaction and inclusiveness in planning and environmental management. Following on from the classic example of Arnstein (1969), the concept has also been applied to MSP, such as the ladder by Kidd and McGowan (2013) or the PartiSEApate handbook (Matczak et al. 2014) to characterise the intensity of transboundary collaboration. Reed (2008, p. 2419), analysing participation typologies in environmental management, distinguishes four types of approaches to analysing participation based on the following dimensions:

a. different degrees of participation on a continuum (ladder types);
b. the nature of participation, identified by the direction of communication flows;
c. theory, distinguishing between normative and/or pragmatic participation; and
d. the objectives of participation (e.g. research or development-driven, planner or people-centred, diagnostic and learning-centred).

In order to identify suitable starting points for MSP, we review a selection of relevant ladders and related metaphors (Table 10.1), highlighting their conceptual development over time. Our selection focuses mostly on types (a) and (b), but not exclusively so. The table contains a short description of the ladders, extracting key dimensions that also seem important for MSP.

Our chosen starting point in the chronological list is the ladder by Arnstein (1969), which laid the groundwork for many later ladders or similar metaphors. It considers participation mostly from the perspective of power and interaction, starting from what can be called the "dark" side of planning, ascending towards increasing inclusion. The most basic distinctions are non-participation (unable to have a say but being treated or even manipulated into something), tokenism (receiving something but not necessarily exerting influence, described as informing, consulting, placation) and citizen power (implying real influence, by inclusion in a partnership, having power delegated or being in full control). One problem is that Arnstein's ladder appears to be one-dimensional, but it actually blends the power dimension with methods and functions of participation as well as a value judgement, considering the higher steps on the ladder more desirable. The "dark" side may also be overemphasised: information and consultation may in fact be a more neutral way of interaction between authorities and citizens rather than necessarily tokenistic.

The perspective of power and the ability to influence decisions has remained central in subsequent ladders such as those of Pretty (1995) and Selle (1996). Selle uses the metaphor of a stairway of participation in planning, where the

Table 10.1 Participation ladders and related metaphors and important dimensions raised

Author, title/ metaphor	Main content	Key dimensions of participation raised
Ladders	**Urban / land planning**	
Arnstein 1969 *"A Ladder of Citizen Participation"*	"Classic" ladder with eight rungs ranging from two types of non-participation (1. manipulation, 2. therapy), to graded tokenism (3. informing, 4. consultation, 5. placation), to three degrees of citizen power (6. partnership, 7. delegated power, 8. citizen control) (our numbering).	• Power: unequally distributed in planning. • Need to empower and include citizens in planning. • Forms of participation (including valuation). • Dark side and misuse of participation processes (placation, manipulation, tokenism).
Selle 1996 *Stairway of participation in planning*	Stairway of four steps building on each other, also mirroring a shift in participation paradigms over decades: 1. Information of affected (1960s); 2. Information of general public (1970s); 3. Activating participation (1980s); 4. Collaborative participation (1990s).	• Purposes of participation. • Degrees and forms of involvement in terms of power sharing. • Complementary steps: all steps are necessary, may differ in terms of target group and timing during a planning process. • Development over time and paradigm shift.
	Development	
Pretty 1995 *"A Typology of partici-pation: How people participate in development programs and projects"*	Seven degrees of people's involvement in development project work: 1. Manipulative participation (pretence); 2. Passive participation (being told what has been decided or already happened); 3. Participation by consultation; 4. Participation for material incentives (content decided externally); 5. Functional participation (interactive, shared decision-making, but major decisions already taken externally); 6. Interactive participation (joint analysis, participation as a right and not just a means to achieving project goals, more control over resources and outcomes); 7. Self-mobilisation (initiative independent from external institutions, taking advice, full control over use of resources).	• Power and purposes of participation • Roles and activities • Local knowledge • Power and empowerment
Beyond simple ladders	**Climate change adaptation**	
Collins & Ison 2009 *"Jumping off Arnstein's ladder: Social Learning as new policy paradigm"*	Four embedded ovoids connected in one place instead of a ladder from smaller innermost to larger encompassing: 1. Information 2. Consultation 3. Participation 4. Social learning. (our numbering)	• (Social) learning: need to think in terms of open communication and learning of larger groups, governance systems or whole societies. • Participation is not just about more power and authority. • No linear relation or hierarchy between levels of participation. • Change over time of policy problems, roles and responsibilities, requiring flexible structures and learning processes.

(continued)

Table 10.1 (continued)

Merging and meeting ladders	Natural resource management & planning	
Berkes 1994 *Levels of co-management in fisheries* Pomeroy & Berkes 1997 *"A hierarchy of co-management arrangements"* Carlsson & Berkes 2005 Berkes 2009 *"Evolution of co-management: the role of knowledge generation, bridging organisations and social learning"*	Seven levels of co-management (from bottom to top, our numbering): 1. Informing (community informed about decisions already made); 2. Consultation (start of face-to-face contact, community input heard, but not necessarily heeded); 3. Co-operation (community starts to have input into management, e.g. use of local knowledge, research assistants); 4. Communication (start of two-way information exchange; local concerns begin to enter management plans), (divided in two in Pomeroy & Berkes 1997: 4. Communication, 5. Information exchange) 5.(6) Advisory Committees (partnership in decision-making starts; joint action on common objectives); 6. (7) Management Boards (community is given opportunity to participate in developing and implementing management plans); 7. (8) Partnership/Community Control (partnership of equals; joint decision-making institutionalised, power delegated to community where feasible); 9. Inter-area coordination (geographical expansion) (Pomeroy & Berkes 1997)	• Rights and roles (decision, use, ownership) in relation to sustainable resource management: multiple forms of graded property and use rights in relation to natural resources, including communal management. • Local and scientific knowledge need to be combined for efficient management. • Need for central government-based and community-based resource management to meet and interact (delegation, self-management). Both can need each other – for various reasons (support, knowledge, acceptance, legitimacy, long-term commitment to sustainable resource use). • Embedded systems of management require organisational cross-communication (Carlsson & Berkes 2005): exchange system - joint organisations - nested system. • Turn to adaptive co-management after 2008 adds a time and larger scale learning dimension, with emphasis on knowledge generation and exchange, individual and organisational and social learning, evolving management arrangements and adaptive management (Berkes 2009).
Hurlbert & Gupta 2015 *"The split ladder of participation"* see Fig. 1	A split, X or H-shaped ladder with the two rungs of information and consultation (including testing of ideas and seeking advice) as connecting middle rungs. Arnstein's ladder is extended, using a clear analytical structure. The ladder is formed around four logically connected main dimensions: a. the type of problems to be addressed in relation to agreement on values and agreement/certainty on knowledge, b. the level of learning needed, c. the level of trust, and d. the level of power sharing (management or governance). The ladder extends upward over 8-9 degrees of power sharing and left and right in relation to the level of trust and problem solving. The lower rungs indicate low participation, the higher ones high participation. The left side implies low trust and low problem solving and the right side high trust and problem solving. For details, see Fig. 10.1 and text.	• Level of power sharing (steps of ladder/from management to governance): Arnstein's original ladder lies across within moderately structured problems: the lower end on the low-trust side with zero-loop learning and low problem-solving and the higher end on high-trust side with double-loop learning and high problem-solving. • Type of policy problems: MSP development at present and strategic MSP problems reside most likely in the quadrants three and four, implying disagreement in knowledge and values. • Nature of learning in relation to the problems, from zero to triple-loop learning, requires adaptive management/governance. • Trust is important and related to interaction and learning.

(continued)

Table 10.1 (continued)

Ladders of	Marine spatial planning	
Kidd & McGowan 2013 *"A ladder of transnational partnership working to support MSP"*	The first MSP specific ladder, focusing on collaboration across borders and its degree of institutionalisation: from less formalised (bottom) to increasingly institutionalised (top). From bottom (our numbering): 1. Information sharing (building trust, understanding and capacity) ; 2. Administration sharing (creating collaborative advantages); 3. Agreed joint rules (constituting shared rule systems); 4. Combined organisation (changing the institutional order); 5. Combined constitution (changing the political order).	• Reflection on different dimensions: power, partnerships, process of development (and need of adaptation), relative formality (less formal as less aggressive). • Emphasising development of partnerships over time: degree of active interaction and delegation of responsibilities for this purpose to specific organisations. • Emphasising the need for a mix of different stakeholders to interact (including mandated authorities), focusing on organised actors. • Focus on transboundary collaboration and development of organised multi level multi-actor partnerships supporting MSP, less on formally required MSP participation.
Twomey & O'Mahony, 2018 *Continuum of Stakeholder Participation in European MSP*	Four levels of participation (our numbers) on a reversed ladder including a triangle figure to represent main actor types (government on top, industry and civil society on the bottom) and connecting arrows: 1. Informing and awareness raising without feedback (one-way process/top-down/arrows down). 2. Consultation providing feedback through statutory process (one-way process/top-down/arrows go up). 3. Stakeholder engagement working directly with government (two-way dialogue/top-down/arrows both ways). 4. Stakeholder collaboration (multi-sector dialogue/blend of top-down and bottom-up/arrows both ways in triangle) implying partnering with government.	• Focus is on power sharing and on dialogue and interaction between different parts of society (Government, Industry & Civil Society). • A continuum is more likely than ladder. All types of interactions may be needed and build on each other. • Combining and clarifying dimensions through triangles (who interacts with whom), direction of interaction (arrows), quality of interaction (text). • Need to combine top-down and bottom-up with users but also cross-level among authorities. • Does not address cross-border aspects and mandated key stakeholders (those with more strongly defined formal rights), and timing only to some extent.

bottom steps of information and consultation are seen as complementary and rights based. He also suggests that citizen participation has developed over time towards increasingly inclusive and interactive forms, while the bottom steps remain significant as a base. Collins and Ison (2009) add the dimension of learning, emphasising the need to think in terms of non-linear power and open communication and learning in larger groups and whole societies.

The natural resource co-management discourse adds further important dimensions to the ladders (e.g. Berkes 1994) from the perspective of user rights and functions of participation. Their point of departure differs slightly and grounds in natural resources and ecosystems. They also highlight the knowledge dimension, referring to different types of knowledge situated with science and authorities and resource users that need to meet and learning through adaptive co-management (Armitage et al. 2008).

Moving towards more recent examples in Table 10.1, ladder metaphors gradually become more sophisticated as further dimensions are recognised and specified. An important insight is that there is no linear progression between different levels of power. Citizen control, for example, is not necessarily the highest goal, as policy problems change and may require different levels of participation (Bishop and Davis 2002). Furthermore, as highlighted by Collins and Ison (2009), roles and responsibilities—and with them, degrees of power—are not absolute but may change during the participation process. Power is relational (Dyrberg 1997), circulating through all actors rather than distributed in a linear, hierarchical and one-directional manner (Foucault 1980). These aspects are also important in the context of MSP (see Sect. 4).

A particularly interesting recent ladder of participation is the "split ladder of participation", developed by Hurlbert and Gupta (2015) (see Fig. 10.1). Its attraction lies in its multidimensionality, bringing together many dimensions developed in earlier ladders, including power sharing, problem structure and learning. The ladder consists of four quadrants and four "pull factors". The lower quadrants (1 and 2) imply low participation and adaptive management, the higher ones (3 and 4) high participation and adaptive governance. The left-hand quadrants represent low trust and problem-solving capacity, the right-hand quadrants high trust and problem-solving capacity. Each quadrant is further

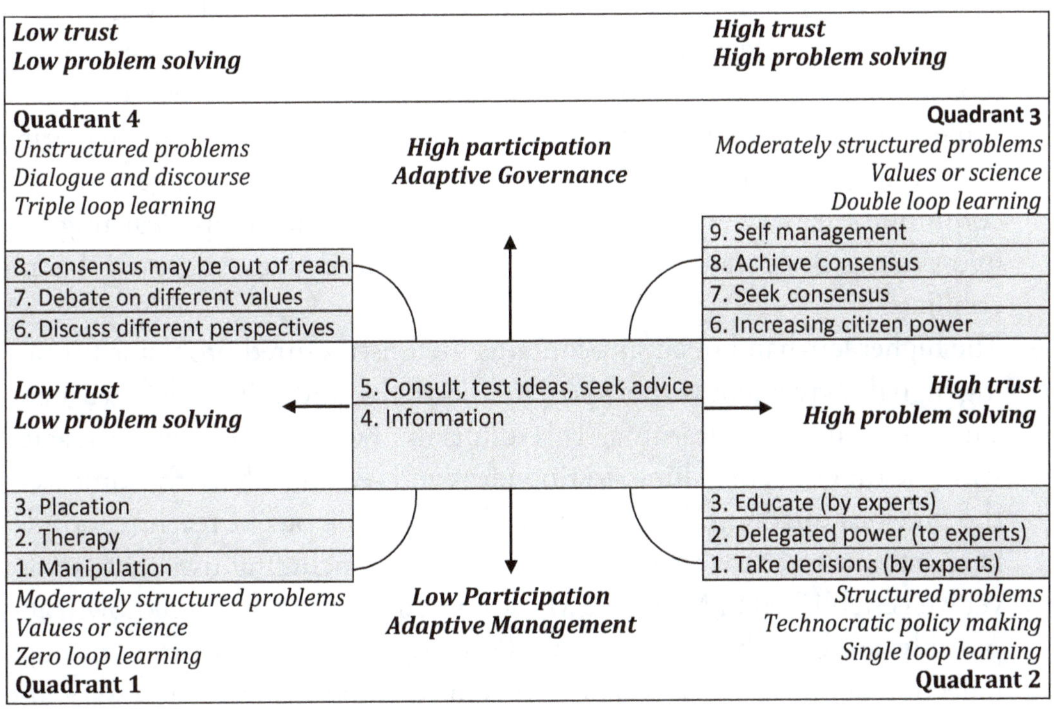

Fig. 10.1 The split ladder of participation (Source: Hurlbert and Gupta 2015, p. 105, adapted)

defined by the structure of problems, dialogue and discourse, and types of learning. The force of the arrows thus pulls the ladder apart into specific expressions of participation. We illustrate the quadrants with examples from MSP.

1. The lower left-hand quadrant covers moderately structured problems with disagreement on values or knowledge, low trust and zero-loop learning of both planners and society. This results in low problem-solving, described in the ladder as manipulation, placation and information. Examples could be controversial top-down planning of risk-filled infrastructure or "old style" conservation and resource management meeting resistance with resource users (e.g. fishers).
2. The lower right-hand quadrant covers structured problems (implying agreement on problems and solutions) and technocratic policymaking, with high trust and single-loop learning. This results in little participation but a high capacity in expert-led problem-solving (educate, delegated power, take decisions). Examples could be construction permits with moderate negative impacts, in situations without major value conflicts, where standard procedures work well, so decisions can in fact be delegated to experts, implying single loop learning in content by experts and public.
3. The higher right-hand quadrant encompasses moderately structured problems (implying disagreement on values or knowledge), high levels of trust and double-loop learning. This results in high problem-solving capacity (consult, test ideas, seek advice, increase citizen power, seek consensus, achieve consensus, self-management). Examples are comprehensive marine or coastal planning with value conflicts, adaptive co-management and community-based resource management. Double loop learning by questioning basic values and goals may be needed to address inherent conflicts.
4. The higher left-hand quadrant contains the unstructured problems (implying high disagreement with respect to knowledge and values), low trust and a need for triple-loop learning. This results in low problem-solving capacity, with the rungs of consulting, testing ideas and seeking advice the same as in the previous quadrant, but consensus here may be out of reach. Examples here are how to adapt coasts to climate change, including uses that are not yet well established in MSP and whose consequences are still unknown, and possibly also cross-border planning. Presently, triple loop learning through reflection about the process and inherent learning is highly necessary and apparently under way, especially in transboundary MSP.

2.2 Elements with Relevance to MSP

From the ladders and metaphors presented, we consider the following to represent key dimensions in analysing and discussing participation in MSP:

- *Problem type and related learning*: dealing with more or less wicked problems with knowledge gaps and value conflicts, requiring more or less direct interaction and learning *across sectors and levels* by individuals, groups, organisations and societies (Collins & Ison)—from zero- to double- to triple-loop learning depending on the type of problems to be addressed (Hurlbert & Gupta).
- *Trust* as promoted by the quality of the process in terms of openness, transparency, legitimacy (ibid.) and the need to interact between different groups (Twomey and O'Mahony 2018).
- The degree of *power sharing* between those in charge (all ladders) and those participating, depending on roles which may change over time (co-management discourse).
- *Functions and objectives* of participation (co-management discourse).

Important dimensions that are not captured sufficiently in the above and need to be added to the metaphors are:

- the *timing* of participation at different stages of MSP and
- the *spatial* context of participation: local, national, transboundary or cross-cutting (Kidd & McGowan). This aspect is particularly important in transboundary settings where cross-border participation may be required.

The following section illustrates the importance of these dimensions by referring to the Baltic Sea perspective.

3 Participation in Transboundary MSP: Conditions and Challenges in the Baltic Sea

The Baltic Sea region (BSR) can be seen as pioneering in institutionalising high level cross-border collaboration on the environment (Helsinki Commission [HELCOM][4]) and planning (Vision and Strategies Around the

[4] HELCOM (Baltic Marine Environment Protection Commission—Helsinki Commission) is the governing body of the Convention on the Protection of the Marine Environment of the Baltic Sea Area (www.helcom.fi).

Baltic Sea [VASAB][5]), including a common MSP Working Group (cf. Kidd and McGowan 2013; Zaucha 2014a). The last decade has seen a chain of transdisciplinary cross-border research and development projects on MSP in the BSR (for a summary, see Zaucha 2014b), where scientists and practitioners have been working together to understand problems and develop practice. A rapid process of institutionalisation of MSP is under way, driven both by planning and coordination needs and the EU's MSP Directive (Zaucha 2014a); this encompasses countries at very different stages of MSP development.

In relation to participation, two out of ten soft-law joint HELCOM-VASAB MSP principles refer to public participation: (1) Principle 5: Participation and Transparency and (2) Principle 7: Transnational coordination and consultation (HELCOM-VASAB MSP WG 2010). Principle 5 states that *relevant authorities and stakeholders*, including coastal municipalities and national and regional bodies, should be involved in MSP initiatives *as early as possible* and that *public participation* should be secured. Principle 7 calls for a *Baltic Sea perspective* in MSP, *pan-Baltic dialogue* in developing maritime spatial plans and *consultation between the BSR countries and the EU* (HELCOM-VASAB MSP WG 2010, *our emphasis*). The related guideline on transboundary consultations, public participation and co-operation (HELCOM-VASAB n.d.) specifies a number of principles departing mainly from an instrumental perspective on participation. Public participation—as defined by these guidelines—should aim to increase quality and acceptance of the public decisions and reduce tensions and disputes over the marine space and marine resources (HELCOM-VASAB n.d.).

However, when translating such general principles into concrete policy processes for the BSR, it becomes apparent that concepts such as "stakeholder", "participation" and "MSP" might not mean the same in each country (Kull et al. 2017). Moreover, the awareness of related problems may vary between countries, sectors and levels (Kull et al. 2017; Morf et al. forthcoming). Although collaboration in the BSR scores high on the collaboration ladder by Kidd and McGowan (2013), stakeholder involvement on a pan-Baltic level has so far tended to be formalist, and it can be difficult for new stakeholders to enter the process (Janßen et al. 2018).

Based on Janßen et al. (2018), Kull et al. (2017), and Morf et al. (forthcoming), the major obstacles and challenges in stakeholder involvement across borders in the BSR include:

[5] VASAB (Visions And Strategies Around the Baltic Sea) is an intergovernmental multilateral co-operation of 11 countries of the Baltic Sea Region in spatial planning and development (www.vasab.org).

1. *Differences in design for participation and in understanding of stakeholders' engagement across borders.* Institutional differences concerning roles and responsibilities in MSP are important when there is a need to collaborate with corresponding authorities or other stakeholders in another juridical system. Both planners and participants need to recognise how MSP operates in the other country, who to ask for information and how to contact a particular type of stakeholder. Governance structures for implementing MSP in the BSR countries vary both vertically (what level of governance is responsible for MSP) and horizontally (what thematic agency or ministry is responsible[6]). Moreover, learning about MSP in other countries takes time and requires financial and other resources. Last but not least, building trust between parties in various countries takes time too.
2. *Limited time and resources of both planners and participants.* Stakeholder participation is usually a lengthy process, and it can be difficult to keep participants engaged. The main challenges on the participants' side are loss of interest (especially if participation in MSP is not considered beneficial initially) and limited personnel and funds. On the planners' side, time and resources are often not available to develop frameworks to support transboundary collaboration and create the networks between planners and other civil servants. Moreover, language differences can impact the process in terms of costs and time (translation and interpretation).
3. *Different timing and time horizons for MSP across borders.* MSP is not synchronised across borders, which impacts stakeholders' awareness and motivation to participate. For example, stakeholders in a country where the process has just started might not have enough knowledge and capacity to meaningfully participate in cross-border proceedings, compared to stakeholders in a country with a more advanced process.
4. *Stakeholders' understanding of MSP and their roles in MSP.* MSP is still a new process and many potential stakeholders are unaware of clear benefits of involvement. There are few capacity-building initiatives to facilitate and empower meaningful stakeholder participation. Stakeholders may have wrong expectations of MSP, risking disappointment in the process, its legitimacy and outcomes.

---

[6] In Latvia MSP is supervised by the Ministry of Environment and Regional Development, while in Estonia the Ministry of Finance is in charge. In Germany the responsibility is delegated to the Federal Maritime and Hydrographic Agency (part of the Federal Ministry of Transport and Digital Infrastructure), while in Sweden it is the Swedish Agency for Marine and Water Management (under the Ministry of Environment and Energy) and in Denmark the Danish Maritime Agency (under the Ministry of Business and Growth).

5. *Difficulties to involve certain types of stakeholders.* Although many stakeholders initially do not see many benefits of participating in MSP, some groups appear to be especially difficult to mobilise at present: regional and local authorities (especially politicians), highly differentiated sectors (e.g. tourism), sectors with little trust in MSP (often the fisheries sector), and the general public.

Both the BaltSpace and the Baltic SCOPE projects include evidence of multilevel governance issues. There is a need to think in terms of multilevel governance and related challenges (Marks 1993; Hooghe and Marks 2003; Jessop 2003; Piattoni 2010; Kull 2014),[7] as MSP has to reach across jurisdictional boundaries and administrative levels and include non-governmental actors. According to legislation, MSP is a top-down exercise; at the same time, it has to rely on endogenous, place-based knowledge and needs to include the corresponding knowledge bearers.

Summing up, with MSP at an initial stage of development in the BSR, there is a need to (1) raise awareness and understanding among marine users, authority stakeholders and the public at large; (2) develop capacity for participatory processes among responsible authorities and stakeholders; (3) develop a better understanding of MSP processes in various BSR countries and (4) evaluate and compare MSP settings and processes across borders. A framework to systematically and group specifically analyse participation in MSP in transboundary settings can be useful to develop participation further, also in national contexts.

4 Towards a Framework for Analysing Participation in MSP in Transboundary Settings

Based on the earlier theoretical analysis and the current situation in the Baltic, we now move towards a framework that could be used to analyse and develop participation in MSP in transboundary settings. We suggest the following dimensions for consideration in such a framework:

(1) reasons and purpose of involvement (why);
(2) depth and breadth of involvement (who);

[7] For example, coordination challenges (e.g. Bache et al. 2012; Kull and Tatar 2015) which require careful consideration of how, why and when to include different types of actors.

(3) intensity of involvement and influence in relation to roles (how much); and
(4) methods, timing and frequency of involvement (how, when, how often).

4.1 Reasons and Purpose of Involvement

As set out above, it is crucial for the organisers of the MSP processes to better understand various purposes and forms of stakeholder participation. Only when the objectives are clearly defined is it possible to decide who should be involved in the process and what tools would best facilitate this involvement (e.g., Reed 2008; NRC 2008). The answer to "why" thus influences "who" and "what"; answering all three is necessary for MSP that is legitimate, widely accepted, open to community and sector values and priorities, and empowering to the whole spectrum of stakeholders.

Purpose-wise, there are two main types of participation: instrumental and transformative. Instrumental participation in MSP aims to enhance the efficiency of the planning processes and their outcomes through proportionate allocation of marine space or other marine resources and mitigation of existing or foreseen conflicts (Stirling 2008). Transformative participation focuses on the process of public communication and reasoning, so that the outcomes of the process are relatively less important compared to giving a voice to society. Transformative participation aims to involve all interested and affected groups and individuals and often attempts to challenge existing relations of power (Jansen et al. 1998; NRC 2008; Stirling 2008).

Both types of participation have to fulfil three basic functions (1) improving the quality of decisions and plans, (2) enhancing legitimacy and (3) capacity building (NRC 2008). Evidently, (1) is vital for instrumental, while (2) and (3) are central for transformative participation.

Importantly, the purpose of participation and the expectations of planners and stakeholders can change throughout a planning process, requiring the who and the how to be adapted as well.

4.2 Depth and Breadth of Involvement

Participants in MSP are usually persons, organisations and groups affected by the plan. Apart from breadth and depth, it is also the interrelationships among participants and in relation to the plan that matter.

Empirical results from the Baltic indicate many different types of stakeholders and a need for a highly differentiated approach to different groups and subgroups (Morf et al. forthcoming). Based on roles and needs at least four main groups need to be involved: (1) authority stakeholders from different levels (often with special mandates and rights), (2) specific stakeholder or user groups related to marine interests, (3) the public at large (a diffuse group, but usually with participatory rights) and (4) stakeholders from across the border (where rights and mandates are less clear). Even within one user group there can be considerable variety (e.g. fishers: vessel size, target species, gear types, harbours), and views and needs can differ considerably. Planners therefore need to avoid over-simplification.

These different types of participants have varying positions, interests, values and basic needs to account for—with the latter important for long-term conflict management but often difficult to address (see Morf 2006). Double- and possibly triple-loop learning[8] is required across groups. For this to take place on equal terms, differences in power need to be taken account of, resulting in a differentiated approach to awareness raising, empowerment and participation methods. Lastly, political decision-makers still need to be mobilised and involved more, as they play important roles both for legitimacy and to dispatch the necessary resources for MSP.

4.3 Intensity of Involvement and Influence in Relation to Roles

In terms of how much influence is exerted by different stakeholders, it is important to consider direct and indirect representation and who has what mandate. MSP processes imply both representative and direct participation of stakeholders (e.g. participation by writing letters or coming to a hearing). For efficiency reasons, however, also in the Baltic, often only representatives of important stakeholders are actively invited to more interactive forms (e.g. advisory boards, steering groups, workshops).

Questions regarding representation need to be linked to consideration of the intensity of involvement and the influence that can and should be exerted by different stakeholders. Not everybody needs to have decision-making power about everything for a process to be considered legitimate and leading

[8] Double-loop learning questions views and goals and leads to different ways of framing a situation (questioning the rules and thinking out of the box). Triple-loop learning asks questions about how we learn, for example, by discussing the dynamics of a meeting, what learning was produced and how it was produced, see also Hurlbert and Gupta (2015).

to well-informed decisions. Equally, not everyone may need to be involved at the same level of intensity at all times. MSP often includes a mix of different forms and intensities of participation depending on the situation, the purpose of the process, and the needs of the actors involved. Moreover, the distribution of responsibilities and roles varies, impacting on the influence participants can have on process and outcomes. Often, actors with formal roles have more influence than others (such as veto rights); it is common that planning authorities are orchestrating participation and have both facilitator and decision-making roles based on sector-specific legislation.

The setting also influences possibilities for exerting influence: In transboundary MSP, participation is likely to be less intense and influential than what might be possible in domestic processes. Involvement are more likely to be less interactive and on a strategic level rather than affecting operational planning content.

Overall, participants' activities could include (1) contributing to problem definition and conflict mapping, (2) providing knowledge for creating planning evidence, (3) proposals/views on how and when to use certain areas, (4) proposals/views on what needs to be protected and how, (5) proposals on how the own user group could contribute to the implementation of the plan, (6) contributions to monitoring and evaluation and (7) criticising and giving input on the process and actively contributing to making it more inclusive and transparent. Unless stakeholders are involved at an early stage of a planning process, or the whole process is open and reactive to their input, there is a risk that stakeholder participation is limited to basic instrumental (or even functional/tokenist) purposes.

4.4 Facilitating Participation

In addition to the principles set out above, methods of facilitating participation are an important added consideration. Currently, there is no standardised procedure helping planners to select suitable participation techniques (Luyet et al. 2012). Planners will usually need to follow existing legislation and consider other factors, such as available resources, including time, facilitation experience and capacity of both the planning team and participants. They also need to take into account the differences in knowledge, formal education and social status, cultural and social norms, experiences in similar managerial initiatives, history of relationships and power disparities (Rowe and Frewer 2000; NRC 2008; Luyet et al. 2012). Usually, whoever designs the process (perhaps with the help of experts) defines the necessary degree of involvement based on their knowledge, experience, feelings or expectations (Daniels et al. 1996).

There are numerous methods and techniques to stimulate interactions between planners, authorities and stakeholders. Public participation methods include (1) formalised approaches, such as referenda, public hearings, presentations, questionnaires and surveys, and periods for comments; (2) more collaborative approaches, like citizen panels, advisory committees, citizen juries, multicriteria analysis, cognitive maps or online deliberations and (3) techniques for which high engagement is central, such as joint fact-finding, policy dialogues, negotiated rulemaking, community partnerships or consensus conferences (Rowe and Frewer 2000; Lynam et al. 2007; NRC 2008; Luyet et al. 2012). None of these methods are by definition better than the other, although of course some are more suitable for specific positions on the MSP participation ladder. The assessment or usefulness of these tools depends on the process evaluation criteria (Rowe and Frewer 2000; Oels 2006; Blackstock et al. 2007; Luyet et al. 2012), which in turn are dependent on the formulation of policy and participation goals.

4.5 Towards a Combined Ladder of Participation in MSP

Based on the previous sections, we now present a combined ladder to analyse participation in transboundary MSP contexts (Table 10.2). Many dimensions could have been included, resulting in very complex constructs. To keep the framework simple, our focus is on elements that can readily be observed. We concentrate on (1) the degree of power sharing (visible, e.g. in the distribution of formal and informal roles), (2) the intensity of communication and learning (e.g. one-way or two-way, listening and acknowledging and the potential for double and triple loop learning) and (3) responsibilities for concrete planning and management tasks (functions). All three increase from the bottom to the top tiers of the ladder. These dimensions are necessarily interlinked and can vary depending on each MSP context. In principle, each step on the ladder could also be associated with a number of techniques for participation, depending on the desired purpose and type of communication and learning (not presented in detail here). We also formulate our ladder to reflect the context of current MSP constraints—for example, legal requirements that demand a dedicated authority formally in charge of the MSP process.

The ladder is structured along the degree of influence that can be exerted and includes the intensity of communication and responsibilities (left-hand column). The other two columns specify what this implies from the perspec-

Table 10.2 A ladder or stairway of MSP participation. Steps build on each other and do not reflect the "dark" manipulative and technocratic sides of participation

Increasing degree of power sharing → / Increasing delegation of functions → / Increasing intensity of communication and learning →	Influence (characteristics)	Authorities	Participants / interaction among stakeholders Note: key stakeholders can have special formal roles (e.g. veto right, implementation)
	Process responsibility (formal and informal, legally based or as complement, recurrent).	Process leadership partially or entirely delegated to participants but keeping overall responsibility.	Process leadership to some extent delegated to (key) stakeholders, within some type of overall mandate/ legislation (e.g. leadership over a local process, responsibility within own sector).
	Decision-making (formal, legally based or as complement, recurrent or at pre-defined stages).	Process in hands of authority/political or decision-making/break-off right. Decisions have to be followed.	Veto right/right to vote/break-off point in relation to specific items defined by authority/legislation.
	Collaboration (on planning process, concrete tasks, partially informal, recurrent, depending on activities).	*Process and decision-making in principle in the hands of authority. Consensus and needs-based collaboration.*	*Collaboration on tasks defined together, based on consensus and available resources or voluntary contributions based on invitation by the authority in charge.* *Right to contribute to the definition of activities.*
	Deliberation: dialogue & learning (partially informal, requires openness, recurrent interaction and mutual accommodation).	*Mutual exchange and learning, recurrent. Authority keeps power to adapt process and content, without formal obligation to accommodate insights.*	*Mutual exchange and learning, without formal obligation for neither part to take in and accommodate lessons learnt.* *Right to have a say and be listened to.*
	Consultation (legally based, two-way).	Obligation to listen. Keeps all other rights related to structure and content of planning process.	Active participation. Right to provide views and be listened to.
	Information (legally based, one-way).	Obligation to inform. Keeps all rights related to process and content of planning.	Passive participation. Right to be informed about issues and process and decisions.

tive of responsible authorities and participants. The term "ladder" should not be understood as linear, nor does it imply that one step is seen as superior over another. Rather, different steps might be complementary or need to differ based on the specific planning stage and context.

Lastly, the ladder should really be understood as three-dimensional, as the steps in *italics* are not part of the power dimension (based on specified roles) but belong to a more open and concurrent *interaction and deliberation* dimension with more shifting roles that can be imagined as superimposed on the power dimension. We also emphasise that this ladder does not include the "dark" or manipulative or technocratic side of participation.

The provision of *information* has become a common basic step in national and transnational MSP processes, expected both by authorities and by stakeholders. Usually, the process is directed by the authorities in terms of both content and spread. Rather than information as such, the issue is whether the information reaches all relevant stakeholders. This is a particular concern in early stage MSP and especially in transboundary MSP when awareness of MSP and its requirements is still limited. Language can act as an added barrier: Examples from the Baltic (e.g. Germany and Poland) highlight a need to translate plans into other languages and to find ways of approaching stakeholders in the other country.

Consultation has also become a standard feature of MSP in most countries due to legal requirements. Typical examples are the consultation processes recently carried out on draft MSP plans in Germany, Poland and other Baltic countries, involving varying ranges of stakeholders. For authorities, this brings with it the obligation to take into account the views put forward by stakeholders, while stakeholders have a right to be listened to. The planning process itself, however, including authority over content and structure, is still in the hands of the authority.

Deliberation broadens the scope of consultation and any subsequent steps on the ladder, although the authority retains the power to adapt the process and content of MSP. There is no formal obligation to accommodate any insights gained from deliberation. This step is, therefore, not directly related to power but more to the degree of interaction between stakeholders and opportunities offered for mutual learning. An example from the Baltic are stakeholder workshops organised in Latvia and Lithuania as part of MSP pilot projects, which were an opportunity for different stakeholder groups to meet and exchange views to be fed back into the MSP process.

Collaboration, like deliberation, is more related to the degree of interaction (and functions of participation), although power is shifting. The power over the process remains in the hands of the authority in charge, but participants may have a right to contribute to defining activities, at least during some stages of the process. Collaboration can be formal or informal and is usually based on available resources or voluntary contributions made on the invitation of the authority in charge (e.g. with limited planning resources, Denmark might need to rely on this).

Decision-making is another form of engagement, where the process is in the hands of the authority but there is a veto right for stakeholders. This gives stakeholders the power to break off a process, albeit still in line with a framework set out by the authority or legislation. For example, in Poland and Latvia, municipalities have much stronger rights, which can be used for this purpose.

Process responsibility, lastly, implies that process leadership has been partially or entirely delegated to participants. The authority retains overall responsibility for the process as defined by its legal mandate. Here, process leadership may be delegated to stakeholders with a particular mandate. Examples can be delegated leadership over a local/sector process, such as in nested approaches to MSP with national and lower-level plans (e.g. Germany).

The ladder is designed as an analytical tool that can reveal the position of certain processes or stakeholders along the three selected dimensions. As such, it does not imply a quality judgement of participation. Each step on the ladder has its justifications and none is automatically inferior to another—although each can be done well and successfully or badly and unsuccessfully.

We note that irrespective of the desired intensity, participation does not happen automatically. Even the provision of information requires active and target group-specific facilitation. An additional question is therefore what methods can be employed to successfully take each desired step. As stated above, suitable methods could easily be added to each step of the ladder.

Another aspect that is often overlooked concerns the timing of participation. If the timing of participation in the planning cycle is bad and the scope for stakeholder input is communicated wrongly, even the best participative process may not be able to achieve the desired outcomes. For example, the kind of input that can be provided by participants and the possibilities to affect outcomes usually decrease as planning proceeds. Each row and step on the ladder would therefore ideally be complemented further by the aspects of methods and timing.

5 Conclusions and Outlook

Based on theoretical reflections and empirical insights from the Baltic Sea, this chapter has attempted to distil key dimensions of participation to develop a framework for analysing and comparing participation in MSP in trans-boundary settings. The framework suggested should be understood as complementary to other relevant ladders, namely:

- The split ladder by Hurlbert and Gupta (2015), which can assist overall diagnosis, but may be difficult to use for finer analysis in cross-border MSP settings.
- The ladder by Kidd and McGowan (2013), which emphasises cross-border collaboration but does not explicitly analyse the roles and influence of individual stakeholders.
- The ladder by Twomey and O'Mahony (2018) which helps to analyse interaction, even if it is less distinctive in relation to rights and roles of specific stakeholder types and subgroups.

Testing and further refining of the proposed concept are required to facilitate both research-based analysis and practical comparison of participatory approaches. A possible future development might be to design a "climbing wall" of parallel ladders for different types of MSP stakeholders, helping to compare the many different dimensions of stakeholder participation in MSP. More work is also required on how to include important time and methodological dimensions that have not yet been developed here.

In practice, as MSP in the Exclusive Economic Zone is a top-down exercise, usually mandated at national or regional levels, participation in MSP will likely be based on formal components. Still, even highly formal types of participation (such as consultation) can benefit from enhanced interaction and mutual learning across levels and sectors, illustrated by the dimensions of double and triple-loop learning. Moreover, top-down driven processes also need to engage with bottom-up processes (e.g. local MSP projects, scenarios, mapping exercises), meaning formal approaches need to give room to informal ones in their process design and vice versa. At the same time, expectation management is important, and authorities should be transparent with respect to the available headroom in designing and implementing participatory processes including rules for decision-making. Here, the ladders could help define the respective status of informal and formal processes and set out ways for how they could enhance each other.

Cross-border and transboundary MSP implies a need to involve neighbouring authorities in MSP processes, but also non-authority stakeholders from adjoining countries. In the BSR at least, processes for doing so have not yet been developed; this could be supported by a more analytical use of the ladder. The ladder could also be helpful in identifying the current national state of play of MSP and—especially in countries where MSP is in the initial stages—needs for capacity building and awareness raising. In some countries coastal spatial planning has had decades of practice (e.g. Sweden in municipal planning, Finnish regional planning or the German Federal States), but, unless a way is found to analyse and compare these approaches, available practices and experiences will not necessarily be easy to incorporate and learn from. This does not imply benchmarking, which may be difficult, as MSP is multidimensional and different countries have different systems and views of participation (e.g. some countries will be striving for more collaborative approaches than others). At the same time, cross-comparison and evaluating the benefits of different types of participation on the basis of set criteria can promote learning from each other. As MSP continues to develop as a new professional field, there is a need to remain attentive to the different purposes of participation and to make constructive use of analytical dimensions when organising and analysing participation in MSP

Acknowledgements The development of the chapter took place under the project "Economy of maritime space" funded by the Polish National Science Centre. The Open Access fee of this chapter was provided by this project. This work was also supported by the BONUS BALTSPACE project, which has received funding from BONUS (Art 185), funded jointly from the EU's Seventh Programme for research, technological development and demonstration, and from Baltic Sea national funding institutions. Furthermore, research by Nordregio, within the authority-driven MSP project Baltic SCOPE (Towards coherence and cross-border solutions in Maritime Spatial Planning; 2015-17), co-financed by the European Maritime and Fisheries Fund and the project partners (national authorities and regional organisations) provided important results. We warmly thank our informants and colleagues in these projects, documented in among others Kull et al. (2017) and Morf et al. (forthcoming).

References

Armitage, D., Marschke, M., & Plummer, D. (2008). Adaptive Co-management and the Paradox of Learning. *Global Environmental Change, 18*, 86–98.

Arnstein, S. R. (1969). A Ladder of Citizen Participation. *Journal of the American Institute of Planners, 35*, 216–224.

Bache, I., Bartle, I., & Flinders, M. (2012). *Unravelling Multilevel Governance: Beyond the Binary Divide.* Paper Delivered at the Workshop "Governance and Participation Research". University of Sheffield, June 6.

Berkes, F. (1994). Co-management: Bridging the Two Solitudes. *Northern Perspectives, 22*, 18–20.

Berkes, F. (2009). Evolution of Co-management: Role of Knowledge Generation, Bridging Organizations and Social Learning. *Journal of Environmental Management, 90*, 1692–1702.

Bingham, L., Nabatchi, T., & O'Leary, R. (2005). The New Governance: Practices and Processes for Stakeholder and Citizen Participation in the Work of Government. *Public Administration Review, 65*, 547–558.

Bishop, P., & Davis, G. (2002). Mapping Public Participation in Policy Choices. *Australian Journal of Public Administration, 61*, 14–29.

Blackstock, K. L., Kelly, G. J., & Horsey, B. L. (2007). Developing and Applying a Framework to Evaluate Participatory Research for Sustainability. *Ecological Economics, 60*, 726–742.

Borrini-Feyerabend, G., Pimbert, M., Farvar, M. T., Kothari, A., & Renard, Y. (2004). *Sharing Power. Learning by Doing in Co-management of Natural Resources Throughout the World.* Tehran: IIED and IUCN/CEESP/CMWG, Cenesta.

Bryson, J. M., Quick, K. S., Slotterback, C. S., & Crosby, B. C. (2012). Designing Public Participation Processes. *Public Administration Review, 73*, 23–34.

Carlsson, L., & Berkes, F. (2005). Co-management: Concepts and Methodological Implications. *Journal of Environmental Management, 75*, 65–76.

Collins, K., & Ison, R. (2009). Jumping Off Arnstein's Ladder: Social Learning as a New Policy Paradigm for Climate Change Adaptation. *Environmental Policy and Governance, 19*, 358–373.

Daniels, S. E., Lawrence, R. L., & Alig, R. J. (1996). Decision-Making and Ecosystem-Based Management: Applying the Vroom-Yetton Model to Public Participation Strategy. *Environmental Impact Assessment Review, 16*, 13–30.

Dyrberg, T. B. (1997). *The Circular Structure of Power: Politics, Identity, Community*. New York: Verso.

Fainstein, S. S., & Fainstein, N. (1996). City Planning and Political Values: An Updated View. In S. Campbell & S. S. Fainstein (Eds.), *Readings in Planning Theory* (pp. 265–287). Oxford: Blackwell.

Foucault, M. (1980, May). *Power/Knowledge. Selected Interviews and Other Writings, 1972–1977* (1st ed.). Harvester Press.

HELCOM-VASAB. (2010). Baltic Sea Broad-Scale Maritime Spatial Planning Principles. Retrieved from http://www.helcom.fi/action-areas/maritime-spatial-planning/msp-principles.

HELCOM-VASAB. (2016). Guidelines on Transboundary Consultations, Public Participation and Co-operation. Adopted by the MSP Working Group in June 2016. Retrieved from http://www.helcom.fi/Documents/Action%20areas/Maritime%20spatial%20planning/Guidelines%20on%20transboundary%20consultations%20public%20participation%20and%20co-operation%20_June%202016.pdf.

Hooghe, L., & Marks, G. (2003). Unraveling the Central State, but How?—Types of Multi-level Governance. *American Political Science Review, 97*, 233–243.

Hurlbert, M., & Gupta, J. (2015). The Split Ladder of Participation: A Diagnostic, Strategic, and Evaluation Tool to Assess When Participation Is Necessary. *Environmental Science & Policy, 50*, 100–113.

Jansen, F. M., Jensen, P. M., Vindig, D., Ellemann-Jensen, H., & Steffen, L. C. (1998). *"We Are Strong Enough": Participatory Development in Practice*. Copenhagen: DanChurchAid.

Janßen, H., Varjopuro, R., Luttmann, A., Morf, A., & Nieminen, H. (2018). Imbalances in Interaction for Transboundary Marine Spatial Planning: Insights from the Baltic Sea Region. *Ocean and Coastal Management, 161C*, 201–210.

Jentoft, S., & Chuenpagdee, R. (2009). Fisheries and Coastal Governance as a Wicked Problem. *Marine Policy, 33*, 553–560.

Jessop, R. (2003). *Governance, Governance Failure, and Metagovernance*. Paper Presented at an International Seminar Held at the Università della Calabria, November 21–23.

Kidd, S., & McGowan, L. (2013). Constructing a Ladder of Transnational Partnership Working in Support of Marine Spatial Planning: Thoughts from the Irish Sea. *Journal of Environmental Management, 126*, 63–71.

Kull, M. (2014). *European Integration and Rural Development—Actors, Institutions and Power*. London: Routledge.

Kull, M., & Tatar, M. (2015). Multi-level Governance in a Small State—A Study on Involvement, Participation, Partnership, and Subsidiarity. *Regional & Federal Studies*. https://doi.org/10.1080/13597566.2015.1023298.

Kull, M., Moodie, J., Giacometti, A., & Morf, A. (2017). *Lessons Learned: Obstacles and Enablers When Tackling the Challenges of Cross-Border Maritime Spatial Planning—Experiences from Baltic SCOPE*. Stockholm: Espoo and Gothenburg—Baltic SCOPE. Retrieved from http://www.balticscope.eu/content/uploads/2015/07/BalticScope_LL_WWW.pdf.

Luyet, V., Schlaepfer, R., Parlange, M. B., & Buttler, A. (2012). A Framework to Implement Stakeholder Participation in Environmental Projects. *Journal of Environmental Management, 111*, 213–219.

Lynam, T., de Jong, W., Sheil, D., Kusumanto, T., & Evans, K. (2007). A Review of Tools for Incorporating Community Knowledge, Preferences, and Values into Decision Making in Natural Resources Management. *Ecology and Society, 12*(1), 5.

Marks, G. (1993). Structural Policy and Multilevel Governance in the EC. In A. Cafruny & G. Rosenthal (Eds.), *The State of the European Community* (pp. 391–411). New York: Longman.

Matczak, M., Przedrzymirska, J., Zaucha, J., & Schultz-Zehden, A. (2014). *Handbook on Multi-level Consultations in MSP*. PartSeaPate. Retrieved from http://www.partiseapate.eu/wp-content/uploads/2014/09/PartiSEApate_handbook-on-multi-level-consultations-in-MSP.pdf.

Morf, A. (2006). *Participation and Planning in the Management of Coastal Resource Conflicts: Case Studies in West Swedish Municipalities*. Ph.D. Dissertation. Gothenburg: School of Global Studies, University of Gothenburg. ISBN 91-975290-3-6.

Morf, A., Strand, H., Gee, K., Gilek, M., Janßen, H., Hassler, B., Luttmann, A., Piwowarczyk, J., Saunders, F., Stalmokaite, I., & Zaucha, J. (forthcoming). BONUS BALTSPACE Deliverable D2.3: Exploring Possibilities and Challenges for Stakeholder Integration in MSP. Swedish Institute for the Marine Environment Report Series. Gothenburg: Swedish Institute for the Marine Environment, University of Gothenburg.

NRC (National Research Council). (2008). *Public Participation in Environmental Assessment and Decision Making*. Washington, DC: The National Academies Press.

Oels, A. (2006). Evaluating Stakeholder Dialogues. In S. Stoll-Kleemann & M. Welp (Eds.), *Stakeholder Dialogues in Natural Resources Management. Theory and Practice* (pp. 117–151). Berlin Heidelberg: Springer–Verlag.

Piattoni, S. (2010). *The Theory of Multi-level Governance: Conceptual, Empirical, and Normative Challenges*. Oxford: Oxford University Press.

Pomeroy, R., & Berkes, F. (1997). Two to Tango: The Role of Government in Fisheries Co-management. *Marine Policy, 21*, 465–480.

Pretty, J. (1995). Participatory Learning for Sustainable Agriculture. *World Development, 23*, 1247–1263.

Quick, S., & Bryson, J. (2016). Public Participation. In C. Ansell & J. Torfing (Eds.), *Handbook on Theories of Governance*. Cheltenham: Edward Elgar.

Reed, M. S. (2008). Stakeholder Participation for Environmental Management: A Literature Review. *Biological Conservation, 141*, 2417–2431.

Rittel, H., & Webber, M. (1973). Dilemmas in a General Theory of Planning. *Policy Sciences, 4*, 155–169. Elsevier Scientific Publishing Company, Inc.: Amsterdam. [Reprinted in] Cross, N. (Ed.), Developments in Design Methodology. 1984. J. Wiley & Sons: Chichester, pp. 135–144.

Rowe, G., & Frewer, L. (2000). Public Participation Methods: A Framework for Evaluation in Science. *Technology and Human Values, 25*, 3–29.

Sandercock, L. (1998). The Death of Modernist Planning: Radical Praxis for a Postmodern Age. In M. Douglass & J. Friedmann (Eds.), *Cities for Citizens: Planning and the Rise of Civil Society in a Global Age* (pp. 163–184). Chichester and New York: John Wiley and Sons.

Selle, K. (Ed.). (1996). *Planung und Kommunikation: Gestaltung von Planungsprozessen in Quartier, Stadt und Landschaft: Grundlagen, Methoden, Praxiserfahrungen.* Wiesbaden und Berlin: Bauverlag.

Stirling, A. (2008). "Opening Up" and "Closing Down". Power, Participation, and Pluralism in the Social Appraisal of Technology. *Science, Technology, & Human Values, 33*, 262–294.

Stoker, G. (1998). Governance as Theory: Five Propositions. *International Social Science Journal, 50*, 17–28.

Twomey, S., & O'Mahony, C. (2018). Stakeholder Processes in Marine Spatial Planning: Ambitions and Realities from the European Atlantic Experience. In J. Zaucha & K. Gee (Eds.), *Marine Spatial Planning—Past, Present, Future.* Palgrave Macmillan.

Zaucha, J. (2014a). Sea Basin Maritime Spatial Planning: A Case Study of the Baltic Sea Region and Poland. *Marine Policy, 50*, 34–45.

Zaucha, J. (2014b). *The Key to Governing the Fragile Baltic Sea. Maritime Spatial Planning in the Baltic Sea Region and Way Forward.* Riga: VASAB Secretariat.

11

Taking Account of Land-Sea Interactions in Marine Spatial Planning

Sue Kidd, Hannah Jones, and Stephen Jay

1 Introduction

As the opening chapter in this book reveals, since the first major international conference on marine/maritime spatial planning (MSP) organised by the Intergovernmental Oceanographic Commission of United Nations Educational, Scientific, and Cultural Organization (UNESCO) in 2006, MSP has emerged as a significant new direction in global governance with systems of MSP being established for the first time in coastal states across the world. It therefore appears that, for many, MSP is seen as a key mechanism to achieve more effective planning and management of human relationships with the sea. One of the leading definitions of MSP as 'a public process of analysing and allocating the spatial and temporal distribution of human activities in marine areas to achieve ecological, economic and social objectives that have been specified through a political process' (Ehler and Douvere 2009) provides an insight into the nature of these MSP developments which focus on improving governance of human activities in the marine environment. There is no doubt such efforts are needed not only to respond to growing human use of the sea and rising demands on marine space and potential for

S. Kidd (✉) • S. Jay
Department of Geography and Planning, University of Liverpool, Liverpool, UK
e-mail: suekidd@liverpool.ac.uk

H. Jones
School of Environmental Sciences, University of Liverpool, Liverpool, UK

J. Zaucha, K. Gee (eds.), *Maritime Spatial Planning*,
https://doi.org/10.1007/978-3-319-98696-8_11

conflict between different marine activities but also to growing understanding of the scale of deterioration in the health of marine ecosystems and its connection with current patterns of human development. However, while recent MSP advances must be welcomed, the purpose of this book is to critically reflect on this experience and explore future directions for this embryonic form of governance. This chapter does this from a Land-Sea Interaction (LSI) perspective and raises questions about the role and limitations of MSP in addressing sustainable development of the world's oceans as many of the issues it is concerned with are inextricably connected to activity on the land. The chapter starts by identifying LSI considerations that are evident in some of the key documents that are guiding the establishment of MSP across the world. It then sets out a framework for understanding LSI and explores the different dimensions identified with particular reference to examples from European experience where LSI issues have become a renewed focus of concern in recent times. The chapter ends with an exploration of how LSI matters might inform future directions for MSP and may be heralding a new era of Territorial Spatial Planning (TSP), which spans both land and sea.

2 LSI and MSP: Directions from International Law and Guidance on MSP

A key reference point for all those involved in the development of MSP is the 1982 United Nations Convention of the Law of the Sea (UNCLOS) as this is one of the most significant international legal frameworks guiding human relations with the world's oceans (Maes 2008; Soininen and Hassan 2015). A reading of UNCLOS from an MSP and LSI perspective provides some interesting insights as it is evident that related considerations permeate key aspects of the convention. For example, the economic significance of the sea to landward communities is reflected in the preamble to the document which states that the convention aims 'to contribute to the realization of a just and equitable international economic order which takes into account the interests and needs of mankind as a whole and, in particular, the special interests and needs of developing countries, whether coastal or land-locked' (United Nations General Assembly 1982, preamble). This aim is perhaps reflected most prominently in the sections related to protection of rights of innocent passage for ships of all nations. Recognition of key shipping routes and port infrastructure and the location of other sea uses in a way which is consistent with their continuing operation is a central concern for MSP (Nautical Institute and World Ocean Council 2013), and its significance cannot be overstated. It is important because over 80 per cent of global trade is carried on board ships and seaborne trade is inextricably connected to the activities and well-being of the world economy (United Nations Conference on Trade and Development 2017) and indeed most aspects of human life in all countries. LSI perspectives are also evident in the sections in UNCLOS related to protection and preservation of the marine environment. For example, Article 194 requires states to take measures to deal with all sources of pollution of the marine environment including those from land-based sources. Many see protection of the marine environment as another critical aspect of MSP (Douvere 2008; Foley et al. 2010; Ehler 2018), but UNCLOS raises interesting questions about the nature of MSP's role and its limitations in dealing with this agenda which extends well beyond ocean shores.

It should be noted that UNCLOS predates the development of the modern era of MSP which evolved at least in part from experimentation from the 1970s onwards with Integrated Coastal Zone Management (ICZM sometimes called Integrated Coastal Management (ICM)). Its emergence as a significant new field of global governance in its own right was reflected in the publication in 2009 of a step-by-step guide to MSP by UNESCO (Ehler and Douvere 2009). This document remains an important global reference point for MSP (Ehler 2014; Pınarbaşı et al. 2017), and it therefore also seems relevant to reflect on its mention of LSI issues here. These are first addressed in its discussion of ICZM which it acknowledges played a key part in demonstrating the need for integrated planning and management of human relationships with the marine environment. However, it notes that ICZM at that point in time tended to be focussed on a narrow coastal strip—both landward and seaward—and rarely extended inland to cover, for example, coastal watersheds or seawards to include all of the territorial seas or Exclusive Economic Zones (EEZs) where significant human interactions also occurred. In contrast to ICZM, the guide saw MSP as focusing on the human use of *marine* spaces and envisaged it as 'the missing piece that can lead to truly integrated planning from coastal watersheds to marine ecosystems' (Ehler and Douvere 2009, p. 21). This line of thinking is developed further in the guide's discussion of MSP and Ecosystem-based Management which it described as an integrated approach to management that considers the entire ecosystem, including humans with the goal being to maintain an ecosystem in a healthy, productive and resilient condition so that it can provide the goods and services humans want and need. In this context, attention is drawn to the need to recognise the interconnectedness among systems, such as among air, land and sea. In this way, the guide is a useful reminder of early thinking about the scope of MSP with its marine focus and its place within a wider governance architecture addressing LSI issues.

More recently, for countries within the European Union (EU), the 2014 MSP Directive has become a key document guiding MSP development. It is also emerging as a reference point for MSP practice in other parts of the world as the European Commission joins forces with UNESCO to promote the roll out of MSP process worldwide (UNESCO and European Commission 2017). Here too LSI considerations are evident. For example, Article 6 of the Directive indicates that one of the minimum requirements of MSP is that LSI should be taken into account, while Article 7 says that Member States may achieve this through the MSP process itself or by other formal or informal processes, such as ICZM in which case, the outcome must be reflected in the maritime spatial plans. Beyond this, the Directive is significant in noting that MSP has an important role in promoting coherence with other relevant processes related to LSI and in this way it establishes a legal basis for MSP authorities to make connections inland (EC 2014). It is perhaps for these reasons that LSI issues have been a source of much interest in recent MSP discussions in Europe which have prompted action across a range of fronts. This chapter draws upon this experience to help develop a closer understanding of the connections between LSI, MSP and wider systems of 'territorial' governance.

3 A Framework for Considering Land-Sea Interactions

However, before discussing recent European experience related to LSI, it is important to acknowledge that LSI-related work is by no means new or indeed unique to this part of the world. One illustration of this is the Land-Ocean Interactions in the Coastal Zone (LOICZ) project which was established in 1993 as a core element of the International Geosphere-Biosphere Programme (IGBP). This has produced many helpful LSI-related outputs over the years (Ramesh et al. 2015). With an initial focus on the biology, chemistry and physics of the coastal zone, it has more recently extended its research scope to include social, political and economic sciences to better address the human dimensions of LSI. Since 2015, LOICZ has become a core project of the new Future Earth initiative under the new name of Future Earth Coasts (Future Earth Coasts 2018) and this is likely to be a key point of reference for those with an interest in LSI in years to come.

Beyond research associated with the LOICZ project, LSI-related activities have been longstanding in Europe and an early expression of this was also focussed around the coastal zone where interactions are arguably at their most obvious and intense. This took the form of activity related to ICZM and included projects which formed part of the EU Demonstration Programme on ICZM which ran from 1996 to 1999. This informed the development of eight ICZM principles (EC 1999) which at first sight can seem surprisingly lacking in specific reference to LSI. However, closer examination reveals that core areas for LSI consideration are identified including interactions within and between natural systems and human activity (e.g. Principles 1 and 5) and in governance arrangements—involving relevant administrative bodies at different levels (Principle 7) and making use of a combination of instruments to facilitate coherence (Principle 8). In this way, the ICZM principles mirror to some extent the LSI research areas identified in the LOICZ programme and seem to confirm these as central aspects to consider in scoping LSI concerns. Beyond this however, it is worth noting that the ICZM principles set out a number of operational points which also seem to be of relevance to those involved in addressing LSI matters. These include the need to take a long-term perspective (Principle 2), adopt an adaptive management approach (Principle 3), recognise local specificity (Principle 4) and involve all parties (Principle 6).

More recently, various EU institutions have supported further investigation into LSI matters. For example, in 2013, The European Observation Network for Territorial Development and Cohesion (ESPON) published the findings of its ESaTDOR project which examined territorial development opportunities and risks in European Seas (University of Liverpool 2013). Through an analysis of EU-wide data sets related to economic activity, energy and pipelines and cables, and transport and environment, the study developed a typology of European maritime regions reflecting the varying intensity of LSI. The typology distinguished European Core, Regional Hub, Transition, Rural and Wilderness regions (see Fig. 11.1). The study is of interest in setting out a methodology to enable comparable analysis of land and sea data and using this to identify LSI hotspots and cold spots covering both land and sea areas. Equally, some may find helpful its exploration of the potential policy implications arising from the definition of maritime regions with different levels of LSI intensity.

ESPON is funding a follow-up project on LSI called MSP-LSI, which is extending understanding in particular of landward economic linkages of key maritime sectors drawing upon the value chain analysis outlined below. The project will produce guidelines for both MSP and terrestrial planning agencies on how best to manage LSI, The results will be published in 2019.

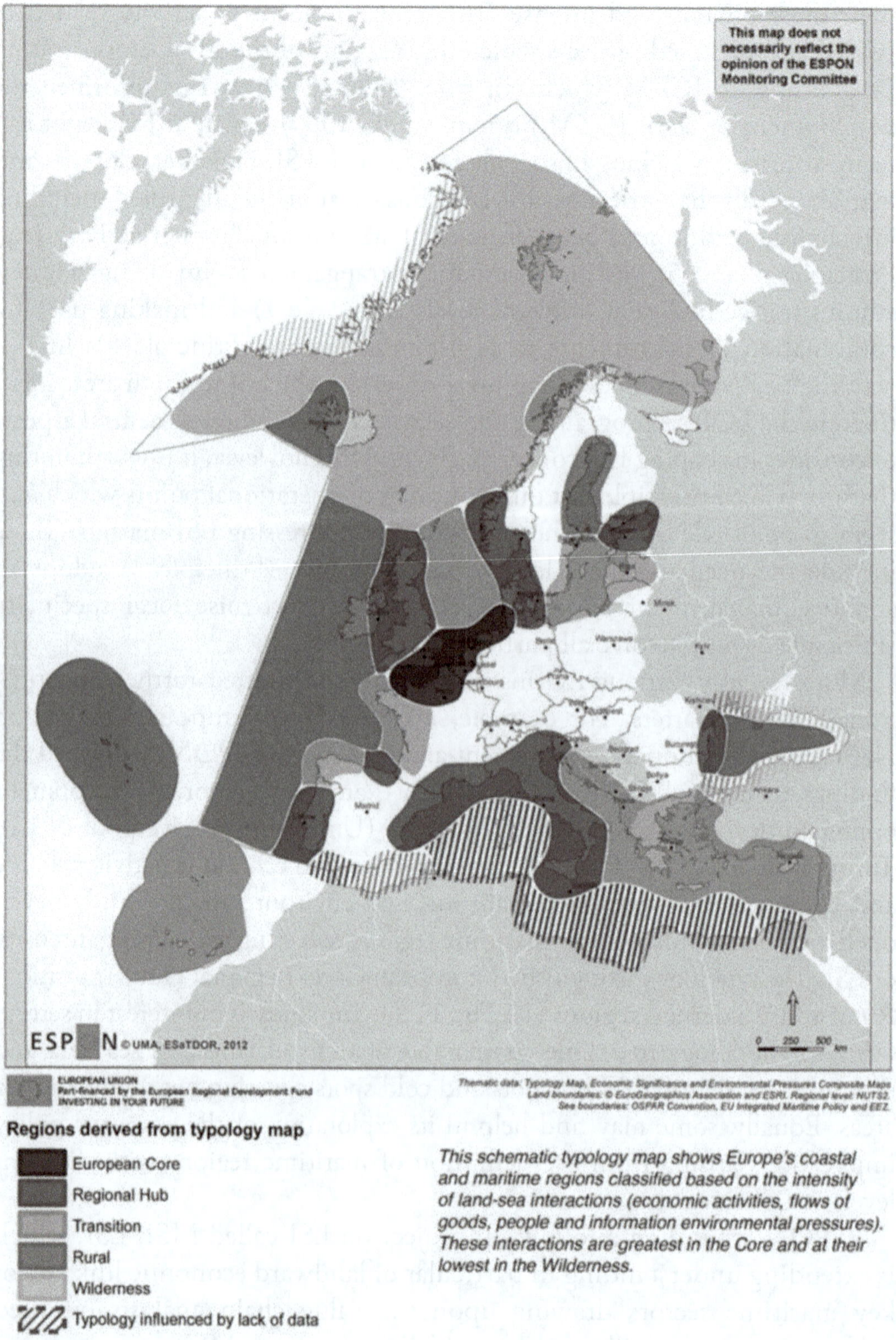

Fig. 11.1 ESaTDOR (European Seas Territorial Development and Risks) LSI typology of European maritime regions. Source: Based on University of Liverpool (2013, p. 6)

Another strand of LSI-related activity has been supported by DG Environment and is reflected in the publication of an LSI guide for MSP (DG Environment 2018). This includes a scoping of environmental, socio-economic and technical LSI issues in relation to eight key marine development sectors: Aquaculture, desalination, fisheries, marine cables and pipelines, minerals and mining, ports and shipping, tourism and coastal recreation, and offshore energy. It then examines how those engaged in MSP might respond to these in various stages of their plan making. In this way, the guide is a helpful step forward in operationalising LSI understanding within MSP.

A third area of activity is, however, the main source of discussion in the remainder of this chapter, and this relates to work undertaken by the European MSP Platform which is supported by DG Mare. This work was prompted by the MSP Expert Group which advises the European Commission on the roll out of the MSP Directive. In 2017, the expert group identified LSI as an area of particular concern for MSP practitioners, and in response, a conference to examine MSP and LSIs was held in Malta in June 2017, which was attended by over 70 stakeholders from across European seas. This initiative provided a valuable forum to discuss current understanding of LSI issues among European MSP practitioners.

In preparation for the event, the authors of this chapter developed a framework to examine the topic (European MSP Platform 2017a) (see Fig. 11.2). This reflected previous studies and recognised that LSI is a complex phenomenon, involving both natural processes across the land-sea interface and the interactions with human activities on both land and sea. To address LSI the framework proposes that MSP authorities and other stakeholders should, first, seek to understand the dynamics involved, and, second, find institutional mechanisms that are most suited to managing LSI within their particular governance context. The framework acknowledges that there may be a range of options available, involving different types and spatial scales of intervention. The different dimensions of the framework are explained below.

3.1 The Dynamics of Land-Sea Interactions

Within the framework, interactions between the land and sea are broadly grouped into two categories—bio-geochemical processes and socio-economic activities—which are closely interrelated and dynamic in their character and expression.

Of the two categories bio-geo-chemical processes in particular have been the subject of a significant number of European research projects and associ-

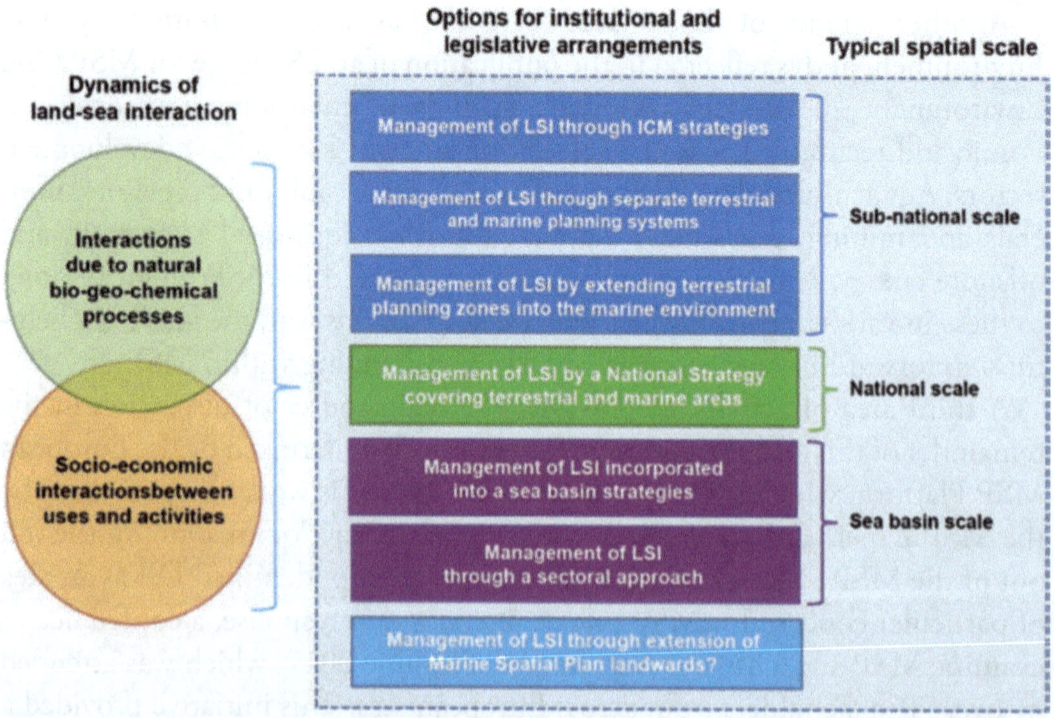

Fig. 11.2 Framework for addressing LSI. Source: Based on European MSP Platform (2017a)

ated efforts to inform the practices of planning and management stakeholders. One example is work undertaken in relation to Danish Marine Waters in response to obligations under the Oslo-Paris (OSPAR) Convention, which involved an assessment of factors and parameters that cause, control or respond to eutrophication of the sea. This presents in a simple graphic way key landward sources of marine pollution (see Ærtebjerg et al. 2003, p. 11).

Another example is the Celtic Seas Partnership LIFE funded project which brought together marine and landward stakeholders, governments and the scientific community within the Celtic Seas to find workable ways of supporting the implementation of the Marine Strategy Framework Directive (MSFD) and its ambition to achieve Good Environmental Status of European Seas. One of the outputs of this project was a set of guidelines for those engaged in terrestrial planning explaining how they might support MSFD efforts. The guidelines are useful in providing a simple written account of how landward development can impact on the health of the marine environment (University of Liverpool 2016a).

For example, they explain that the sea is the major sediment and nutrient sink for the land and that pollutants and sediments from land-based activities released into water and air are ultimately likely to find their way to the sea, creating pressures on the functioning of marine and coastal ecosystems. In terms of the water environment, the guidelines highlight that marine life relies on good water quality and habitat integrity to live and function well and to provide many of the ecosystem services humans rely on, such as seafood and waste regulation. However, marine water quality can be significantly affected by the effluents and sediments from land-based sources that reach the sea from rivers and direct runoff. Equally, atmospheric emissions (including carbon dioxide) from landward activities are a key cause of ocean acidification as these get transported over the oceans where they fall with rain into the sea. The sea absorbs carbon dioxide, but in the process, this makes the sea more acidic. This can have an impact on many animals in the sea which are adapted to less acidic conditions. In particular, animals that have shells or external skeletons, such as coral reefs, are at risk. This can also impact upon marine industries such as shellfish aquaculture.

In addition, the guidelines emphasise that human-induced climate change mainly from land-based activity has major implications for marine ecosystems. For example, increased storm activity combined with sea-level rise can have important implications for land use due to coastal flooding and erosion. At the same time, an increase in sea temperature will affect the natural range of species which are adapted to colder or warmer temperatures. The sea level will also change, leading to a change in coastal habitats, meaning that current intertidal habitats may reduce in extent or be lost and replaced by different habitats. Any landward activity that contributes to global warming will therefore also have implications for the sea.

Finally, the guidelines note that landward development can result in disturbance to and loss of critical coastal and marine habitats which provide many services such as regulating coastal erosion, flood protection, food production and opportunities for recreation and leisure. They indicate that poorly planned coastal development can have direct, indirect and unintentional impacts on these natural services, the negative impacts of which can significantly outweigh the benefits of the original development.

To assist terrestrial planning stakeholders in assessing the impact of their activities on the marine environment, the guidelines include a checklist of pressures on marine ecosystems that can arise from landward development, which shows how these can impact on different MSFD Good Environmental Status descriptors (See Fig. 11.3).

The discussion above provides some examples of the complex interactions between bio-geochemical processes and socio-economic activities and their impacts on the marine environment. However, in order to develop a rounded understanding, the LSI framework indicates that it is also important to examine relationships from a socio-economic led perspective as well.

MSFD / Pressure Descriptor	Biodiversity	Non-indigenous species	Commercial Fish & Shellfish	Foodwebs	Eutrophication	Seafloor Integrity	Hydrographic condition	Contamination (ENV)	Fish and Seafood Contaminants	Marine Litter	Energy introduction (incl. noise)
Abrasion (physical disturbance to habitats)	X			X		X		X			
Barrier to species movement	X		X	X							
Change in wave exposure (alteration of normal regime)	X		X	X		X	X				
Changes in Siltation (above natural levels)	X		X	X		X					
Electromagnetic changes	X		X	X			X				
Emergence regime changes (alteration of natural regime)	X			X		X	X				
Input of organic matter (above natural levels)	X		X	X	X	X			X		
Introduction of microbial pathogens (above natural levels)	X		X	X		X			X		
Introduction of Non-Indigenous Species and Translocations	X	X	X	X	X	X					
Introduction of Non-synthetic compounds (e.g. heavy metals, oils) (above natural levels)	X		X			X		X	X		
Introduction of Radionuclides (above natural levels)	X		X	X		X		X	X		
Introduction of Synthetic compounds (e.g. pesticides, pharmaceuticals)	X		X	X		X		X	X		
Introduction of Litter (all types)	X		X	X		X		X	X	X	
Nitrogen and phosphorus enrichment	X		X	X	X	X			X		
pH changes (alteration of normal pH regime)	X		X	X		X	X				
Salinity changes (alteration of normal salinity regime)	X		X	X		X	X				
Habitat change (due to sealing with new materials (e.g. concrete) or loss to land (land reclaim))	X		X	X		X					
Selective extraction of fauna and flora (e.g. due to fishing, collecting, recreational harvesting, loss on cooling inlets)	X		X	X		X					
Smothering of flora and/or fauna (due to addition of materials onto natural habitat where there is change in the properties of the habitat but the habitat is not lost)	X			X		X					
Thermal changes (alteration of natural temperature regime)	X		X	X	X	X	X				
Underwater noise (outside of natural levels of noise)	X		X	X							X
Water flow rate changes (alteration of normal water flow regime)	X		X	X	X	X	X				
Emissions (leading to changes in environmental drivers like temperature or acidity resulting from climate change)	X		X	X	X	X	X				

Fig. 11.3 Land-sea pressure impact matrix. Source: Based on University of Liverpool (2016a, p. 16)

In this respect, the LSI and MSP brochure published by DG Environment (DG Environment 2018) provides a helpful reference. For example, its scoping of socio-economic interactions of marine cables and pipeline development identifies potential benefits to landward communities in the form of employment and income regionally and nationally with respect to direct and ancillary activities. However, it also raises questions about potential displacement effects which might have socio-economic consequences. It notes that fishing vessels may be excluded from cable corridors or landfall sites or these could require changes in fishing activity such as shifts in fishing gear which may have negative impacts on landings, fishermen's income, jobs and fishing communities. A similar socio-economic scoping undertaken for marine mining and minerals highlights that the sector could also bring socio-economic benefits in the form of employment and income regionally and nationally including that related to ancillary sectors such as exploration services and ship building and secondary activities such as construction. In addition, it suggests that tourism and recreation might benefit from such activity if it is directed to the provision of material for beach recharge and coast protection. Similarly there could be wider socio-economic benefits to coastal communities if the activity is associated with the development of flood defence structures. On the other hand, possible negative consequences of marine mining and minerals activity could relate to its potential restriction on offshore energy development, as this may need to be excluded from extraction areas for the duration of an extraction licence. From these scoping examples it can be seen that the socio-economic impacts of human uses of the marine environment can be both positive and negative and the brochure is helpful in indicating the type of assessments that might aid MSP decision-making taking account of socio-economic LSI concerns.

It is interesting to observe however, that this LSI analysis with its MSP focus tends towards a seaward perspective and that a complementary view on LSI which is arguably more landward in its orientation is perhaps evident in the European Commission initiatives related to Blue Growth. Since the inception of the EU's Blue Growth Strategy in 2012, it has been clear that European seas and oceans are increasingly seen as one of the important drivers for the European economy and that MSP is regarded a key tool for achieving sustainable Blue Growth. In a review published in 2017, it was estimated that Europe's maritime industries employed over 5 million people and generated almost EUR 500 billion a year for the European economy. The potential to create many more jobs was also noted, with Organization for Economic Cooperation and Development (OECD) forecasts suggesting that the value of the global ocean economy could more than double by 2030.

Significant growth rates were already evident in some sectors in Europe, for example, in the rapid development of offshore wind farms which have since 2012 emerged as a major contributor to European employment accounting for around 150,000 jobs (EC 2017a). Increasing recognition of the economic significance and potential for growth in maritime sectors is prompting a new phase of LSI research with a more direct socio-economic and landward orientation. This is reflected in the Blue Growth strand of Horizon 2020 research programme (EC 2017b) as well as research directly commissioned by different arms of the European Commission such as DG Mare. An example that is helpful to the current discussion is a study of Scenarios and Drivers for Sustainable Growth from the Oceans, Seas and Coasts (Ecorys 2012). Out of an identified 27 maritime economic activities of significance in Europe, this study examined 11 sectors which were considered to offer the most growth potential (short sea shipping, marine aquatic products, blue biotechnology, oil and gas, offshore wind, ocean renewable energy, marine minerals, coastal tourism, cruise tourism, coastal protection, maritime security and surveillance and environmental monitoring). Value chain analysis formed a key part of the investigation. A summary example of its application to short sea shipping together with an outline of the approach is shown in Fig. 11.4. Value chain analysis explores the landward

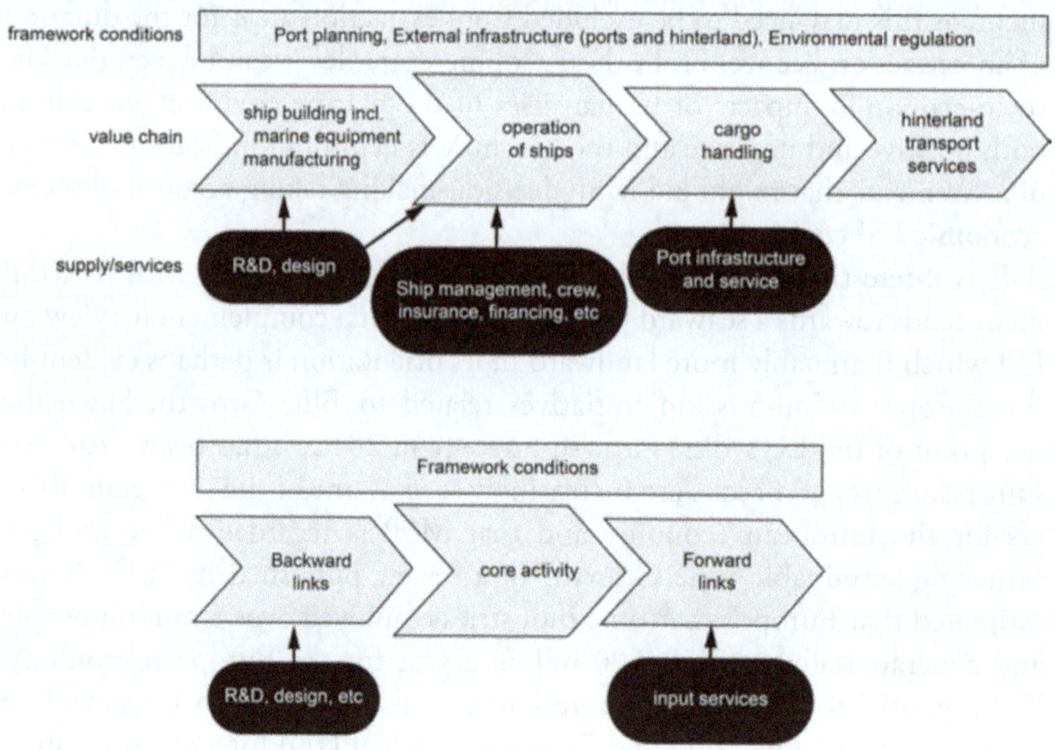

Fig. 11.4 Example of LSI value chain analysis. Source: Based on Ecorys (2012, pp. 32, 55)

implications of maritime sectors including direct employment associated with the core activity and indirect employment associated with backward and forward sector linkages. These aspects are all affected by surrounding framework conditions that provide the context for the maritime economic activities to develop (e.g. inland port and transport infrastructure, and training and research institutions). It is perhaps important to note that many aspects of these framework conditions lie beyond the control of MSP but are influenced by the policy and activities of terrestrial agencies including those related to terrestrial planning. Further development of this value chain analysis is a central component of the ESPON funded MSP-LSI project mentioned above and it is envisaged that it will be a core element in the LSI guidelines which will be produced by the project to help both MSP and terrestrial planning regimes understand and address LSI issues.

Although value chain analysis seems to be emerging as key tool in LSI investigation in relation to socio-economic activities, some might argue that the social aspects of LSI are still under-represented. A possible means of responding to this is by incorporating ecosystems services perspectives in LSI analysis. Ecosystem Services are defined as the benefits human beings can obtain from ecosystems (Millennium Ecosystem Assessment 2005, p. v.) and traditionally are separated into four categories: Provisioning, Regulating, Supporting and Cultural Services. Although not without its critics and limitations, an Ecosystem Services approach does offer the potential to look deeper into LSI considerations not least through its identification of Cultural Services. These are generally considered to be those non-material benefits people obtain from interaction with their surroundings and can take many different forms, from aesthetic appreciation of the natural environment to artistic inspiration, using different spaces or settings for activities such as leisure, education, improving health and well-being, spiritual enrichment, appreciation of symbols, history and diverse cultures. In this regard, the Nature's Services and the Sea resource pack which was also an output of the Celtic Seas Partnership project (University of Liverpool 2016b) could provide inspiration for how social aspects of LSI might be given more attention.

3.2 Legislative and Institutional Arrangements to Address Land-Sea Interactions

Having provided guidance on the dynamics of land-sea interaction, the LSI framework shown in Fig. 11.2 then indicates that these may be addressed through a variety of legislative and institutional arrangements which may have different spatial configurations in terms of land-sea coverage and operate at different spatial scales. A key point to emphasise is that local circumstances will determine which option might be most appropriate and the choice will be determined by a wide range of factors including existing institutional and legislative structures, cultural norms and practices, and geography. The following section outlines different LSI governance options with reference examples in Europe.

One option that is available is for LSI interactions to be managed through ICM initiatives. For example, Croatia is developing a Joint Management Strategy for Marine Environmental and Coastal Zone Areas and a related Action Programme. The strategy links to its obligations under the 1995 Barcelona Convention for the Protection of the Marine Environment and the Coastal Region of the Mediterranean and its ICM Protocol which was adopted in 2010. LSI considerations are fundamental to this protocol which reflects an understanding that the preservation and sustainable development of the Mediterranean requires a specific integrated approach at the level of the Mediterranean basin as a whole and of its coastal States including their inland areas. The protocol calls for ICZM activities at both sea-basin and national scales and specifically mentions connection to land policy and for states to put in place economic, financial and/or fiscal instruments intended to support local, regional and national initiatives for the integrated management of coastal zones. This emphasis on ICZM approaches to LSI is particularly relevant to the Mediterranean as sea borders between states are still to be agreed, meaning that the scope for MSP (sea focussed) activity is restricted. However, from an LSI perspective, this means that the Mediterranean experience merits close attention, as this is a region where LSI issues are a particular focus of attention.

Some countries have chosen to maintain separate terrestrial and marine planning systems whilst providing for land-sea interactions to be taken into consideration. This is the case in England where marine planning and terrestrial planning are separate but with an overlapping area of jurisdiction in the intertidal zone. Despite this, LSI issues are addressed in a number of ways. First, the UK Marine Policy Statement, which is the key guidance for marine planning in England, is also identified as guidance for terrestrial planning. Equally, the National Planning Policy Framework for inland planning authorities includes sections on maritime matters and is also a relevant guidance for marine planners. There is also a formal duty to cooperate between the two systems, and at a local level, a mechanism is available to put in place a formal 'coastal concordat' coordinating the consenting processes for developments within the coastal zone. In addition, the planning inspectorate's checklist related to the 'test of soundness' of terrestrial development plans now includes a range of requirements related to marine matters (University of Liverpool 2016c).

Another option is for local and regional scale terrestrial plans to extend into the marine environment with a view to addressing LSI within these areas. In France, for example, since 2000, an instrument called the *Schéma de Cohérence Territorial* (SCOT) has been available as an optional plan prepared by communes. The SCOT sets out strategic goals for the area it covers and it has been possible to include a specific chapter which can stand as a *schéma de mise en valeur de la mer* (SMVM) and integrate maritime concerns into the plan. (University of Liverpool 2016c). In Germany also for some time the Lander authorities have had a planning remit which extends over their adjoining sea areas. An example is Mecklenburg-Vorpommern where work started on the first marine spatial plan in 2002 when the concept was very new. The Mecklenburg-Vorpommern plan came into force in 2005 and was revised in 2016. In the recent version, the approach has been to extend the terrestrial plan into the sea (12 nm) in order to create one holistic plan with a common legal framework. In preparing the plan, a process of screening LSIs was undertaken in order to establish what kind of interactions were taking place, what data was available and who was responsible for managing them. As a result of this approach, most of the stipulations in the updated State Development Programme are regulations that deal with LSIs (European MSP platform 2017b).

The emergence of integrated planning approaches as a means of addressing LSI is also seeing the creation of national strategies which encompasses both terrestrial and marine areas. This is the approach taken by the Netherlands, for example, in the National Policy Strategy for Infrastructure and Spatial Planning (Ministry of Infrastructure and Environment 2011). This includes a National Spatial Structure map which extends over areas of Dutch jurisdiction in the North Sea, and the document develops a comprehensive vision for the development of Dutch territory looking to 2040. An interesting feature of the document is that it doesn't separately consider marine areas, but LSI matters are interwoven throughout the plan which aims to support the ambition for the Netherlands to develop in a competitive, accessible, liveable and safe manner. Malta takes a similar approach through their Strategic Plan for Environment and Development. This is an overarching plan covering both land and sea and also acts as the national Maritime Spatial Plan. Again LSI issues permeate the document, but it does include a separate chapter on the Coastal Zone and Marine Area (Government of Malta 2015).

LSI can also be managed on a larger, sea-basin scale and a prominent example of this is the Baltic Sea Region (BSR). This region has much experience with transboundary cooperation with an appreciation of LSI as a central driving force. For example, within the frame of the Baltic Marine Environment Protection Commission (or Helsinki Commission [HELCOM]), a Baltic Sea Action Plan (HELCOM 2007) for environmental protection has been developed with, for example, landward responses to eutrophication concerns a major focus of attention. The region also has developed the VASAB Long-term Vision for the BSR (Szydarowski and VASAB Committee on Spatial Development 2009), which recognises the Baltic Sea as a unifying factor and a shared resource and proposes a list of actions to stimulate territorial development potentials in the region related to urban networking and urban rural relations, accessibility and management of the Baltic Sea. LSI issues of a more socio-economic nature are prominent here. As is discussed in other chapters in this volume, activities related to HELCOM and VASAB have spawned a range of Baltic Sea-wide endeavours related to MSP with LSI a common thread running through them.

LSI can also be managed within sectors themselves, such as oil and gas, and tourism, sometimes operating at a sea-basin scale. For example, the INTERREG MED-funded project CO-EVOLVE, which started at the beginning of 2017, is analysing and promoting the co-evolution of human activities and natural systems in coastal tourist areas in the Mediterranean, allowing for the sustainable development of tourist activities, based on the principles of ICZM and MSP. CO-EVOLVE recognises that a key challenge for sustainable coastal and maritime tourism development is the strengthening of cooperation among regions and the joint development and transferring of approaches, tools, guidelines and best practices. It brings together an analysis at a Mediterranean scale of threats and enabling factors for sustainable tourism with local studies on seven representative pilot areas. The aim is to demonstrate through pilot actions the feasibility and effectiveness of an ICZM/MSP-based planning process.

The framework presented in Fig. 11.2 also notes that it is technically possible that LSI could be addressed by extending the remit of MSP inland, in contrast to extending a terrestrial planning area seawards. However, this is not an approach that appears to have been adopted so far.

It is clear from the above examples that LSI can be addressed at a variety of spatial scales. These include:

- Local areas, such as ICZM partnerships and economically driven initiatives, involving municipalities and other local interests
- Sub-national planning territories, such as maritime plan areas, involving MSP authorities working in collaboration with coastal and maritime stakeholders

- National territories, where a national strategy or plan, covering the whole of the nation's waters, and possibly its land area as well, may guide LSI efforts
- Sea basins/transnational regions, where transnational cooperation may produce a strategy or protocol for guiding national LSI efforts and ensuring ongoing cross-border cooperation

These scales are not mutually exclusive. For example, a higher-level strategy may be implemented or supplemented at a sub-national or local level by other instruments. A key notion underlying the framework presented in Fig. 11.2 is that alternative governance approaches to addressing LSI are available, and what is the most suitable in a particular context must be informed by local factors including existing institutional arrangements for spatial planning and management.

4 Some Reflections on the LSI Framework

This penultimate section of the chapter draws upon the discussions of the different aspects of the framework at the conference on LSIs held in Malta, June 2017 (European MSP platform 2017b). Summary points related to each European sea basin are presented in Tables 11.1, 11.2, 11.3, 11.4 and 11.5 and the key findings from the workshop are then outlined below.

4.1 Context Matters

Although many LSI issues are common to all European seas, as Tables 11.1, 11.2, 11.3, 11.4 and 11.5 reveal, the detailed experience of LSI varies in significant ways between countries and sea basins. This reflects differences in physical/human geography and legal/administrative arrangements as well as wider culture norms and perspectives. For example, the LSI experience of small islands was considered to be very different from that of large countries with small areas of coastline. This suggests that while there is much scope for developing common understanding and collaborative approaches to LSI, localised variations highlight the value in fostering diversity in LSI-related practices.

Table 11.1 LSI issues and arrangements in the European Atlantic

Key LSI issues	Institutional and legislative arrangements
• Coastal erosion • Climate change • Air quality associated with ports and related activities • Landscape and visual impacts of windfarms	**OSPAR** OSPAR: Protecting and conserving the Northeast Atlantic and its resources is the mechanism for governments to cooperate on the implementation of the Convention for the Protection of the Marine Environment of the Northeast Atlantic. OSPAR works on a number of fields including biodiversity and ecosystems, hazardous and radioactive substances, human activities and offshore industries. OSPAR had installed a dedicated working group on MSP that is inactive at this time. **Atlantic Strategy** The Atlantic Strategy as it stands provides directions for investment and funding relevant to LSI issues. As it is high level, its influence is rather intangible and bottom-up interaction is limited at present. **Atlantic Arc Commission** The Atlantic Arc Commission is one of the Conference of the Peripheral Maritime Region's (CPMR) six geographical Commissions. In the general work of the CPMR, LSI is being looked at in terms of implementation of the MSP Directive with reference to Articles 4, 6 and 7, but also Article 9, which includes a requirement for consulting with other relevant parties and stakeholders. **Voluntary and Sectoral Fora** For example, Fisheries Advisory Councils provide mechanisms for discussion and knowledge and experience sharing concerning a range of LSI issues. There are different levels of involvement in different Member States across the Atlantic Ocean.

4.2 Integrated Perspectives Are Important

The workshop discussions confirmed that interactions between land and sea and between environmental, socio-economic and governance elements are highly complex. While disaggregation of different LSI elements can aid understanding and help direct action, it was felt that integrated 'whole system' perspectives and approaches are required to address LSI in an effective way. The need to develop a broad-based understanding of LSI issues among both terrestrial and marine stakeholders and foster integrated 'territorial' approaches to planning and management across land and sea was an important overarching message.

Table 11.2 LSI issues and arrangements in the Baltic Sea

Key LSI issues	Institutional and legislative arrangements
• Strategic alliances between mechanisms and projects feeding into practices include LSI aspects • Pilot plans have fostered MSP processes and support strong collaboration between stakeholders • Eutrophication of marine waters • Increased shipping activity	**VASAB** VASAB is an intergovernmental multilateral cooperation of 11 countries of the Baltic Sea Region (BSR) on spatial planning and development. Its current work is guided by the 2009 'VASAB Long-Term Perspective for the Territorial Development of the Baltic Sea Region' strategic document, which considers MSP as a key instrument for the alleviation of potential sea use conflicts. **HELCOM** HELCOM is the governing body of the Convention on the Protection of the Marine Environment of the Baltic Sea Area (Helsinki Convention). In 2007, HELCOM developed the new Baltic Sea Action Plan (BSAP), which introduced MSP as a process aiming at more coherent management of human activities in the Baltic Sea. HELCOM is an excellent coordinator for LSI activities. **HELCOM—VASAB MSP working group** The joint Baltic Sea MSP Working Group, established by HELCOM and VASAB, is a forum for intergovernmental discussions on MSP. The Working Group hosts dialogues on recent and future developments in the field of MSP in the BSR. **Networks and Organisations** The European Union Strategy for the Baltic Sea Region (EUSBSR) is one of numerous organisations which consider LSI issues. The objectives of the European Strategy for the BSR are to save the sea, increase prosperity and connect the region. Spatial planning is here seen as one of four major tools to achieve these objectives. The strengths of the strategy are transnational cooperation and tools developed to implement the strategy on vertical and horizontal levels. However, there is still a lack of continuous implementation and huge potential to be used by Member States. Other examples include the Council of the Baltic Sea States (CBSS) and the SUBMARINER Network: a unique platform that brings actors and initiatives from the BSR together to actively promote innovative and sustainable uses of marine resources. It integrates perspectives from local to international scale, different science disciplines as well as policy and economic stakeholders. Current work on MSP-related LSI concerns marine aquaculture, maritime cultural heritage as well as development of new tools to consider LSI. **The CPMR Baltic Sea Commission** The CPMR Baltic Sea Commission is one of six Geographical Commissions, which comprise the Conference of Peripheral Maritime Regions of Europe (CPMR). The Baltic Sea Commission contributes to CPMR reflections and policy positions and acts as a lobby and think tank for the regions around the Baltic Sea. The thematic working group, Maritime Working Group, monitors developments on MSP. **BSSSC** The Baltic Sea States Sub-regional Co-operation (BSSSC) is a political network consisting of regional authorities from the 10 littoral states of the Baltic Sea. The network stresses the importance of coordinating different EU and national policies across the BSR and has installed a working group on Maritime Policy.

Table 11.3 LSI issues and arrangements in the Black Sea

Key LSI issues	Institutional and legislative arrangements
• Mass coastal tourism and related environmental impacts • Lack of accurate and up to date data • Geopolitical barriers to addressing LSI • Lack of legislation and strategies to deal with LSI	**Black Sea Basin Programme** The programme provides opportunities to extend existing European experience to the Black Sea and is particularly useful concerning the development of transboundary cooperation and improving LSI practices through networking. **Black Sea Commission** The Commission on the Protection of the Black Sea against Pollution provides an inventory of data, partnership and governance of relevance to the environmental dimensions of LSI, as well as challenges related to political issues. It could take a lead in data standardisation and monitoring of environmental aspects.

Table 11.4 LSI issues and arrangements in the Black Sea

Key LSI issues	Institutional and legislative arrangements
• Transnational management of supply chains linking shipping, port & inland transport infrastructure, import/ export industries • Offshore renewable energy developments and impacts on shipping/port accessibility • Difficulties in transnational management of LSI as many issues are country specific	**OSPAR** OSPAR is an international cooperation organisation with the potential to take responsibility for transnational LSI issues; however, it is questioned whether or not there is a mandate for this and if the correct management systems are in place. **North Sea Commission** The North Sea Commission is a political cooperation platform for regions across the North Sea. The aim is to promote common interests, especially concerning EU institutions, national governments and other organisations that deal with issues relevant to the North Sea, including LSI. One of the focus areas of the North Sea Region 2020 Strategy is MSP. One of the thematic working groups, 'Marine Resources', includes exchange of best practice on ICZM and MSP across the North Sea.

4.3 LSI Challenges and Opportunities

Discussions also highlighted that efforts to address LSI should focus not only on the challenges raised by LSI but also on finding beneficial synergies and realising the opportunities that LSI can bring.

Table 11.5 LSI issues and arrangements in the Baltic Sea

Key LSI issues	Institutional and legislative arrangements
• Coastal erosion • Climate change • Intense transport/ traffic • Urbanisation and impacts on wider marine environment • Institutional fragmentation	**Barcelona Convention ICZM Protocol** The Protocol is directly concerned with LSI and provides for exchange of experience, but there are different stages of application between countries. **EUSAIR** The macro-regional strategy provides a common political agreement for the Adriatic, which is of key relevance to LSI issues, but there is a need to improve the operationability. There is also a need to develop good practices regarding the integration of ecological and economic parts. **UNEP MAP/Regional Activity Centres** (RACs) In the Mediterranean RACs disseminate information on areas of special protection and marine litter for example and provide informal institutional settings (although there is a desire for more formalised settings) and facilitate networking through events and conferences **BLUE MED** This initiative strengthens cooperation on Mediterranean issues of relevance to LSI, but the long-term sustainability of the initiative may be a challenge

4.4 The Value of Diversity in Approaching LSI

It was felt that the LSI framework provides a useful way to structure discussion of different approaches to addressing LSI issues. The conference confirmed that a 'one-size-fits-all' approach to tackling LSI is not appropriate. Variations in context mean that what might be relevant and work well in one area might not be appropriate in another. However, it was felt that there was much merit in sharing different experiences.

4.5 Learning from ICZM

A recurring theme was that Europe's experience of ICZM, which had been developed to address many LSI issues, remained a valuable source of inspiration and in some instances could provide established mechanisms that could be built upon in finding new ways to integrate maritime and terrestrial planning and address LSI issues in contemporary times.

4.6 Opportunity for Cooperative Sea-Basin Approaches to LSI

In all sea basins, delegates identified established transnational institutional and legislative arrangements that could help to address LSI. These ranged from institutions associated with international conventions such as HELCOM and OSPAR; EU-supported sea-basin strategies and regional development programmes/projects; mechanisms associated with the delivery of European Directives including the MSFD, Water Framework and MSP Directives; and various other transnational fora ranging from the formal to the informal. However, it was noted that many of these organisations were only partial in their coverage of LSI issues and/or their land/sea responsibilities and that the scope for improved synergy and joined up action to better address LSI at a regional sea scale was great.

4.7 Connecting Strategic and Local Level Action

Equally, it was noted that all European sea basins have agencies and projects of various sorts at national and local levels for addressing LSI issues. Again it was noted that in many cases these were also partial in their scope/remit. It was also recognised that there was often a gap in understanding between the various levels that hampered effective responses to LSI. More generally, it was considered that there is a need to increase knowledge of LSI among all stakeholders and that this requires improved LSI-related data collection.

4.8 The Importance of Sea-Basin-Scale Approaches

Overall, it was felt that the MSP Directive and new MSP arrangements presented new opportunities to address LSI issues that have been investigated for many years. However, at the same time, it was important to acknowledge that MSP is only one of many mechanisms which can be used to address LSI. For example, DG MARE has recognised the importance of stepping back from national MSP efforts and has provided support to look at maritime issues at sea-basin or sub-sea-basin scale. Other forms of European funding have also been made available to support LSI-related initiatives, and consequently there are many examples of successful projects that have addressed LSI. Whilst further EU funding for projects can provide an avenue for continuing this work, it was felt that this was not a sustainable mechanism in the long term. Instead, the Baltic Sea HELCOM-VASAB collaboration was put forward as a notable example showing how countries can work together on an ongoing basis to address LSI at a transnational scale, and it was felt that this provided a useful model that other sea basins could follow.

5 Conclusions

This chapter has explored the relationship between MSP and LSI and in so doing has hopefully provided some useful reminders about the origins and scope of MSP but also offered some inspiration for future directions for MSP and for ocean governance more generally. By looking back at some of the core documents and experience that has framed the development of MSP, we can see that LSI considerations were prominent in its evolution and remain central to MSP activities today. However, consideration of LSI issues related to bio-geochemical processes highlights that MSP has only a small part to play in addressing many of the environmental challenges facing the world's oceans. Equally, investigation of LSI issues related to socio-economic activities indicates that MSP is only one element in a wider governance and economic and social system that influences sea-based activities and that can help to deliver sustainable patterns of Blue Growth. These reflections suggest the need for realism about what MSP can deliver and for more extensive debate about where it sits within existing structures of governance on both land and sea. While the chapter suggests the need to qualify what MSP can achieve, it also perhaps reveals the valuable role that it is and can play in providing an arena for discussion about human relationships with the world's oceans and in highlighting how continuing innovation in governance arrangements seem to be needed. One area of innovation in particular emerges from the commentary presented here which connects to notions of integrated governance, ecosystem-based management and the Ecosystem Approach as well as LSI. This relates to the development of a new era of what might be termed Territorial Spatial Planning—integrated planning which is place based and spans land and sea. This is a recurring feature in many efforts to address LSI presented above, and examples are evident at all levels and in different regions. It is as yet developing in an embryonic and seemingly haphazard manner, but a groundswell seems to be emerging that this is a key way forward. Whether TSP approaches develop and are helpful in addressing LSI challenges and opportunities and what this might mean for the evolution of MSP will be key subjects of enquiry in the years ahead.

Acknowledgements We are grateful to the project 'Economy of maritime space' funded by the Polish National Science Centre for contributing the Open Access fee for this chapter and facilitating our discussions and preparation of the book.

References

Ærtebjerg, G., Andersen, J. H., & Hansen, O. S. (Eds.). (2003). *Nutrients and Eutrophication in Danish Marine Waters. A Challenge for Science and Management.* Copenhagen: Danish National Environmental Research Institute.

DG Environment. (2018). *Land-Sea Interactions in Maritime Spatial Planning.* Brussels: European Commission.

Douvere, F. (2008). The Importance of Marine Spatial Planning in Advancing Ecosystem-Based Sea Use Management. *Marine Policy, 32*(5), 762–771.

Ecorys. (2012). *Blue Growth Scenarios and Drivers for Sustainable Growth from the Oceans, Seas and Coasts: Final Report.* Brussels: European Commission, Directorate General for Maritime Affairs and Fisheries.

Ehler, C. (2014). *A Guide to Evaluating Marine Spatial Plans.* IOC Manuals and Guides No. 70, ICAM Dossier 8. Paris: UNESCO, Intergovernmental Oceanographic Commission UNESCO IOC, 96 pp.

Ehler, C. (2018). Marine Spatial Planning: An Idea Whose Time Has Come. In *Offshore Energy and Marine Spatial Planning* (pp. 6–17). Routledge.

Ehler, C., & Douvere, F. (2009). *Marine Spatial Planning: A Step-by Step Approach Towards Ecosystem-based Management.* Manual and Guides No 153 ICAM Dossier No 6. Paris: Intergovernmental Oceanographic Commission UNESCO IOC, 99 pp.

EC. (1999). *Towards a European Integrated Coastal Zone Management (ICZM) Strategy General Principles and Policy Options.* Luxembourg: Office for Official Publications of the European Communities.

EC. (2014). *DIRECTIVE 2014/89/EU Establishing a Framework for Maritime Spatial Planning.* Brussels: European Commission.

EC. (2017a). *Report on the Blue Growth Strategy Towards more Sustainable Growth and Jobs in the Blue Economy.* Brussels: European Commission.

EC. (2017b). *Research EU Results Pack All Aboard for Better Marine Stewardship Through Research and Innovation.* Luxembourg: The Community Research and Development Information Service (CORDIS).

European MSP Platform. (2017a). *Maritime Spatial Planning: Addressing Land-Sea Interaction a Briefing Paper.* Brussels: European Commission, DG Mare.

European MSP Platform. (2017b). *Maritime Spatial Planning Conference Addressing Land-Sea Interactions Conference Report.* Brussels: European Commission, DG Mare.

Foley, M. M., Halpern, B. S., Micheli, F., Armsby, M. H., Caldwell, M. R., Crain, C. M., Prahler, E., Rohr, N., Sivas, D., Beck, M. W., & Carr, M. H. (2010). Guiding Ecological Principles for Marine Spatial Planning. *Marine Policy, 34*(5), 955–966.

Future Earth Coasts. (2018). *Our Coastal Futures: A Strategy for the Sustainable Development of the World's Coasts*. Cork: Future Earth Coasts.

Government of Malta. (2015). *Strategic Plan for Environment and Development*. Santa Venera: Government of Malta.

HELCOM. (2007). *HELCOM Baltic Sea Action Plan*. HELCOM: Helsinki.

Maes, F. (2008). The International Legal Framework for Marine Spatial Planning. *Marine Policy, 32*(5), 797–810.

Millennium Ecosystem Assessment. (2005). *Ecosystems and Human Well-Being Synthesis*. Washington, DC: Island Press.

Ministry of Infrastructure and Environment. (2011). *Summary National Policy Strategy for Infrastructure and Spatial Planning: Making the Netherlands Competitive, Accessible, Liveable and Safe*. The Hague: Ministry of Infrastructure and Environment.

Nautical Institute and the World Ocean Council. (2013). *The Shipping Industry and Marine Spatial Planning: A Professional Approach*. London: Nautical Institute.

Pınarbaşı, K., Galparsoro, I., Borja, Á., Stelzenmüller, V., Ehler, C. N., & Gimpel, A. (2017). Decision Support Tools in Marine Spatial Planning: Present Applications, Gaps and Future Perspectives. *Marine Policy, 83*, 83–91.

Ramesh, R., Chen, Z., Cummins, V., Day, J., D'Elia, C., Dennison, B., Forbes, D. L., Glaeser, B., Glaser, M., Glavovic, B., & Kremer, H. (2015). Land–Ocean Interactions in the Coastal Zone: Past, Present & Future. *Anthropocene, 12*, 85–98.

Soininen, N., & Hassan, D. (2015). Marine Spatial Planning as an Instrument of Sustainable Ocean Governance. In *Transboundary Marine Spatial Planning and International Law* (pp. 3–20).

Szydarowski, W., & VASAB Committee on Spatial Development. (2009). *VASAB Long-Term Perspective for the Territorial Development of the Baltic Sea Region*. Riga: Vasab Secretariat.

UNESCO and European Commission. (2017). Joint Road Map to Accelerate Marine/Maritime Spatial Planning Processes World Wide. Retrieved July 4, 2018, from http://www.unesco.org/new/fileadmin/MULTIMEDIA/HQ/SC/pdf/Joint_Roadmap_MSP_v5.pdf.

United Nations Conference on Trade and Development. (2017). *Trade and Development Report 2017: Beyond Austerity: Towards A Global New Deal*. Geneva: UNCTD.

United Nations General Assembly. (1982). *Convention on the Law of the Sea*. Retrieved July 4, 2018, from http://www.refworld.org/docid/3dd8fd1b4.html.

University of Liverpool. (2013). *ESTaDOR European Seas Territorial Development Opportunities and Risks: Executive Summary*. Liverpool: University of Liverpool.

University of Liverpool. (2016a). *Marine Proofing for Good Environmental Status of the Sea: Good Practice Guidelines for Terrestrial Planning*. Cardiff: WWF, Celtic Seas Partnership.

University of Liverpool. (2016b). *Nature's Services and the Sea: A Resource Pack for Marine and Coastal Stakeholders*. Cardiff: WWF, Celtic Seas Partnership.

University of Liverpool. (2016c). *Marine Proofing for Good Environmental Status of the Sea: Good Practice Guidelines for Terrestrial Planning: Country Fact Sheets*. Cardiff: WWF, Celtic Seas Partnership.

12

Linking Integrated Coastal Zone Management to Maritime Spatial Planning: The Mediterranean Experience

Emiliano Ramieri, Martina Bocci, and Marina Markovic

1 Introduction to the Mediterranean Basin

The Mediterranean Sea is the largest semi-enclosed sea in the world, stretching 4,000 km from east to west, with a maximum width of 800 km. Its coastline is approximately 46,000 km long, with nearly 19,000 km of island coastline (UNEP/MAP-Plan Bleu 2009). The mean depth of the Mediterranean Sea is 1,370 m, while the maximum is around 5,267 m (recorded at Calypso Deep, Greece). It is connected to the Atlantic Ocean through the Strait of Gibraltar. The Dardanelles, Marmara Sea and the Bosporus Strait connect it to the Black Sea, and the Suez Canal connects it to the Red Sea. The main rivers bringing significant water flow to the Mediterranean Sea are the Rhone, Po, Nile and Ebro Rivers (Saliot 2005).

Marine waters fall into different legal regimes, as defined by the United Nations Convention on the Law of the Sea (UNCLOS): internal waters, territorial sea, contiguous zone, Exclusive Economic Zones (EEZs), continental

E. Ramieri (✉) • M. Bocci
Thetis SpA, Venice, Italy
e-mail: emiliano.ramieri@thetis.it

M. Markovic
UNEP/MAP PAP/RAC, Split, Croatia

J. Zaucha, K. Gee (eds.), *Maritime Spatial Planning*,
https://doi.org/10.1007/978-3-319-98696-8_12

shelf and high seas. Delimitation of maritime boundaries in the Mediterranean is complex, mainly due to geographical, geopolitical and economic reasons, and requires complex agreements among neighbouring states. This results in several yet unsolved issues even with respect to territorial sea borders. Most Mediterranean states have established a 12-mile territorial sea, while declaration of EEZs through the adoption of the national legislation has been carried out only by some Mediterranean states (MRAG, IDDRA, and LAMANS 2013; DOALOS 2018). In addition, "EEZ derived zones", such as fisheries zones, fisheries protection zones, ecological protection zones and ecological and fishery protection zones, have been declared through national legislation by a number of states; however, these zones encompass only some of the rights that can be exercised within the EEZ. It should also be noted that claim for EEZs and "derived" zones, based on adopted national legislation, does not automatically lead to their full validation and implementation. Although the first formal step for the establishment and delimitation of a maritime zone is the adoption of legislation in the form of law, a number of additional steps are necessary before the final validation of the claimed zone and its boundary (MRAG, IDDRA, and LAMANS 2013; DOALOS 2018). Unlike the EEZ, a coastal state does not need to declare its continental shelf, as its existence is inherent. However, its delimitation (in line with art. 77 of UNCLOS) is often done in agreement with the neighbouring states. In the Mediterranean, there are some delimitation issues still pending also related to the continental shelf (Chevalier 2004). Taking into account the above-mentioned status of EEZs and "derived" zones, more than 20% of the marine waters in the Mediterranean fall under a high seas regime governed by international norms (Cinnirella et al. 2014). This limits interventions of coastal states in economic and environmental maritime affairs and calls for strong cooperation at the regional level.

Cooperation is particularly relevant for the preservation of natural and environmental conditions that are the basis for various economic activities and social benefits of Mediterranean states. Known as a biodiversity hotspot, the Mediterranean is rich in endemic flora and fauna, with biodiversity representing between 4% and 18% of all the marine species known worldwide (Piante and Ody 2015). Richness of species and habitats lead to outstanding aesthetic value which (apart from other values such as cultural heritage) represents a vital resource for tourism development. However, intensified coastal and maritime activities (including tourism) are often responsible for loss of biodiversity. To date, nearly 19% of assessed species are considered threatened by extinction (UNEP/MAP-Plan Bleu 2009).

Moreover, according to the analysis carried out by the MEDTRENDS project (Piante and Ody 2015), almost all Mediterranean maritime sectors (such as tourism, shipping, aquaculture, offshore oil and gas), except professional fisheries, are expected to grow during the next 15 years. Emerging sectors, such as renewable energy, seabed mining and biotechnology, are expected to grow even faster, although in absolute terms they will be less relevant than more traditional uses also in the future, and there is greater uncertainty on their possible evolution. Such growing development can increase existing conflicts between sectors and generate new ones; in addition, it will represent additional pressure on already stressed Mediterranean ecosystems. It also calls for strengthening collaboration among the countries, in order to ensure:

- reduction of overfishing and improvement of sustainable management of fish resources;
- management of maritime traffic specifically in congested or strategically important areas (e.g. the Adriatic Sea, the Aegean Sea and the connection to the Black Sea, the routes connecting to the Suez Canal, Gibraltar Strait);
- reduction of risk of ship collisions and environmental accidents; and
- management of conflicts that might arise from the exploitation of submarine natural gas and oil resources.

Having in mind the existing threats to the marine environment and the migratory nature of marine species, collaboration between the Mediterranean countries is particularly important for achieving Good Environmental Status (GES) of the sea. Therefore, the ongoing shift from habitat conservation approaches to biodiversity and ecosystem functioning approaches, beyond national boundaries, reflects much better the rationale which sustains the management and conservation of marine ecosystems. This shift calls for holistic, integrative and ecosystem-based frameworks (UNEP/MAP 2017).

Beyond this introduction, this chapter is structured into four sections. The following one illustrates policies supporting coastal and marine planning in the Mediterranean Sea, referring, in particular, to the cooperation framework of the Barcelona Convention. Section 3 discusses links between Integrated Coastal Zone Management (ICZM) and Maritime Spatial Planning (MSP) in this sea basin and the important role played by land-sea interactions (LSI); examples of practices are provided in Sect. 4. Finally, some elements that can support the future integration of ICZM and MSP in the Mediterranean Sea are provided. Part of the contents of this chapter is based on initial results and outputs of the ongoing SUPREME and SIMWESTMED projects (co-funded by the European Union [EU] through the EC-DG Maritime Affairs and Fisheries) to which the authors of the chapter directly contributed.

2 The Policy Frame for Coastal and Marine Planning and Management in the Mediterranean

The Convention for the Protection of the Marine Environment and the Coastal Region of the Mediterranean (Barcelona Convention; signed in 1976; amended in 1995) is the main policy achievement of the Mediterranean Action Plan (MAP) of the United Nation Environment Programme (UNEP). The contracting parties to the Barcelona Convention are 21 countries bordering the Mediterranean Sea (Albania, Algeria, Bosnia and Herzegovina, Croatia, Cyprus, Egypt, France, Greece, Israel, Italy, Lebanon, Libya, Malta, Monaco, Montenegro, Morocco, Slovenia, Spain, Syria, Tunisia and Turkey) together with the EU. The Barcelona Convention is the only regional, legal and regulatory framework for the protection of the entire Mediterranean marine and coastal environment providing for objectives and obligations agreed by all the contracting parties:

- "to prevent, abate, combat and to the fullest extent possible eliminate pollution of the Mediterranean Sea Area" and
- "to protect and enhance the marine environment in that area so as to contribute towards its sustainable development" (Barcelona Convention, art. 4).

The Barcelona Convention is complemented by seven protocols (Land-Based Source Protocol, Hazardous Wastes Protocol, Prevention and Emergency Protocol, Dumping Protocol, Offshore Protocol, Specially Protected Areas/Biological Diversity Protocol and ICZM Protocol) and a number of strategies and plans (UNEP/MAP 2015). In addition to the legal framework, MAP contributed in setting out an institutional framework for cooperation addressing common marine and coastal challenges. The MAP Coordinating Unit and its Regional Activity Centres (RACs) are acting as a technical mechanism assisting the Mediterranean governments to implement their respective commitments for the protection of the marine and coastal environment. Standing out, for more than 40 years, as a coherent legal and institutional framework of cooperation, the Barcelona Convention system is a platform that contributes to building trust among Mediterranean countries in the joint actions towards planning and management of marine and coastal activities.

The Ecosystem Approach (EcAp) represents the overarching guiding principle to all policy implementation and development undertaken under the auspices of the Barcelona Convention. EcAp is to be integrated in all of the Convention's policies and activities, as it makes explicit the links between the status of natural resource systems and the services they provide. It also seeks to maintain the integrity and functioning of ecosystems as a whole, and recognises that the impacts of human activities are a matter of social choice. In the context of MAP, EcAp refers to a specific process, as the contracting parties have committed to implementing the EcAp with the ultimate objective of achieving GES of the Mediterranean Sea and coast. They do so through informed management decisions, and based on integrated quantitative assessment and monitoring of the marine and coastal environment of the Mediterranean. Decision IG.21/3 (UNEP(DEPI)/MED IG.21/9; the so-called EcAp Decision) expresses the agreement on regionally common targets and lists of indicators to achieve GES in the Mediterranean.

The EcAp process under the Barcelona Convention shares many commonalities with the process of implementation of the EU Marine Strategy Framework Directive (MSFD; Directive 2008/56/EC) (EC 2008a): for example, achieving GES and Healthy Environment which are independent of national jurisdictional waters. Both aim to establish a Programme of Measures to achieve their respective goals by 2020. The subregional initial assessment prepared by MAP under the EcAp framework has been directly relevant to Mediterranean EU member states in their initial assessment required under MSFD. Even if MSFD is not applicable to the entire Mediterranean, its philosophy and principles could, nonetheless, be applied to the whole marine Mediterranean domain through the development of a shared vision via MAP. Both the MSFD and the MAP EcAp processes are committed to seeking mutual collaboration for the protection of the Mediterranean marine environment. However, there are important differences in the capacity for implementing specific measures or initiatives, with the implementation of such goals driven by different visions and concerns between different jurisdictions.

Planning of coastal and maritime activities is clearly taken on board by the Barcelona Convention and some of its protocols, primarily the Protocol on ICZM in the Mediterranean (UNEP/MAP/PAP 2008). Entered into force in 2011, the ICZM Protocol was a major innovation being the first (and still only) supranational legal instrument for coastal zone management (Rochette et al. 2012). ICZM is defined by the Protocol as a "dynamic process for the sustainable management and use of coastal zones, taking into account at the same time the fragility of coastal ecosystems and landscapes, the diversity of activities and uses, their interactions, the maritime orientation of certain activities and uses and their impact on both the marine and land parts." (ICZM Protocol, art. 2f). Spatial planning of coastal zones is an essential component of the ICZM Protocol, as one of the main objectives of ICZM is to "facilitate, through the rational planning of activities, the sustainable development of coastal zones by ensuring that the environment and landscapes are taken into account in harmony with economic, social and cultural development" (ICZM Protocol, art. 5).

In addition, the Protocol provides, for the first time, a common geographical criterion for the definition of coastal zones. Contrary to a common perception of the coastal zone as only the landward part from the coastline, the definition provided by the Protocol (art. 3) clearly includes the marine component as well; the coastal zone is the area between:

- the seaward limit of the coastal zone, which shall be the external limit of the territorial sea of parties; and
- the landward limit of the coastal zone, which shall be the limit of the competent coastal units as defined by the parties.

ICZM is therefore depicted as an integrated management approach, acknowledging that the coastal area is a whole system formed by both its land and sea components, with interdependent human uses and coastal resources. It, therefore, implies taking into account the interrelationships that exist between coastal uses and the environment they potentially affect. As elaborated within MedOpen, a permanent virtual training course on coastal management in the Mediterranean, ICZM requires integration at different levels, that is, across zones, time, sectors and disciplines. Still according to the Protocol, ICZM calls for reinforcement of institutional coordination, integration of sectoral policies and management approaches, as well as adoption of a participatory process facilitating horizontal and vertical dialogue, agreements and compromises between all parties involved in the use and management of coastal resources. There is no uniform approach to coastal management, and therefore there is no single way to apply ICZM in the Mediterranean. The experiences vary, reflecting the diversity of geographic conditions, policy priorities and specific concerns related to coastal areas. Therefore, agreed principles and methodological approach need to be respected, but also adapted to the country's national and local contexts.

While MSP is not expressly mentioned in the ICZM Protocol and can be considered a relatively new term within the frame of the Barcelona

Convention, the above makes clear that planning of marine space is a concept already taken on board by the Protocol. Specifically, spatial planning of the coastal zone is mentioned by the Protocol, with the sea clearly referred to as a component of the coastal zone.

As reported in the MAP Mid-Term Strategy 2016–2021 (UNEP(DEPI)/MED IG.22/28), the contracting parties of the Barcelona Convention at their 18th Ordinary Meeting (December 2013, Istanbul, Turkey) recommended to strengthen MAP activities on MSP as part of ICZM, in order to contribute to the GES of the Mediterranean Sea, investigate in more detail connections between land and sea areas and propose coherent and sustainable land and sea-use planning. Moreover, the opportunity to apply MSP is mentioned several times in the Mediterranean Strategy for Sustainable Development (MSSD) 2016–2025 (UNEP/MAP, 2016) and, in particular, under MSSD Objective 1, strategic direction 1.2: "Establish and enforce regulatory mechanisms, including Maritime Spatial Planning, to prevent and control unsustainable open ocean resource exploitation". Given these premises and following two years of work coordinated by MAP Priority Actions Programme Regional Activity Centre (PAP/RAC), the 20th Ordinary Meeting of the contracting parties to the Barcelona Convention, held in December 2017 in Tirana (Albania), adopted the "Conceptual Framework for Marine Spatial Planning" in the Mediterranean Sea (UNEP(DEPI)/MED IG.23/23). This is recognised as a guiding document to facilitate the introduction of MSP under the Barcelona Convention and, in particular, link it to ICZM, as well as to provide a common context to contracting parties for implementing MSP in the Mediterranean Region.

MSP, compared to land planning, is a fairly new and emerging process in the Mediterranean Region. In general, the process is at its initial stage and is highly influenced by differences among countries. These particularly relate to their institutional and legal framework and to some extent the availability of a reliable knowledge base (Policy Research Corporation 2011). The EU Directive on MSP (Directive 2014/89/EU) (EC 2014) is a key enabling factor (Zerkavi 2015) that has triggered concrete actions towards MSP implementation in EU member countries. All EU countries in the Mediterranean have finalised the transposition of the MSP EU Directive into national legislation and identified the competent MSP national authorities. Coordination mechanisms exist or are being created to improve cross-sector integration within MSP, and EU countries are busy developing other MSP-related activities, such as data collection and structuring, elaboration of guidelines, development of MSP methodologies, stocktaking of maritime uses and activities, elaboration of overarching vision/strategic elements and/or identification of the number of expected MSP plans and related geographic scope. Some initial actions have also been taken in some non-EU countries— for example, the advisory/strategic level "Israel Marine Plan" (Portman 2015) or the design and testing of a methodology for marine vulnerability assessment based on EcAp in Boka Kotorska Bay (Montenegro) with the explicit aim of supporting MSP (see Sect. 4). Nevertheless, MSP initiatives are still unbalanced between the two shores (northern and southern) of the Mediterranean Sea.

The Mediterranean context can rely on a wide number of cross-border projects—a few of them also involving non-EU countries—focusing on MSP or indirectly dealing with related aspects (e.g. MEDTRENDS, SHAPE, ADRIPLAN, SUPREME, MSP Med—Paving the Road to MSP in the Mediterranean, THAL-CHOR, SIMWESTMED and POCTEFEX-ALBORÁN "Cross-border Space of Nature Shared Management"). These projects have delivered a valuable set of MSP practices and tools.

Together with the EU Directive on MSP, the Conceptual Framework for MSP is expected to support dissemination of the MSP concept and further foster its implementation in the Mediterranean Sea in close interaction with ICZM.

3 Linking ICZM and MSP: The Importance of LSI in the Mediterranean Basin

In the Mediterranean context, there is an evident overlap of the geographical scope of ICZM, as defined by the Protocol on ICZM, and MSP as defined by Directive 2014/89/EU (EC 2014): both include the territorial sea. From this perspective, MSP can be seen as one of the main tools for implementing ICZM in the marine part of the coastal zone, also to avoid this overlap becoming an obstacle for their joint implementation. Considering the definition of the coastal zones in the ICZM Protocol (see Sect. 2), almost all other Protocols of the Barcelona Convention are related to this in one way or another. ICZM can therefore support the implementation of several of these Protocols; vice versa, the relevant objectives and provisions of these Protocols should be taken into account in all ICZM projects, plans and strategies. Given these links, the application of MSP within the framework and the geographic scope of the ICZM Protocol can contribute to the goals set by other protocols, as in the case of identification, planning and management of protected areas according to the Protocol concerning Specially Protected Areas and Biological Diversity in the Mediterranean (SPA/BD) or the protection of the

Mediterranean Sea against pollution resulting from exploration and exploitation of the continental shelf and the seabed and its subsoil (referring to the so-called Offshore Protocol).

ICZM and MSP share common principles, for example, sustainable management and development of coastal-marine areas, sustainable use of natural resources, importance of stakeholder participation and so on. Figure 12.1 highlights links among the principles identified by art. 6 of the ICZM Protocol and the MSP principles first included in the EC Roadmap (EC 2008b) and subsequently embedded in the EU MSP Directive. For example, the ICZM Protocol highlights the importance of adequate and timely participation in a transparent decision-making process by stakeholders concerned with the coastal zones (principle C4), which clearly matches MSP principles M4—Stakeholder participation; and M3—Developing MSP in a transparent manner.

Notwithstanding these evident commonalities, they are different processes, which need to be complementary and coherently implemented. According to the Protocol for the Mediterranean, ICZM essentially aims to ensure the sustainable management of coastal zones. It stresses the need for integration/cooperation among different governance bodies and policy sectors dealing with and active on the coast, as well as informed participation and cooperation of all stakeholders. The same can be applied to MSP as regards the sustainable management of marine areas. ICZM may result in strategies and management plans and might lead to the allocation of space to specific activities (through spatial planning), in the way that MSP does for the sea.

Both processes acknowledge the importance of applying the EcAp; Fig. 12.1 also highlights the main links between MSP/ICZM and EcAp principles (the latter as defined by UNEP/CBD/COP/5/23—Annex III).

Links between ICZM and MSP are particularly evident in the Mediterranean Sea. Some Mediterranean countries have not claimed EEZ or “derived” zones, which they might be entitled to establish under the international law (UNCLOS), while for some of the claimed EEZ or “derived” zones, full validation and implementation is still pending (Suarez de Vivero 2010; MRAG, IDDRA, and LAMANS 2013). This implies that in these countries MSP implementation focuses or will focus mainly on the territorial sea, which is also part of the geographic scope of the ICZM Protocol. Maritime activities tend to concentrate in coastal waters, and leading and emerging maritime sectors in the basin (such as shipping and port activities, aquaculture, small-scale fisheries and coastal tourism) have significant interactions with the land territory. Pure offshore activities in the Mediterranean are still limited. With

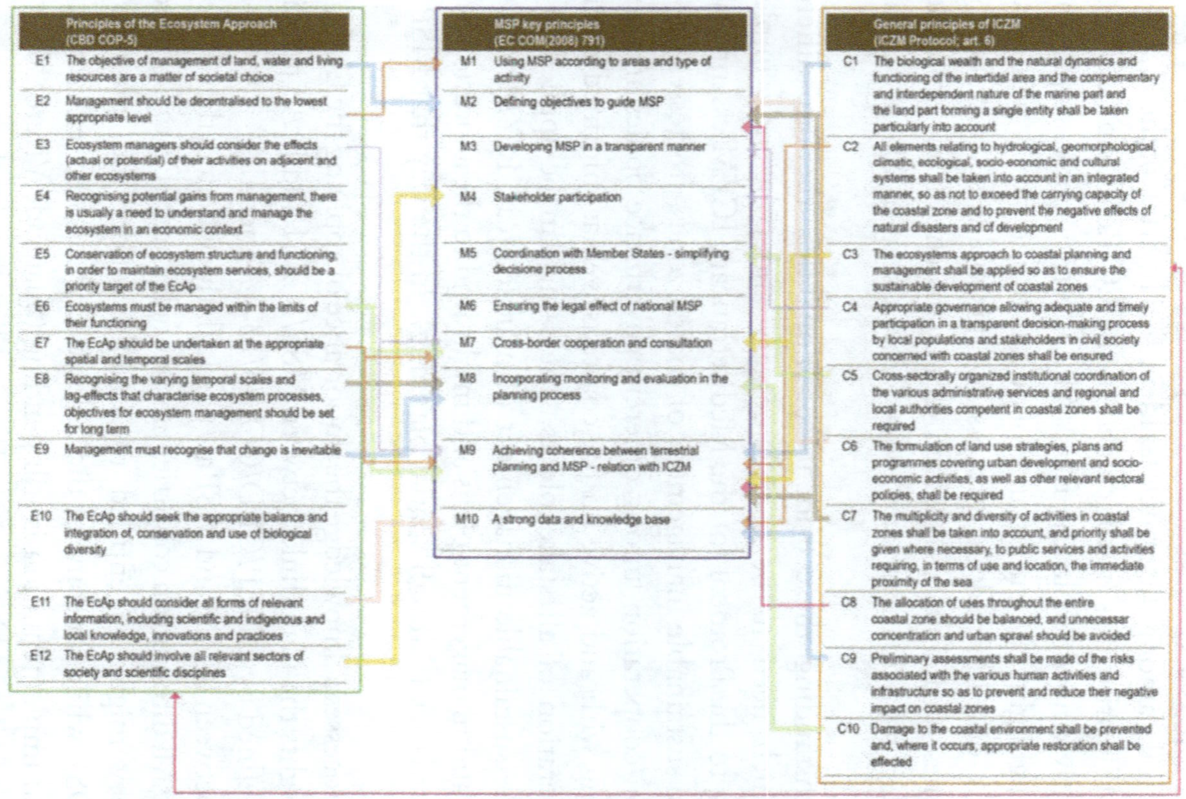

Fig. 12.1 Links among EcAp, MSP and ICZM principles (source: UNEP(DEPI)/MED IG.23/23)

the rapid expansion of maritime economy, these connections are becoming more and more relevant (Piante and Ody 2015).

The concentration of maritime activities along the coast and relevance of LSI are also related to some geographic features of the Mediterranean basin: a large number of islands, countries with a long and indented coastline (e.g. Italy, Croatia and Greece), a high concentration of people along the coast and the presence of important land-sea transition systems (e.g. deltas of Nile, Po, Rhone and Ebro or the numerous coastal lagoons; Cataudella et al. 2015).

According to the study "ESaTDOR—European Seas and Territorial Development, Opportunities and Risks" (ESPON and University of Liverpool 2013), the marine areas close to the coasts of Spain, France, Italy, Malta, Slovenia and northern Croatia are those with greater LSI intensity in the Mediterranean and can be classified as a "regional hub" of LSI, characterised by strong land-sea interactions, high maritime activities and employment (although less than those of the European Core for LSI, i.e. the English Channel and the southern coast of the North Sea), but also significant environmental pressures. Other hotspots emerge at a more detailed scale, as in the case of the Suez Canal, Athens and Piraeus port, the Strait of Gibraltar or the coastal area of Cyprus. Great parts of the Eastern Mediterranean can be considered as area of transitional LSI intensity, with medium environmental pressures and more narrow or localised concentration of maritime economy. The Alboran Sea is characterised in a similar way. Still according to the ESaTDOR study, the southern rim of the Mediterranean is categorised as rural area (with the exception of the Strait of Gibraltar and the Suez Canal), with relatively low environmental pressure but also low levels of maritime-related activities and employment, dominated by primary production and tourism.

Most relevant LSI challenges in the Mediterranean basin are linked to both socio-economic and environmental aspects, also considering that good environmental quality plays an essential role in sustaining important economic activities, such as coastal tourism and fishery. These challenges tend to vary within the basin and are specific at sub-basin level. They include climate change adaptation and disaster risk reduction (including both natural risks, e.g. coastal erosion and flooding, and technological risks, e.g. ship collision and oil spills); planning and management of connections between land and sea-borne transportation; coastal urbanisation and littoralisation; booming of coastal tourism; land-based impacts on the marine environment such as eutrophication, chemical contamination and plastic pollution along hotspot areas; degradation/transformation of land-sea transition systems; difficulties in establishing proper protection of vulnerable and high-value coastal-marine systems; and limited connection between coastal-marine and rural development.

All the elements described earlier call for a common implementation of ICZM and MSP, especially since they also share a number of procedural steps, for example, the creation of a strong and reliable data and knowledge base (corresponding to principles M10–C2, C9 of Fig. 12.1), the elaboration of a common and long-term vision and identification of strategic objectives (corresponding to principles M2–C6, C7), cross-sector and vertical integration (corresponding to principles M5–C5), stakeholder engagement (corresponding to principles M3, M4–C4) and so on. Indeed, the Barcelona Convention system, including its Protocols and specifically the ICZM Protocol, provides a common framework agreed at the level of the entire Mediterranean within which MSP implementation can be anchored and possibly spread beyond EU countries. From this perspective, ICZM and MSP are expected to work jointly in addressing common issues such as local socio-economic development of coastal communities or nature protection across land and sea. One of the major challenges affecting this integration is still the fragmentation of competences, which characterises both the land and sea components of the coastal area and which is even magnified when the two are considered together. The issue of competence fragmentation has been recognised as particularly relevant by the Mediterranean states since the phase of negotiation of the ICZM Protocol, when the specific request to establish appropriate coordination mechanisms to improve integration was emphasised (UNEP(DEC)MED WG. 270/5). By 2016, nearly half of the countries reported (to the ICZM Protocol/Reporting Questionnaire, based on COP Decision IG22/16) the establishment of coordination mechanisms, with a 60% increase compared to 2013 (PAP/RAC 2013). Where existing and operationally fully implemented, these mechanisms could provide a suitable platform for coordinated governance with MSP as well.

4 Practices from the Mediterranean Experience

Although the process of MSP implementation is at an initial stage in most of the Mediterranean countries, a wide range of project-based experiences are available, including some demonstrating the importance of encompassing coast and sea in marine planning and management, as outlined in Sect. 3. Linking ICZM process with MSP can significantly contribute to the effectiveness of plans, leading to easier overall planning processes and improving efficiency of implementation. Experiences with ICZM provide insights on several key MSP issues like considering land-sea interactions and applying the EcAp. In many cases, cross-sector dialogue mechanisms were identified and started in the Mediterranean under ICZM processes, as well as multilevel cooperation experiences. Stakeholder engagement and public participation are also widely practised within ICZM. All these activities can be efficiently capitalised on within MSP. Some examples of Mediterranean practices are illustrated in this section.

Applying the EcAp. As outlined in Sect. 2, the EcAp represents a framing element for ICZM in the Mediterranean. The Montenegrin experience on Boka Kotorska Bay is a good example for how this approach and related indicators can provide a basis for the MSP process. A pilot study was developed within the Project "Defining the methodological framework for marine spatial planning in Boka Kotorska Bay (Montenegro)" (PAP/RAC and MSDT 2017), focusing on Boka Kotorska Bay, which is one of the most vulnerable zones of the Montenegrin coastal area. The pilot study designed and tested an EcAp-based methodology for marine vulnerability assessment, considering the EcAp Ecological Objectives and using related EcAp indicators. The potential use of this approach to inform the MSP and ICZM processes was also tested. The EcAp-based vulnerability assessment included three main steps:

- Identification and mapping of data related to EcAp indicators, including indicators of the environmental state of the marine and coastal area (biodiversity and landscape features, such as habitat distributional range, population abundance of selected species and alike) and indicators of existing pressures (e.g. eutrophication, contamination, physical disturbance of the coastline).
- Attribution of values to the current state (i.e. value index) and pressures on the marine areas (i.e. impact index). By using different criteria (e.g. conservation status, rareness, endemism), the value index is applied to different components of the environment. The impact index reflects the intensity of the impact on the marine environment and is defined based on criteria related to exposure to and sensitivity of the marine environment to the pressures coming from existing human activities.
- Assessment of vulnerability, which depends on the current state of the marine environment (value index), the current intensity of pressures (impact index), characteristics of future activities and resilience of the marine environment to future activities (i.e. its capacity to absorb additional pressures). Based on expert opinion on the resilience of the marine environment to each individual future activity, a vulnerability value was assigned on a scale of 1–10 for each spatial unit.

Results of the vulnerability assessment pointed out the areas where proper management of coastal and maritime activities is needed, e.g. in terms of relocation of specific activities and/or the need to seek alternative solutions for marine uses (Fig. 12.2). The results of the vulnerability assessment can also underpin the identification of technological improvement needs or other measures needed to reduce the impacts of specific activities on the marine environment.

Engaging stakeholders. The Cypriote MSP pilot experience elaborated in 2014–2015 for the coastal and marine area of Limassol (south of Cyprus) in the framework of the project Cross-border Cooperation for Maritime Spatial Planning Development THAL-CHOR (ΘΑΛ-ΧΩΡ)[1] provides an example of joint planning for land/coastal-related activities and maritime sectors, which was based on stakeholder engagement across the entire process. The pilot experience developed tailored tools in order to communicate spatial information relevant for the plan and facilitate informed dialogue and cooperation. The process faced issues of great interest for ICZM and MSP and their integration: resolution of spatial conflicts between different uses of the sea and coastal areas, better coordination between different stakeholders and creation of conditions for achieving sustainable development in line with the strategy "Europe 2020" were among THAL-CHOR objectives.

The analysis performed in the MSP pilot experience identified a high concentration of coastal and maritime activities in the Limassol area, including shipping, ports activities, fisheries, aquaculture, tourism, military use, cables and pipelines, and securing freshwater supply. Oil and gas exploitation and offshore renewable energy production were also considered as potential future activities. Conflicts and compatibilities among these activities were analysed. Spatial data were structured in a common Web-GIS system, which was made available via the THAL-CHOR project website to share results and support stakeholder engagement. Great emphasis was placed on stakeholder engagement during pilot plan elaboration: results of the conflict analysis were shared with local stakeholders through consultation workshops. Despite the Limassol plan being a pilot MSP plan and, therefore, not legally binding, it represents a valuable MSP example embedding relevant aspects of ICZM. This MSP pilot experience and similar ones conducted in Lesvos and Rhodes (Greece) in the frame of the same project enabled designing and testing a methodology for the development of MSP plans.

Promote institutional coordination and integrated governance. The Coastal Area Management Programme (CAMP; UNEP/MAP 1999) funded

[1] www.mspcygr.info; accessed on 22 June 2018.

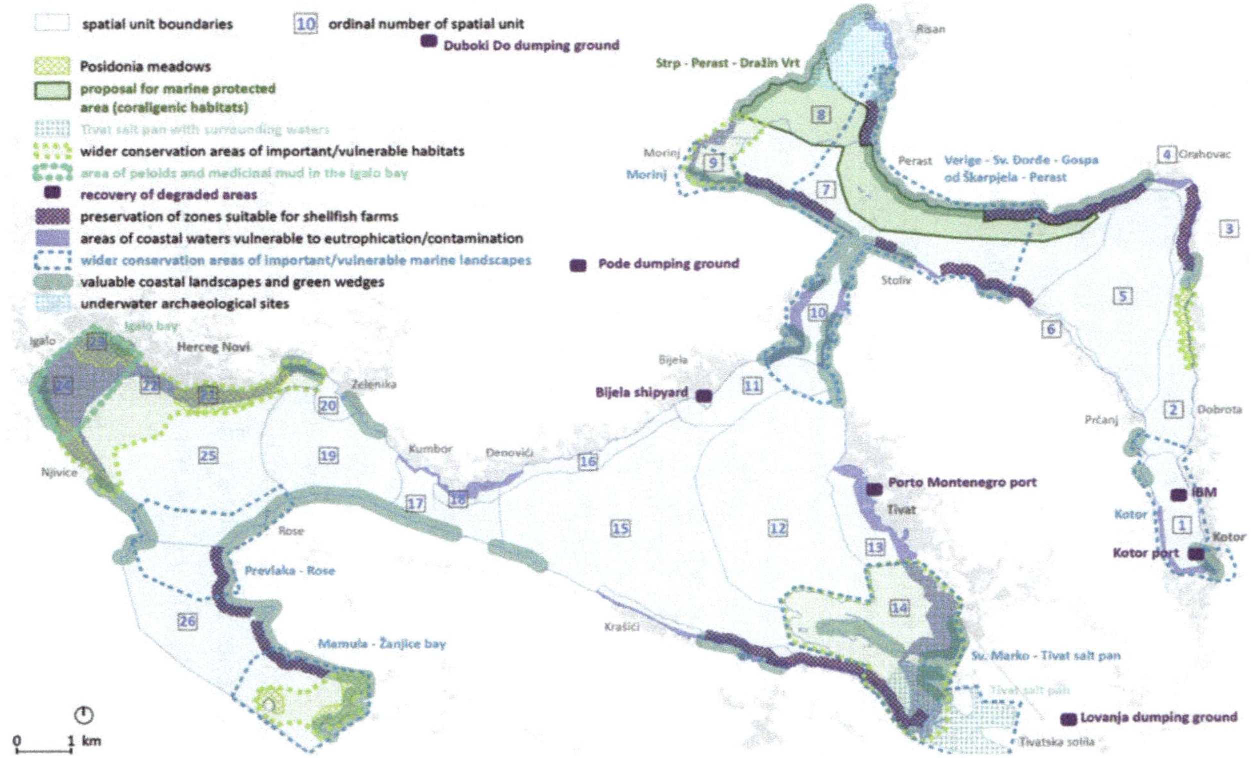

Fig. 12.2 Recommendations for marine and coastal planning in Boka Kotorska Bay (Montenegro) deriving from environmental vulnerability assessment (source: PAP/RAC and MSDT 2017)

by MAP, with co-financing from the respective countries, and coordinated by PAP/RAC, is oriented at the implementation of practical coastal management projects in selected Mediterranean coastal areas, applying ICZM as a major tool. CAMP projects have been implemented in most Mediterranean countries (UNEP/MAP/PAP 2015).

The CAMP project implemented in France (2014–2017) focused on the Var Department in Provence-Alpes-Côte d'Azur Region, covering an area of about 6,000 km^2, with 432 km of coastline, including ten islands and islets (70 km are occupied by military establishments, 40 km are urban areas and ports, while 92 km are beaches and 230 km are rocky shores). Objectives of the project were to identify and support local initiatives contributing to ICZM Protocol implementation; to facilitate and encourage more effective and coherent governance of the Var territory with its multiple layers of policies and regulations at different scales; and to develop transferable methodologies which can benefit other territories. Specifically, the actions of the project included:

- setting up of a consultative body, named *Terre et Mer Var Forum*, organising informal mediation meetings that facilitated exchanges between the different stakeholders of the Var coastal area;
- development of an evaluation study of sea and coastal management and planning policies in the Var, with regard to their relevance for the implementation of the ICZM Protocol; and
- development of operational actions, in partnership with civil society and institutions. These included concrete initiatives such as the management of ports and navigation basins but also educational initiatives aiming at developing a cultural "brand" for the Mediterranean islands (named "Archipelago of excellence"), like, for example, the production of films on some of the small Mediterranean islands and their surrounding maritime area, for example, the Specially Protected Areas of Mediterranean Importance (SPAMI), namely, the Port-Cros National Park and the archipelago of Embiez/Six-Fours.

The *Terre et Mer Var Forum* was set up to be a place of mediation and governance at the local scale of the Var Department, supporting the actions of key local actors such as the *Conservatoire du littoral*. The latter is a state institution managing public coastal land for conservation purposes and ensuring free access to the public. Biodiversity, aesthetics and cultural identity of the littoral are protected, also with direct engagement of local communities and associations. The Forum was conceived as a laboratory to stimulate a dynamic dialogue among coastal stakeholders, to advance in the development of an integrated governance mechanism and to highlight the successes and difficulties encountered in the implementation of ICZM policies. Although created within the CAMP France project, the Forum continues to operate, facilitating discussions related to coastal and marine management stimulated by other projects as well, like SIMWESTMED.

Ensure integrated managed development of coastal and marine areas. The proposals for ICZM-MSP for the Emilia Romagna Region (Italy) formulated within the Italian National Project RITMARE (Barbanti and Perini 2018) provide an example of integrated scenario analysis for the coast and the sea, aiming to address and guide the growth and development of coastal and maritime activities. The analysis consisted of the following steps: (1) update of existing assessments; (2) analysis of expected trends of maritime activities in the area; (3) analysis of conflicts, synergies and cumulative impacts; (4) definition of a common vision and related objectives; (5) identification of a portfolio of spatial measures; and (6) elaboration of an integrated scenario. Sectors, Departments and Services of the Emilia Romagna Region were engaged along the different steps.

A vision was developed where tourism is seen as pivotal for economics, ensuring it also acts as promoter of other economic sectors and does not compromise natural resources, thus supporting the regional economy. A portfolio of spatial measures was identified referring to six major uses: coastal defence, decommissioning of oil and gas offshore platforms, development of offshore wind farms, fishery and aquaculture, environmental protection and management of areas used for military purpose. The set of measures has the following objectives:

- Sustainable development through establishment of new uses, promoting Blue Growth in the area by overcoming existing barriers but safeguarding the uses already in place
- Reduction of conflicts and increase of synergies among uses
- Reduction of environmental impacts, particularly in the area between 0 and 6 nm, and increase the level of protection of relevant habitats and species

An integrated "managed development" scenario was finally developed which considers all the measures implemented at the same time and aims to pursue the above objectives in an integrated way. This led to an assessment of the possible overall reduction in use conflicts and cumulative impacts as a result of applying the proposed measures.

Integrate different specific spatial policy for coastal and marine areas. The variety and heterogeneity of coastal and marine environments in the Mediterranean call for a tailored approach for the specificities of different areas. In order to enable the implementation of specific policy measures, the Israel Marine Plan (Portman 2015), an initiative of a group of researchers and planners at the Faculty of Architecture and Town Planning's Center for Urban and Regional Studies at the Technion,[2] proposes to divide Israeli marine space into five functional marine areas that are spatially distinguished from each other. The distinction between the functional marine areas lies within the priorities and reciprocal relationships between uses or actions in each of them. Each of the marine areas is therefore an exploration area that offers a variety of opportunities and yet enables managing conflicts and attaining synergies in accordance with the goals set by the plan. Decisions made according to the distinctive guidelines for each one of the marine areas would thus enable the social, economic and environmental functioning of the entire marine space. The areas are identified as follows:

- "Marine protected areas" considered the most protected areas among all those included in the plan. the areas. They are the main exploration areas for designations related to nature and landscape preservation and protection. The policy proposed here is a proactive policy for high priority location and approval of marine protected areas at various levels.
- "Marine shared areas" spread out between the "horizon line" as delineated in the plan and the boundary of the coastal shipping corridor that stretches to its west, and between strips of sections of the "Marine protected area". They are development-oriented areas and are the main exploration areas for intervention and development designations. Here, too, a proactive policy is suggested for locating, at a high priority, sustainable human usages.
- "Marine horizon areas" are visible from densely populated urban shores and extend in front of them; therefore, they are very sensitive from a social point of view and constitute a significant public resource. Decision-makers will act to preserve an open landscape, to decrease the risk from existing and future infrastructure facilities and to utilise this area for purposes of leisure and recreation. This typology of area also includes vulnerable habitats and a main concentration of heritage values linked with the shore; it also includes harbours and other coastal infrastructures. Proximity to the shore also implies proximity to sources of pollution.

[2] msp-israel.net.technion.ac.il; accessed on 22 June 2018.

- "Marine shared-protected area" spreads over the back of the horizon space and between the strips of the marine protected area up to the edge of the continental slope. The policy proposed for this area is a reactive policy intended to preserve this area as an open marine space and to view it as a secondary exploration area for additional protected areas, as well as a future secondary exploration area for limited constrained human uses.
- "Deep sea area" spreads out from the edge of the continental slope up to the limit of the economic waters. The proposed policy for this area will give high priority to exploration intended for human usage.

5 Ways Forward

The EU MSP Directive is one of the main enabling factors for MSP implementation in some of the Mediterranean Sea countries. However, its application is limited to EU member states. Nevertheless, the principle of sincere cooperation, including cooperation with non-EU countries, is fundamental for the implementation of such a Directive. Embedding MSP in the ICZM process defined by the Barcelona Convention can provide a wider, common and legally based framework for MSP implementation in the entire Mediterranean. From this perspective, the adoption of the Conceptual Framework for Marine Spatial Planning in the Mediterranean Sea represents an initial step in this direction. Integration of ICZM and MSP is an important component of the activities that MAP is carrying out in the biennium 2018–2019. This specifically includes the ongoing elaboration of the Common Regional Framework on ICZM, within which the Conceptual Framework for MSP is to be integrated. Together with the essential role the EU MSP Directive plays for the EU countries, this process is expected to contribute to the dissemination and implementation of the MSP concept in the coming years.

The importance of applying integrated ICZM-MSP in the Mediterranean Sea also stems from the high relevance of environmental, social and economic land-sea interactions which characterise this sea basin. Indeed, integrating ICZM and MSP would also seem highly relevant in other semi-enclosed basins, such as, in particular, the Black Sea (Golumbeanu and Nicolaev 2015).

Notwithstanding the relevance of a pan-Mediterranean approach to MSP, it is also important to acknowledge that this sea basin is characterised by subregional specificities. As highlighted by the Conceptual Framework for MSP, a multi-scalar approach is recommended to tailor a common approach to specific characteristics. The scale of the entire Mediterranean might be relevant for defining common strategic goals and approaching transboundary challenges; some of them might assume specific significance at the subregional level. With the gradual introduction of MSP, the Barcelona Convention can provide an institutional framework for cooperation in the Mediterranean also for marine planning and management. At the same time, at the subregional level, other cooperation mechanisms can play a relevant role, for example, the EU Strategy for the Adriatic Ionian Region (EUSAIR). The multi-scalar approach is completed by the national and, in some cases, subnational levels which are expected to implement the statutory MSP processes.

Although challenging, the establishment of new EEZs and the full application of existing ones would extend the area of MSP implementation, providing opportunities for both the managed exploitation of marine resources and space as well as for improved conservation (Katsanevakis et al. 2015). The resolution of pending disputes on maritime borders would also help in defining a clear legal basis for MSP implementation in the Mediterranean.

Project-based experience on (cross-border) MSP in the Mediterranean is rather rich and keeps growing, focusing also on integration between marine and coastal planning. This has delivered a variety of practices (such as data-sharing infrastructures, tools, methodologies, handbooks, guidelines, recommendations and pilot plans) that can be transferred and used in the formal MSP processes, considering the necessary phases of testing and adaptation. Although the uptake of project outcomes still needs to be improved, the main challenge lies in the unbalanced distribution of experiences between EU and non-EU countries, also affecting data availability and accessibility. Some recently started or upcoming initiatives will contribute to filling this gap, for example, the MSP project in Albania and Montenegro (Implementation of the EcAp in the Adriatic Sea through MSP) funded by the Global Environmental Facility, which was officially launched in May 2018, or the pilot project on cross-border MSP to be launched in the Western Mediterranean according to the "Joint Roadmap to accelerate Maritime/Marine Spatial Planning processes worldwide" adopted on 24 March 2017 by the Intergovernmental Oceanographic Commission of UNESCO and the Directorate-General for Maritime Affairs and Fisheries of the European Commission. CAMP projects implemented up to now (18 plus 1 in preparation, each in a different country) have resulted in the successful spread of the ICZM concept around the entire Mediterranean and in testing the application of different provisions of the ICZM Protocol. A second round of CAMP projects could focus on integrating spatial planning of the sea with the overarching ICZM framework at the national or subnational scale, thus further contributing to increased capacity building in MSP in non-EU countries.

Acknowledgements The authors acknowledge that part of the contents of this chapter derives from their contribution to the ongoing EU co-funded SUPREME and SIMWESTMED projects. In addition, the authors thank Ms Željka Škaričić (PAP/RAC) and Mr Sylvain Petit (PAP/RAC) for their suggestions and final reading of the chapter. We are grateful to the project "Economy of maritime space" funded by the Polish National Science Centre for contributing the Open Access fee for this chapter and facilitating our discussions and preparation of the book.

References

Barbanti, A., & Perini, L. (Eds.). (2018). Between Land and Sea: Analysis and Proposals for Maritime Spatial Planning in Emilia-Romagna Region (*in Italian*: Fra la terra e il mare: Analisi e proposte per la pianificazione dello Spazio Marittimo in Emilia Romagna). https://doi.org/10.5281/zenodo.1184364.

Cataudella, S., Crosetti, D., & Massa, F. (2015). Mediterranean Coastal Lagoons: Sustainable Management and Interactions Among Aquaculture, Capture Fisheries and the Environment. General Fisheries Commission for the Mediterranean, Food and Agriculture Organisation of the United States, Studies and Reviews No. 95. Retrieved from http://www.fao.org/3/a-i4668e.pdf.

Chevalier, C. (2004). Governance in the Mediterranean Sea. Legal Regime and Prospective. IUCN Centre for Mediterranean Cooperation. Retrieved from https://cmsdata.iucn.org/downloads/legalspects_en_1.pdf.

Cinnirella, S., Sardà, R., Suárez de Vivero, J. L., Brennan, R., Barausse, A., Icely, J., Luisetti, T., March, D., Murciano, C., Newton, A., O'Higgins, T., Palmeri, L., Palmieri, M. G., Raux, P., Rees, S., Albaigés, J., Pirrone, N., & Turner, K. (2014). Steps Toward a Shared Governance Response for Achieving Good Environmental Status in the Mediterranean Sea. *Ecology and Society, 19*(4), 47. https://doi.org/10.5751/ES-07065-190447.

DOALOS—Division for Ocean Affairs and the Law of the Sea, Office of Legal Affairs, United Nations Secretariat. Maritime Space: Maritime Zones and Maritime Delimitation. Retrieved June, 2018, from http://www.un.org/Depts/los/LEGISLATIONANDTREATIES/index.htm.

EC. (2008a). Directive 2008/56/EC of the European Parliament and of the Council of 17 June 2008, establishing a framework for community action in the field of marine environmental policy (Marine Strategy Framework Directive). Retrieved from https://eur-lex.europa.eu/legal-content/EN/TXT/PDF/?uri=CELEX:32008L0056&from=EN.

EC. (2008b). EC COM(2008)791 final. Communication from the Commission: Roadmap for Maritime Spatial Planning: Achieving Common Principles in the EU. Retrieved from https://eur-lex.europa.eu/LexUriServ/LexUriServ.do?uri=COM:2008:0791:FIN:EN:PDF.

EC. (2014). Directive 2014/89/EU of the European Parliament and of the Council of 23 July 2014 Establishing a Framework for Maritime Spatial Planning. Retrieved from https://eur-lex.europa.eu/legal-content/EN/TXT/PDF/?uri=CELEX:32014L0089&from=EN.

ESPON & University of Liverpool. (2013). *ESaTDOR European Seas and Territorial Development, Opportunities and Risks*. Applied Research 2013/1/5. Final Report, Version 15/4/2013. Retrieved from https://www.espon.eu/programme/projects/espon-2013/applied-research/esatdor-european-seas-and-territorial-development.

Golumbeanu, M., & Nicolaev, S. (Eds.). (2015). *Study on Integrated Coastal Zone Management*. 454 pp. Ex Ponto Publishing House, 2015. ISBN: 978-606-598-397-7

Katsanevakis, S., Levin, N., Coll, M., Giakoumi, S., Shkedi, D., Macklewotrh, P., Levy, R., Velegrakis, A., Koutsoubas, D., Caric, H., Brokovich, E., Ozturk, B., & Kark, S. (2015). Marine Conservation Challenges in an Era of Economic Crisis and Geopolitical Instability: The Case of the Mediterranean Sea. *Marine Policy, 51*, 31–39. https://doi.org/10.1016/j.marpol.2014.07.013.

MRAG, IDDRA, & LAMANS. (2013). *Costs and Benefits Arising from the Establishment of Maritime Zones in the Mediterranean Sea*. Final Report. Retrieved from https://ec.europa.eu/maritimeaffairs/sites/maritimeaffairs/files/docs/body/maritime-zones-mediterranean-report_en.pdf.

PAP/RAC. (2013). Ensuring Appropriate Coordination: An Explanatory Report of Article 7 of the ICZM Protocol. Project SHAPE. Priority Actions Programme – Regional Activity Centre, Split. 34 pp. Retrieved from https://www.pap-thecoastcentre.org/pdfs/explanatory_article_7_outline_final_feb13.pdf.

PAP/RAC & MSDT. (2017). Vulnerability Assessment of the Marine Environment in the Boka Kotorska Bay. Methodological Guidelines (Analiza ranjivosti morske sredine u Bokokotorskom zalivu. Metodološke smjernice). Priority Actions Programme Regional Activity Centre; Ministry of Sustainable Development and Tourism (Montenegro). Podgorica, 2017.

Piante, C., & Ody, D. (2015). *Blue Growth in the Mediterranean Sea: The Challenge of Good Environmental Status*. MEDTRENDS Project. WWF-France. 192 pp.

Policy Research Corporation. (2011). *Exploring the Potential for Maritime Spatial Planning in the Mediterranean Sea*. Final Report. Study Carried Out on Behalf of the European Commission Directorate-General for Maritime Affairs and Fisheries. Retrieved from https://ec.europa.eu/maritimeaffairs/documentation/studies/study_msp_med_en.

Portman, M. E. (2015). Marine Spatial Planning in the Middle-East: Crossing the Policy Planning Divide. *Marine Policy, 61*(2015), 8–15. https://doi.org/10.1016/j.marpol.2015.06.025.

Rochette, J., Wemaëre, M., Billé, R., & du Puy-Montbrun, G. (2012). A Contribution to the Interpretation of Legal Aspects of the Protocol on Integrated Coastal Zone Management in the Mediterranean. UNEP/MAP, Priority Actions Programme—Regional Activity Centre, Split, 72 p. + annexes. Retrieved from https://www.pap-thecoastcentre.org/regional_medpartnership_workshop/documents/ICZM%20Protocol_Legal%20aspects.pdf.

Saliot, A. (Ed.). (2005). *The Mediterranean Sea. The Handbook of Environmental Chemistry*. Springer-Verlag Berlin Heidelberg. 410 pp.

Suarez de Vivero J. L. (2010). *Jurisdictional Waters in the Mediterranean and Black Seas*. Study Carried Out for European Commission Directorate-General for Internal Policies of the Union—Policy Department B: Structural and Cohesion Policy—Fisheries, 134 pp.

UNEP(DEC)MED WG. 270/5. Draft Protocol on Integrated Coastal Zone Management in the Mediterranean. Meeting of the MAP Focal Points. Athens, Greece, 21–24 September 2005. Retrieved from https://wedocs.unep.org/bitstream/handle/20.500.11822/5515/05wg270_5_eng.pdf?sequence=1&isAllowed=y.

UNEP(DEPI)/MED IG.21/9—Annex II—Thematic Decisions. Decision IG.21/3 on the Ecosystems Approach including adopting definitions of Good Environmental Status (GES) and targets. Retrieved from https://wedocs.unep.org/rest/bitstreams/8206/retrieve.

UNEP(DEPI)/MED IG.22/28. Decision IG.22/1: UNEP/MAP Mid-Term Strategy 2016–2021. Retrieved from https://wedocs.unep.org/rest/bitstreams/8364/retrieve.

UNEP(DEPI)/MED IG.23/23. Decision IG.23/7: Implementation of the Integrated Coastal Zone Management Protocol: Annotated Structure of the Common Regional Framework for Integrated Coastal Zone Management and Conceptual Framework for Marine Spatial Planning. Retrieved from http://wedocs.unep.org/bitstream/id/74412/17ig23_23_2307_eng.pdf.

UNEP/CBD/COP/5/23—Annex III. Report of the Fifth Meeting of the Conference of the Parties to the Convention on Biological Diversity. Annex III: Decisions Adopted by the Conference of the Parties to the Convention on Biological Diversity at its Fifth Meeting, Nairobi, 15–16 May 2000. Retrieved from https://www.cbd.int/doc/meetings/cop/cop-05/official/cop-05-23-en.pdf.

UNEP/MAP. (1999). *Formulation and Implementation of CAMP Projects: Operational Manual*. MAP-PAP/RAC, Athens-Split. 86pp. Retrieved from https://pap-thecoastcentre.org/itl_public.php?public_id=28.

UNEP/MAP. (2015). The Mediterranean Action Plan. Barcelona Convention and its Protocols. Overview. United Nations Environment Programme/Mediterranean Action Plan. Retrieved from https://wedocs.unep.org/rest/bitstreams/1298/retrieve.

UNEP/MAP. (2016). *Mediterranean Strategy for Sustainable Development 2016–2025*. Valbonne: Plan Bleu, Regional Activity Centre. Retrieved from https://planbleu.org/sites/default/files/upload/files/MSSD_2016-2025_final.pdf.

UNEP/MAP. (2017). *2017 Mediterranean Quality Status Report*. UNEP(DEPI)/MED IG.23/Inf.10/rev.1.

UNEP/MAP/PAP. (2008). *Protocol on Integrated Coastal Zone Management in The Mediterranean*. Priority Actions Programme—Regional Activity Centre, Split, 118 pp. Retrieved from https://www.pap-thecoastcentre.org/pdfs/Protocol_publikacija_May09.pdf.

UNEP/MAP/PAP. (2015). *Assessment of Coastal Area Management programme (CAMP) Projects*. Mediterranean Action Plan, Priority Actions Programme Regional Activity Centre. Split, Croatia. 76 pp. Retrieved from https://pap-thecoastcentre.org/itl_public.php?public_id=28.

UNEP/MAP–Plan Bleu. (2009). *State of the Environment and Development in the Mediterranean—2009*. Plan Bleu, Athens, Greece, 200 pp. Retrieved from https://planbleu.org/sites/default/files/publications/soed2009_en.pdf.

Zerkavi, A. (2015). Introducing Maritime Spatial Planning Legislation in the EU: Fishing in Troubled Waters? *Maritime Safety and Security Law Journal*, 1/2015, 95–114. Retrieved from http://www.marsafelawjournal.org/wp-content/uploads/2015/09/Issue1_Zervaki_Article.pdf.

13

Stakeholder Processes in Marine Spatial Planning: Ambitions and Realities from the European Atlantic Experience

Sarah Twomey and Cathal O'Mahony

1 Introduction

Bordered by two oceans and four seas, the European Union (EU) has the largest maritime territory in the world with marine regions accounting for over five million jobs and generating 40% of its gross domestic product (GDP) (EC 2017). The EU's political economy is inexorably linked to the marine environment and ensuring the health of marine ecosystems is necessary for the future of ocean biodiversity and sustaining maritime development. The scale and diversity of coastal and marine activities across Europe's regional seas thus present huge challenges for governance and policy frameworks. Governance sets the stage within which management occurs (Olsen 2003) and its success is key to dealing with conflict and escalating pressures on the marine environment. The principles of good environmental governance are well documented: openness; participation; transparency; and, accountability (Wingqvist et al. 2012; Lockwood et al. 2010; Heldaweg 2005). In particular, effective governance goes beyond information provision and consultation by governments; it requires the active participation of stakeholders (Colvin et al. 2016; Reed 2008). Stakeholders represent a host of

S. Twomey (✉) • C. O'Mahony
MaREI Centre for Marine and Renewable Energy, Environmental Research Institute (ERI), University College Cork, Cork, Ireland
e-mail: s.twomey@ucc.ie

J. Zaucha, K. Gee (eds.), *Maritime Spatial Planning*,
https://doi.org/10.1007/978-3-319-98696-8_13

marine activities operating in the seas comprising diverse statutory, regulatory, commercial and societal perspectives; they are gatekeepers to a vast amount of experience, knowledge, values and interests and play a pivotal role in contemporary marine governance.

Blue Growth is an EU long-term strategy to harness the untapped potential of Europe's oceans, regional seas and coasts for jobs and growth (EC 2012). Specific activities have been earmarked for additional effort—aquaculture, coastal tourism, marine biotechnology, ocean energy and seabed mining. The success of the Blue Growth agenda is contingent on ecosystem health, and current and future activities need to be carefully planned in relation to each other and the surrounding environment (i.e. the ecosystem approach). Marine Spatial Planning (MSP) is recognised as a key mechanism for achieving these goals and applies the ecosystem approach to conduct integrated, forward-looking and strategic decisions on human uses of the sea (Ehler and Douvere 2009).

2 Aims and Objectives

This chapter presents a contribution to the MSP literature that is practice-based and is primarily targeted towards planners and a general audience. While the majority of the existing Europe-focused literature has originated from experiences in the semi-enclosed Baltic Sea and the North Sea, this chapter focuses on a selection of coastal nations bordering Europe's Atlantic sea basin—coastal nations which heretofore have not featured prominently in the literature were chosen so as to provide new insight to the challenges along the European Atlantic coastline. The aim of this chapter is to illustrate how different state-based MSP settings affect the type and degree of stakeholder participation in practice. In particular, the focus is on recent trends in stakeholder participation in MSP across different geographic, ecological and socio-political contexts from the island of Ireland (i.e. Ireland and Northern Ireland) and the Iberian coast (i.e. Spain and Portugal). The chapter outlines the complexities and practical challenges associated with the ambitions and realities of delivering multi-sector participatory MSP processes. Insights are drawn from multiple case studies of stakeholder processes including research-based transboundary MSP pilot projects from the northern and southern European Atlantic, and statutory initiatives at different stages of MSP implementation.

3 Methodology

The methodological approach comprised a desk-study analysis of peer-review and grey literature (e.g. project reports, policy statements) relevant to the EU Atlantic region. The literature analysis included outputs from projects which were informed by activities focused on the engagement of stakeholders through workshops and interviews to ascertain their views and opinions on the implementation of MSP in different jurisdictional settings. Findings to emerge from additional semi-structured interviews with stakeholders engaged in MSP processes, and the recent advancement in implementation, were also incorporated into analysis for this chapter.

4 Conceptualising Stakeholder Processes

Multiple distinctions relating to the term 'stakeholder' can be found throughout relevant literature. Definitions are not consistently used and can mean numerous things in different management and regulatory contexts (Long 2012). The phrase first emerged in the realm of corporate governance in the 1930s (Preston and Sapienza 1990). In recent decades, it has become widely used in the field of environmental governance and particularly in the marine and maritime sphere (Fig. 13.1).

In MSP, the term 'stakeholder' refers to *any individual, group, or organisation that are or will be affected, involved or interested (positively or negatively)* and can be classified into the following three broad categories:

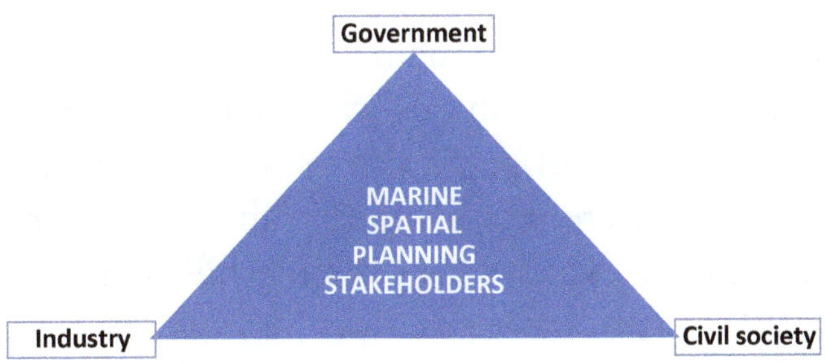

- Government decision-makers at various levels (i.e. government stakeholders including ministries, state agencies, municipalities and local government);
- Commercial or industry stakeholders representing the key marine sectors operating in the area;
- Civil-society stakeholders represented by the research community, citizen and community-based organisations, non-governmental organisations (NGOs), and conservation groups.

Fig. 13.1 Widely accepted definition of stakeholders and their categories in MSP and marine governance (based on Pomeroy and Douvere 2008; Long 2012; Roxburgh et al. 2012; Flannery et al. 2015; Jay 2015, Jay et al. 2016)

The terms participation, engagement and consultation are regularly used interchangeably to signify a process by which individuals and groups (i.e. stakeholders) converge to communicate, interact, exchange information, provide input or share in decision-making. Although these words are often used synonymously in policy documents and academic literature, they have different meanings. In particular, participation can mean many things to different people. It is frequently used as an umbrella term to describe activities ranging from information provision, public consultation, discussions with the public, or co-decision-making and partnerships.

It is important to take into account that the scope and extent of stakeholder participation differs greatly across regions and from country to country. The level of involvement will also largely depend on the political or legal requirements for participation that already exist in a country or region. In particular, various countries in Europe have used different ways to involve stakeholders in their MSP initiatives (Fig. 13.2).

5 Institutional Context for Participation in Marine Spatial Planning

This section outlines they key legal, policy and institutional frameworks that govern contemporary approaches to MSP and the obligations that exist for involving stakeholders.

Early and effective stakeholder participation is a fundamental aspect of the Ecosystem Approach and is also a legal requirement under a host of other different international and European instruments presented in Table 13.1.

The Rio Declaration on Environment and Development (1992) is a document widely regarded as the founding charter of sustainable development. The following principles are of particular relevance: Principle 10 emphasises that environmental issues are best handled with the participation of all concerned citizens; Principle 20 advocates for the full participation of women; while, Principle 22 refers to indigenous peoples and their communities. The Rio conference also led to the approval of Agenda 21, a comprehensive blueprint of action for the twenty-first century to be implemented globally, nationally and locally by UN organisations and the world's governments. The text of Agenda 21 is an extensive 351-page document with multiple references to participation and participatory mechanisms (Charnoz 2009).

Under the Convention on Biological Diversity (CBD), Annex of COP 6 Decision VI/19 specifies the need to ensure the participation of major stakeholders from different sectors in sustainable development and biodiversity

STAGE ONE: ***INFORMING***	Communicating information and raising awareness of the MSP process outside of government institutions with industry and civil society stakeholders with no avenue for receiving feedback	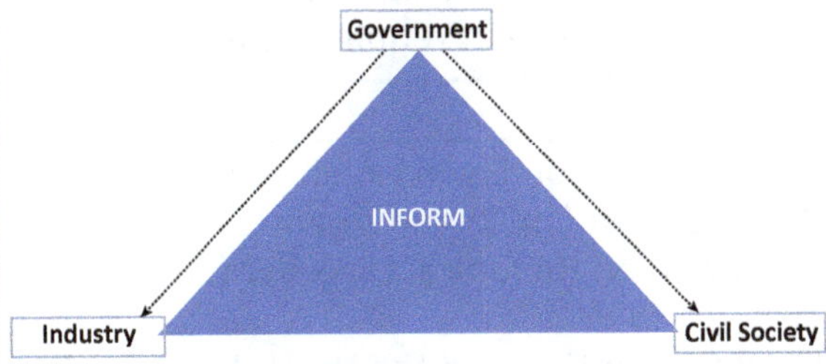 **One-way process/ Top-down**
STAGE TWO: ***PUBLIC CONSULTATION***	Industry and civil society stakeholders providing feedback to government decision-makers on potential decisions and alternatives through a formal statutory process.	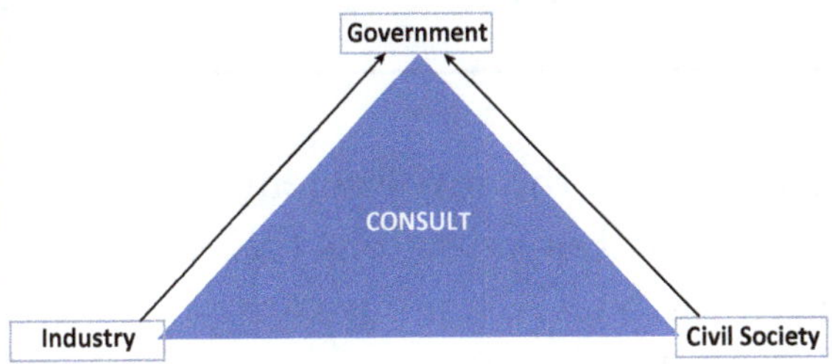 **One-way process/ Top-down**
STAGE THREE: ***STAKEHOLDER ENGAGEMENT***	Working directly with government ***throughout*** the process (and especially before the consultation phase) to ensure that stakeholder concerns are understood and considered in decision-making processes.	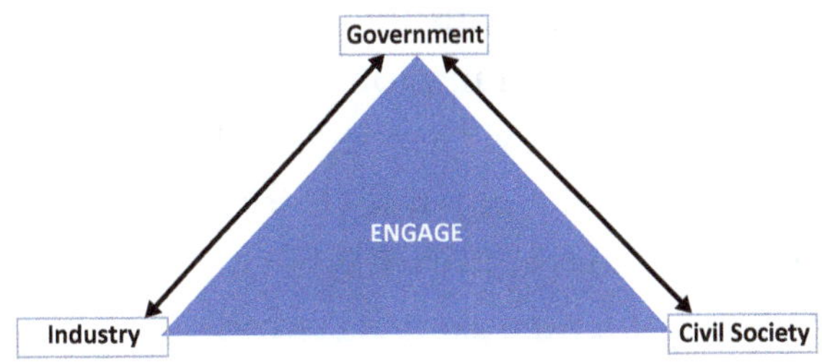 **Two-way dialogue/ Top-down**
STAGE FOUR: ***STAKEHOLDER COLLABORATION***	Partnering with government in each aspect of the decision-making process including the development of alternatives and the identification of the preferred solution (Blend of top-down and bottom-up)	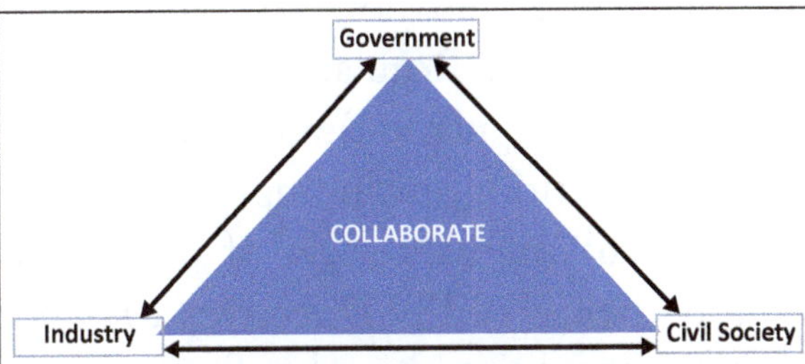 **Multi-sector dialogue/ Blend of Top-down and Bottom-up**

Fig. 13.2 The Continuum of stakeholder participation (using the categories of industry, civil society and government—the latter can include different levels of authority from local, regional to national) in European MSP with various stages ranging from information provision to collaboration between all categories of stakeholders. The arrows represent the flow of information and the direction of interactions between stakeholders (Adapted from Arnstein 1969)

Table 13.1 List of key international and European instruments relevant to the European Atlantic that require stakeholder participation

International	European Union
Rio Declaration on Environment and Development (1992) and Agenda 21	Atlantic Strategy and Action Plan
Convention on Biological Diversity (CBD)	Directive 2014/89/EU on Maritime Spatial Planning
OSPAR Convention and the North-East Atlantic Environment Strategy	Directive 2003/4/EC on Public Access to Environmental Information
Aarhus Convention	Directive 2003/35/EC on Public Participation

conservation. The involvement of environmental interest groups and non-governmental organisations (NGOs) are integral, and the distinct role sign-posted for NGOs is perhaps the most major innovation of the Convention (Lee and Abbot 2003).

Adopted in 1998 under the auspices of the United Nations Economic Commission for Europe, the Aarhus Convention, is the first comprehensive effort at the supranational level at implementing Principle 10 of the Rio Declaration, providing legally binding obligations on three pillars: public access to environmental information; decision-making; and, justice. The European Community has been a Party to the Aarhus Convention since 2005 and has implemented it via two EU Directives and a Regulation. The Directives address public access to environmental information (Directive 2003/4/EC) and public participation in environmental decision-making (Directive 2003/35/EC). The Regulation (Regulation 1367/2006, hereafter referred to as the Aarhus Regulation) addresses the application of the provisions of the Aarhus Convention, including that enabling NGOs meeting certain criteria to request an 'internal review' of administrative acts or omissions.

Under the EU's MSP Directive (2014/89/EU), MSP is defined as a process by which the relevant Member State's authorities analyse and organise human activities in marine areas to achieve ecological, economic and social objectives (EC 2014). There is a legal obligation to establish and implement MSP plans by 2021 that apply an ecosystem approach; consider economic, social and environmental aspects; and promote the coexistence of relevant activities and uses. Inherent in this is the provision of opportunities for stakeholders to participate throughout the process (Article 6). In addition:

> *Member States shall establish means of public participation by* ***informing all interested parties*** *and by* ***consulting the relevant stakeholders and authorities, and the public concerned, at an early stage*** *in the development of maritime spatial*

plans, in accordance with relevant provisions established in Union legislation. Member States shall ***also ensure that the relevant stakeholders and authorities, and the public concerned, have access to the plans once they are finalised*** *(Article 9).*

The EC's Communication on Developing a Maritime Strategy for the Atlantic Ocean Area (COM(2011)782) identified a number of themes of relevance to marine stakeholders in the Atlantic sea basin—implementing the ecosystem approach; reducing Europe's carbon footprint; sustainable exploitation of the Atlantic seafloor's natural resources; responding to threats and emergencies; and socially inclusive growth. An Action Plan for a Maritime Strategy in the Atlantic area was subsequently adopted: delivering smart, sustainable and inclusive growth (COM(2013)279) which sets out priorities for research and investment to advance the 'blue economy' in the Atlantic area. The Action Plan was developed through consultations conducted in the Atlantic Stakeholder Forum which consisted of representations from each of the five Atlantic Member States, the European Parliament, regional and local authorities, civil society and industry.

6 MSP Guidance and Recommendations

MSP aims to achieve multiple objectives (social, economic and ecological) and should therefore reflect as many expectations, opportunities or conflicts occurring in the MSP area. This section summarises (Table 13.2) the conceptual reasons for encouraging stakeholder participation in MSP, which reflect good practice, guidance and associated principles (e.g. EU Roadmap).

According to Principle 4 of the EU Roadmap for MSP (COM(2008)791), in order to achieve broad acceptance, ownership and support for implementation, it is important to involve all stakeholders at the earliest possible stage in the planning process. Stakeholder participation is also reported as a source of knowledge that can significantly improve the quality of MSP (Ehler and Douvere 2009).

Moving towards greater stakeholder participation to holistically address interactions among multiple sectors and communities within coastal and marine areas requires fresh thinking and new approaches. Public consultation alone is no longer appropriate. Implementing effective MSP entails the adoption of inclusive participatory planning processes that move beyond traditional top-down approaches. In addition, a host of guidance documents report that MSP requires active engagement with stakeholders throughout the entire planning process from preparatory, drafting and implementation to evaluation phases (e.g. Ehler and Douvere 2009; U.S. Institute for Environmental Conflict Resolution 2011; Pentz 2012; Agardy et al. 2011, 2012).

While the rationale for engagement of stakeholders in MSP is well established (Table 13.2) and is reflected in the guidance documentation available to practitioners, there is a need to consider: firstly, how appropriate are the mechanisms being used and secondly, how different stakeholders react to engagement opportunities is influenced by: the resources and power they have at their disposal, and the design of the actual engagement mechanisms employed in any given situation. Planners should be aware that differences exist between stakeholders in terms of power and influence, and MSP processes should look to mitigate these differences in order to deliver more equitable and democratic approaches (Flannery et al. 2018).

Using the case study material (research- and practice-based) that forms the basis of this chapter, the theoretical considerations on stakeholder participation from the MSP literature are mapped onto each example to illustrate where divergence occurs between theory and practice and the drivers behind this divergence (e.g. geographical, socio-political influences).

Table 13.2 Rationale for actively involving stakeholders in MSP (NOAA Coastal Services Centre 2007; EC 2008; Ehler and Douvere 2009; EC 2014)

Rationale for actively involving stakeholders in MSP:
• Encourages ownership of the plan, engenders trust among stakeholders and decision-makers and voluntary compliance with rules and regulations.
• Improves understanding of the complexity (spatial, temporal) and human influences of the marine management area.
• Develops a mutual and shared understanding about the problems and challenges in the management area.
• Increases understanding of underlying (often sector-oriented) desires, perceptions and interests that stimulate and/or prohibit integration of policies in the management area.
• Examines existing and potential compatibility and/or conflicts of multiple use objectives of the management area.
• Aids the generation of new options, consensus and solutions that may not have been considered individually.
• Expands and diversifies the capacity of the planning team, in particular through the inclusion of secondary and tertiary information (e.g. local knowledge and traditions).

7 Geographical Context for the European Atlantic Case Studies

The European Atlantic region (or sea basin) broadly refers to the coasts, territorial and jurisdictional waters of five EU Member States: Ireland; France; Spain; Portugal and, until March 2019, the UK[1] (i.e. all of Northern Ireland and Wales, the western parts of England and Scotland). Unlike other European marine regions that are somewhat sheltered and semi-enclosed seas, the European Atlantic countries look outwards to an exposed open ocean (Fig. 13.3).

The countries showcased in this chapter are considered as geographically peripheral nations of the EU which in turn provides socio-economic advantages and disadvantages. Ireland, Spain and Portugal have only recently emerged from the economic recession brought about by the European Debt Crisis of 2009. In terms of sea uses, fishing is a major sector within the Atlantic, whilst coastal tourism and shipping are of great importance to all Member States bordering this area. Given the Atlantic region's geographic position, it is considered a gateway to continental Europe (O'Hagan 2018). There is limited oil and gas production, but the region has high potential for the development of offshore renewable energy given its favourable physical and climatic conditions for wind, tidal and wave energy devices (Pérez-Collazo et al. 2015; Magagna and Uihlein 2015).

8 Stakeholder Processes in Reality

The following section outlines experiences and outcomes from a research-based transboundary MSP project with two pilot studies from northern and southern European Atlantic contexts. This is followed by an examination of four statutory initiatives at different phases of MSP implementation in Ireland and Northern Ireland (which is one of the devolved administrations of the UK) on the island of Ireland (Fig. 13.4), and Spain and Portugal on the Iberian coast (Fig. 13.5).

[1] Following the results of an EU referendum in 2016, on 29 March 2017, the UK notified the European Council in accordance with Article 50(2) of the Treaty on European Union of their intention to withdraw from the EU in 2019.

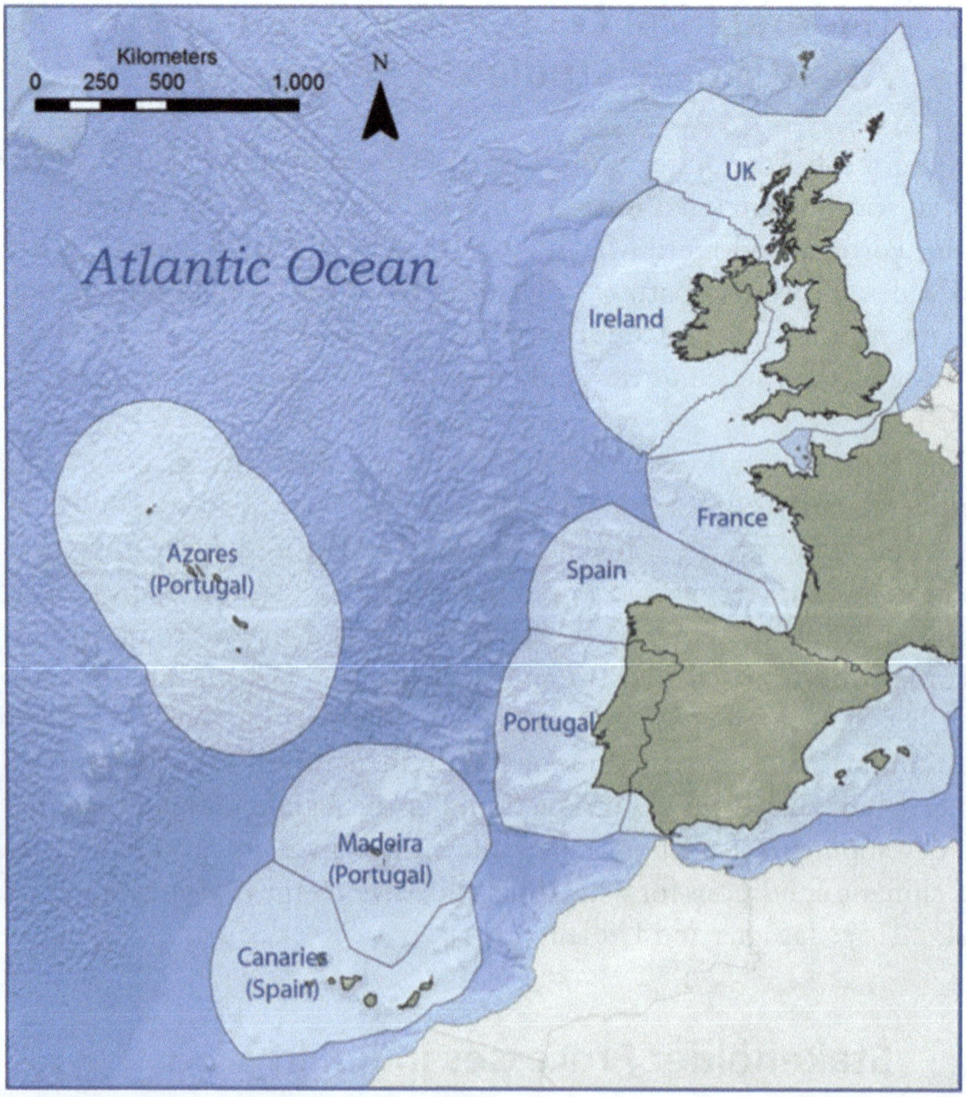

Fig. 13.3 Map of EU Member States bordering the Atlantic Ocean and the extent of their respective Exclusive Economic Zones (EEZs). Data sources: EEA and EMODNET

8.1 Research-Based Applications

Although MSP is a national task, the nature of the marine environment and the activities taking place in the sea mean that cooperation in MSP across borders is essential. This is recognised in the MSP Directive, which requires Member States to cooperate with respect to transnational issues with the aim of ensuring that maritime spatial plans are coherent and coordinated across the marine region concerned.

The Transboundary Planning in the European Atlantic (TPEA) project focused on two pilot areas; one on the island of Ireland in the Irish Sea between Ireland and Northern Ireland, UK; the second, in the Gulf of Cadiz between

Fig. 13.4 Map of the island of Ireland. Data sources: EEA and EMODNET

Portugal and Spain (Fig. 13.7). It was a civil-society-led project coordinated by a research institution and involved a consortium of six governmental and four research partners across the region. Its primary aim was to explore cross-border (transboundary) MSP exercise and three key pillars provided a framework for planning activities (Jay et al. 2016; Almodovar et al. 2014; TPEA 2013) (Fig. 13.6):

1. Participation of multi-sector stakeholders (from government, industry and civil society groups) throughout the entire process as a means to inform, guide and validate the activities and outputs at all stages of the process (e.g. pre-planning, developing the vision and objectives, establishing the current context, developing scenarios, etc.) (Stage 3 and 4 on the Stakeholder Participation Continuum);

Fig. 13.5 Map of Spain and Portugal with the Iberian coast to the west. Data sources: EEA and EMODNET

2. Analysis of legal, policy and governance frameworks; and
3. Use of Geographic Information Systems (GIS) and geospatial technology.

The authors of this chapter coordinated the stakeholder engagement aspects of the project including the development of a strategy that established the objectives of stakeholder engagement throughout the planning exercise process indicating how and when stakeholders were to be engaged at each stage of the preparation, planning and dissemination process (Fig. 13.7). The central mechanism for participation was the organisation of facilitated stakeholder workshops (three in each pilot area), where participants were invited to explore

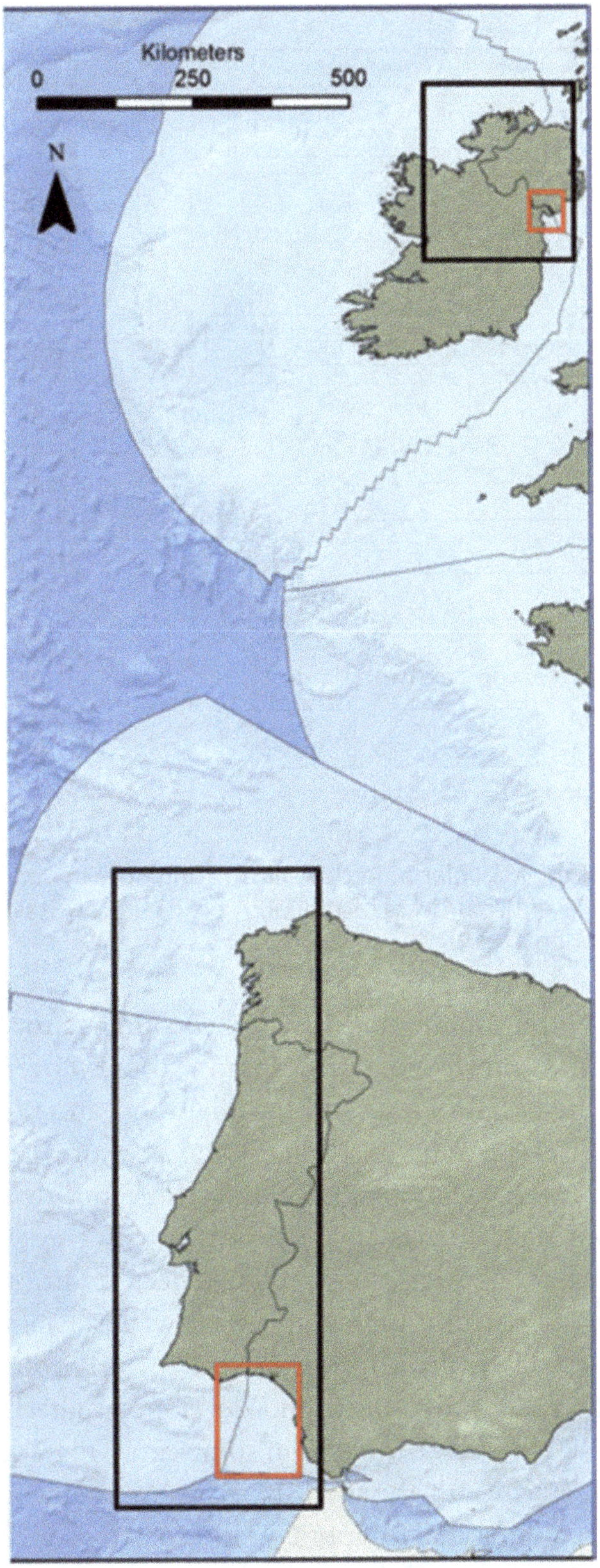

Fig. 13.6 Map illustrating the location of the two MSP pilot areas within the European Atlantic

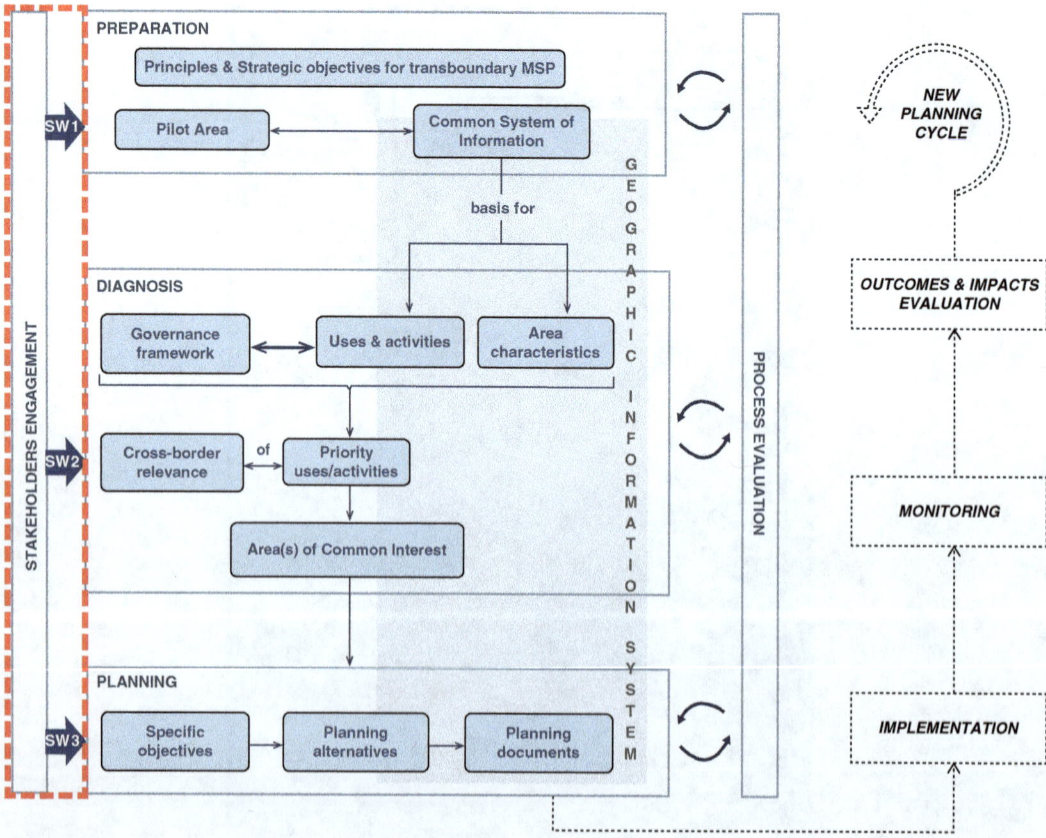

Fig. 13.7 Phases of the MSP planning cycle illustrating how the participation of stakeholders informed the entire process of the TPEA project. (SW=Stakeholder Workshop) (Twomey and O'Mahony 2014)

different aspects of transboundary MSP, and encouraged to share their experiences, expectations, knowledge and opinions. Topics covered during workshops included methods for establishing planning areas; data sharing and harmonisation across jurisdictions; identifying pressures and opportunities; agreeing specific and strategic planning objectives; and development of scenarios (Almodovar et al. 2014).

The TPEA workshops were designed to reflect traditions in stakeholder participation and consisted of presentations, as well as facilitated group work and interactive exercises. In advance of each workshop, TPEA partners planned the format and programme and developed materials specific to the participants and activities on the day—in some cases, materials were provided to participants in advance of the workshop as part of introductory information. Participants were asked to appraise all aspects of each workshop (e.g. content, facilitation), and this information was used to plan subsequent events over the course of the project (Jay 2015; TPEA 2014).

During the first round of workshops, stakeholders were asked to comment on their preferred means of stakeholder engagement and which methods they felt would be most appropriate for the MSP process. A variety of methods were proposed—representing the opinions of participants—but also indicating that when embarking on a stakeholder engagement process, planning teams will have to be mindful of the need to tailor messages (e.g. their content and how they are communicated) according to the requirements of different interested parties—either individuals or groups (Table 13.3)

While workshops were an important element of stakeholder participation, they were not the only means of incorporating stakeholder input. Recognising that many stakeholders could make a valuable contribution to the process outside of the workshops, members of the project team actively sought stakeholder input over the course of the project, facilitated through meetings, presentations at industry and NGO events, use of formal and informal networks—this flexible and adaptive approach to engagement ensured the MSP process benefitted from the participation of a wide range of stakeholders across government, industry and civil society.

Some difficulty was experienced in the southern pilot area (Spain and Portugal), particularly in the early stages of the process. For example, despite efforts to ensure equal representation, the first workshop was attended by an overwhelming majority of Portuguese stakeholders. It then became apparent that this situation had transpired as a result of national differences. In 2013, the concept of MSP was largely unknown in Spain coupled with a weaker tradition of public engagement, whereas Portugal had already gained experience of a non-statutory national MSP study (i.e. Planning and Ordering of Maritime Space (POEM)).

Table 13.3 Participatory mechanisms proposed by the TPEA stakeholders and justifications for their use in an engagement process

Participatory Mechanism	Justification
Identify organisational champions	Committed individuals with access to extensive networks can support engagement efforts of planning team
Involve politicians	Builds trust and allows for in-depth discussion
Use of social media	Twitter and LinkedIn are the new media of choice for many professionals
Stakeholder forum	Potential to cater for numerous interest groups
Public campaign	Means of raising awareness and encouraging involvement
Public meetings	Provides participants with a voice and an opportunity to contribute, and for planning team to cover any technical aspects of plan
Visualisation (GIS/MSP Game)	Can be novel tools to initiate and facilitate discussion
Online forum	A means of communicating FAQs and preferred media for many
Roadmap/strategy for engagement	Sets out where, when and how stakeholders can get involved

Although Northern Ireland had already embarked on its statutory MSP process in 2012 (and Ireland had not yet), the challenges experienced in the southern pilot area with stakeholder representation were not shared in the northern pilot area. Perhaps this can be attributed to stakeholders on the island of Ireland sharing a common language in addition to similar traditions of public participation (Table 13.4).

8.2 Statutory-Based Applications

This section provides an overview of four case studies of real-life statutory MSP developments from the island of Ireland; Ireland and Northern Ireland, and the Iberian coast; Spain and Portugal. The focus of the analysis is on the aspirations for stakeholder participation and the progress to date (where relevant) as documented by relevant government publications.

Table 13.4 High-level summary of the stakeholder mechanisms employed and trends in representation across different categories of stakeholders at the TPEA multi-sector workshops (2012–2014) in the northern and southern European Atlantic pilot areas

Location	Stakeholder mechanisms employed	Representation
Ireland and Northern Ireland	Three multi-sector transboundary stakeholder workshops. Stakeholders were seated in a series of roundtables (i.e. cabaret style room set-up) to ensure representatives from different sectors could interact and hear diverse perspectives on the topics. Experienced facilitators at each table helped to guide semi-structured group discussions and ensure all voices had an opportunity to contribute. Pro-active outreach and communications with government, industry and NGO at workshops and conferences, and government meetings.	Government: 44% Industry: 26% Civil Society: 30%
Spain and Portugal	Three multi-sector transboundary stakeholder workshops. Stakeholders self-selected their own seats around a U-shape/boardroom style table and the event followed a formal meeting-style process. The Chair introduced various topics and asked for feedback and comments at various points. Additional single-sector meetings were essential in Spain to raise awareness of MSP and encourage attendance.	Government: 72% Industry: 8% Civil Society: 20%

Ireland

As an island nation, Ireland has over 7500 km of coastline and sovereign rights to an Exclusive Economic Zone (EEZ) of 880,000 km^2—over ten times its land mass. Over 50% of the population reside on the Irish coast, the inshore and offshore waters contain some of the largest fisheries resources in Europe; are the western gateway for shipping to European and international seaports; and are amongst the most valuable and accessible marine renewable resources (wind, wave and tidal) globally.

Legislation	Ireland has transposed the Directive through the EU (Framework for Maritime Spatial Planning) Regulations 2016, signed into law on 29 June 2016. The competent authority for MSP is the Department for Housing, Planning and Local Government (DHPLG).
Status of MSP	A Government-led Inter-Departmental Marine Coordination Group was established in 2009 followed by the launch of a high-level policy document, *Harnessing Our Ocean Wealth* (HOOW) in 2012. This sets out the vision, high-level goals and key actions to enable Ireland's marine potential to be realised. A roadmap for Ireland's first plan was published in 2017- *Towards a Marine Spatial for Ireland.*
EU Projects	TPEA, SIMCELT
Stakeholder Participation	HOOW was developed following a period of public consultation in 2012. Over this consultation period, 192 responses were received from a variety of stakeholders including NGOs, Trade and Professional associations, Small or Medium-sized Enterprises (SMEs), Higher Education Institutions and others. The results of this consultation fed into the final version of Harnessing Our Ocean Wealth. *Towards a Marine Spatial for Ireland* provides details on the proposed public participation and consultation processes from 2018 to 2020. The aim of this document is to describe how, when and what DHPLG will do with the outcomes of stakeholders' views. The plan will be guided by the following principles for engaging in MSP: • involve people early on in the decision-making process and in developing specific policy within the framework provided by HOOW; • engage with interested people and organisations at the appropriate time using effective engagement methods and allowing sufficient time for meaningful consultation; • be adaptable, recognising that some consultation methods work better for some people and some issues and that a one-size-fits-all approach will not work; • respect the diversity of people and their lifestyles and give people a fair chance to have their voice heard regardless of gender, age, race, abilities, sexual orientation, circumstances or wherever they live;

	• be clear in the purpose of any engagement and how stakeholders may contribute and let people know how their views have been taken into account within agreed timescales; • make documents publicly available on the Department's website; and • communicate clearly with people using plain English and avoiding jargon. An Advisory Group has been established to ensure the participation of relevant NGOs, professional bodies and technical experts in the process. In advance of and running alongside subsequent formal public consultation, the DHPLG will also seek to engage stakeholders through the following means: geographic or sector-based groups workshops; web portal; one-to-one meetings; exhibitions and drop-in sessions; attendance at stakeholder meetings; questionnaires; web updates; and, newsletters.
Stakeholder Participation Continuum	Stages 1 and 2 and a plan has been prepared for Stage 3 activities (as of time of writing).
Key challenges encountered	• developing capacity and resources for different forms of stakeholder participation beyond public consultation and information sessions (discussed in interviews) and • geographical extent of coastline and marine area.
Links	http://www.ouroceanwealth.ie/publications http://www.housing.gov.ie/search/topic/marine-spatial-planning http://msp-platform.eu/countries/ireland

Northern Ireland

The Marine Plan for the Northern Ireland Plan Area will cover an area of approximately 12,350 km^2 and include 650 km of coastline. A number of large cities and towns are located along the coastline and the marine area is a key asset in terms of biodiversity, recreation, tourism and the transportation of goods and services by sea. The marine waters of Northern Ireland also support industries such as aquaculture and fishing and there is significant potential for economic growth from tidal and offshore wind energy.

Legislation	The MSP Directive has been transposed and the competent authority for MSP is the Department of Agriculture, Environment and Rural Affairs (DAERA). In line with the UK's Marine & Coastal Access Act (2010), the Marine Act (Northern Ireland 2013) sets out a new MSP framework which applies to the inshore region (i.e. the territorial sea out to twelve nautical miles).
Status of MSP	DAERA are in the process of finalising their first plan. However, whilst the plan exists and has gone through the Sustainability Appraisal process, it is still in draft format and cannot be officially adopted as Northern Ireland has been without a government since February 2017.

EU Projects	TPEA, SIMCELT
Stakeholder Participation:	A Statement of Public Participation (SPP) was published in 2012. The aim was to: • set out how and when people can be involved in the preparation of Northern Ireland's first Marine Plan; • indicate the associated time frame leading to its publication for public consultation; and • Invite comment as to the matters to be included in the proposed Marine Plan. This document is for 'interested persons', anyone or any group likely to be interested in or affected by policies or proposals to be contained in the plan. This will involve those who live or work near the sea, those who derive their livelihood from the sea, as well as those who enjoy it, care about it or manage it in some way. It will include individuals as well as stakeholder groups and larger organisations. According to the SPP, key methods of engagement and communication will include sector-based workshops, geographic-based workshops, one-to-one meetings, attendance at stakeholder group meetings, provision of information through a designated website, newsletters, public meetings and drop-in sessions. Key stakeholder engagement to date has included extensive series of stakeholder events throughout 2012, including schools and the wider public in coastal areas, sectoral meetings and meetings with NGOs; Multi-stakeholder event (2013); Sustainability Appraisal Scoping Workshop (2014); Ongoing engagement with Northern Ireland and UK statutory bodies to ensure their respective responsibilities are accurately reflected; ongoing cooperation with neighbouring authorities that share the regional seas.
Stakeholder Participation Continuum	Stages 1 and 2 and a plan was prepared for Stage 3 (as of time of writing).
Key challenges encountered	• Managing expectations of stakeholders with regard to anticipated scope and level of participation (discussed in interviews). • Influence of political inertia on implementation process.
Links	https://www.daera-ni.gov.uk/articles/marine-plan-northern-ireland https://www.daera-ni.gov.uk/publications/marine-plan-statement-public-participation http://msp-platform.eu/countries/united-kingdom

Spain

Spain has a coastline of over 4680 km bordering the Atlantic Ocean. The key maritime sectors are coastal and marine tourism, fisheries, maritime transport and mariculture. Plans for offshore wind farms in Spain, the world's fourth largest producer of wind energy, have been set up and will be established in its territorial sea.

Legislation	Spain adopted the Royal Decree 363/2017 on 8 April 2017 establishing a framework for MSP. The competent authority has been designated as the Ministry of Agriculture, Food and Environment.
Status of MSP	Spain has just recently transposed MSP into national law and is embarking on its preparatory phase for implementation. A number of MSP-related initiatives (e.g. MPAs, Natura2000, renewable energy plans) have been carried out, but so far, no multi-sector MSP initiatives have been developed at the national level. The focus over recent years has been on the implementation of the EU's Marine Strategy Framework Directive (MSFD). Marine strategies are under development for Spain's five marine subregions to reach the good environmental status of the marine environment. Each of these subregions will also benefit from MSP.
EU Projects	SIMNORAT, SIMWESTMED, TPEA, POCTEFEX-ALBORAN
Stakeholder Participation:	At this stage, it is unclear what opportunities will be available for stakeholders to participate in Spain's forthcoming process. In terms of statutory stakeholder involvement, the Inter-Ministerial Commission on Marine Strategies (CIEM) was created in 2012 and is responsible for coordination between Ministerial Departments of the Central State Administration. Competencies on maritime and coastal affairs are shared between central and regional governments, including those facing the Atlantic. According to the EU MSP Platform, it is likely that stakeholder participation will be similar to that of the MSFD process, whereby top-down engagement between high-level government stakeholders was supplemented with a series of targeted stakeholder workshops with technical experts from the science community and NGOs.
Stakeholder Participation Continuum	N/A
Key challenges encountered	• Involving non-statutory stakeholders (e.g. industry, NGOs, local communities, etc.) in participatory processes for MSP. • Geographical footprint in multiple sea-basins.
Links	http://www.marineplan.es/en/ATLAS_13_06_11_EN.pdf http://msp-platform.eu/countries/spain

Portugal

With an Atlantic coastline of 942 km, Portugal has one of the largest maritime areas in Europe with an EEZ of 1,700,000 km^2 encompassing both continental Portugal and two large insular regions surrounding the Azores and Madeira. Traditional economic activities include fishing, aquaculture and maritime transport, whilst new emerging sectors such as deep-sea mining, biotechnology and ocean energy are begin developed.

Legislation	Portugal has approved legislation for MSP for all Portuguese maritime space. In 2014, the first Portuguese MSP law was enacted, followed by Order No. 11494/2015 in 2015 which paved the way for the development of the Situation Plan, (Portuguese MSP). The competent authority is the Ministry of the Sea.
MSP Status	From 2008 to 2012, the government led a (non-statutory) multidisciplinary MSP study—Planning and Ordering of Maritime Space (POEM). A new agency was then established, the Ministry for Agriculture, Sea, Environment and Spatial Planning (MAMAOT), with responsibility for both terrestrial planning and MSP. Several agencies were reorganised or disbanded, including the agency responsible for the development of POEM.
EU Projects	TPEA, SIMNORAT, GPS Azores
Stakeholder Participation:	The draft version of POEM was published online for public consultation for 12 weeks in late 2011. This was supplemented by a series of sector-specific workshops that were also organised in different coastal locations, focusing on transport and navigation, fisheries and aquaculture, coastal tourism and leisure, maritime defence and scientific research. However, the stakeholder process has been described as 'tokenistic' with very limited public consultation over a short period despite the complexity of the plan (Portman 2011; Calado et al. 2010). The National Ocean Strategy 2013–2020 (NOS) was developed through public debate following a large number of public meetings both in mainland Portugal and in the Autonomous Regions of the Azores and Madeira. Over 100 contributions were received from public and private entities and also from civil society. The NOS aims to promote the 'effective participation of everyone from a central, regional and local level—involving public and private entities and civil society as key partners for the identification and evaluation of threats and pursuing opportunities, ensuring reflection and production of strategic thinking' (Government of Portugal 2014: 56).
Stakeholder Participation Continuum	Stages 1 and2.
Key challenges encountered	• Absence of a detailed plan for stakeholder participation. • Involving industry stakeholders in participatory processes for MSP (Portman 2011; Calado et al. 2010). • Geographical footprint in multiple sea-basins.
Links	http://www.portugal.gov.pt/pt/ministerios/mm.aspx http://www.dgpm.mam.gov.pt/Pages/POEM_PlanoDeOrdenamentoDoEspacoMarinho.aspx https://www.dgpm.mm.gov.pt/enm-en http://msp-platform.eu/countries/portugal

9 Comparative Analysis: Stakeholder Processes Across the European Atlantic

In comparison to other European marine regions such as the Baltic or North Sea, MSP is still in its infancy in the European Atlantic. Whilst it is clear that MSP implementation across the region is at different stages in terms of implementation, some preliminary trends and issues have been identified. At a country level, the general approaches to planning are diverse as a result of varying underlying social and political contexts. MSP in Portugal is currently in its second cycle (although the first plan was more of a research-based exercise and not a statutory plan); Northern Ireland is at the latter stages of pre-implementation of its first plan; Ireland has recently embarked on the preparatory stages of their first plan, and Spain has transposed the relevant legislation but has not yet published any further details on their proposed plan. This variance has implications for the comparative analysis of stakeholder processes across the region; however, projects such as TPEA provide empirical evidence that can be used for the purposes of regional assessment.

Even in adjoining jurisdictions, there will be differences of approach to stakeholder participation and representation by sectors, possibly reflecting different political (e.g. political inertia in Northern Ireland) socio-economic conditions (e.g. emergence from recession), cultures and organisational structures. Unlike the government-led statutory processes presented in the previous section, TPEA was a bottom-up civil society-led process which tested a participatory approach to MSP and aimed to employ all four stages on the Continuum of Participation model (Fig. 13.3). An interesting distinction between the two pilot areas was the design of the stakeholder workshops, particularly in terms of the seating arrangements and (non)-use of trained facilitators. On the island of Ireland, stakeholders were seated at a series of round tables with representatives from all different sectors (i.e. government, industry and civil society) to encourage interaction and multi-perspective discussions (consistent with Stages 3 and 4 on the Stakeholder Participation Continuum). This was based on the assumption that the way in which you design stakeholder interactions at an event will inevitably influence the type of input you receive and smaller groups (e.g. six to eight individuals) at a number of round tables is conducive to more inclusive and meaningful conversations between stakeholders. Experienced facilitators were also appointed at each table to guide the group through a series of semi-structured group discussions; ensure that every stakeholder had an opportunity to meaningfully contribute; and no one voice could dominate the group. A more formal approach was favoured on the Iberian coast with stakeholders self-selecting their seats in a U-shaped meeting format without the assistance of facilitators.

Resources were specifically assigned to ensure high levels of representation across sectors. Evidence from these research-based case studies indicate that stakeholder representation across different categories was evenly distributed in the Northern Atlantic countries (Table 13.4). Government decision-makers and statutory stakeholders were well represented across all four countries and over-represented in Spain and Portugal (72%). Civil society stakeholders from NGOs, the science community and local community groups also played an active role (i.e. 30% in on the island of Ireland and 20% on the Iberian coast). However, stakeholders from industry have been under-represented, particularly in Spain and Portugal (8%).

In terms of statutory processes, of those that have published in-depth information on stakeholder participation (i.e. Northern Ireland and Ireland), the planning authorities indeed claim to be moving beyond traditional consultation methods to more inclusive and participatory mechanisms of engagement. According to these policy documents from Ireland, the ambitions for stakeholder participation fall within Stage 3 of the Stakeholder Participation Continuum—(Engagement)—presented in Table 13.2. Portugal has yet to release specific details of their plans for stakeholders but has indicated that they intend to engage with 'everyone from a central, regional and local level'. However, unlike Ireland and Northern Ireland, they have not published any detailed plans (e.g. a Stakeholder Engagement Strategy or SPP) on when and how they will provide these opportunities. Similarly, with Spain, it is unclear at this point whether the planning authority will apply a participatory approach to MSP (Stage 3 of the Stakeholder Participation Continuum) or merely rely on public consultation (Stage 2 of the Stakeholder Participation Continuum) to obtain input from stakeholders.

10 Key Challenges

The case studies reveal valuable insights into the complexity and practical challenges associated with delivering multi-stakeholder MSP processes. Whilst acknowledging that the experiential data are limited, the overarching findings indicate that the policy guidance on participation on MSP has been interpreted in different ways across the European Atlantic. As a result, the cases presented in this chapter from research-based pilot exercises to statutory processes demonstrate that there are variable ambitions for participation in

MSP. This chapter thus contributes a European Atlantic perspective to an emerging body of literature that is critical of the realities of MSP implementation in other parts of the EU and beyond (e.g. Jones et al. 2016; Flannery and Ellis 2016; Flannery et al. 2018).

According to existing MSP theoretical frameworks, once planning authorities follow a step-by-step approach characterised by interactive multi-stakeholder participation, a strategic plan can determine where and when human activities occur in marine spaces. The European Atlantic case studies indicate that in reality only the research-based initiative employed mechanisms to engage and involve stakeholders at an early stage and continuously throughout the process. Evidence emerging from the interviews and literature indicate that although Northern Ireland and Portugal had policy ambitions to conduct pro-active and inclusive stakeholder processes in line with good practice, in actuality, the approaches to date have not matched the expectations of stakeholders from industry and civil society.

The diverse and often conflicting activities and perspectives of humans are therefore very much at the heart of MSP. Everyone has some type of stake and, in reality, it is primarily a socio-political process which strives to balance the demands of powerful stakeholders with robust scientific data through strategic trade-offs. Despite these high stakes, evidence from both research and statutory initiatives indicate that industry stakeholders have engaged in lower numbers than those representing government and civil society interests. This is a critical point for MSP and is something that needs to be factored into subsequent planning cycles. Perhaps a tailored engagement strategy with this sector is necessary to clarify the value and importance of their role in terms of providing a unique perspective and tacit knowledge from those operating in an offshore setting.

In theory, the greater the scale of the planning area and the more the marine activities that have to be considered, the greater the number of stakeholders that need to be engaged throughout the process. This complexity inevitably adds to costs in both time and financial resourcing for planning authorities. For example, Northern Ireland has the smallest planning area in all of the case studies, yet it has a larger MSP team than that of Ireland and Portugal.

11 Practical Recommendations for Planning Authorities

The following recommendations to address challenges for stakeholder participation in MSP are based on lessons learnt from a synthesis of: the good practice guidelines developed by the TPEA project (Almodovar et al. 2014); discussions with stakeholders in interviews; and findings from the case studies presented in this chapter.

1. **Transparency**: Mutual respect and fairness with a transparent process was raised as a key starting point for any engagement exercise.

   - Transparency needs to extend to the objectives, outcomes, roles, expectations and limits of any MSP process. For some stakeholders who have had no previous experience with MSP, the concept of MSP can be intangible. Sometimes communicating what MSP *is not* is more important than explaining what *it is*. Likewise planning authorities should be explicit on what will be involved in the plan, as well as justifying what won't be covered.
   - Stakeholder expectations need to be managed. Honesty goes a long way—to avoid potential conflict with stakeholders, be realistic and don't raise expectations by proposing high levels of engagement if you can't deliver on these promises. If it's going to be mostly public consultation, communicate that from the beginning but aim to improve participatory opportunities in future MSP cycles (e.g. by organising single-sector or multi-sector stakeholder meetings and workshops).

2. **Early and ongoing inclusive engagement mechanisms**: Engagement with stakeholders outside of government organisations should not be an after-thought.

   - Develop a comprehensive profile of stakeholder interests and contacts (e.g. in the form of a database) at the earliest possible stage in the process is essential to identify a wide pool of stakeholders before deciding which particular stakeholders that need to be considered and then targeted at different stages when necessary. The sectoral interests and the associated organisations or groups should be shared with stakeholders (e.g. at meetings or workshops) for validation and to plug any gaps in representation across the various sectors. This database needs to be maintained and updated throughout the planning process in order to ensure that all relevant stakeholders are being engaged and to avoid a situation where stakeholders have unintentionally been excluded.

- Many planning authorities are familiar with statutory and regulatory stakeholders, but extra consideration needs to be given to identify other key stakeholders such as industry representatives (e.g. traditional sectors, new emerging industries, environmental consultancies, trade unions); the science community including socio-economic researchers, marine social scientists, GIS specialists; and NGOs at the earliest stage (i.e. preparatory phase) in the MSP process.
- Don't expect stakeholders to initiate contact with you. Outreach by the planning authority is essential, for example, by attending industry and NGO events, conferences, community meetings and so on.
- Strive to find a balance between consultation (i.e. one-way communication or no participation) and trying to engage everyone throughout the process (complete participation) is key. This is a difficult task, but it must be the aim in any MSP process.
- The importance of face-to-face contact through interactive workshops should not be underestimated and over-reliance on websites and newsletters should be avoided. Use participation mechanisms that encourage dialogue and interaction between different stakeholder groups such as the facilitated multi-sector workshops outlined in this chapter. Strive to find a balance between consultation (i.e. one-way communication or no participation) and trying to engage everyone throughout the process (complete participation) is key.
- Stakeholders will often need to justify their attendance at MSP meetings or workshops. Dates and locations may not always suit them. Be flexible and promote an open-door policy to allow stakeholders to drop in and out of the process.

3. **Promoting inclusiveness and developing capacity for stakeholders to participate effectively in the process**: A diverse group of stakeholders from different professions and backgrounds should be encouraged but be cognisant of power and resource imbalances.

   - In the early stages of MSP, not all stakeholders will be familiar with the concept or what the process entails. MSP and time and resources need to be allocated to raise awareness so that (as much as possible) stakeholders are on a level playing field.
   - Technical information should be tailored to different audiences and communicated clearly in basic terms. Communication professionals are

invaluable resource in these situations as they have the skills to bridge these gap science and policy gaps with different sectors.

- The process needs to be balanced and designed in a manner that ensures overly vocal stakeholders do not have a disproportionate influence over the process. Using facilitators in a workshop setting is a strategy to overcome this imbalance as various strategies can be applied to promote an environment where no one stakeholder can dominate a discussion

12 Conclusion: *Participation Is a Contested Concept That Often Fails to Live Up to Its Promise*

On a theoretical level, the motivation and rationality for the integration of stakeholders throughout the MSP cycle is unquestionable. However, in practice, the case studies presented in this chapter highlight varying degrees of disconnect between the conceptual underpinnings of MSP theory and the realities of recent stakeholder processes in MSP. The ultimate challenge is to map out ways in which the processes and outcomes of stakeholder processes can align more realistically with the policy aspirations of national planning authorities. All countries profess to be implementing inclusive and multi-dimensional stakeholder processes in addition to long-established mechanisms such as formal public consultation. However, it is clear from the European Atlantic experience that definitions of stakeholder participation, and exactly what it should entail, vary greatly. The true meaning of 'early and effective engagement' seems open to interpretation and sectors have different opinions on how and what it should look like. In order to contribute to the co-production of the knowledge base upon which Marine Plans are developed, rather than just being consulted, it is vital that stakeholders are allowed flexible opportunities to participate in all stages of MSP.

Acknowledgements This material is based upon works supported by Science Foundation Ireland (SFI) under Marine and Renewable Energy Ireland (MaREI) Centre (12/RC/2302). The Open Access fee of this chapter was provided from the same source.

References

Agardy, T., Davis, J., Sherwood, K., & Vestergaard, O. (2011). Taking Steps Toward Marine and Coastal Ecosystem-based Management: An Introductory Guide. https://doi.org/10.17605/OSF.IO/AKH93

Agardy, T., Christie, P., & Nixon, E. (2012). Marine Spatial Planning in the Context of the Convention on Biological Diversity: A Study Carried Out in Response to CBD COP 10 Decision X/29. Secretariat of the Convention on Biological Diversity.

Almodovar, M., de Armas, D., Lopes Alves, F., Bentes, L., Fonseca, C., Galofré, J., Gee, K., Gómez-Ballesteros, M., Gonçalves, J., Henriques, G., Jay, S., O'Mahony, C., Rooney, A., & Twomey, S. (2014). *TPEA Good Practice Guide: Lessons for Cross-border MSP from Transboundary Planning in the European Atlantic*. Liverpool: University of Liverpool.

Arnstein, S. R. (1969). A Ladder of Citizen Participation. *Journal of the American Institute of Planners, 35*(4), 216–224.

Calado, H., Ng, K., Johnson, D., Sousa, L., Phillips, M., & Alves, F. (2010). Marine Spatial Planning: Lessons Learned from the Portuguese Debate. *Marine Policy, 34*(6), 1341–1349.

Charnoz, O. (2009). *The Global Discourse on 'Participation' and its Emergence in Biodiversity Protection* (No. 83). Working Paper [online]. Retrieved March 20, 2018, from https://www.afd.fr/sites/afd/files/imported-files/083-document-travail-VA.pdf.

Colvin, R. M., Witt, G. B., & Lacey, J. (2016). Approaches to Identifying Stakeholders in Environmental Management: Insights from Practitioners to Go Beyond the 'Usual Suspects'. *Land Use Policy, 52*, 266–276.

EC. (2008). Roadmap for Maritime Spatial Planning: Achieving Common Principles in the EU. [online]. Retrieved March 19, 2018, from http://eur-lex.europa.eu/legal-content/EN/TXT/PDF/?uri=CELEX:52008DC0791&from=EN.

EC. (2012). Blue Growth Opportunities for Marine and Maritime Sustainable Growth. [online]. Retrieved March 18, 2018, from http://eur-lex.europa.eu/legal content/EN/TXT/PDF/?uri=CELEX:52012DC0494&from=EN.

EC. (2014). Directive 2014/89/EU of the European Parliament and of The Council of 23 July 2014 Establishing a Framework for Maritime Spatial Planning. [online]. Retrieved March 16, 2018, from http://eur-lex.europa.eu/legal-content/EN/TXT/PDF/?uri=CELEX:32014L0089&from=EN.

EC. (2017). Report on the Blue Growth Strategy Towards More Sustainable Growth and Jobs in the Blue Economy. [online]. Retrieved March 18, 2018, from https://ec.europa.eu/maritimeaffairs/sites/maritimeaffairs/files/swd-2017-128_en.pdf.

Ehler, C., & Douvere, F. (2009). *Marine Spatial Planning: A Step-by Step Approach Towards Ecosystem-based Management*. Manual and Guides No 153 ICAM Dossier No 6. Paris: Intergovernmental Oceanographic Commission UNESCO IOC, 99 pp.

Flannery, W., & Ellis, G. (2016). Exploring the Winners and Losers of Marine Environmental Governance (Edited Interface Collection). *Planning Theory and Practice, 17*(1), 121–122.

Flannery, W., Healy, N., & Luna, M. (2018). Exclusion and Non-participation in Marine Spatial Planning. *Marine Policy, 88*, 32–40.

Flannery, W., O'Hagan, A., O'Mahony, C., Ritchie, H., & Twomey, S. (2015). Evaluating Conditions for Transboundary Marine Spatial Planning: Challenges and Opportunities on the Island of Ireland. *Marine Policy, 51*, 86–95.

Government of Portugal. (2014). National Ocean Strategy: 2013–2020. [online]. Retrieved March 26, 2018, from https://www.dgpm.mm.gov.pt/enm-en.

Heldaweg, M. (2005). Towards Good Environmental Governance in Europe. *European Environmental Law Review, 14*(1), 2–24.

Jay, S. (2015). *Transboundary Marine Spatial Planning in the Irish Sea*. Transboundary Marine Spatial Planning and International Law, p. 174.

Jay, S., Alves, F. L., O'Mahony, C., Gomez, M., Rooney, A., Almodovar, M., Gee, K., de Vivero, J. L. S., Gonçalves, J. M., da Luz Fernandes, M., Tello, O., Twomey, S., Prado, I., Fonseca, C., Bentes, L., Henriques, G., & Campos, A. (2016). Transboundary Dimensions of Marine Spatial Planning: Fostering Inter-jurisdictional Relations and Governance. *Marine Policy, 65*, 85–96.

Jones, P. J. S., Lieberknecht, L. M., & Qiu, W. (2016). Marine Spatial Planning in Reality: Introduction to Case Studies and Discussion of Findings. *Marine Policy, 71*(September), 256–264.

Lee, M., & Abbot, C. (2003). The Usual Suspects? Public Participation Under the Aarhus Convention. *The Modern Law Review, 66*(1), 80–108.

Lockwood, M., Davidson, J., Curtis, A., Stratford, E., & Griffith, R. (2010). Governance Principles for Natural Resource Management. *Society and Natural Resources, 23*(10), 986–1001.

Long, R. (2012). Legal Aspects of Ecosystem-based Marine Management in Europe. In *Ocean Yearbook*. Hijhoff.

Magagna, D., & Uihlein, A. (2015). Ocean Energy Development in Europe: Current Status and Future Perspectives. *International Journal of Marine Energy, 11*, 84–104.

NOAA Coastal Services Center. (2007). Introduction to Stakeholder Participation. Retrieved April 06, 2018, from https://coast.noaa.gov/data/digitalcoast/pdf/stakeholder-participation.pdf.

O'Hagan, A. M. (2018). Regulation and Planning in Sea Basins—NE Atlantic. In K. Johnson, G. Dalton, & I. Masters (Eds.), *Building Industries at Sea: 'Blue Growth' and the New Maritime Economy*. Gistrup and Delft: River Publishers.

Olsen, S. B. (2003). Frameworks and Indicators for Assessing Progress in Integrated Coastal Management Initiatives. *Ocean & Coastal Management, 46*(3–4), 347–361.

Pentz, T. A. (2012). *Stakeholder Involvement in MSP*. BaltSeaPlan Report, 24.

Pérez-Collazo, C., Greaves, D., & Iglesias, G. (2015). A Review of Combined Wave and Offshore Wind Energy. *Renewable and Sustainable Energy Reviews, 42*, 141–153.

Pomeroy, R., & Douvere, F. (2008). The Engagement of Stakeholders in the Marine Spatial Planning Process. *Marine Policy, 32*, 816–822.

Portman, M. E. (2011). Marine Spatial Planning: Achieving and Evaluating Integration. *ICES Journal of Marine Science, 68*(10), 2191–2200.

Preston, L. E., & Sapienza, H. J. (1990). Stakeholder Management and Corporate Performance. *Journal of Behavioral Economics, 19*(4), 361–375.

Reed, M. S. (2008). Stakeholder Participation for Environmental Management: A Literature Review. *Biological Conservation, 141*(10), 2417–2431.

Roxburgh, T., Dodds, L., Ewing, J., Morell, T., Sutton, E., Garcia Varas, J. L., Viada, C., Teleki, K., Siciliano, D., Vallet, E., O'Mahony, C., & Twomey, S. (2012). *Towards Sustainability in the Celtic Sea – A Guide to Implementing the Ecosystem Approach Through the Marine Strategy Framework Directive* (p. 47). Cardiff: World Wildlife Fund.

TPEA. (2013). Conceptual Framework: Transboundary Planning in the European Atlantic. Retrieved March 20, 2018, from http://www.tpeamaritime.eu/wp/wpcontent/uploads/2013/09/TPEA_CONCEPTUALFRAMEWORK_FINAL.pdf

TPEA. (2014). Stakeholder Engagement Factsheet. [online]. Retrieved March 20, 2018, from http://www.tpeamaritime.eu/wp/wp-content/uploads/2013/09/TPEA-Factsheet-Stakeholder-Engagement_Updated_2014.pdf.

Twomey, S., & O'Mahony, C. (2014). Stakeholder Engagement in Transboundary Maritime Spatial Planning. TPEA Project Showcase: European Maritime Day, Bremen, Germany. [online]. Retrieved September 6, 2018, from http://www.researchgate.net/publication/281092036_Stakeholder_Engagement_in_Transboundary_Marine_Spatial_Planning.

U.S. Institute for Environmental Conflict Resolution. (2011). Principles for Stakeholder Involvement in Coastal and Marine Spatial Planning. [online]. Retrieved April 04, 2018, from http://projects.ecr.gov/cmspstakeholderengagement/pdf/PrinciplesforStakeholderInvolvementinCoastalandMarineSpatialPlanning_0.pdf.

Wingqvist, G. O., Drakenberg, O., Daniel Slunge, D., Martin Sjöstedt, M., & Ekbom, A. (2012). The Role of Governance for Improved Environmental Outcomes-Perspectives for Developing Countries and Countries in Transition. Swedish Environmental Protection Agency [online]. Retrieved April 06, 2018, from http://www.swedishepa.se/Documents/publikationer6400/978-91-620-6514-0.pdf?pid=3823.

14

Scenario-Building for Marine Spatial Planning

Lynne McGowan, Stephen Jay, and Sue Kidd

1 Introduction

The use of scenarios in strategic (terrestrial) spatial planning has been widely accepted for a number of years (ESPON 2007; Haughton et al. 2010). A range of scenario-building techniques has also been applied within marine management in order to support decision-making (Van Hoof et al. 2014; Lukic et al. 2018). Scenario-building is now attracting some attention within the specific context of marine/maritime spatial planning (MSP) too. This chapter explores the extent to which scenario-building is starting to be introduced and presents a scenario-building exercise carried out in relation to a transboundary MSP exercise for the Celtic Seas region. This example of scenario-building takes previous practice as its starting point but aims to produce a narrative that focuses more directly on two of the most critical issues for MSP in transboundary spaces—namely evolving patterns of spatial development in the marine area and the need for increased cooperation between MSP authorities.

Scenario-building is one of a set of terms being used to suggest a future-oriented, strategic dimension to planning; visions, forecasts, strategies, prospective road maps and action plans also suggest forward-looking tools to support plan-making. For simplicity, in this chapter, we focus on the notion of scenarios. The concept has its origins in military strategy and business plan-

L. McGowan • S. Jay (✉) • S. Kidd
Department of Geography and Planning, University of Liverpool, Liverpool, UK
e-mail: Stephen.Jay@liverpool.ac.uk

J. Zaucha, K. Gee (eds.), *Maritime Spatial Planning*,
https://doi.org/10.1007/978-3-319-98696-8_14

ning (Lindgren and Bandhold 2009) but has been adopted within public administration. Whilst there is no single definition of a scenario, one useful definition from the Intergovernmental Panel on Climate Change (2001) states:

> A scenario is a coherent, internally consistent and plausible description of a possible future state of the world. It is not a forecast; rather, each scenario is one alternative image of how the future can unfold.

Therefore, any process that examines a scenario or scenarios involves the creation of alternative images of the future and evaluating them against some kind of goal or set of values. In doing so, the purpose of using scenarios is inextricably linked to the question of *what do we want to know about the future?* At a general level, van Hoof et al. (2014) suggest that scenarios "can contribute to policy decision making by identifying and anticipating developments (desirable and undesirable) and information gaps and inconsistencies" that help to focus attention on causal processes and decision points that can be used in making better strategies.

2 Existing Scenario-Building Practice

The United Nations Educational, Scientific and Cultural Organization (UNESCO) MSP guide (Ehler and Douvere 2009) suggests that an MSP process should include the consideration of alternative spatial scenarios, of which one should be selected as the goal of the plan. Indeed, there are a number of existing examples of scenario-building within the context of coastal and marine planning and management. These have been partly experimental in nature, related to pilot projects, but also include some official processes. A frequently quoted example is that of the Belgian GAUFRE project (Maes et al. 2005). This was a research project that developed a visionary approach for the marine environment, applying certain land-use planning concepts and methodologies.

Here, we present more detailed examples from the UK, France and the wider Celtic Seas region. Firstly, in 2004, the UK's Department for the Environment, Food and Rural Affairs commissioned a study on Alternative Futures for Marine Ecosystems (AFMEC), (Pinnegar et al. 2006). This aimed to create a set of scenarios for use in strategic planning over a 20–30-year time frame. This resulted in a four-quadrant, two axes *possibility space*, which helped to define four scenarios. The two axes were the driving forces of the scenarios: societal values (from individual to community) along the horizontal axis, and distribution of power (autonomy to interdependence) along the vertical axis. Incorporating other key parameters, such as gross domestic product (GDP), demographic change and water consumption, it was then possible to build narratives related to the four separate quadrants of the possibility space. These narratives were given summary names: World Markets, Global Commons, Fortress Britain and Local Stewardship. The scenarios were then applied to a range of activity domains such as climate, fisheries and aggregates to demonstrate how possible trends may play out.

Secondly, within the UK's English marine planning process, there are key stages where future uses of the sea are being considered. In the initial plan preparation phase, these include "identifying issues" and "gathering evidence". In these two stages, the Marine Management Organisation (MMO) in conjunction with stakeholders gathered information about the plan area. In the "options development" stage, the MMO considered different ways of achieving the plan objectives and vision. Options were compared to a *business as usual* scenario, which considered how the marine area might develop in the absence of a MSP. Scenarios were again incorporated into the "plan policy development" stage. The MMO commissioned research to review past trends and current drivers and develop future projections for selected industry sectors active in the plan areas (MMO 2017). The scenarios used in this exercise were developed as part of the Celtic Seas Partnership's Future Trends project (described later) and consisted of *Business as Usual*, *Nature @ Work* (maximising ecosystem services) and *Local Stewardship* (local decision-making and differentiation) scenarios. Changes in activity for each sector were mapped and plotted according to the most appropriate unit of activity (e.g. MW of energy generated, Gross Value Added (GVA), freight tonnage). Potential trade-offs between sectors and the environment were identified in each of the marine plan areas.

Thirdly, in France, the North Atlantic-Western Channel Façade is piloting the implementation and monitoring of strategic planning for maritime space and coastal areas (*façades*). A guide to the process by which it will be produced was recently published (Ministère de l'Environnement, de l'Énergie et de la Mer 2017 and Direction Interrégionale de la Mer Nord Atlantique-Manche Ouest (DIRM-NAMO) 2017). In the first stage, the existing conditions of the *façades* and emerging issues and risks will be identified. This will be followed by the definition of a Vision for 2030, priority objectives for the *façades* and the selection of indicators to measure progress against the objectives. In defining the Vision for 2030, a scenarios method will be adopted that builds in different socio-economic, institutional and environmental factors to develop contrasting pathways and visions and enables different points of view and actors to be brought together for collective reflection.

Finally, the Celtic Seas Partnership's Future Trends work examined the future of the Celtic Seas region, with reference to what this means for the achievement of good environmental status (GES) and the need for an integrated, ecosystem-based approach to marine management (ABPmer and ICF International 2016). The period considered was approximately 20 years from 2016 and covered 10 maritime sectors including conservation. In this project a set of three scenarios was used to project and map spatial development and highlight potential opportunities and spatial conflicts that may need to be resolved through cooperation.

It is also worthwhile to note that within the context of the European Union's framework for MSP, Directive 2014/89/EU (the MSP Directive, EC 2014) does not directly refer to scenarios, but Article 4(5) states:

> Member States shall have due regard to the particularities of the marine regions, relevant *existing and future activities and uses* [emphasis added] and their impacts on the environment, as well as to natural resources, and shall also take into account land-sea interactions.

In addition to this, maritime spatial plans should "identify the spatial and temporal distribution of relevant existing and future activities and uses in their marine waters" (Art. 8) to support the sustainable development and growth of the maritime sector. In so doing, Member States should take into consideration relevant interactions of activities and uses, such as aquaculture areas, fishing areas, installations and infrastructures for energy, transport routes and so on. Hence the need to take into account future uses and activities across a range of sectors may provide for the consideration of alternative options or scenarios. Furthermore, where maritime spatial plans are likely to have significant effects on the environment, they are subject to Directive 2001/42/EC on the assessment of the effects of certain plans and programmes on the environment (the SEA Directive, EC 2001). This requires that in thinking about policy responses, "reasonable alternatives taking into account the objectives and geographical scope of the plan" should be considered, thereby ensuring plan-making authorities explore differing futures in some way.

3 Developing a Typology of Scenarios

These examples of scenario-building for marine management illustrate different ways of thinking about the future and the different types of scenario that may be used to answer questions about pathways for development. Borjeson et al. (2006) provide a simple distinction between scenario types based on the principal questions that a user may want to pose about the future:

- What will happen?
- What can happen?
- How can a specific target be reached?

Normative scenarios address the question "how can a specific target be reached?" and are most frequently used when a desired end state is known, with the user wanting to determine how that state can be reached by working backwards. Back casting from an end state can help to identify incremental steps that should be taken to achieve the desired goal. Back casting can also identify the factors that may prevent achievement of the end goal.

Predictive scenarios attempt to answer the question of "what will happen?" In this case, information about the past and present is projected forward to a future point to see what the situation might be, that is, changes are determined by forecasting. For example, predictions of coastal erosion around the UK coast have been used to develop Shoreline Management Plans that respond to potential risks over 20-, 50- and 100-year periods.

Exploratory scenarios consider "what can happen?" given a set of plausible futures. They are often used to understand developments over a longer time horizon or more strategic issues (Borjeson et al. 2006, 727). An example of this is a project which aimed to strengthen the preparedness and adaptive capacity of communities within the Hudson River watershed in the face of climate change (Roberts 2014). Here four scenarios (Procrastination Blues, Stagflation Rules, Nature be Damned! and Give Rivers Room!) were used to determine the consequences of different paths of action and the likelihood that different response options would be taken up under each scenario.

The pathways explored by each of the three types of scenario are illustrated in Fig. 14.1. Visualising scenarios in this way, normative scenarios may be seen as *inward bound* as they work backwards to see how a desired future might grow from the present. In contrast, predictive and exploratory scenarios might be described as *outward bound* as they extrapolate trends into the future or ask "what if?" or "what can happen? questions to arrive at a range of possibilities.

There are instances when different types of scenario (exploratory, normative etc.) can be used in conjunction with each other. For example, in the Water Scenarios for Europe and for Neighbouring Countries (SCENES) project, exploratory scenarios for freshwater management were first developed to provide a specific "end point" that set a socio-economic and institutional

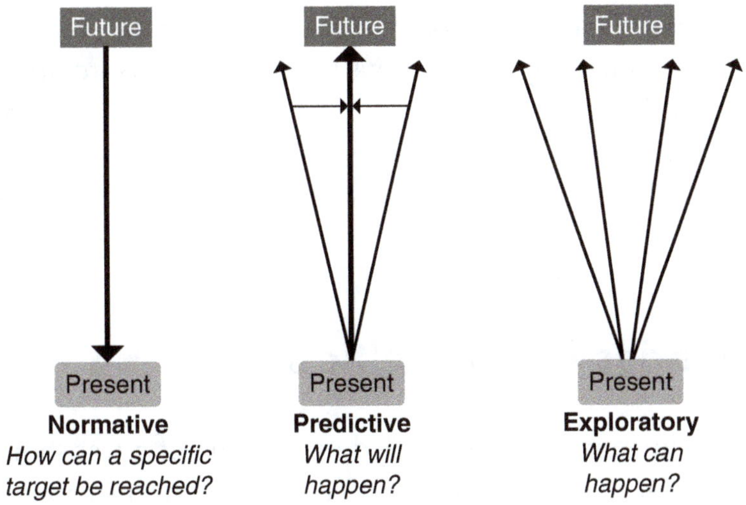

Fig. 14.1 Types of scenario

context for water management. Then a back casting (normative) method was used to identify interim objectives, policy actions and strategies to achieve this vision (Kok et al. 2011).

4 Principles of Scenarios

Based on these examples and wider literature, a set of principles for developing scenarios for use in MSP can be outlined. These are:

1. ***Scenario-building should be participatory:*** Scenarios should be created with stakeholder input, either in the creation of the initial narrative, defining focus/scope, or in checking plausibility and potential outcomes (Herry and Winder 2015).
2. ***An appropriate time frame should be adopted:*** this may vary depending on the nature of driving forces. This should be at least 5 years where change happens quickly but up to 50 years where change may be more slow or uncertain; at least 10 years is typical (Pinnegar et al. 2006, 16).
3. ***Plurality is required:*** two to four scenarios are considered to be the optimal number for exploring a range of potential futures.
4. ***The scenarios developed should have plausibility:*** whilst scenarios are not intended to be accurate forecasts of the future, they should be constructed in such a way that users can see the scenarios as possible futures.
5. ***Scenarios should have internal consistency:*** the building blocks (or drivers) that are used to create each scenario should be joined together in an explainable and logical manner (Haines-Young et al. 2011; Van Hoof et al. 2014).

6. ***Scenarios should have resonance with their users:*** the scenarios produced should have sufficiently distinct narratives for users to understand the varying conditions and drivers to be considered. They should tell a story that is convincing. Dramatic or extreme scenarios, using memorable names, are helpful in this instance (Joint Research Commission 2008).

5 Scenarios for the Celtic Seas

A scenario-building exercise was carried out as part of a transboundary MSP project that sought to understand possible future patterns of spatial development in the Celtic Seas region and what this might mean for transboundary cooperation on MSP (see McGowan et al. 2018; and www.simcelt.eu). This region incorporates national waters from France, Ireland and the UK and involves seven administrations with MSP responsibilities.

The development of scenarios in this exercise followed a four-stage process:

- Background material on key maritime sectors was collected and used to produce a set of sectoral Briefing Notes. These covered policies and MSP processes in relation to the specific sector and identified a series of drivers (political, economic, technological, etc.) that may be critical to the future development of each sector across the Celtic Seas.
- Based on previous examples of scenario development, a *possibility space* was developed as a framework to shape four distinct scenarios. This was shaped around two axes, representing two dimensions of particular importance to transboundary cooperation.
- The drivers for change identified in the Briefing Notes were mapped on to the new possibility space in order to create more in-depth narratives or *pen pictures* for each scenario.
- The scenarios were tested by stakeholders in a workshop setting, where they considered different sectoral trajectories for growth, what this might mean for integration and identified key issues where transboundary working would become more important.

5.1 Chosen Sectors

The sectors included in this exercise were deliberately limited to a small number due to the time-limited nature of the project and to make it possible to examine each sector in depth. They were selected using two criteria:

- The sector has a distinct transnational dimension, in terms of movement across transnational space or fixed patterns of spatial development (or structures) that span national borders or
- The sector is known to have growing spatial demands, that is, it is an expanding sector that must be taken into account in the development of maritime spatial plans

The sectors chosen were:

- Cables and Pipelines
- Ports and Shipping
- Offshore Wind Energy
- Wave and Tidal Energy
- Aquaculture

5.2 The Development of Scenarios: The Possibility Space

Following previous examples of scenario development, the scenarios were developed using the four-quadrant or *possibility space* approach with two main variables used to construct the horizontal and vertical axes. The axes represented to key dimensions, as follows.

Footprint: Spatial Diffusion Versus Efficiency (Horizontal Axis)

Changing spatial footprint was represented by a continuum from spatial diffusion to spatial efficiency. Whereas previous scenario exercises have tended to use environmental concerns or green approaches against economic development as a proxy for changing spatial footprint, this approach recognised how the activities of many new maritime sectors are shaped by technological advances and the drive to decarbonise the economy, providing *greener* or more sustainable patterns of development, for example, energy generation from offshore wind turbines or cleaner, more fuel-efficient ship design. Given the different stages of economic growth that can be attributed to different maritime sectors, some activities can be expected to expand in terms of spatial distribution (e.g. the development of new offshore wind farms) and/or resource use (e.g. more intensive aquaculture). Conversely, other maritime sectors could be expected to decrease their spatial footprint (e.g. when oil and gas fields are exhausted and rigs are decommissioned).

In this case, **spatial diffusion** is used to describe a situation where different marine users or sectors:

- take up the maximum amount of marine space that is available to them;
- use that space exclusively (i.e. do not coexist or co-locate with other marine users); and
- use marine resources both expansively and most intensively to maximise exploitation of the marine resource available to them.

Spatial efficiency, on the other hand, occurs when users or sectors:

- take up a smaller amount of marine space;
- use the same space—coexisting or co-locating with other compatible activities;
- use limited resources or use marine resources in a more sustainable manner.

Cooperation: Autonomy Versus Cooperation (Vertical Axis)

This axis reflected the degree of cooperation that takes place between MSP authorities. At the bottom end of the scale, *autonomy* refers to minimal levels of cooperation between authorities (at national or international scales) and the maintenance of "hard" boundaries around a given entity's maritime space. At the opposite end of the vertical axis, *cooperation* refers to strong relationships between planning authorities that span national borders, more permeable boundaries (whilst respecting national sovereignty) and a recognition of shared responsibility for maritime regions. This may manifest itself in the development of regional cooperation, new models of governance, ecosystem-based management or more integrated forms of planning (van Tatenhove 2013).

The Possibility Space

By combining the two axes, a possibility space is created containing four quadrants or possible outcomes (scenarios), depending on different combinations of footprint and cooperation (Fig. 14.2). This allows the development of more detailed scenarios. In the top left-hand quadrant, Scenario 1 represents a situation where cooperation may be high between authorities and patterns of development also show high levels of spatial diffusion. Moving to the right,

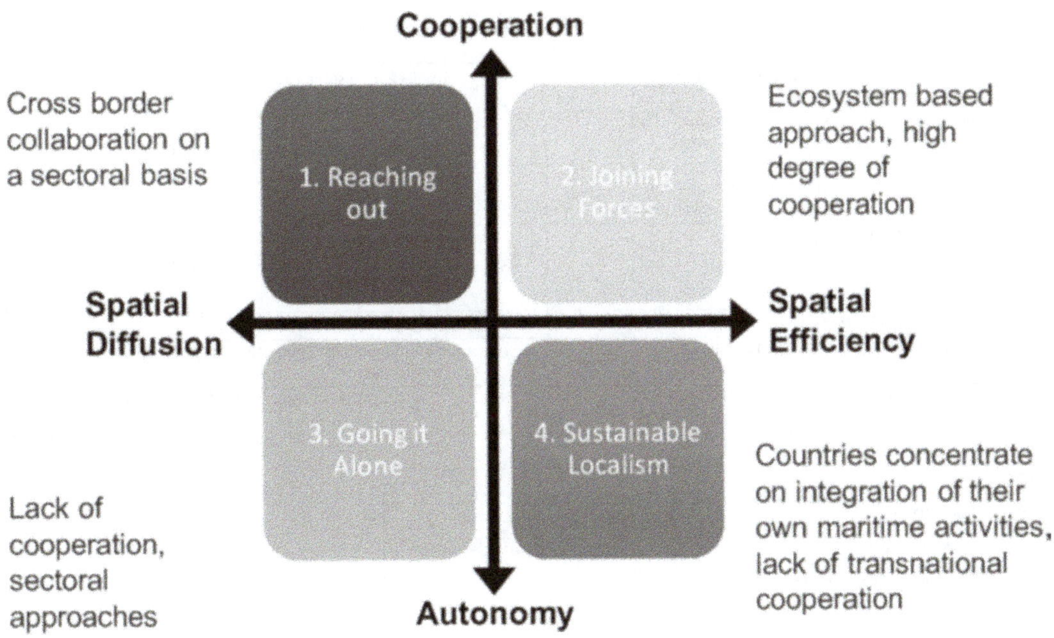

Fig. 14.2 The SIMCelt possibility space

Scenario 2 also displays high levels of cooperation but a high level of spatial efficiency. In certain circumstances this may be considered the *ideal* scenario as it represents the greatest level of cooperation between authorities and most efficient use of marine resources. In contrast, the bottom left-hand quadrant (Scenario 3) represents a situation of little cooperation and high levels of spatial diffusion, inferring uncoordinated and expansive resource use. Finally, Scenario 4 (bottom right-hand quadrant) refers to a situation of little trans-boundary cooperation but more efficient resource use within individual jurisdictions. Full descriptions for each scenario are given in Fig. 14.4.

5.3 Mapping Drivers onto the Possibility Space

Having defined the four different scenarios, the next step was to develop a storyline for each scenario. This was expressed in a set of four *pen pictures*. To do this:

1. The drivers for change identified in the Briefing Notes were collated in a table for each sector, with additional columns for spatial efficiency, increasing cooperation and justification,
2. For each driver or planning issue, a judgement was made whether it would lead to increasing or decreasing spatial efficiency and increasing or decreasing levels of cooperation between authorities. The results of this were recorded in a table (see example in Fig. 14.3) with justification.

Sector: Aquaculture		Spatial Efficiency	Cooperation	Justification
A8	National ambitions to increase aquaculture production	-	+	Will require some expansion of sites, but also increased stakeholder engagement
A21	Development of integrated multi-trophic aquaculture (IMTA), or polyculture, where different species such as shellfish, seaweed and fish are cultivated together to enable the recycling of nutrients through the food chain	+	=	Positive impacts in terms of spatial efficiency. May not require transboundary cooperation
Sector: Offshore Wind Energy				
O24	Increasing size and generation capacity of wind turbines	+	-/=	Possibility to concentrate higher output in smaller areas.
Sector: Ports and Shipping				
P5	Reduction of CO_2 emissions and pollution by shipping	=	+	Enforcement may require cooperation between authorities
P8	Continued development of the TEN-T network	=/+	++	Concentration of shipping traffic through key routes to enable accessibility of all regions

KEY			
++	High increase	=	Neutral
+	Some increase	-/=	Neutral – tending to decrease
=/+	Neutral-tending to increase	-	Decreasing

Fig. 14.3 Example of mapping cooperation and spatial impacts

3. Except where the resultant impact for each driver was judged to be neutral, the driver was mapped onto the possibility space.

5.4 Developing Storylines

Mapping Impacts on to the Possibility Space

Having identified the likely impacts on sectoral drivers and planning issues across all the sectors in the Briefing Notes, these were mapped on to the possibility space in order to create storylines for each scenario, as follows:

1. Each driver/issue was numbered (e.g. O1, O2 for offshore wind, C1, C2 for conservation).
2. The possibility space was further divided up into 7 × 7 grid squares for each quadrant.
3. Based on the likely impacts of each driver/issue in the table, a decision was made about where this would fit within the possibility space using the two axes as a guide. For example, where spatial efficiency was seen to be increasing, a marker was placed on the right-hand side (Scenario 2 or 4), or if it was decreasing on the left (Scenario 1 or 3).
4. Level of cooperation was then considered. If this was reckoned to be increasing, the marker would be moved towards the top half (Scenario 1 or

2), or if decreasing in the bottom half (Scenarios 3 and 4). For moderate or low changes, markers would be placed closer to the centre of the corresponding axes, and for extreme changes markers would be further out.

5. This process was repeated with each driver until they had all been placed within the possibility space. Multiple markers were allowed in each grid square (Fig. 14.4).
6. The drivers in each quadrant were then assembled into a pen picture with illustrative examples of how different sectors may develop up to the year 2050.
7. Each scenario was given a title that conveys its main characteristics (Fig. 14.5).

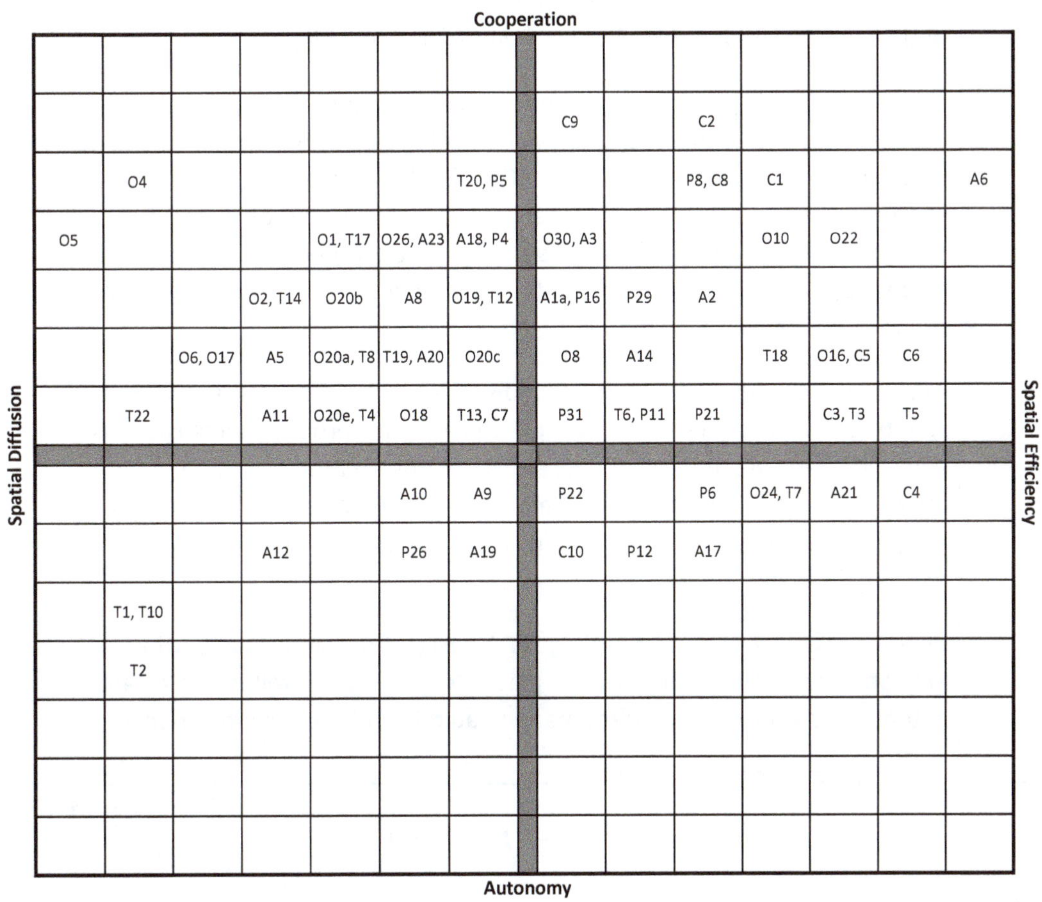

Fig. 14.4 Mapping individual drivers onto the possibility space

Scenario 1. Reaching Out

Key features: Cross border collaboration on a sectoral basis

International and national climate change targets and pollution controls are key drivers of change.

These lead to countries making greater efforts to deploy **marine renewables** in coastal areas and further offshore. More areas are zoned for the primary purpose of renewable energy growth both in coastal areas and further out to sea, creating competition for space between energy interests and other sea users such as aquaculture and shipping and increasing cumulative impacts. Transnational energy infrastructure is put in place to support the distribution of green energy.

Sharing of **information** within sectors is seen as a way to increase coordination, e.g. E-navigation, maritime service portfolios and development of the Common Information Sharing Environment for shipping.

Within the **shipping** sector international agreements on pollution are also key drivers of change, with more Emission Control Areas being designated and a much greater number of ships using LNG fuels. The seasonal opening of Arctic sea routes takes place but is dependent on high levels of international cooperation to maintain safety and security. Motorways of the Sea continue to develop along key routes and into more remote areas to connect with Arctic routes and growing renewable energy zones.

Ambitions for **aquaculture** production remain high across Celtic Seas countries as consumer demand for aquaculture product increases. As aquaculture moves further offshore this creates greater competition with other sea users. Climate change impacts such as increases in sea water temperature and increasing storminess also make large-scale production more challenging.

Increased sharing of data regarding **MPA** designations and collaboration on environmental monitoring takes place, e.g. using satellite data and autonomous vehicles to monitor marine habitats and species movements.

(*continued*)

Scenario 2. Joining Forces

Key features: Ecosystem based approach, high degree of governmental cooperation

This scenario affords the highest level of protection to the **marine environment**, with regards to international requirements such as CBD and MSFD. Countries cooperate on decisions about new MPAs, including some in international waters. At the national level, there is greater clarity and direction in the way that MPAs are designated and managed.

Tight environmental constraints mean that countries think more strategically about the location of maritime activities and there is a strong drive towards **colocation** of marine renewables with activities such as coastal defences, tourism, fisheries and aquaculture.

International **shipping** activity continues to increase, with larger ships being used to take advantage of economies of scale. In EU Member State waters, reduced customs formalities increase the efficiency and volume of goods moved through ports. Upgrades to port facilities and connectivity to ports hinterlands are implemented to take advantage of both international and local shipping movements. In areas where multiple marine users are active, protection of navigational safety is considered a priority.

Aquaculture growth is managed through the allocation of space in maritime spatial plans. Continued financial support from the EU and other institutions helps to deliver new operations that use innovative methods such as multi-use platforms shared with offshore wave energy and monitoring stations.

As well as developing colocation with aquaculture, fisheries and environmental monitoring, **renewable energy** continues to grow in two main areas. **Offshore wind** energy moves further out to sea, as technology for deeper waters (including floating platforms) becomes more viable both technologically and financially. A limited number of **tidal lagoons** are built, primarily for energy generation, but also supporting new leisure and tourism activities.

(*continued*)

Scenario 3. Going It Alone

Key features: Minimal cooperation, expanding sectoral approaches

Under this scenario, countries work independently to pursue their own **Blue Growth** targets, expanding and maximising exploitation of their maritime resources across marine territories. Coordination and cooperation on MSP is minimal. Competition within maritime sectors becomes fiercer, leading to distinct winners and losers, for example bigger **ports** using economies of scale and their connectivity to capture more shipping trade compared to smaller ports.

Efforts to protect the **marine environment** are limited as countries seek greater levels of economic exploitation, e.g. using waters more intensively for aquaculture, fishing and producing energy.

In terms of **aquaculture**, increasing demand for farmed products and the need to combat impacts of climate change such as increased seawater temperatures lead to the use of genetically modified alternatives to fishmeal, and GM species that grow faster.

To ensure security of energy supplies, existing sources of hydrocarbons continue to be extracted whilst new sources are explored. **Offshore wind, wave and tidal energy** continue to expand, with devices deployed in coastal waters and further offshore. Large tidal lagoons and barrages are built where these do not interfere with key navigational routes, resulting in some loss of habitats.

(continued)

Scenario 4. Sustainable Localism

Key features: Countries concentrate on developing their own maritime activities but there is a lack of transnational cooperation.

Under this scenario economic growth in traditional industries is slow but there is accelerated growth in green and high-tech sectors. Smart **specialisation** within the maritime sector helps regions to develop unique strengths and capacities. New technologies also help to integrate different sectors using the same space as shared platforms, monitoring systems and less polluting ways of doing things are found.

Conservation and environmental objectives focus on the reinforcement of existing management and regulation measures. Where new MPAs are considered for designation, there is a strong emphasis on additional socioeconomic benefits that can be provided through designation.

To use space more effectively, the **aquaculture** sector adopts a polyculture approach and multi-trophic species. High quality, niche aquaculture products with greater added value and traceability throughout supply chains are developed for local markets.

Diversification occurs within the **port** sector due to the slow growth of international trade, for example specialised shipbuilding services and innovations in logistics through greater use of IT and real-time tracking. Facilities servicing the offshore energy industries are adopted by some ports to compensate for the decrease in international cargos. In other ports, **short sea shipping** experiences a modest increase for specialised cargos such as liquid bulk.

Wave and tidal energy is increasingly favoured over offshore wind as technologies improve and both small and large-scale projects become more financially viable. Tidal lagoons are built in locations for the dual purposes of energy generation and protecting areas vulnerable to flood risk.

Fig. 14.5 The four scenarios

5.5 Stakeholder Preferences

The four scenarios were tested in a workshop with 35 participants representing the different administrations of the Celtic Seas, consultants, researchers, ecologists, planers and industry representatives from the energy, fisheries and shipping sectors. Interactive sessions were used to explore the scenarios. Participants were organised into groups representing the key sectors involved. They then explored, firstly, *sectoral ambitions* up to 2050. Secondly, they considered *sectoral interactions*, looking at other sectors' ambitions for 2050 and what this might mean in terms of potential competition for space or new synergies that might arise. Thirdly, significant cross-border MSP issues were identified and discussed with a view to *promoting cross-border cooperation.*

5.6 Sectoral Ambitions

In the first session, participants were asked to consider where their sector would be by the year 2050 in terms of the degree of transboundary cooperation that might take place and whether the sector would increase its spatial efficiency. Their views were recorded on the possibility space (Fig. 14.6). For all sectors, there was an aspiration to move towards greater spatial efficiency:

- For conservation, some permitted activities may develop within Marine Protected Areas (MPAs), but this may be on an ad hoc basis
- For offshore wind energy, location would be influenced by potential supergrids and interconnectors

With regard to levels of cooperation, there was more variation in participants' views and across different sectors. In some cases this may have been due to uncertainties and speculation surrounding the UK's intended exit from the

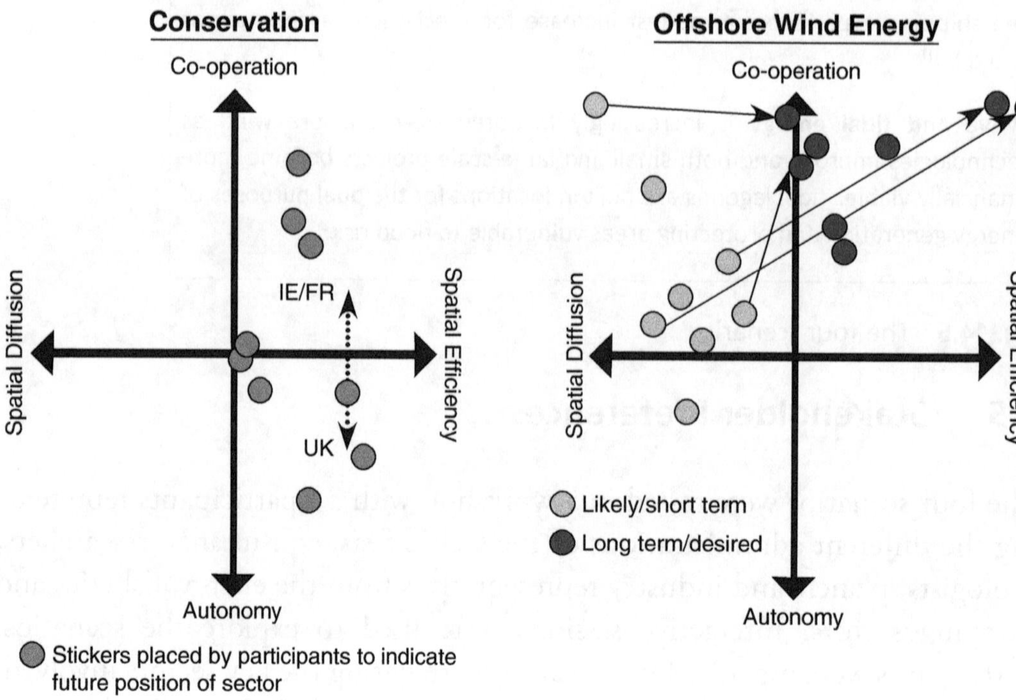

Fig. 14.6 Future directions for selected sectors

European Union and potential implications for existing regulations and mechanisms, such as commitments to the Marine Strategy Framework Directive or the maritime transport space without barriers. Other reasons for changing levels of cooperation included:

- For aquaculture, existing low levels of cooperation expected to continue,
- For conservation, other regional cooperation mechanisms such as OSPAR may facilitate cooperation more than planning authorities,
- For ports and shipping, the need to ensure navigational safety may lead to increased cooperation and
- For wave and tidal energy, large-scale projects such as tidal lagoons can only be successfully implemented with sufficient buy-in from local communities, developers and planners.

5.7 Sectoral Interactions

In the second session, participants were asked to consider the positions taken by other sectors within the possibility space and consider what this would mean for their sector. Some of the key points emerging were:

- *Aquaculture* and *conservation* have a mutual interest in maintaining good water quality.
- The co-location of *aquaculture* and offshore *energy* was identified as a key opportunity; however, some big questions remain about the possibility of co-location as aquaculture areas may not be suitable for energy installations (and *vice versa*). Similarly, it was noted that the case for economic viability and societal benefits has not been made so developers may be unwilling to take risks.
- *Wave energy* appears to offer the greatest opportunities for co-design that can incorporate wider community benefits.
- Co-location between *ports* and *aquaculture* is unlikely to take place, but ports may benefit from the spatial management of aquaculture as they can influence location to protect navigational safety.
- *Ports* may need to evolve in order to keep pace with logistical demands from larger *wave, tidal energy* and *offshore wind* developments.

5.8 Actions to Promote Cross-Border Working

Following the discussions of sectoral interactions, the two top issues from each table were identified by facilitators for elaboration of problems and possible solutions. The issues identified were:

1. Biosecurity and shipping
2. Conservation and offshore wind
3. Co-location of aquaculture and conservation areas
4. Co-location of aquaculture and offshore wind
5. Transnational energy grids and storage facilities
6. Co-location of aquaculture and ocean renewable energy (further offshore)
7. Port diversification
8. Designation of new shipping lanes

Participants then considered these issues in more detail and think of ways they could be addressed by MSP authorities. For each issue, discussion helped to identify the transnational nature of the issue, possible solutions and the resources or mechanisms that would need to be put into place in order to improve the existing situation.

5.9 Outcomes

The scenarios' workshop exercises helped to confirm the predictions that all of the sectors under consideration would continue to grow in terms of their activities over the period up to 2050. With regard to the possibilities of more spatially efficient forms of development, a number of key trends were discerned (Fig. 14.7).

6 Reflections

The use of a four-quadrant or *possibility space* has mirrored the approach to developing scenarios used in other exercises, such as the Millennium Ecosystem Assessment and the AFMEC project (Pinnegar et al. 2006). However, a critical difference in this case was in regards to the two variables used to construct the axes of the possibility space, namely autonomy/cooperation and spatial diffusion/efficiency. The use of these two axes or dimensions, together with the scenario pen pictures, provided for a broad range of possibilities in terms

- Aquaculture may increase spatial efficiency through better integration into marine plans and use of new technologies such as multi-trophic systems; however co-location with other sectors is unlikely to become large scale
- Continuing development of offshore energy infrastructure (cabling), with some cross border interconnectors coming into service
- Designation of MPAs will continue, but management will be more challenging
- Ports and shipping will remain a critical focus for MSP. Diversification of ports and cooperation with other sectors such as offshore wind energy may provide some spatial efficiency
- Offshore wind will continue to have a growing spatial footprint in the Celtic Seas. Some spatial efficiencies may be achieved through technological improvements such as increased generation capacity of turbines
- Upscaling of wave and tidal energy deployment will lead to increased spatial footprint. Additional socio-economic benefits may be gained through co-design and consultation with local communities where they are sited

Fig. 14.7 Key outcomes of the scenarios workshop

of the spatial footprint of future maritime activities and cooperation between planning authorities to be represented. By having contrasting scenarios, this brings into focus the extremes of what might be the most or least desirable futures. For example, the "Joining Forces" scenario represents the ideal in terms of promoting integration between uses, co-location and a high degree of transboundary cooperation in MSP, whilst "Going it Alone" represents the opposite. Use of these extremes also provides for reality checks to occur, as although the ideal situation may be integration or co-location of different maritime activities, there will always be some that require exclusive use of space, such as for navigational safety. Therefore, the scenarios can highlight what may be more feasible within the scope of MSP or specific plans going forward.

In testing these scenarios with participants, their feedback provided points for reflection.

- The presentation of scenarios including drivers and targets in the scenarios could be adapted to the subnational level at which MSP is taking place in many contexts.
- The definition of a baseline position for each sector on the possibility space was important for context setting and demonstrating the geographic specificities of development for each sector.
- Within larger maritime sectors, there are specific sub-sectors for which the more desirable future may differ quite considerably from the bigger picture. For example, for ports and shipping, cooperation and spatial effi-

ciency may be of less importance to the recreational boating sector as it operates in a different way to the commercial (freight) shipping sector. The example of conservation (as shown in Fig. 14.6) also demonstrates that different scenarios may be preferred or be more likely depending on geographical specificities and country contexts. This demonstrates how although it would be expedient to have one "agreed" scenario to help in the formation of marine spatial plans, there will always be alternative views and potential outcomes that may need to be accommodated in the plan-making and implementation stages.

Although the scenarios were not intended to provide an accurate prediction of the future, the process of scenario-building can promote debate about the direction that plans might take. This can allow for more creativity and opportunities for learning about the potential of MSP to facilitate particular outcomes. Specifically, looking at where each sector may be (or would like to be) in the future can reveal the aspirations of the sector towards cooperation and spatial efficiency. Comparing these aspirations across sectors may then help to show where there are likely to be spatial conflicts as different sectors strive for integration or co-location in limited space, or where sectors may be more resistant to integration and cooperation. For MSP authorities, understanding the direction of travel for different sectors and their aspirations for the future is critical. If this information can be recorded, such as through scenario-building exercises, then plans and policies may be better informed.

7 Conclusions

Within terrestrial settings, there has been significant interest in future-oriented approaches to planning (Albrechts 2004; Haughton et al. 2010; Nadin 2002). This has been mostly at a strategic level of planning, where there is greater scope for considering a range of broad possibilities, reflecting different overall objectives, than may be the case at a more localised, project-specific level of planning. Similar approaches have been adapted to a small extent in marine settings too, as exemplified in this chapter. Arguably, the potential and need for these exploratory approaches is greater in the context of MSP, where planning spaces are geographically vast, the possibilities for human interaction are diverse and priorities for action are far from settled.

Developing alternative scenarios, such as by the method presented here, or engaging in some other future-oriented exercise, can be a productive way of envisaging possible trajectories and shaping preferred lines of travel over the coming years. It can help those involved in plan making to set out alternatives that higher-level policymakers can consider and stakeholders can deliberate. This may assist in preventing MSP from becoming too narrowly focused on meeting the immediate spatial needs of the most demanding activities and failing to consider broader, long-term objectives and foreclosing opportunities which may become more important with time.

The UNESCO MSP guide (Ehler and Douvere 2009) suggests that identifying alternative spatial scenarios is an integral part of an MSP process. Various decision-making criteria may then lead to the selection of a preferred scenario which then becomes the goal of the subsequent steps of the process, and which the plan aims to deliver. However, it is unlikely that a single, preferred scenario would be easy to agree; differing, competing scenarios may persist, at least in the background, throughout the process. And even if one scenario is formally selected as a goal to be reached, it is unlikely to remain completely fixed but may evolve and be adjusted in the light of realities and changing priorities that come to the fore as plan-making proceeds, not to mention during efforts to implement a plan once completed.

It is perhaps more productive to develop a range of scenarios, through a process such as that outlined earlier, and to allow them to live throughout an MSP process, acting as points of reference as more definitive aspects of planning are carried out. The questions then become, in relation to individual planning decisions: Which of the scenarios does this lead us towards? Is this desirable? Or should we act more in favour of heading towards a different scenario? This is not dissimilar to Hillier's argument for broad visions to be set in the background ("planes of immanence") and more specific plans and projects then to be brought into the foreground ("planes of organisation") (Hillier 2010, 454). One can imagine an oscillation between these two dimensions of planning activity; priorities and criteria of one kind or another for the use of a sea space are shaped by the scenario(s) judged to be preferable, and the scenarios themselves may be revisited in light of the hard facts of establishing those priorities and criteria. The *possibility space* offered by a range of scenarios is thus kept open throughout, so that the MSP process can seek out desirable futures, but remain open to opportunities, so that a clearly defined end point is never quite in view (Boelens and de Roo 2015).

Acknowledgements We are grateful to the project "Economy of maritime space" funded by the Polish National Science Centre for contributing the Open Access fee for this chapter and facilitating our discussions and preparation of the book.

References

ABPmer & ICF International. (2016). *Future Trends in the Celtic Seas: Scenarios Report*. ABPmer Report No. R.2584d produced Celtic Seas Partnership, ABPmer, Southampton.

Albrechts, L. (2004). Strategic (spatial) Planning Re-examined. *Environment and Planning B: Planning and Design, 31*(5), 743–758.

Boelens, L., & de Roo, G. (2015). Planning of Undefined Becoming: First Encounters of Planners Beyond the Plan. *Planning Theory, 15*(1), 42–67.

Borjeson, L., Höjer, M., Dreborg, K. H., Ekvall, T., & Finnveden, G. (2006). Scenario Types and Techniques: Towards a User's Guide. *Futures, 38(7), 723–729.*

Direction Interrégionale de la Mer Nord Atlantique-Manche Ouest (DIRM-NAMO). (2017). *Document Stratégique de Façade*. Retrieved February 5, 2018, from http://www.affaires-maritimes.pays-de-la-loire.developpement-durable.gouv.fr/document-strategique-de-facade-dsf-r188.html.

EC. (2001). *Directive 2001/42/EC of the European Parliament and of the Council of 27 June 2001 on the Assessment of the Effects of Certain Plans and Programmes on the Environment (The Strategic Environmental Assessment Directive)* OJ L 197, 21.7.2001, 30–37.

EC. (2014). *Directive 2014/89/EU of the European Parliament and of the Council of 23 July 2014 Establishing a Framework for Maritime Spatial Planning* OJ L 257, 28.8.2014, 135–145.

Ehler, C., & Douvere, F. (2009). *Marine Spatial Planning: A Step-by Step Approach Towards Ecosystem-based Management*. Manual and Guides No 153 ICAM Dossier No 6. Paris: Intergovernmental Oceanographic Commission UNESCO IOC, 99 pp.

ESPON. (2007). *Scenarios on the Territorial Future of Europe*. Report of ESPON Project 3.2, ESPON, Luxembourg.

Haines-Young, R., Paterson, J., Potschin, M., Wilson, A., & Kass, G. (2011). *The UK NEA Scenarios: Development of Storylines and Analysis of Outcomes*. The UK National Ecosystem Assessment Technical Report. UK National Ecosystem Assessment, UNEP-WCMC, Cambridge

Haughton, G., Allmendinger, P., Counsell, D., & Vigar, G. (2010). *The New Spatial Planning: Territorial Management with Soft Spaces and Fuzzy Boundaries*. London: Routledge.

Hillier, J. (2010). Strategic Navigation in an Ocean of Theoretical and Practice Complexity. In J. Hillier & P. Healey (Eds.), *The Ashgate Research Companion to Planning Theory* (pp. 447–480). Aldershot: Ashgate.

IPCC. (2001). Developing and Applying Scenarios. In J. McCarthy, O. Canziani, N. Leary, D. Dokken, & K. White (Eds.), *Climate Change 2001: Impacts, Adaptation and Vulnerability, Contribution of Working Group II to the Third Assessment Report of the Intergovernmental Panel on Climate Change*. Cambridge:

Cambridge University Press. Retrieved from http://www.ipcc.ch/ipccreports/tar/wg2/index.php?idp=126.

Joint Research Commission. (2008). *Methodology: Exploratory versus Normative Methods*. FOR-LEARN Online Foresight Guide. Retrieved from http://forlearn.jrc.ec.europa.eu/guide/0_home/index.htm.

Kok, K., van Vliet, M., Dubel, A., & Sendzimir, J. (2011). Combining Participative Backcasting and Exploratory Scenarios Development: Experiences from the SCENES Project. *Technological Forecasting and Social Change, 78*(5), 835–851.

Lindgren, M., & Bandhold, H. (2009). *Scenario Planning: The Link Between Future and Strategy*. Palgrave Macmillan.

Lukic, I., Schutz-Zehden, A., & de Grunt, L. S. (2018). *Handbook for Developing Visions in MSP*. Technical Study under the Assistance Mechanism for the Implementation of Maritime Spatial Planning.

Maes, F., Schrijvers, J., & Vanhulle, A. (2005). *A Flood of Space: Toward a Spatial Structure Plan for Sustainable Management of the North Sea*. Brussels: Belgian Science Policy.

Marine Management Organisation. (2017). *MMO1127. Futures Analysis for the North East, North West, South East and South West Marine Plan Areas, June 2017*. Retrieved January 9, 2018, from https://www.gov.uk/government/uploads/system/uploads/attachment_data/file/650895/Futures_analysis_for_the_North_East__North_West__South_East_and_South_West_marine_plan_areas__MMO_1127_.pdf.

McGowan, L., Jay, S. A., & Kidd, S. J. (2018). *Overview Report on the Current State and Potential Future Spatial Requirements of Key Maritime Activities* (D3c) EU Project Grant No.: EASME/EMFF/2014/1.2.1.5/3/SI2.719473 MSP Lot 3. Supporting Implementation of Maritime Spatial Planning in the Celtic Seas (SIMCelt). University of Liverpool. 130 pp.

Ministère de l'Environnement, de l'Énergie et de la Mer. (2017). *Documents Stratégiques de Façade et de Bassin Maritime Volet Stratégique Guide d'élaboration: Volume 2—Méthodologie*. Retrieved January 10, 2018, from http://www.affaires-maritimes.pays-de-la-loire.developpement-durable.gouv.fr/IMG/pdf/guide_dsf_vol_2_cle25ae5b.pdf.

Nadin, V. (2002). Visions and Visioning in European Spatial Planning. In A. Faludi (Ed.), *European Spatial Planning: Lessons for North America* (pp. 121–137). Cambridge: Lincoln Institute.

Pinnegar, J. K., Viner, D., Hadley, D., Dye, S., Harris, M., Berkout, F., & Simpson, M. (2006). *Alternative Future Scenarios for Marine Ecosystems*. Technical Report. Cefas Lowestoft, 109 pp.

Roberts, E. J. (2014). *Exploratory Scenario Planning: Lessons Learned in the Field*. Lincoln Institute of Land Policy Working Paper WP14ER1. Retrieved from https://www.lincolninst.edu/sites/default/files/pubfiles/roberts_wp14er1.pdf.

van Tatenhove, J. P. M. (2013). How to Turn the Tide: Developing Legitimate Marine Governance Arrangements at the Level of the Regional Seas. *Ocean and Coastal Management, 71*, 296–304.

Van Hoof, L., Hendriksen, A., & Bloomfield, H. J. (2014). Sometimes You Cannot Make It on Your Own; Drivers and Scenarios for Regional Cooperation in Implementing the EU Marine Strategy Framework Directive. *Marine Policy, 50*(2014), 339–346.

15

Managing Risk Through Marine Spatial Planning

Roland Cormier and Andreas Kannen

1 Introduction

Marine spatial planning (MSP) is ultimately the allocation of spatial and temporal measures to ensure that human activities or, more specifically, sector and socio-economic development in the sea can take place in a sustainable manner (Cormier et al. 2015). Planning is a management process to establish objectives, in order to reach strategic goals set by a governance process. In this context, the planning process has to allocate space to address development objectives to reach the goals set by the policymaking process of the political system (Douvere and Ehler 2009; Cormier et al. 2017). In this discussion, the political system governs the process of policymaking: The planning process is a management function that follows the direction set by the governance processes (Anthony and Dearden 1980).

Planning does not occur in complete isolation from ongoing activities and existing legislation or policies (Maes 2008). Although MSP may often be confused with conservation planning within an ecosystem approach (Ansong et al. 2017), environmental, health and safety considerations also have to be integrated into the spatial allocation to achieve the development objectives of the sectors seeking opportunities (Christie et al. 2014). The European Maritime Spatial Planning Directive (MSPD) (EC 2014) is primarily socio-economic legislation that has to integrate the other European environmental directives

R. Cormier (✉) • A. Kannen
Helmholtz Zentrum Geesthacht, Geesthacht, Germany
e-mail: roland.cormier@hzg.de

J. Zaucha, K. Gee (eds.), *Maritime Spatial Planning*,
https://doi.org/10.1007/978-3-319-98696-8_15

such as the Water Framework Directive (EC 2000) and the Marine Strategy Framework Directive (EC 2008) (Moss 2008; Junier and Mostert 2015). It also has to integrate health and safety requirements that are set in legislative and regulatory frameworks of sector activities such as international and national legislation for shipping safety and safety buffer zones around marine wind farms (Aps et al. 2015) which trumps other environmental and socio-economic considerations.

Regulatory planning (on the essence of regulatory planning, see Chap. 1) is not an output per se. It is a process for which the output is a marine spatial plan (Cormier et al. 2015). From a regulatory perspective—and recognizing that the success of MSP or a planning process may be defined differently—the success of the regulatory planning process is the production of a plan. The success of the regulatory plan, in turn, is the implementation of its spatial allocation in the daily operations of the industry sectors and other human activities. Indeed, it is the implementation of the marine spatial plan in the regulatory approval processes of the sectors that will carry into effects the objectives stipulated by the plan to reach the development goals set by the political system (Cormier et al. 2017).

Based on the risk management standards of International Organization for Standardization (ISO) 31000 (ISO 2018), risk is the effect of uncertainty on achieving objectives. In risk management, processes, procedures, controls, tasks and reporting are used to reduce the uncertainties of achieving the objectives (Cormier et al. 2015). Thus, the objectives of managing the risks in a planning process are to reduce the uncertainties of producing a marine spatial plan. The objective of managing the risks in the plan is to reduce the uncertainties of achieving environmental, social and economic objectives once implemented. Therefore, risk management in the regulatory planning process and the plan is not dealing with the same risks. For example, an ill-managed planning process could lead to mistrust from stakeholders because they feel that their cultural sensitivities are not being acknowledged (Gee et al. 2017), or industry feels that they are being fingered as the problem. The planning process could fail to deliver a plan that has specific, measurable, achievable, realistic and time-bounded objectives (e.g. SMART objectives), or deliver a plan that cannot be implemented effectively and efficiently to achieve development and conservation objectives (Rice et al. 2005; Cormier and Elliott 2017). The plan may not incorporate pre-established regulatory requirements of given sectors in order to provide space for another activity such as shipping safety regulation and environmental standards and regulations (Aps et al. 2015). Scientific, management and operational uncertainties may have been inadvertently missed during the planning process (Cormier et al. 2015).

This chapter discusses the nuances of risk management between the regulatory planning process and the resulting plan, providing insight into perceived mismatches between stakeholders, science and policy. It also contrasts differences and the purpose of policymaking within a governance context, planning within a management context and implementation within a regulatory context. International risk management and quality management standards are also introduced to demonstrate how these standards can help address and integrate such varied risks.

2 The Confusing Jargon of Policy Development and Implementation

There is a very broad understanding of what is meant by management. It is sometimes confused as policy development initiatives, assessments of environmental concerns or managing human activities, all of which are elements of management (Chun and Rainey 2005; Cormier et al. 2019; Loehle 2006; Mingers and White 2010). In environmental realms, assessment and monitoring activities are often viewed as management, or the mere knowledge of scientific facts is viewed as good management practices (Browman and Stergiou 2004). The underlying problem is in understanding the differences between policymaking, planning processes and implementation of measures.

In management disciplines (Anthony and Dearden 1980; Green 2015), policymaking is a function of the governance system for setting long-term goals (Ackoff 1990). In MSP, the political system assumes that role through international collaboration and national political processes (Anderson 2011) that sets development and conservation goals expressed in either conventions, legislation or policies. Examples are the United Nations Convention on the Law of the Sea (UNCLOS 1982) as well as the European MSPD (EC 2014), the Marine Strategy Framework Directive (EC 2008) or the Integrated Maritime Policy (EC 2011). In addition to setting long-terms goals and in some cases shorter-term objectives, they also set the scope and the context for planning initiatives. For example, once adopted as national legislation, the competent authority delegated under the MSPD has to achieve the objectives outlined in Article 5 to reach the goals of Article 1. In contrast, planning is a function of the management process undertaken by administrations and departments that were delegated to lead such a process (Anthony and Dearden 1980). In MSP, the competent authority assumes the management role of the planning process through consultation and advisory processes to produce a marine spatial plan within the scope and context established by the conventions, legislation and policies established by the political system (Gavaris 2009; Cormier et al. 2017). The competent authority or the person delegated the job of leading the planning processes has to manage the stakeholder, scientific and technical consultation and advisory processes that are needed to identify the spatial and temporal allocations needed to address development and environmental objectives as well as stakeholder concerns. So-called policy integration has more to do with recognizing the constraints of existing regulatory frameworks and stakeholder concerns that can be addressed by a marine spatial plan (Long et al. 2015; Cormier et al. 2015; Creed et al. 2016). In some cases, there may be environmental quality requirements, socio-economic issues or health and safety concerns that cannot be addressed solely by a spatial plan. These may be identified in the plan as additional requirements that are being addressed by other mechanisms and jurisdictions. Implementation, however, is a function of the operational processes driven by the regulators that ultimately are delegated the authority to license the sector to undertake their activities (Girling 2013; Hupe and Hill 2016). Here, the regulators need to include the spatial and temporal allocations of the marine spatial plan for their sector as conditions of authorizations, licences or permits. It is this last step that ultimately implements the plan into the operations of sector activity.

The confusion may stem from concerns regarding the goals and objectives of the marine spatial legislation and policies, and the hope that these can be addressed through the planning process. Given that the planning process is conducted within the scope and context of the same legislation, the planning process cannot necessarily address these concerns without going back to the political system. The manager of the planning process still has to follow the direction provided by the legislation and policies. As an example, a regulator for a specific sector may initiate a planning process for a marine area in isolation of the other activities occurring in that area. Such an approach would ultimately have an impact on development and conservation objectives, but these would be outside the span of authority of that particular regulator. In most cases, the confusion lies in the lack of understanding of the respective roles of governance processes of the political system, management processes of MSP and the operational processes of regulatory approvals (Green 2015).

Without standardized processes and harmonized vocabulary, the various approaches and processes used in integrated oceans management and MSP continue to propagate a broad range of definitions, concepts and understandings that are most often implied, not explicitly defined and therefore provoke misunderstandings between planners from different countries and/or sectors as well as between planners and stakeholders. Abspoel et al. (forthcoming) conceptualize language and communication problems in MSP, concluding that the "lingua franca" of marine planning is evolving but not fully developed yet. According to Abspoel et al. (forthcoming) "few planners master the language already. For stakeholders, the language is unclear and not yet accepted". In addition, taking into account the increasing need for transnational cooperation in MSP, this "lingua franca" needs to bridge also different planning cultures and terminology of planning in different countries. Generally, a common language will need to evolve from practice and root deeply into the community and also be emphatic, fair and related to practical experiences and languages of stakeholders (Abspoel et al. forthcoming). For scientific (risk) assessments to be supportive in policymaking and policy implementation, a suitable language must therefore be found, not least to show how data are linked to the decision context. In practice, scientific understanding needs to be transferred into what it means for the specific decision-making context; it must also be related to policy or management language. In the context of risk assessment, this means that risk assessments for policy formulation and risk assessments for policy implementation may have to address different types of questions, need to use a different analytical approach and require a different way of interpretation, presentation and communication.

3 Assessing Versus Managing Risk

Historically, the ecosystem approach to management has relied on significant contributions from the ecological sciences (Christensen et al. 1996; McLeod et al. 2005; Browman and Stergiou 2004). Over time, this has spawned a variety of ecological assessment frameworks and state of the environment reporting activities that have been tailored to specific ecological and management contexts (Borja et al. 2009; Paetzold et al. 2010). Both environmental impact assessments and strategic environmental assessment are considered the hallmark of decision-making regarding specific projects or industry sector ecological considerations in legislation today (MacKinnon et al. 2018). However, the diversity in the approaches used in various assessments is most often cited as an impediment in making such knowledge usable in decision-making. Assessments and monitoring without a policy context is only an assessment that reflects the concerns of the person or the stakeholders doing the assessment (Holsman et al. 2017). In addition, MSP requires an assessment of a much broader set of concerns (Cormier et al. 2015). In planning, there are cultural, social, economic and liability concerns that have to assessed systematically given that the plan has to address a broader set of objectives than a strict focus on marine ecosystem impacts. The policy context also helps identify the concerns that can be managed by a marine spatial plan and the concerns that fall outside the span of control of such a plan. Keeping the planning process on the policy is a very important aspect in planning to avoid the so-called scope creep where assessments spawn more assessments and discussions, resulting in stakeholder fatigue and loss of credibility in the process.

Uncertainty is also considered an impediment to decision-making with the view that reducing scientific uncertainty would help the uptake of assessment knowledge in decision-making (Leung et al. 2016; Uusitalo et al. 2015). In fact, there are more than scientific uncertainties that are taken into account in decision-making (DFO 2014). They include management uncertainties that arise from a lack of coordination and vertical integration policies for specific sectors or misinterpretations of legislation and policies by the stakeholders involved in the planning process. There are also operational uncertainties related to the effectiveness of the management measures that are implemented to achieve the objectives as well as the potential of accidental failure (Veland and Aven 2015). Monitoring and adaptive management are typically considered a management approach that will tell us where to make improvements (Behn 2003; Douver and Ehler, 2010; Stelzenmüller et al. 2015). In risk management, the spatial and temporal allocations of a marine spatial plan should reduce the uncertainties of achieving development and conservation objectives.

In planning, assessments have to be conducted to identify the concerns that can be addressed by a marine spatial plan and the concerns that should be addressed by other management regimes. The assessments have to inform the planner and the stakeholders as to their concerns. The outcomes of the spatial and temporal allocations being considered also has to be assessed to evaluate which can best address the objectives of the plan. The uptake of such knowledge depends on the relevance of such information to the context of the decisions being made. Understanding the nuances between risk assessment and risk management would shed light as to some of the mismatch between the knowledge generated by science and the uptake of usable knowledge in decision-making.

4 The Perception and Understanding of Risk in Planning

The scientific and technical communities involved in risk assessment define risk as a function of magnitude and probability (Renn 2008). However, society generally perceives risks in terms of the magnitude of the consequences

(Gardner 2009). Few conceptually appreciate the likelihood aspect of the consequence in terms of the meaning of risk (Slovic 1986). Even though a given consequence is highly unlikely, people generally focus on the severity of the consequences of risk as the basis to drive policy decisions (Aven 2015; Leung et al. 2016). The most obvious example are public debates on the risk of accidents in nuclear power plants and the risks associated with the storage of nuclear waste, which focus mostly on the sheer magnitude of impacts of any risk event independent of its likelihood. The role of science is to explicitly address perceptions based on the current knowledge of a given risk and associated uncertainties (Conrad and Ferson 1999). However, it is the role of policymakers to identify and frame the risks in consultation with the public (Harremoës and Turner 2001). Given that policymaking is a social process (Fletcher 2007; Fletcher et al. 2013), it is the process of policymaking that provides the space and time for policymakers and the public to acquire an understanding of the risks in contrast to their perception of those risks (Pouyat 1999). In such a process, scientific advice informs decisions regarding the selection of a course of action, which is most often expressed through legislation and public policy objectives.

In planning, there is a risk that the marine spatial plan does not address the development and conservation objectives that it was intended to achieve. The negotiations and debate may lead a planner and the stakeholder to lose sight of the initial objectives or may even be influenced by objections and obstructions that may be inadvertently introduced during the process. Given that the planning has to ensure that it is conducted within the legislative and policy context that started the process, the quality of the process itself is as important as the quality of the marine spatial plan at the end of the process (Cormier et al. 2015). The quality management principles of Hoyle (2011) provide a short list of quality elements for a process. Adapted to MSP, the principles provide a checklist to evaluate the outputs of the planning process step as the process is progressing (Table 15.1).

Referring back to the issue of jargon and communication, a well-documented and clearly understandable scientific and technical assessment which refers back to the diverse perceptions of risk helps to create transparency of decision-making. This then helps to avoid mistrust among policymakers, planners, stakeholders and different groups in society, even when not everybody agrees to the policy objectives and legislation resulting from the policy process or agrees with the interpretation of risk that underpins these decisions. For policy implementation and planning, however, the objectives and legislation from the policy process define its route, clarify which issue might get specific priority and which constraints need to be taken into account,

Table 15.1 Hoyle's quality management principles adapted to marine spatial planning from Cormier et al. 2015

Hoyle's Principles	Quality management principles in MSP
Consistency of purpose	The MSP process will deliver the required marine spatial plan when there is consistency between the purpose of outputs of the MSP process steps and the objectives. When this principle is applied, the outputs of the MSP process step in terms of feedback and advice would have been guided and derived from the feedback, advice and expectations of the competent authorities, industry stakeholders and communities of interest.
Clarity of purpose	Clear, measurable objectives with defined outputs for each step of the MSP process establish a clear focus for all actions and decisions and enable the tracking of progress as expected by the competent authorities, industry stakeholders, communities of interest and scientific experts. When this principle is applied, people involved in the MSP process understand what they are expected to provide as feedback and advice and understand what they are trying to achieve and how the plan performance will be measured and reported in addressing the objectives.
Connectivity with objectives	The actions and decisions that are undertaken in the MSP process will be those necessary to achieve the objectives and hence there will be demonstrable connectivity between the two. When this principle is applied, the actions and decisions of the people involved in the planning process will be those necessary to deliver the outputs needed to achieve the objectives and no others as stipulated by the policy context.
Competence and capability	The quality of the MSP process outputs is directly proportional to the competence of the people, including their behaviour. When this principle is applied, people involved in scientific advisory peer review activities and consultation tables should have the competencies that reflect their role at the deliberation tables as well as contribute the view and opinions of the constituency they represent.
Certainty of results	Desired results are more certain when the output of each step of the MSP process has performance indicators and planned periodic reporting requirements. When this principle is applied, people involved in the MSP process and, in some cases, the public will have the knowledge and understanding of the progress and performance of the planning process as stipulated by the policy context and the objectives.
Conformity to best practice	The performance of the MSP process is greatly optimized and efficient when actions and decisions conform to established and recognized practices. When this principle is applied, MSP process activities are performed in the manner intended providing confidence that it is being performed in the most efficient and effective way as stipulated by the policy context.

(*continued*)

Table 15.1 (continued)

Hoyle's Principles	Quality management principles in MSP
Clear line of sight	The MSP process outputs are more likely to satisfy everyone involved when periodic reviews are conducted to verify whether there is a clear line of sight between the objectives and the requirements and expectations of the competent authorities, industry stakeholders and communities of interest. When this principle is applied, the scope or objectives of the MSP process may have to be periodically changed causing realignment of activities and resources, thus ensuring continual improvement of the planning process in light of new developments and knowledge.

often with details left to be sorted out in the planning and implementation process. Different types of assessment, in combination with stakeholder involvement processes, therefore form a significant component of the policy-making process and the resulting process of developing (spatial) plans by creating joint understanding of the associated risks (but not necessarily agreement on the outcomes).

5 The Perception and Understanding of Risk in Policy Implementation

In policy implementation, the role of science does not change significantly as it still has to provide advice based on the current knowledge and uncertainties of a given risk. The risks, however, are expressed in terms of achieving the legislative or public policy objective established by the policymakers (Assmuth et al. 2010; Olagunju and Gunn 2016). The role of management is to identify and structure the issues that need to be managed to achieve the objectives in consultation with stakeholders (Harremoës and Turner 2001). In this situation, the role of science is to address the risk perceptions of stakeholders in terms of the potential impact that management measures will have on their vested interest (Soma and Vatn 2014). In contrast to policymaking, integrated planning and management processes provide the space and time for managers and stakeholders to acquire an understanding of the implementation of measures to reduce the risks of not achieving a policy objective (Vigerstad and McCarty 2000). In such process, the scientific advice informs decisions regarding the selection of management measures most often expressed through regulations, standards and guidelines.

As mentioned earlier, the marine spatial plan has to reduce the uncertainties in achieving the development and conservation objectives established in legislation and policy. As with Hoyle's principles for the MSP process, The ten tenets (Barnard and Elliott 2015) of environmental management for the successful and sustainable development of environmental management strategies provide for comprehensive quality considerations for the marine spatial plan (Cormier et al. 2015).

1. **Environmentally/ecologically sustainable**: That the measures will ensure that the ecosystem features and functioning and the fundamental and final ecosystem services are safeguarded.
2. **Technologically feasible**: That the methods, techniques and equipment for ecosystem protection are available.
3. **Economically viable**: That a cost-benefit assessment of environmental management indicates viability and sustainability.
4. **Socially desirable/tolerable**: That the environmental management measures are as required or at least are understood and tolerated by society as being required; that societal benefits are delivered.
5. **Legally permissible**: That there are regional, national or international agreements and/or statutes, which will enable and/or force the management measures to be performed.
6. **Administratively achievable**: That the statutory bodies such as governmental departments, environmental protection and conservation bodies are in place and functioning to enable successful and sustainable management.
7. **Politically expedient**: That the management approaches and philosophies are consistent with the prevailing political climate and have the support of political leaders.
8. **Ethically defensible**: That the environmental management measures that allow development at the risk of losing ecosystem services upon which people depend on are ethically defensible.
9. **Culturally inclusive**: That the environmental management measures also integrate cultural ecosystem consideration that may not have societal or economic value.
10. **Effectively communicable**: That the environmental management objectives are communicated and understood by all the stakeholders, especially to achieve the vertical and horizontal integration of the other nine tenets.

Therefore, scientific advice and assessments from whatever scientific discipline need to be targeted towards impacts of potentially alternative sets of measures for achieving the policy objectives and at the same time minimizing impacts on vested interests or conflicts between implementation of several policy objectives.

This may imply, for example, the need to identify technical or regulatory measures and analyse them in terms of efficiency of enabling or constraining impacts on different vested interests, sectors and policy objectives. It can refer to regulatory measures such as zoning and the designation of priority areas for specific sectors (thereby constraining other sectors within the same area), including any follow-on conflicts these may trigger. It can refer to regulation of activities in time or regulatory demands for technical mitigation of environmental impacts, which would then become part of approval processes for a particular sector activity. Practical examples are the zoning approach used in the marine spatial plan for the German Exclusive Economic Zone (EEZ) (BSH 2009; Kannen 2014), which spatially separates shipping and offshore wind farms by designating priority areas for both sectors. Other risk mitigation measures include a closure of offshore wind farm areas for fishing activities due to safety reasons (avoiding collisions) or demanding mitigation measures for noise (minimizing risk of disturbing marine mammals) to be part of the construction approval process.

6 The Benefits and Efficiencies of Risk Management Standards

International standards such as the ISO 31000 risk management standard or the ISO 9001 quality management system can be applied to any management situation and policy context (ISO 2008, 2018). Updated in 2018, ISO 31000 provides definitions, performance criteria and a common overarching process for identifying, analysing, evaluating and managing risks within a policy context. These are written by experts in their field and are off-the-shelf processes and procedures. Applying these standards can reduce the start-up time of an MSP initiative by eliminating the need to develop a planning process, including the principles and framework, recognizing there may be reasons to develop or adapt these principles and framework to suit particular contexts. More importantly, they also come with a lexicon of technical terms and definitions that are consistent across the standards provided by the ISO. Adopting such standards also formalizes the planning process and provides a common road map for all parties involved in the planning process (Ciocoiu and Dobrea 2007). Given that the standard can be acquired by anyone, standards can also improve transparency and help align expectations. The parties involved are provided with the step ahead of the planning process starting that allows them to prepare questions and contributions ahead of time. Instead of debating the steps of the planning process, or where to start, ISO 31000 already comes with a structured process starting with policy context and followed by a risk assessment to identify and implement measures to achieve objectives. Cormier et al. (2015) undertook the exercise to link the ISO 31000 risk management approach with the general approaches of applying a policy cycle (Douvere and Ehler 2009; MMO 2014) to MSP.

As MSP is about the allocation of spatial and temporal measures to achieve development objectives, the MSP policy context according to ISO 31000 is the development of objectives for the various sectors. The risk assessment is subsequently used to identify the impediments to achieving those objectives. The aim of such a risk-based approach to planning is to ultimately find solutions to resolve the spatial and temporal conflicts between marine uses and produce a plan. Generally, Cormier et al. (2015) propose to structure the MSP process along the various steps of risk assessment ranging from risk identification and risk analysis to risk treatment, with the latter being the step to define the measures (regulatory or technical) to deal with the risks identified and recognized as relevant in the specific planning context. Furthermore, decision-making in the planning process is accompanied by a process of stakeholder involvement and a (separate) process of scientific advice in each of the risk assessment steps. These are elements to guarantee that the planning process is properly informed by scientific assessments and stays involved in a regular communication with stakeholders in order to provide transparency on the decisions taken.

However, in order to avoid the risk of a failing process, MSP needs more than a structured process to successfully produce a plan. It also needs to have criteria to review the quality of the process and the quality of the plan. Hoyle's process principles (Hoyle 2011) provide the quality management objectives for the planning process as such (Cormier et al. 2015). The principles are criteria to ensure that the planning process maintains consistency and clarity of the MSP purpose while ensuring that the process and the debates stay connected to the objectives of the planning process. The principles recognize the need for competence and capability to deliver the process conducted, thereby providing certainty of the expected results or outcomes of each step of the process. But a well-structured regulatory planning process, even when creating a lot of common understanding and a large amount of agreement from various stakeholders, does not automatically guarantee an implementable plan that achieves the intended outcomes. Therefore, the quality of the plan itself, its outputs and its intended outcomes depend entirely on a different set of criteria (Cormier et al. 2015). The ten tenets of adaptive management and sustainability provide one holistic framework and criteria for understanding and managing the socioecological system (Barnard and Elliott 2015).

These tenets outline the type of stakeholder consultation and feedback as well as scientific and technical advice needed to ensure that any marine spatial plan addresses the objectives, concerns and expectation of the parties involved and is implementable along existing legislative and administrative realities (Cormier et al. 2015). Without a roster of quality objectives for the planning process on one hand and the plan on the other hand, confusion is likely to happen as the participants will focus the planning process and set priorities that reflect their individual agendas and views.

In Fig. 15.1, we created a matrix that combines the ISO 31000 risk management process steps with the ten tenets of environmental management with the MSP elements in line with Hoyle's principles. In summary, "establishing the context" sets the purpose for the planning process, as well as competencies, capabilities and best practices that will support the planning process. The role of "risk identification" and "risk analysis" is to provide clarity and understanding to the perceptions of the risks as to what are the causes that may have an effect on achieving objectives. Based on the "risk analysis", the role of "risk evaluation" is to gain an understanding of the severity of risks using criteria and identify which are the risks that are unacceptable in relation to achieving objectives and that will require management guided by precautionary principles. Based on the "risk evaluation", "risk treatment" is the selection of management measures in the development and implementation of a management plan to achieve the objectives. The row for "Effectively communicate" highlights the information and support functions as well as the oversight, consultation and feedback activities for the entire process. The last two columns have been organized in terms of the "monitoring and review" and "communication and consultation" activities that will be required once the management plan has been implemented. As stipulated by ISO 31000, these activities generate the information that will be needed to evaluate the effectiveness of the plan in the future, enabling improvements to the plan adhering to adaptive management principles. Successful environmental management can only be achieved by environmental and compliance monitoring and review.

7 Conclusion

Even though it may sound very technical, linking risk management structures and quality management objectives with approaches referred to in spatial planning literature and practice may help to develop well-accepted MSP regu-

Quality Objectives 10 tenets	Establishing the Context	Risk Assessment			Risk Treatment	Monitoring and Review	Communication and Consultation
		Risk Identification	Risk Analysis	Risk Evaluation			
Tenet 1	Consistency of Purpose	Clarity of Purpose	Clarity of Purpose	Connectivity with Objectives	Connectivity with Objectives	Certainty of Results	Clear Line of Sight
Environmentally/ ecologically sustainable	Development objectives Environmental objectives	Ecologically and Biologically Significant Areas	Environmental Impact Assessments	Severity of the Consequences	Selection of the Areas to be Protected and Conserved	Environmental Effects Monitoring	Ecosystem Status and Trends Overview Reports
Tenet 2	Competence and Capability	Clarity of Purpose	Clarity of Purpose	Connectivity with Objectives	Connectivity with Objectives	Certainty of Results	Clear Line of Sight
Technologically feasible	Industry Stakeholders Representatives	Driver Activity Significant Pressures	Cause and Effect Pathways	Identification of Pressures that Require Management	Selection of Guidelines, Standards and Procedures	Environmental Monitoring of the Pressures	Human Use and Environmental Atlases
Tenet 3	Competence and Capability	Clarity of Purpose	Clarity of Purpose	Connectivity with Objectives	Connectivity with Objectives	Certainty of Results	Clear Line of Sight
Economically viable	Management Area Economic Activities	Economically Significant Ecosystem Services	Economic Impact Assessments	Severity of the Consequences	Management Plan Costs and Benefits	Economic and Development Monitoring	Economic Ecosystem Services Overview Reports
Tenet 4	Competence and Capability	Clarity of Purpose	Clarity of Purpose	Connectivity with Objectives	Connectivity with Objectives	Certainty of Results	Clear Line of Sight
Socially desirable/ tolerable	Communities of Interest Representatives	Socially Significant Areas	Social Impact Assessments	Severity of the Consequences	Selection of the Areas to be Protected and Conserved	Monitor Trends Social Values	Cultural Ecosystem Services Overview Reports
Tenet 5	Consistency of Purpose			Connectivity with Objectives	Connectivity with Objectives	Certainty of Results	Clear Line of Sight
Legally permissible	National and Trans-National Legislation and Policies			Legislative Legal Liabilities and Repercussions	Management Plan Legislative Tools	Regulatory Compliance Verification	Compliance Reports
Tenet 6	Competence and Capability			Competence and Capability	Connectivity with Objectives	Certainty of Results	Clear Line of Sight
Administratively achievable	Management Area Governance Structure			Institutional Program Costs and Repercussions	Competent Authorities Agreement	Program Performance Monitoring	Program Priorities and Performance Reports
Tenet 7	Consistency of Purpose			Connectivity with Objectives	Connectivity with Objectives	Certainty of Results	Clear Line of Sight
Politically expedient	Development Public Policy Mandate Timeframe			Public Policy Implications	Management Strategy Public Policy Endorsement	Monitor Public Policy Trends and Priorities	Public Policy Announcements
Tenet 8	Conformity to Best Practice			Connectivity with Objectives		Certainty of Results	
Ethically defensible	Charter of Rights ISO 26000: Social Responsibility			Human Rights Social Responsibility Implications		Monitor Ethical Concerns	
Tenet 9	Competence and Capability	Clarity of Purpose	Clarity of Purpose	Connectivity with Objectives	Connectivity with Objectives	Certainty of Results	Clear Line of Sight
Culturally inclusive	Traditional Communities Representatives	Culturally Significant Areas	Cultural Impact Assessments	Severity of the Consequences	Selection of the Areas to be Protected and Conserved	Monitoring Trends Traditional Values	Cultural Ecosystem Services Overview Reports
Tenet 10	Conformity to Best Practice	Connectivity with Objectives			Connectivity with Objectives	Clear Line of Sight	Conformity to Best Practice
Effectively communicable	Terms of References Decision Business Rules Advisory Bodies	Management Oversight Stakeholder Advice and Feedback Scientific and Technical Advice			Governance Oversight Stakeholder Feedback	Planning Process Evaluations and Auditing	Communication and Consultation Reporting Procedures

Fig. 15.1 Activities and outputs of the planning process steps in relation to the ten tenets

latory planning processes as well as the implementation of the resulting plans. It aims to avoid an unstructured "muddling through" and associated unintended consequences by defining clear milestones and competencies, providing criteria for decision-making and supporting transparency.

As Barnard and Elliot highlighted (2015), effective environmental management does not simply rely on science underpinning and a participative planning process to address the sustainability concerns of stakeholders. It relies on the management of human activities by the implementation of management practices and measures that operate under voluntary conformity, industry sector standards or legislative compliance. In management, standards and certification play a huge role in a variety of services and industries, particularly in terms of quality management and risk management where most countries have adopted ISO standards as their own. Although most would argue that each planning initiative is unique to the institutional make-up of governance, stakeholder concerns and ecological considerations of the planning area, ISO standards of framework, process and vocabulary can still be adapted to harmonize environmental management across planning process initiatives. The use of international standards, such as the ones available under ISO, can avoid the need to develop a framework and debate definitions that can consume valuable time and use the scarce resources that are usually allocated for these initiatives. In addition to training in the use of these standards that is already available for most ISO standards, standardized frameworks can facilitate knowledge transfer and lessons learnt between initiatives improving future processes. Finally, ISO also provides a suite of standards that can guide and facilitate effectiveness and performance evaluations.

In the marine environment, MSP could greatly benefit from such standards. As these initiatives are just starting to get under way in Europe, they could facilitate and minimize start-up costs and public investment. Give the widespread use of these standards in various countries, they may enhance public trust in environmental management as well as alleviate concerns through a structured process that can educate and inform as well as consult. By tracing environmental impacts from the effects to the causes combined with the effectiveness of management practices, such process may reduce uncertainty for some decisions while providing justification for further research in others. There are also links to MSP evaluation (see Chap. 18 in this book) and benchmarking.

In the future, there may be a need to develop a standard that would be designed specifically for an ecosystem approach to management, particularly in relation to the ever increasing level of human activities in the marine environment. There may also be a need to develop a new educational approach for graduate and post-graduate students as well as training approach for practitioners (see Chap. 19 in this book). For those wishing to pursue a career in environmental planning and management, course curricula and training workshops could bring a broader set of competencies and skills that are not always acquired by existing academic fields of study in the natural sciences, social sciences and economics.

Acknowledgements We gratefully acknowledge the support received from the BONUS BALTSPACE project. BONUS BALTSPACE project was funded from BONUS (Art 185), itself funded jointly from the European Union's Seventh Programme for research, technological development and demonstration, and from Baltic Sea national funding institutions.

References

Abspoel, L., Mayer, I., Keijser, X., Fairgrieve, R., Ripken, M., Abramic, A., Kannen, A., Cormier, R., & Kidd, S. (forthcoming). *Does the Helmsman Speak English? Improving Communication in Maritime Spatial Planning with the MSP Challenge Games*. Paper submitted to Marine Policy.

Ackoff, R. L. (1990). Strategy. *System Practice, 3*, 521–524.

Anderson, J. E. (2011). *Public Policymaking: An Introduction* (7th ed.). Boston: Wadsworth Cengage Learning, 342 pp.

Ansong, J., Gissi, E., & Calado, H. (2017). An Approach to Ecosystem-Based Management in Maritime Spatial Planning Process. *Ocean & Coastal Management, 141*, 65–81. https://doi.org/10.1016/j.ocecoaman.2017.03.005.

Anthony, R. N., & Dearden, J. (1980). *Management Control Systems* (4th ed.). Homewood, IL: Richard D. Irwin, Inc.

Aps, R., Fetissov, M., Goerlandt, F., Helferich, J., Kopti, M., & Kujala, P. (2015). Towards STAMP Based Dynamic Safety Management of Eco-Socio-Technical Maritime Transport System. *Procedia Engineering, 128*, 64–73. https://doi.org/10.1016/j.proeng.2015.11.505.

Assmuth, T., Hildén, M., & Benighaus, C. (2010). Integrated Risk Assessment and Risk Governance as Socio-Political Phenomena: A Synthetic View of the Challenges. *Science of the Total Environment, 408*, 3943–3953. https://doi.org/10.1016/j.scitotenv.2009.11.034.

Aven, T. (2015). On the Allegations that Small Risks Are Treated out of Proportion to Their Importance. *Reliability Engineering & System Safety, 140*, 116–121. https://doi.org/10.1016/j.ress.2015.04.001.

Barnard, S., & Elliott, M. (2015). The 10-tenets of Adaptive Management and Sustainability: An Holistic Framework for Understanding and Managing the Socio-ecological System. *Environmental Science & Policy, 51*, 181–191. https://doi.org/10.1016/j.envsci.2015.04.008.

Behn, R. D. (2003). Why Measure Performance? Different Purposes Require Different Measures. *Public Administration Review, 63*, 586–606. https://doi.org/10.1111/1540-6210.00322.

Borja, Á., Ranasinghe, A., & Weisberg, S. B. (2009). Assessing Ecological Integrity in Marine Waters, Using Multiple Indices and Ecosystem Components: Challenges for the Future. *Marine Pollution Bulletin, 59*, 1–4.

Browman, H. I., & Stergiou, K. I. (2004). Perspectives on Ecosystem-based Approaches to the Management of Marine Resources. *Marine Ecology Progress Series, 274*, 269–303.

BSH. (2009). Anlage zur Verordnung über die Raumordnung in der deutschen ausschließlichen Wirtschaftszone in der Nordsee (AWZ Nordsee-ROV) vom 21. September 2009: Raumordnungsplan für die deutsche ausschließliche Wirtschaftszone in der Nordsee.

Christensen, N. L., Bartuska, A. M., Brown, J. H., Carpenter, S., D'Antonio, C., Francis, R., Franklin, J. F., et al. (1996). The Report of the Ecological Society of America Committee on the Scientific Basis for Ecosystem Management. *Ecological Applications, 6*, 665–691.

Christie, N., Smyth, K., Barnes, R., & Elliott, M. (2014). Co-location of Activities and Designations: A Means of Solving or Creating Problems in Marine Spatial Planning? *Marine Policy, 43*, 254–261. https://doi.org/10.1016/j.marpol.2013.06.002.

Chun, Y. H., & Rainey, H. G. (2005). Goal Ambiguity and Organizational Performance in U.S. Federal Agencies. *Journal of Public Administration Research and Theory, 15*, 529–557. https://doi.org/10.1093/jopart/mui030.

Ciocoiu, C. N., & Dobrea, R. C. (2007). The Role of Standardization in Improving the Effectiveness of Integrated Risk Management. *Advanced Risk Management*, 1–19.

Conrad, S., & Ferson, S. (1999). Decision Making. *Risk Analysis, 19*, 63–68.

Cormier, R., & Elliott, M. (2017). SMART Marine Goals, Targets and Management—Is SDG 14 Operational or Aspirational, Is 'Life Below Water' Sinking or Swimming? *Marine Pollution Bulletin, 123*, 28–33. https://doi.org/10.1016/j.marpolbul.2017.07.060.

Cormier, R., Elliott, M., & Rice, J. (2019). Putting on a Bow-Tie to Sort Out Who Does What and Why in the Complex Arena of Marine Policy and Management. *Science of the Total Environment, 648*, 293–305. https://doi.org/10.1016/j.scitotenv.2018.08.168.

Cormier, R., Kannen, A., Elliott, M., & Hall. P. (2015). *Marine Spatial Planning Quality Management System*. ICES Cooperative Research Report No. 327. 106 pp.

Cormier, R., Kelble, C. R., Anderson, M. R., Allen, J. I., Grehan, A., & Gregersen, Ó. (2017). Moving from Ecosystem-based Policy Objectives to Operational Implementation of Ecosystem-based Management Measures. *ICES Journal of Marine Science, 74*, 406–413. https://doi.org/10.1093/icesjms/fsw181.

Creed, I. F., Cormier, R., Laurent, K. L., Accatino, F., Igras, J. D. M., Henley, P., Friedman, K. B., Johnson, L. B., Crossman, J., Dillon, P. J., & Trick, C. G. (2016). Formal Integration of Science and Management Systems Needed to Achieve Thriving and Prosperous Great Lakes. *BioScience, 66*, 408–418. https://doi.org/10.1093/biosci/biw030.

DFO. (2014). *Science Advice for Managing Risk and Uncertainty in Operational Decisions of the Fisheries Protection Program*. DFO Canadian Science Advisory Secretariat Science Advisory Report 2014/015.

Douvere, F., & Ehler, C. (2009). New Perspectives on Sea Use Management: Initial Findings from European Experience with Marine Spatial Planning. *Journal of Environmental Management, 90*, 77–88.

Douvere, F., & Ehler, C. (2010). The Importance of Monitoring and Evaluation in Adaptive Maritime Spatial Planning. *Journal of Coastal Conservation, 15*, 305–311.

EC. (2000). European Community. Directive 2000/60/EC of October 23 2000 of the European Parliament and of the Council Establishing a Framework for Community Action in the Field of Water Policy. *Official Journal of the European Communities, L327*, 1–72.

EC. (2008). Directive 2008/56/EC of the European Parliament and of the Council of 17 June 2008 Establishing a Framework for Community Action in the Field of Marine Environmental Policy (Marine Strategy Framework Directive). OJ L 164, 25.6.2008, p. 19.

EC. (2011). Regulation (EU) No 1255/2011 of the European Parliament and of the Council of 30 November 2011 Establishing a Programme to Support the Further Development of an Integrated Maritime Policy. OJ L 321, 5.12.2011, p. 1.

EC. (2014). Directive 2014/89/EU OF THE European Parliament and of the Council of 23 July 2014 Establishing a Framework for Maritime Spatial Planning. OJ L157. 28.82014, p. 135.

Fletcher, S. (2007). Converting Science to Policy Through Stakeholder Involvement: An Analysis of the European Marine Strategy Directive. *Marine Pollution Bulletin, 54*, 1881–1886. https://doi.org/10.1016/j.marpolbul.2007.08.004.

Fletcher, S., McKinley, E., Buchan, K. C., Smith, N., & McHugh, K. (2013). Effective Practice in Marine Spatial Planning: A Participatory Evaluation of Experience in Southern England. *Marine Policy, 39*, 341–348.

Gardner, D. (2009). *The Science of Fear: How the Culture of Fear Manipulates Your Brain*. London: Penguin Books Ltd.

Gavaris, S. (2009). Fisheries Management Planning and Support for Strategic and Tactical Decisions in an Ecosystem Approach Context. *Fisheries Research, 100*, 6–14. https://doi.org/10.1016/j.fishres.2008.12.001.

Gee, K., Kannen, A., Adlam, R., Brooks, C., Chapman, M., Cormier, R., Fischer, C., Fletcher, S., Gubbins, M., Shucksmith, R., & Shellock, R. (2017). Identifying Culturally Significant Areas for Marine Spatial Planning. *Ocean and Coastal Management, 136*, 139–147.

Girling, P. X. (2013). Operational Risk Management: A Complete Guide to a Successful Operational Risk Framework. 1st ed. Wiley & Sons.

Green, P. E. J. (2015). *Enterprise Risk Management: A Common Framework for the Entire Organization*. Elsevier Inc. 240 pp.

Harremoës, P., & Turner, R. K. (2001). Methods for Integrated Assessment. *Regional Environmental Change, 2*, 57–65. https://doi.org/10.1007/s101130100027.

Holsman, K., Samhouri, J., Cook, G. S., et al. (2017). An Ecosystem-based Approach to Marine Risk Assessment. *Ecosystem Health and Sustainability, 3*, e01256. https://doi.org/10.1002/ehs2.1256.

Hoyle, D. (2011). *ISO 9000 Quality Systems Handbook: Using the Standards as a Framework for Business Improvement* (6th ed.). New York: Routledge. ISBN 978-1-8561-7684-2. 802 pp.

Hupe, P. L., & Hill, M. J. (2016). And the Rest Is Implementation. Comparing Approaches to What Happens in Policy Processes Beyond Great Expectations. *Public Policy and Administration, 31*, 103–121. https://doi.org/10.1177/0952076715598828.

ISO. (2008). Quality Management Systems—Requirements. International Organization for Standardization, Geneva. ISO 9001:2008(E).

ISO. (2018). Risk management—Guidelines (2nd ed.). International Organization for Standardization. ISO 31000:2018(E).

Junier, S. J., & Mostert, E. (2015). The Implementation of the Water Framework Directive in The Netherlands: Does It Promote Integrated Management? *Physics and Chemistry of the Earth, 47–48*, 2–10. https://doi.org/10.1016/j.pce.2011.08.018.

Kannen, A. (2014). Challenges for Marine Spatial Planning in the Context of Multiple Sea Uses, Policy Arenas and Actors Based on Experiences from the German North Sea. *Regional Environmental Change, 14*, 2139–2150.

Leung, W., Noble, B. F., Jaeger, J. A. G., & Gunn, J. A. E. (2016). Disparate Perceptions About Uncertainty Consideration and Disclosure Practices in Environmental Assessment and Opportunities for Improvement. *Environmental Impact Assessment Review, 57*, 89–100. https://doi.org/10.1016/j.eiar.2015.11.001.

Loehle, C. (2006). Control Theory and the Management of Ecosystems. *Journal of Applied Ecology, 43*, 957–966. https://doi.org/10.1111/j.1365-2664.2006.01208.x.

Long, R. D., Charles, A. T., & Stephenson, R. L. (2015). Key Principles of Marine Ecosystem-Based Management. *Marine Policy, 57*, 53–60. https://doi.org/10.1016/j.marpol.2015.01.013.

MacKinnon, A. J., Duinker, P. N., & Walker, T. R. (2018). The Application of Science in Environmental Impact Assessment. Routledge.

Maes, F. (2008). The International Legal Framework for Marine Spatial Planning. *Marine Policy, 32*, 797–810. https://doi.org/10.1016/j.marpol.2008.03.013.

McLeod K. L., Lubchenco J., Palumbi S. R., & Rosenberg A. A. (2005). Scientific Consensus Statement on Marine Ecosystem-Based Management. Signed by 221 academic scientists and policy experts with relevant expertise and published by the Communication Partnership for Science and the Sea. 21 pp.

Mingers, J., & White, L. (2010). A Review of the Recent Contribution of Systems Thinking to Operational Research and Management Science. *European Journal of Operational Research, 207,* 1147–1161. https://doi.org/10.1016/j.ejor.2009.12.019.

MMO. (2014). Guidance: Marine Planning and Development. The 12-stage Process on How a Marine Plan Is Made from Selection to Implementation and Monitoring and How You Can Get Involved. Retrieved from https://www.gov.uk/marine-plans-development.

Moss, B. (2008). The Water Framework Directive: Total Environment or Political Compromise? *Science of the Total Environment, 400,* 32–41. https://doi.org/10.1016/j.scitotenv.2008.04.029.

Olagunju, A. O., & Gunn, J. A. E. (2016). Integration of Environmental Assessment with Planning and Policy-Making on a Regional Scale: A Literature Review. *Environmental Impact Assessment Review, 61,* 68–77. https://doi.org/10.1016/j.eiar.2016.07.005.

Paetzold, A., Warren, P. H., & Maltby, L. L. (2010). A Framework for Assessing Ecological Quality Based on Ecosystem Services. *Ecological Complexity, 7,* 273–281 Elsevier B.V.

Pouyat, R. V. (1999). Science and Environmental Policy-making Them Compatible. *BioScience, 49,* 281–286.

Renn, O. (2008). Concepts of Risk: An Interdisciplinary Review. *GAIA, 17,* 50–66.

Rice, J., Trujillo, V., Jennings, S., Hylland, K., Hagstrom, O., Astudillo, A., & Jensen, J. (2005). Guidance on the Application of the Ecosystem Approach to Management of Human Activities in the European Marine Environment. *ICES Cooperative Research Report, 273,* 1–22.

Slovic, P. (1986). Perception of Risk. *Science, 236,* 280–285.

Soma, K., & Vatn, A. (2014). Representing the Common Goods—Stakeholders vs. Citizens. *Land Use Policy, 41,* 325–333. https://doi.org/10.1016/j.landusepol.2014.06.015.

Stelzenmüller, V., Vega Fernández, T., Cronin, K., Röckmann, C., Pantazi, M., Vanaverbeke, J., Stamford, T., Hostens, K., Pecceu, E., Degraer, S., Buhl-Mortensen, L., Carlström, J., Galparsoro, I., Johnson, K., Piwowarczyk, J., Vassilopoulou, V., Jak, R., Louise Pace, M., & van Hoof, L. (2015). Assessing Uncertainty Associated with the Monitoring and Evaluation of Spatially Managed Areas. *Marine Policy, 51,* 151–162. https://doi.org/10.1016/j.marpol.2014.08.001.

UNCLOS. (1982). United Nations Convention on the Law of the Sea of 10 December 1982. Retrieved 2018-06-07 from http://www.un.org/depts/los/convention_agreements/texts/unclos/unclos_e.pdf.

Uusitalo, L., Lehikoinen, A., Helle, I., & Myrberg, K. (2015). An Overview of Methods to Evaluate Uncertainty of Deterministic Models in Decision Support. *Environmental Modelling and Software, 63,* 24–31. https://doi.org/10.1016/j.envsoft.2014.09.017.

Veland, H., & Aven, T. (2015). Improving the Risk Assessments of Critical Operations to Better Reflect Uncertainties and the Unforeseen. *Safety Science, 79*, 206–212. https://doi.org/10.1016/j.ssci.2015.06.012.

Vigerstad, T. J., & McCarty, L. S. (2000). The Ecosystem Paradigm and Environmental Risk Management. *Human and Ecological Risk Assessment, 6*, 369–381. https://doi.org/10.1080/10807030091124518.

16

The Role of the Law of the Sea in Marine Spatial Planning

Dorota Pyć

1 Introduction

Once, for our ancestors, the ocean was a link between Heaven and Earth. Nowadays, the World Ocean is a universal and common space for all humanity. It is difficult to assess whether the ocean will divide us in the future or bring us closer together. Undoubtedly, solving the problems of the ocean, its protection and the rational use of its resources requires effective cooperation at a global, regional and national level.

To some, the sound of the ocean may evoke the harmony of the past flowing into the future. In order not to lose our connection with it, marine spatial planning (MSP) for sustainable marine governance should be put into practice following the principles of equity. Well-defined, flexible and transparent instruments of marine sustainable governance at a regional and national level are key tools towards achieving governance goals concerning the global ocean (Kingsbury et al. 2005).

Global ocean governance (GOG) is a highly complex concept (Dorman Mc 2000) on account of the multidimensionality and dynamics of ocean management on a legal, economic and social as well as political and cultural level. Ocean governance can be defined as an integrative concept which nowadays allows us to distinguish a set of global problems related to the World Ocean (Galletti 2015; Pyć 2016).

D. Pyć (✉)
University of Gdańsk, Gdańsk, Poland
e-mail: dpyc@prawo.ug.edu.pl

J. Zaucha, K. Gee (eds.), *Maritime Spatial Planning*,
https://doi.org/10.1007/978-3-319-98696-8_16

MSP is a practical way to create and implement rational organization in the use of ocean space. It is important to strengthen the interaction between ocean users in accordance with the principles of sustainable development and environmental protection and in connection with the implementation of socio-economic goals (Ehler and Douvere 2009).

The Law of the Sea confirms that it is possible to develop an international legal regime, although the creation of a global regime of the seas and oceans complicates the decentralized nature of the international public law system (Harrison 2011; Pyć 2011). The comprehensive approach expressed in the United Nations Convention on the Law of the Sea (UNCLOS 1982) concerning, inter alia, the protection and preservation of the marine environment, testifies to its constitutional dimension. Taking into account the essence of MSP, the legal norms of UNCLOS—which formulate the obligation to protect and preserve the marine environment—are essentially important. They have already strengthened existing treaty norms and supported solutions developed in the process of creating common law adopted in international practices (Pyć 2011).

2 Marine Spatial Planning in the Law of the Sea

2.1 Propaedeutics of Marine Areas

The Law of the Sea is one of the oldest areas of international public law that regulates the uses of the World Ocean. The hugely influential work of Hugo Grotius—"the Father of the Law of Nations"—is worth mentioning here as it has significantly impacted the development of the Law of the Sea. Grotius created the paradigm which provides the foundation for the modern Law of the Sea. Claiming an established and important role in the doctrine and jurisdictional practices of the coastal States, Hugo Grotius' paradigm, expressed in *Mare Liberum* written in 1609 ("The Freedom of the Seas or the Right which belongs to the Dutch to take part in the East Indian Trade"), is still valid today and confirms the fundamental foundations of the Law of the Sea, namely that (1) the coastal States have the right to exercise jurisdiction in their marine spaces and (2) the ocean and its resources beyond national jurisdiction are open to all States.

The Law of the Sea was codified in the 1950s in four Geneva Conventions (the Convention on the Territorial Sea and the Contiguous Zone, the

Convention on the High Seas, the Convention on Fishing and Conservation of the Living Resources of the High Seas and the Convention on the Continental Shelf) and, afterward, in the UNCLOS. In general, UNCLOS consists of norms regulating the use of the marine environment and its resources in accordance with the norms defining the legal status of different marine spaces, overseeing the fulfilment of the rights and obligations of States in marine areas and providing the basis for creating an ocean governance framework.

The UNCLOS states in its preamble that "the problems of ocean space are closely interrelated and need to be considered as a whole". This statement is an important starting point for discussions on ocean governance and MSP. The preamble to UNCLOS includes a normative justification for recognizing the ecological unity of the World Ocean. This recognition is of great importance for MSP, especially in the adjacent and interacting areas of Exclusive Economic Zones (EEZ) (or the continental shelf) and Areas Beyond National Jurisdiction (ABNJs). The planning of ocean space is the logical advancement of the structuring of obligations and the use of rights granted under UNCLOS as well as a practical tool in assisting State Parties to comply with their obligations. It should be clearly emphasized that UNCLOS does not contain any provisions relating *expressis verbis* to GOG or MSP.

In relevant literature, the marine environment is presented in a multidimensional way—from the processes taking place at the level of the World Ocean to those of a narrower focus such as habitat, species or genetic resources. The ecological unity of the marine environment implies—in terms of, research needs or applying appropriate management tools—a focus on species, habitats and landscapes and their mutual dependencies. Particularly noteworthy are the provisions of the Convention on the Law of the Sea, which are comprehensive and treat the marine environment from the perspective of ecological unity. The concept of marine environment as it is commonly understood refers to the space of sea water, the air above it and the seabed, all of which include various species of fauna and flora which, in turn, contain various other natural and anthropogenic elements. In practice, the marine environment is an area of economic activity. The World Ocean can be considered a synonym of the marine environment (Pyć 2011).

The term "marine areas" (or marine spaces, marine zones) has a purely conventional meaning in the Law of the Sea. On the basis of their legal status, UNCLOS divides marine areas into three categories: (1) marine areas included in the territory of a State, (2) marine areas which are subject to limited jurisdiction and in which a coastal State enjoys sovereign rights and (3) marine areas located beyond national jurisdiction. The marine areas included in the territory of a coastal State are: internal waters (Article 8 UNCLOS), territorial sea, (Article 3–4 UNCLOS) and archipelagic waters (Articles 46–54 UNCLOS).

The internal waters are the waters landward of the baseline of the territorial sea. A coastal State has sovereignty over its internal waters, extending to the air space over the internal waters as well as to their bed and subsoil. Similarly, an archipelagic State has sovereignty over the international waters of the archipelago.

The territorial sea includes a narrow band of water extending seaward from a coastal State's baseline. Every State has the right to establish the breadth of its territorial sea up to a limit which does not exceed 12 nautical miles measured from the baselines. The outer limit of the territorial sea is the line every point of which is at a distance from the nearest point of the baseline equal to the breath of the territorial sea. The external boundary of the territorial sea is the border of the coastal State's territory. The legal status of the territorial sea is subject to the coastal State's sovereign authority which extends to the air space over the territorial sea as well as to its bed and subsoil. Regarding the territorial sea, the legal order of the coastal State is in force. The specificity of the State's maritime territory reflects the compromise resulting from the idea of freedom of the seas, the provision of a number of rights to foreign ships in the territorial sea belonging to the coastal State and the sovereignty and territorial authority of the coastal State over its territorial sea.

The marine areas under limited jurisdiction in which the coastal State has sovereign rights include the EEZ (Articles 55–75 UNCLOS), the continental shelf (Articles 76–85 UNCLOS) and the contiguous zone (Article 33 UNCLOS).

The EEZ is an area beyond and adjacent to the territorial sea which does not extend beyond 200 nautical miles from territorial sea baselines, and it is subject to a special legal status (Article 55 of UNCLOS). Within EEZ, the coastal State has the right to exploit the water column, seabed and subsoil.

The EEZ is not a part of the State territory. The coastal State's rights in those area are functional, not territorial in nature. It is a special, *sui generis* kind of area which belongs neither to a territorial sea nor to High Seas. While the coastal State has sovereign rights over the resources of the zone and its economic use, it does not exercise sovereignty over the zone itself. Only those rights which, in accordance to the purpose and character of the zone, are related to conducting economic activity in it are qualified as sovereign. The coastal State is not obliged to make these resources available to other States, even if it does not take advantage of them. However, the principle of rational use of living resources, also called the principle of optimal use of living resources, stating that if a coastal State cannot obtain all acceptable catches, it should allow other countries to fish within certain limits, still applies.

The sovereign rights granted to the coastal State in the EEZ were limited in two ways. First, the State exercises these rights only for the purpose of exploiting, researching, protecting and managing the natural resources of the zone, and second, when exercising these rights, the coastal State should duly take into account the rights and obligations of other States and should act in accordance with the provisions of the Convention on the Law of the Sea.

The coastal State in the EEZ zone also has jurisdiction in the establishment and use of artificial islands, installations and structures, marine scientific research and the protection of the marine environment. The consequence of the application of the freedom of the seas principle in the EEZ is the application of provisions on the High Seas, provisions which regulate and form part of the legal status of the EEZ, with restrictions resulting from the sovereign rights of coastal States. Freedom of navigation may be limited by the rights of the coastal State in the scope of the marine environment's protection, for example, against pollution from ships. However, these powers do not give the coastal State complete freedom of action. In order to protect the interests of other States, laws and regulations issued to prevent, reduce and control pollution from ships, the coastal State must act in compliance with generally accepted international standards and principles.

UNCLOS provides that in the EEZ the coastal State has jurisdiction with regard to the protection and preservation of the marine environment (Article 56(1)(b)(iii)). In exercising this jurisdiction, the coastal State is empowered to enact laws and regulations for the prevention, reduction and control of vessel-source pollution in the EEZ. In accordance with Article 211(5) of UNCLOS, such laws and regulations must conform to and give effect to generally accepted international rules and standards established through the competent international organizations.

The contiguous zone provides a buffer consisting of an additional 12 nautical miles beyond the territorial sea. Thus, the outer limit of the contiguous zone does not exceed 24 nautical miles from territorial sea baselines. Within this zone, a State has the right to enforce its customs, fiscal, immigration or sanitary laws and regulations within its territory or territorial sea.

Marine ABNJs include the High Seas (Articles 86–115 UNCLOS) and "the Area" (deep seabed, 133–155 UNCLOS). The High Seas is the water column beyond the EEZ. It is neither subject to any sovereign power nor appropriated, open to the common use of all States, in accordance with the principle of freedom of the seas. From a legal standpoint, the High Seas is not subject to the sovereignty of any State and its use is free for all States. The principle in force regarding the freedom of the seas, specifically the High Seas, means that all States can use this area. Certainly, the use of the High Seas must be carried out in such a way as not to affect the interests of other States. The High Seas is *res usus publicum* (Pyć 2011).

The seabed, which is either the continental shelf or "the Area", that is, the seabed and Deep Ocean beyond national jurisdiction, is not a part of the High Seas. The High Seas, however, includes airspace, and all States have the right to rationally use this space. Although subject to certain regulations, within this ocean space, all States have equal rights in terms of essentially enjoying freedom of navigation, freedom of overflight, freedom to lay submarine cables and pipelines, freedom to construct artificial islands and other installations permitted under international law, freedom of fishing and freedom of scientific research (Attard and Mallia 2014).

"The Area" is the seabed, ocean floor and subsoil beyond national jurisdiction and has special legal status. "The Area" and its resources are the common heritage of mankind (CHM). No States shall claim or exercise sovereignty or sovereign rights over any part of "the Area" or its resources, nor shall any State or natural or juridical person appropriate any part thereof. All rights in the resources of "the Area" are vested in mankind as a whole. "The Area" is intended only for the use of peaceful aims. Activities related to exploration and use of "the Areas" are managed by the International Seabed Authority (ISA), a special management unit established for this very purpose. All State Parties to UNCLOS are ipso facto members of the ISA. The ISA is the organization through which State Parties organize and control activities in "the Area", particularly with a view to administering the latter's resources (Article 133). "The Area" will ensure a fair distribution of benefits to all States, taking into account good faith (e.g. Articles 157 and 300). UNCLOS regulates the issue of "the Area's" legal status under part XI.

2.2 The Protection and Preservation of the Marine Environment

For many centuries, the division of the seas and oceans was based on the assumption that marine resources are infinite and, even if not, far greater than humanity's needs. Yet, empirical research confirms the degradation of the World Ocean's ecological condition. Global threats include, inter alia, sea-level rise, accumulation of pollutants in the marine environment, deterioration of the self-cleaning capacity of closed or semi-closed seas, climate change resulting in ocean acidification and overfishing. The results of the negative changes affecting seas and oceans accumulate over time and space. The future of humanity depends on the health of the oceans which should translate into a careful maintenance of their natural balance, including biodiversity. When considering the state of the marine environment, it is often emphasized that protection of the marine environment is effective when entities operating in this environment act in accordance with obligations resulting from international laws (Harrison 2017).

The Convention on the Law of the Sea, otherwise known as the "constitution of the seas and oceans", pays special attention to international law on the protection of the marine environment (Franckx 1998). UNCLOS confers the power on coastal States to adopt laws and regulations on the safety of navigation and the regulation of maritime traffic in its territorial sea, in respect to, inter alia, the conservation of the sea's living resources (Article 21(1)(d)), the preservation of the coastal State's environment and the prevention, reduction and control of pollution (Article 21(1)(f)).

UNCLOS refers to the rights and obligations of the participatory States regarding the protection and preservation of the marine environment and the prevention of marine pollution not only in the territorial sea, but also in the EEZ and the High Seas. These provisions should be interpreted alongside those included in Part XII, which deals exclusively with the protection and preservation of the marine environment from different sources of pollution (Molenaar 1998).

In order to prevent, reduce and control pollution, UNCLOS obliges its States to create legal rules, standards and recommendations, both at the global and regional level (Articles 207–208, 210, 212). The agreement refers to the relationship between international regulations and internal legislation (domestic law), with the aim of unifying the law and, as a result, increasing maritime safety and security.

UNCLOS contains legal norms aimed at the effective protection of the marine environment, for example, the obligation of States to prevent transboundary pollution, including pollution from or through the atmosphere, the introduction of the environmental impact assessment, the concept of the protection of marine biological diversity or the creation of marine protected areas (MPAs) (e.g. clearly defined areas, Article 211(6a)). Some of these norms are particularly important for MSP.

The coastal State may adopt special mandatory measures for the prevention of vessel-source pollution in specific clearly defined areas of its EEZ. To justify the adoption of such measures, evidence must indicate that the existing international rules and standards are inadequate for the special circumstances of the area concerned. The area must be "clearly defined" and the adoption of special measures must be required for recognized technical reasons regarding the oceanographical and ecological conditions as well as the utilization or protection of the resources and the particular character of the traffic of the area concerned. Article 211(6)(a) and (b) include specific conditions for the adoption of special mandatory measures: the coastal State should conduct appropriate consultations through the competent international organization (e.g. International Maritime Organization (IMO)) with other States concerned. It should also submit a communication to the organization regarding special mandatory measures, supported by scientific and technical evidence and information on reception facilities; the organization, within 12 months of receiving the communication, shall determine whether the conditions in the proposed area justify the adoption of special mandatory measures; following a decision by the organization, the coastal State may adopt laws and regulations implementing such international rules and standards or navigational practices as are made applicable, through the organization, for special areas. These laws shall not apply to foreign vessels until 15 months after the submission of the communication to the organization. The coastal State shall publish the limits of the area where the special mandatory measures are to be enforced.

The coastal State has sovereign rights in the EEZ in the field of exploration and exploitation of natural resources, but these rights should be interpreted in conjunction with the responsibilities for the protection and rational management of these resources. The coastal State acts as "the resource manager" in its EEZ. According to the provisions of the Convention on the Law of the Sea, which concern the living resources of the High Seas and, in particular, highly migratory species, anadromous and catadromous stocks whose protection in UNCLOS has been specifically regulated and referenced in Part XII of UNCLOS, it is clear that its purpose is to protect and preserve the marine environment.

2.3 Global and Regional Cooperation

The natural unity of the World Ocean can be protected through the effective cooperation of all actors of the international community. The duty to cooperate is a fundamental norm in the legal context of the marine environment's protection.

Observations from the last decade illustrate the efforts of both the international society (e.g. by international organizations: IMO, Intergovernmental Oceanographic Commission [IOC]) and regional communities to develop and implement solutions using various ocean governance instruments. UNCLOS prescribes that States shall cooperate on a global or regional basis, directly or through competent international organizations, in formulating and elaborating international rules, standards and recommended practices and procedures for the protection and preservation of the marine environment, taking into account characteristic regional features (Article 197).

Science-based, integrated, adaptive, strategic and participatory approaches are all core values that the IOC promotes in the context of MSP. With a view towards building the technical and institutional capacities of nations around the world, the IOC integrated its MSP initiative as part of the Integrated Coastal Area Management Strategy that was endorsed by the IOC Assembly in 2011. Since then, the IOC has continued to document the international practice of MSP around the world, synthesizing lessons learnt and updating technical guidance in various aspects of MSP design and implementation. Ten years after the first MSP conference in Paris, the IOC contribution in the MSP field culminated with the organization of the second International Conference on MSP in March 2017 at IOC/United Nations Educational, Scientific, and Cultural Organization (UNESCO) in tandem with the European Commission's Directorate-General for Maritime Affairs and Fisheries (DG Mare). This Conference helped consolidate the international network of MSP practitioners and assessed the contribution of MSP towards sustainable Blue Growth and marine ecosystem conservation, as well as identified priorities for the future of MSP. Also, the IMO cooperates, perhaps not directly on MSP, but in tandem with the Regional Seas Programme of the United Nations Environment Programme (UNEP). In particular, the IMO has played a key role in the establishment of international conventions (e.g. International Convention for the Prevention of Pollution from Ships (MARPOL), International Convention for the Safety of Life at Sea (SOLAS), International Convention for the Control and Management of Ship' Ballast Water and Sediments (BWM), International Convention on the Control of Harmful Anti-fouling Systems on Ships (AFS)), as well as regional arrangements for combating marine pollution (Molenaar 1998). The degree of the international acceptance of the IMO norms, standards and recommended practices is decisive in establishing the extent to which State Parties to UNCLOS are under the obligation to implement them (Harrison 2011). This factor is important, bearing in mind that international shipping has undergone tremendous changes in the last few decades. These changes are related not only to the growing tonnage of the world fleet but also to technical progress and new technologies which are changing the face of the shipping industry. Efforts to introduce even higher standards in terms of the protection of the marine environment, especially through the establishment of obligatory standards for the prevention of marine pollution from ships, are and will be increasingly stronger. The necessity of strong international cooperation and coordination between States is already visible.

Nowadays, work is being carried out more intensively than ever before on improving the effectiveness of international and regional cooperation for the implementation of GOG as well as MSP objectives (Zaucha 2014). These improvements are aimed at developing the cross-sectoral organization of national work (Kroepelien 2007).

The doctrine indicates the need to continuously improve international cooperation which, in turn, facilitates the development of ocean governance methods. Across the world, within international and non-governmental organizations, researchers in various fields conduct both individual and joint research on changes occurring in the marine environment and the design of instruments necessary for its effective protection (Juda 1996; Friedheim 2000; Kimball 2003).

The ecosystem approach, holistic and integrated, as well as the experience gained from network cooperation at regional levels suggest that the transferring of regional cooperation mechanisms to the global level is possible. Work on global administration and management of the marine environment has already begun. The effectiveness of the legal regime of the Law of the Sea in the protection of the World Ocean depends on the level of commitment and will of the international community.

3 An Effective Approach to Ocean Governance

An integrated, interdisciplinary, cross-sectoral and ecosystem approach to ocean governance, in conjunction with the legal framework included in the Convention on the Law of the Sea and the objectives of Chap. 17 of Agenda 21 (Agenda 21), is not only desirable but necessary and of fundamental importance to humanity. The need to introduce integrated management is mentioned, referring especially to the implementation of management at the regional level.

In general, the Law of the Sea refers to maritime human activity, taking into account particular categories of marine areas and their legal status. This approach is referred to as a sectoral approach or zonal approach. In response to the weakness of the sectoral approach, a cross-sectoral approach has been developed. The basis for promoting and implementing the cross-sectoral approach is cooperation, in particular cross-border cooperation (Tanaka 2004; Gilek et al. 2015).

A complete dismissal of the sectoral approach is unreasonable. Instead, the sectoral approach used to solve the problems of the World Ocean should be complemented or supported by a holistic and integrated approach to management. The aim of combining the potential of the sectoral approach with the integrated approach in the management of seas and oceans is the identification of environmental problems in the complexity of socio-economic and political conditions and the design of proper solutions. Although the lack of financial resources is generally considered a basic problem, the main barrier is setting priorities when allocating available funds for economic development (global economic policy) with environmental problems pushed into the background (Pyć 2011).

GOG policy, based on integration and coordination, must take into account interdependencies that closely and in a multidimensional way link mankind with the ocean. Striving for effective global and regional cooperation requires integration into functional ocean management, in particular regarding global shipping, the management and protection of endangered species and their habitats, sustainable development of technologies, marine scientific research and tourism. The same applies to global problems: climate change, sea-level rise, reduction of biodiversity, the disposal and storage of hazardous waste at the bottom of the sea and under the seabed. The right approach to ocean governance must reflect the idea of a peaceful use of the seas and the harmonious coexistence of nations regarding the maintenance of international security.

The Manado Ocean Declaration adopted in Indonesia in May 2009 includes important findings for GOG. States have declared their willingness to achieve the long-term conservation, management and sustainable use of living marine resources and coastal habitats through a precautionary and ecosystem approach and to implement long-term strategies with internationally agreed sustainable development goals (SDGs), including those outlined in the UN Millennium Declaration regarding the marine environment, thereby strengthening the global partnership for development. The Declaration stressed the need of implementing national strategies for the sustainable management of coastal and marine ecosystems, in particular mangrove forests, wetlands, grassland clusters, estuaries and coral reefs, protective zones that minimize the negative effects of climate change on one hand and, on the other hand, resources. Countries have also referred to the introduction of integrated coastal zone management and ocean management, including maritime and coastal zoning, in order to minimize and reduce the risk of adverse climate change in coastal communities (critical infrastructure) (Manado Ocean Declaration 2009).

It is worth paying attention to the ten principles of open sea management (10 Principles for High Seas Governance) developed by the International

Union for Conservation of Nature (IUCN), which are increasingly supported by the literature of the subject and practice. These are conditional freedom of activity on the High Seas, protection and preservation of the marine environment, international cooperation, a science-based approach to management, public availability of information, transparent and open decision-making processes, a precautionary approach, an ecosystem approach, sustainable and equitable use, and responsibility of States as stewards of the global marine environment.

The implementation of the Sustainable Development Goal on Ocean (SDG#14), which is one of the 17 goals of the UN Agenda for Sustainable Development 2030 and takes into account MSP and generally ecosystem-based management, provides an effective framework for guiding the sustainable development of coasts and oceans. The UN's vision regarding MSP is based on the use of interdisciplinary sciences for better policymaking and management, for example, to strengthen socio-economic analysis; plan for the local context—"No one size fits all"; combine single-sector and multi-sector area-based approaches; advance the cross-border use of MSP, integrated coastal zone management (ICZM) and MPAs; harmonize the legal and regulatory frameworks across borders; ensure full benefit sharing among stakeholders; develop practical trade-off analyses for realistic planning; use risk analysis and investment scenarios for the engagement of the private sector.

4 The Usefulness of Marine Spatial Planning

The first international meeting devoted to MSP was held in 2007 by the IOC (of UNESCO). Then, as a way of improving the decision-making and implementation process, the definition of MSP was formulated based on an ecosystem approach in managing human activities in the marine environment. The inclusion of MSP in the planning process enables an integrated, forward-looking and consistent decision-making regarding the use of the sea by humans (Ehler and Douvere 2009).

When addressing the concept of integrated management, two doubts need to be resolved. First, the selection of elements which should be integrated in this approach, and second, the extent to which the foundations of this approach are truly supported by the contemporary international Law of the Sea and the international environmental law. It is commonly accepted that, although there is no unified definition of integrated management in international law, the primary goal of this approach is to effectively solve problems that cannot be effectively addressed using traditional instruments. In solving the problems of the World Ocean, a certain degree of integration is required, at least on three levels: axiological, normative and functional. The necessity of integrated management is already visible in the axiological dimension through moral obligation and the development of preventive responsibility for marine and normative protection. This requires the implementation of jointly designed standards included in international agreements, providing them with mechanisms of law monitoring and coordination, including improving existing weaknesses resulting from the sectoral approach (Pyć 2011).

MSP has been defined by the IOC (of UNESCO) in 2009 as "a public process of analyzing and allocating the spatial and temporal distribution of human activities in marine areas to achieve ecological, economic, and social objectives that are usually specified through a political process" (Ehler and Douvere 2009).

The IOC guide "Marine Spatial Planning: A Step-by-Step Approach Toward an Ecosystem-Based Management" has been used as the reference document for developing the policy context in the European Union in the Directive 2014/89/EU establishing the framework for MSP (EC 2014).

MSP is a process that aims to reconcile the diverse group of entities in disagreement over terms of interests and expectations. The different legal status of marine areas, the diverse types and effects of human activities in the marine environment, multifaceted activities and measures aimed at the protection and conservation of marine ecosystems, as well as many other related factors amount to a highly complex web which planning has to solve. In practice, the implementation of MSP may be burdened with ballast resulting from the sectoral approach and well-established habits when it comes to designating the competence of the administrative bodies responsible for maritime affairs.

It is worth noting that many coastal States have introduced instruments into their domestic law that are used to manage maritime space, in order to meet the environmental protection obligation laid down in Article 192 of UNCLOS.

Management as a decision-making process is implemented at many levels of an organization, and it is assumed that it ensures the elimination of detected threats, the use of opportunities and the organization's effective fulfilment of all the functions necessary to achieve the set goal (Ehler 2014). Literature pertaining to the field of management uses the term "management by control". In a complex management process, control plays a key role. It is assumed that "there is no management without control". The control activity aims at eliminating, before the end of each stage of a specific process, phenomena that may negatively affect the final result. One should take into consideration the following criteria: purposefulness, economy, reliability and legality as well as organizational efficiency, meaning correct and effective directions of action and appropriate means to accomplish set tasks. Control, understood as a fundamental management method, must be based on recognition of the problem's identity and result from a thorough analysis of the problem. This, in turn, may produce a universal and flexible procedure that can be used in different circumstances. This procedure should be easy to interpret, particularly in unpredictable situations.

5 Marine Spatial Planning as a Tool of Integrated Maritime Policy

MSP is an instrument of maritime policy, both at the national and regional levels. In the Baltic Sea region, the development of common principles pertaining to MSP, such as holistic, ecosystem and precautionary management, is associated with the involvement of all relevant entities and bodies (Zaucha 2014; Backer 2015). MSP in the Baltic Sea is of interest to international organizations and institutions, including the European Union and the Baltic Marine Environmental Protection Commission (HELCOM). The HELCOM Action Plan of 2007 contained a commitment addressed to the State Parties to the Helsinki Convention regarding the Protection of the Marine Environment of the Baltic Sea Area and, more specifically, it required of them the joint development of general cross-sectoral MSP principles based on an ecosystem approach in cooperation with other international bodies.

It is worth noting that the HELCOM-VASAB Joint Group on MSP defined the following ten principles of MSP: sustainable management, an ecosystem approach, long-term perspective, the precautionary principle, participation and transparency, high-quality data and information bases, transnational coordination and consultation, coherent terrestrial and MSP, planning adapted to characteristics and special conditions at different areas and continuous planning.

Marine management is based on MSP decision-making and integrated management, that is, making decisions and constantly improving planning procedures. From a legal point of view, maritime management (the marine environment and its resources) operates within two areas: legal and institutional. For maritime management, the legal aspect, i.e. the substantive and formal normative dimension of law, is as important as the institutional level, that is, the executive level, which covers all governmental and non-governmental organizations and international institutions that carry out activities directed at environmental management, or whose activity has specific effects on the environment.

Marine environmental management includes multidimensional and integrated planning of human activity based on the most up-to-date scientific knowledge of ecosystems and their dynamics. It also requires knowledge on any activities that are essential to maintaining ecosystem health, as well as ensuring sustainable use of resources, including maintaining ecosystem integrity and ecosystem services.

The entire management process is essential: from planning, through decision-making, to executing management activities in practice. Management and responsibility for the protection of the marine environment in individual areas should be clearly, consistently, flexibly and comprehensively defined. The precautionary principle and the ecosystem approach determine the current framework for spatial planning in marine areas and regulate various human activities in the marine environment with a view to protecting marine and coastal ecosystems and biological diversity (Söderström 2017; Ansong et al. 2017; Pyć 2017).

This structure will avoid overlapping competences of administrative bodies and other entities (agencies) which set goals for implementation. A great amount of hope relates to monitoring of compliance as an element of more effective law enforcement and an important tool in the effective protection of the World Ocean.

Analysis of the legal status of marine areas confirms the important role of coastal States in creating MPAs and ensuring their effective functioning. All entities of international law are obliged to cooperate in the protection and preservation of the marine environment. The duty to cooperate follows the Convention on the Law of the Sea and is applied to each of the marine areas, including coastal areas. It is strengthened by the provisions of many other international agreements concerning the protection of the environment and natural resources.

An extremely important task is the constant improvement of scientific research regarding the World Ocean. It should be added that building ecological awareness in the community, which consists of explaining the impact of the ocean on people's lives and the impact of human activity on the functioning of the World Ocean and climate, is essential. This type of knowledge translates into more thoughtful behaviour of States, other entities as well as individual people. It allows individuals to participate and make the most appropriate and easy-to-implement decisions which will allow for good quality of life with the environment and nature.

The designed MSP framework must include control and surveillance instruments. Marine planners should also be clearly aware of the importance of "marine domain awareness" and the applicable legal norms for the use of the sea. To achieve these objectives, collection of relevant data on the use of the sea is required. MSP procedures, marine environmental control and data collection must meet the requirements of compliance with international law and, from a European perspective, with EU law, both as part of national cooperation with other States as well as at a regional and global level.

Integrated and independent actions introducing solutions to new global problems contrast with the possibilities of existing organizations. Although these institutions seem to be independent, they are characterized by fragmentation and relatively narrow competences as part of their mandates, which leads to the isolation of decision-making processes. Entities responsible for the management of natural resources and environmental protection are institutionally separated from those responsible for economic management. Isolating economic systems from those related to the environment does not support the desired exchanges within the institutions, and the policies pursued by the State are also negatively affected.

Three imperatives for GOG included in the World Commission on Environment and Development (WCED) report of 1987 still retain their relevance. First, the unity of the ocean requires an effective global management system; second, common resources specific to a given sea require a mandatory regional management system; and third, the main threats to the ocean, which originate on land, require effective national actions, undertaken by States and based on the idea of international cooperation (WCED 1987).

Based on the definition of the MSP Directive, MSP involves the identification of possible uses of marine resources and their rational distribution, as well as the provision of sustainable activity in terms of the ecosystem, all of which is performed in the marine environment in order to achieve the economic, social and environmental objectives arising from regional and national policies. These themselves correspond to international rules and standards, recommended practices and procedures for the protection and preservation of the marine environment (Deidun et al. 2011; Santo De 2015).

MSP understood as a purely technical process serves as an instrument of maritime policy at both regional and national level for the implementation of the European Union's integrated maritime policy. This policy focuses on an integrated approach to maritime affairs, referring to all available research methods used in the field of identifying and solving problems arising from the use of the sea by humans. The reasoning supporting the introduction of an integrated approach is recognition of the "maritime dimension" and the establishment of a link with the competitiveness of maritime industries and job creation, maritime fisheries and aquaculture, international maritime trade, maritime transport and logistics, access to energy sources, the effects of climate change and counteracting them, ensuring a high level of environmental protection and maintaining biodiversity, marine research and innovations.

6 Conclusions

For nearly two decades, there has been a tendency to focus international legal instruments on an integrated approach to ocean governance. GOG includes the way in which the international community sets priorities, goals and systems for the cooperation and coordination of activities within international institutions. The essence of this approach is recognition of the intersection of international, regional and national levels at institutional levels.

Issues related to integrated ocean governance are also clearly derived from international law. Analysing the application of international law instruments leads to the conclusion that even the basic assumptions of the concept of integrated management are arbitrarily interpreted by various interested entities.

The impact of the institutions involved in ocean governance is influenced by the holistic approach adopted in the Convention on the Law of the Sea, which states that "the problems of ocean space are closely interrelated and need be considered as a whole". This one sentence in the preamble to UNCLOS is of particular importance. There are many economic, social, political, as well as scientific factors, among others, which must be considered in the development of policy and law in the context of ocean governance. This particularly applies when formulating principles and specific legal norms for achieving GOG objectives and maintaining their integrity. In addition, the biological diversity of resources is important. Management will need to be carried out with particular attention to biodiversity, not only individually but especially in the context of managing other resources.

International lawyers are considering whether it is possible to design global ocean management programmes at the institutional level based on the assumptions underlying the UNCLOS' concept of mankind's common heritage. The introduction of instruments for the management of the seas and oceans has set a new perspective in international law, particularly regarding the Law of the Sea and its practice. The focus has shifted onto the law of the World Ocean as a dialectical system located between a sectoral approach and an integrated approach to managing marine resources. Recognizing that the contemporary Law of the Sea is essentially still based on a sectoral approach, this view exerts a definite influence on interpretive changes in the Law of the Sea. Bearing in mind the achievement of MSP's objectives, from a functional point of view, it is of utmost importance to apply mechanisms of integrated management to the practice of the Law of the Sea in order to create a long-term, reasonable administration of ocean resources in a sustainable manner.

MSP is a process that serves to ensure the introduction of spatial order in seas and oceans. The main goal of MSP is the division of sea space with the purpose of fairly distributing marine areas and their resources between various entities, including coastal States and legal and natural persons. This process may require restrictions on the use of maritime space (e.g. temporary or zonal) and, in justified cases, with the aim of avoiding conflicts between different users of the environment and improving the management of their activities. Capacity building within administrative bodies and other entities in the field of maritime management is also of utmost importance.

MSP is a process which aims to distribute space dynamically for many types of sea use. As such, it also introduces time constraints and even exclusions in order to avoid conflicts between the various users of the environment and improve the management of human activities. MSP should be based on a holistic approach which assumes the existence of multidimensionality and interdependencies of interactions in the marine environment occurring as a result of carrying various activities undertaken in it, including economic activity. In the European Union, specifically those Member States that have developed MSP instruments, the implementation thereof remains at the national level and is carried out by the authorities of those States. The planning process is subject to the analysis of the use of the marine environment and its resources, necessary for decision-making.

Acknowledgements We are grateful to the project "Economy of maritime space" funded by the Polish National Science Centre for contributing the Open Access fee for this chapter and facilitating our discussions and preparation of the book.

References

Agenda 21. Retrieved June 16, 2018., from https://sustainabledevelopment.un.org/content/documents/Agenda21.pdf.

Ansong, J., Gissi, E., & Calado, H. (2017). An Approach to Ecosystem-based Management in Maritime Spatial Planning Process. *Ocean & Coastal Management, 141*, 65–81.

Attard, D., & Mallia, P. (2014). The High Seas. In D. J. Attard, M. Fitzmaurice, & N. A M. Gutiérrez (Eds.), *The IMLI Manual on International Maritime Law, vol. I: The Law of the Sea* (pp. 243–248). Oxford University Press.

Backer, H. (2015). Marine Spatial Planning in the Baltic Sea. In D. Hassan, T. Koukkanen, N. Soininen (Eds.), *Transboundary Marine Spatial Planning and International Law* (pp. 132–133). Routledge.

Deidun, A., Borg, S., & Micallef, A. (2011). Making the Case for Marine Spatial Planning in the Maltese Islands. *Ocean Development & International Law, 42*(s), 137.

Dorman Mc, T. L. (2000). Global Ocean Governance and International Adjudicative Resolution. *Ocean & Coastal Management, 43*, 256.

EC. (2014). Directive 2014/89/EU of the European Parliament and of the Council of 23 July 2014 Establishing a Framework for Maritime Spatial Planning, OJ L 257, 28.8.2014, pp. 135–145.

Ehler, C. (2014). *A Guide to Evaluating Marine Spatial Plans*. IOC Manuals and Guides No. 70, ICAM Dossier 8. Paris: UNESCO, Intergovernmental Oceanographic Commission UNESCO IOC, 96 pp.

Ehler, C., & Douvere, F. (2009). *Marine Spatial Planning: A Step-by Step Approach Towards Ecosystem-based Management*. Manual and Guides No 153 ICAM Dossier No 6. Paris: Intergovernmental Oceanographic Commission UNESCO IOC, 99 pp.

Franckx, E. (1998). Regional Marine Environment Protection Regimes in the Context of UNCLOS. *The International Journal of Marine and Coastal Law, 13*(3), 310–312.

Friedheim, R. (2000). Designing the Ocean Policy Future: An Essay on How I Am Going To Do That, "Ocean Development & International Law", pp. 183–195.

Galletti, F. (2015). Transformation in International Law of the Sea: Governance of the "Space" or "Resources"? In A. Maroco, P. Prouzet (Eds), *Governance of Seas and Oceans* (pp. 2–33). ISTE.

Gilek, M., Hassler, B., & Jentoft, S. (2015). Marine Environmental Governance in Europe: Problems and Opportunities. In M. Gilek, K. Kern (Eds.), *Governing Europe's Marine Environment. Europeanization of Regional Seas or Regionalization of EU Policies?* (pp. 255–257) Ashgate.

Grotius, H. The Freedom of the Seas or the Right which belongs to the Dutch to take part in the East Indian Trade, Batoche Books Limited, Kitchener 2000. Retrieved

June 17, 2018., from https://socialsciences.mcmaster.ca/econ/ugcm/3ll3/grotius/Seas.pdf.

Harrison, J. (2011). *Making the Law of the Sea, A Study in the Development of International Law* (pp. 238, 257–258). Cambridge.

Harrison, J. (2017). *Saving the Oceans Through Law* (pp. 17–41). Oxford: The International Legal Framework for Protection of the Marine Environment.

Juda, L. (1996). *International Law and Ocean Use Management: The Evolution of Ocean Governance* (pp. 285–289). Routledge.

Kimball, L. A. (2003). *International Ocean Governance, Using International Law and Organizations to Manage Marine Resources Sustainability* (pp. 1–123). Gland: IUCN.

Kingsbury, B., Kirsch, N., & Stewart, R. B. (2005). The Emergence of Global Administrative Law. *Law and Contemporary Problems, 68*, 18.

Kroepelien, K. F. (2007). The Norwegian Barents Sea Management Plan and the EC Marine Strategy Directive: Some Political and Legal Challenges with an Ecosystem—Based Approach to the Protection of the European Marine Environment. *RECIEL, 16*(1), 31.

LEG/MISC/3/Rev.1. Implications of the United Nations Convention on the Law of the Sea for the International Maritime Organization, LEG/MISC/3/Rev.1, 06.01.2003. Retrieved June 17, 2018., from http://www.sjofartsverket.se/pages/13880/LEG-MISC-3-rev1.pdf.

Manado Ocean Declaration. (2009). Retrieved June 30, 2018, from http://cep.unep.org/manado-ocean-declaration.

Molenaar, E. J. (1998). Coastal State Jurisdiction Vessel-Source Pollution. *Kluwer Law International pp., 25-27*, 103–110.

Pyć, D. (2011). *Prawo Oceanu Światowego* (pp. 97–105). Gdańsk: Res usus publicum.

Pyć, D. (2016). Global Ocean Governance. *TransNav—The International Journal on Marine Navigation and Safety of Sea Transportation, 10*(1), 159–162.

Pyć, D. 2017. *The Polish Legal Regime on Marine Spatial Planning, Maritime Law of the Polish Academy of Science* (vol. XXXIII). Gdańsk.

Santo De, E. (2015). The Marine Strategy Framework Directive as a Catalyst for Maritime Spatial Planning: Internal Dimensions and Institutional Tensions. In M. Gilek & K. Kern (Eds.), *Governing Europe's Marine Environment. Europeanization of Regional Seas or Regionalization of EU Policies?* (p. 95). Ashgate.

Söderström, S. (2017). *Regional Environmental Governance and Avenues for the Ecosystem Approach to Management in the Baltic Sea Area*. Linköping.

Tanaka, Y. (2004). Zonal and Integrated Management Approaches to Ocean Governance: Reflections on a Dual Approach I International Law of the Sea. *The International Journal of Marine and Coastal Law, 19*, 453–514.

UNCLOS. (1982). The United Nations Convention on the Law of the Sea. Retrieved June 16, 2018, from http://www.un.org/depts/los/convention_agreements/texts/unclos/unclos_e.pdf.

WCED. (1987). Retrieved June 30, 2018, from http://www.un-documents.net/our-common-future.pdf.

Zaucha, J. (2014). Sea Basin Maritime Spatial Planning: A Case Study of the Baltic Sea Region and Poland. *Marine Policy, 50*, 34–45.

17

The Need for Marine Spatial Planning in Areas Beyond National Jurisdiction

Susanne Altvater, Ruth Fletcher, and Cristian Passarello

1 Introduction

1.1 Why Would Marine Spatial Planning Be Undertaken in Areas Beyond National Jurisdiction?

There are a number of sectors potentially active in Areas Beyond National Jurisdiction (ABNJ) including fishing, shipping and cable laying. In addition to these, mining concessions have been leased in a number of locations although, to date, these are only at the exploration phase. These sectors all have individual frameworks in which they are managed. For example, fishing is managed regionally through Regional Fishery Management Organisations (RFMOs), whereas shipping is supported by various Conventions under the International Maritime Organization (IMO), and underwater mining areas are leased through the International Seabed Authority (ISA). Nonetheless, coordination between the different sectors is currently limited, which challenges the conservation of natural resources in ABNJ, although there is potential for cross-sectoral coordination for the purposes of biodiversity conservation (Gjerde et al. 2016).

S. Altvater (✉) • C. Passarello
s.Pro—Sustainable Projects GmbH, Berlin, Germany
e-mail: sal@sustainable-projects.eu

R. Fletcher
UN Environment World Conservation Monitoring Centre, Cambridge, UK

J. Zaucha, K. Gee (eds.), *Maritime Spatial Planning*,
https://doi.org/10.1007/978-3-319-98696-8_17

It would potentially be valuable to use Marine Spatial Planning (MSP) as it provides a framework for coordinated spatial management, especially in data-poor situations characterised by high uncertainty. Moreover, the enforcement of spatial controls could result to be more cost-effective than other management measures (FAO 2007). Transboundary MSP can help with fishing, shipping and cultural heritage (Soininen and Hassan 2015) and can also be useful to implement the Integrated Maritime Policy (IMP) as well as the Marine Strategy Framework Directive (MSFD) (Becker-Weinberg 2017).

Box 17.1 Explanation of the Two Main Concepts of the Chapter

Marine Spatial Planning

Marine spatial planning (MSP) is a public process of analysing and allocating the spatial and temporal distribution of human activities in marine areas to achieve ecological, economic and social objectives that are usually specified through a political process (Ehler & Douvere 2009).

Areas Beyond National Jurisdiction

The areas beyond the limits of national jurisdiction are defined according to the UN Convention of the Laws of the Seas (UNCLOS):

1. *The water column beyond the Exclusive Economic Zone (EEZ), or beyond the Territorial Sea where no EEZ has been declared, called the High Seas (Art. 86) and*
2. *The seabed which lies beyond the limits of the continental shelf, established in conformity with Art. 76 of the Convention, designated as 'the Area' (Art. 1).*

Commonly called the high seas, no one nation has the sole responsibility for management. Everyone has the freedom to navigate, overflight, exploration and exploitation of natural resources (except mineral resources), and others (Part VII of UNCLOS).

The Area has the status of 'common heritage of mankind'. The ISA is the body entitled to act on behalf of the mankind as a whole (UNCLOS, art. 137(2)) and to give concrete content to the principle of the common heritage of mankind foreseeing the international management of mineral resources (Part XI of UNCLOS).

ABNJ account for most of the global ocean and are home to a great amount of biodiversity and natural resources (UNEP-WCMC 2017). Although the remoteness and difficulty of exploiting the resources located in these areas has historically contributed to maintain their preservation, recent shifts in technological capacity and market opportunities allowed humans to expand their interest in ABNJ (Merrie et al. 2014). This interest has resulted in the development of different human activities, which all have the potential to generate significant threats to the marine species and ecosystems of the high seas, also referred to as the Biodiversity Beyond National Jurisdiction (BBNJ) (Kimball 2005; United Nations 2017). Threats include the over-exploitation of resources, habitat degradation, pollution (including those from terrestrial sources such as plastics), exploitation of mineral resources, climate change and climate engineering, ocean acidification and new human activities (Halpern et al. 2008). Because of these pressures, MSP in ABNJ is increasingly needed to ensure the sustainable use of natural resources and the resilience of marine ecosystems in the high seas (Ardron et al. 2008).

Although some sector-specific ABNJ management measures exist, at present there is no overarching mechanism to ensure that important or vulnerable ecosystems in international waters are comprehensively protected (Druel and Gjerde 2014). Efforts are being undertaken to address this challenge through the creation of a new implementing agreement under the United Nations Convention on the Law of the Sea (UNCLOS) for the conservation and sustainable use of BBNJ, referred to as the International Legally Binding Instrument (ILBI). One of the challenges that has been recognised is the need for cross-sectoral coordination of activities in ABNJ (United Nations 2017). Given the limited experience of area-based planning tools for the protection of ABNJ, it is necessary and appropriate to examine the application of spatial planning tools within Exclusive Economic Zones (EEZs) in order to consider their potential for effective use in ABNJ.

Currently, international waters are governed under several sectoral governance regimes to manage specific activities and pressures (Kimball 2005). For example, the IMO governs shipping in the high seas and implements the MARPOL (The International Convention for the Prevention of Pollution from Ships) Convention and Protocol to prevent pollution from shipping. Whereas, the ISA governs 'the Area' (*the seabed and ocean floor and subsoil thereof, beyond the limits of national jurisdiction*) and implements environmental management measures to reduce the potential impacts of deep-sea mining. However, it is argued that the current sectoral framework leaves legal, governance and geographical gaps in management of activities within ABNJ (Druel and Gjerde 2014). In recognition of governance gaps, and in light of the growing anthropogenic pressures, society is slowly realising the importance of supporting the management of current and future activities occurring in international waters, especially if valuable resources, ecosystems, and biodiversity are to be preserved for future generations (Rayfuse 2012). One strand of discussions pertains specifically to the applicability of various area-based management approaches for the conservation and sustainable use of marine resources and biodiversity. Although MSP has the potential to assist states to fulfil their obligations under international agreements—such as UNCLOS and the Convention on Biological Diversity (CBD)—its implementation in ABNJ by single states is not possible within the current governance frameworks. Moreover, international cooperation between various nations is required (Ardron et al. 2008; EC 2009).

1.2 Introduction to Biodiversity Beyond National Jurisdiction

In the Rio Earth Summit outcome document, the 'The Future We Want' importance of the conservation and sustainable use of marine BBNJ was recognised (United Nations 2012). Following the work done by the BBNJ Working Group, and the potential for increasing pressures in ABNJ, the UN General Assembly (UNGA) adopted the BBNJ Working Group's recommendation in Resolution 69/292 (A/RES/69/292) and decided to develop a new implementing agreement under UNCLOS for the conservation and sustainable use of BBNJ. Since 2015, four Preparatory Committee meetings have been held to explore and provide recommendations to the General Assembly on the elements of a draft text for a new instrument. On 24 December 2017, the UNGA adopted Resolution 79/249 and decided to convene an intergovernmental conference to "*consider the recommendations of the Preparatory Committee and to elaborate the text of an international legally binding instrument*" under UNCLOS (A/RES/79/249). The conference will occur over four sessions between 2018 and 2020.

Box 17.2 Processes of the BBNJ Working Group

International Discussions

The challenge of ensuring that marine biodiversity is effectively conserved in ABNJ has been part of extensive discussions for nearly 15 years. In 2004, the UNGA established a "*Ad Hoc Open-ended Informal Working Group to study issues relating to the conservation and sustainable use of marine biological diversity beyond areas of national jurisdiction*, known as Biodiversity Beyond National Jurisdiction (BBNJ) Working Group" to explore these issues (A/RES/59/24). In 2015, the working group provided recommendations (A/69/780*) to develop a new legally binding instrument for the conservation and sustainable use of marine biological diversity of ABNJ, with a particular focus on four overarching issues:

- Marine Genetic Resources (including issues of benefit sharing);
- Area-Based Management Tools (including Marine Protected Areas);
- Environmental Impact Assessments; and
- Capacity building and the transfer of marine technology.

2 Existing Spatial Measures in the High Seas

Following the Geneva Convention on the Law of the Sea (UNCLOS) in 1958, various legal and governance arrangements have been developed globally with the aim of regulating human activities in the marine environment (Merrie et al. 2014). Amongst the various arrangements, the following are the most prevalent regarding the high seas:

2.1 Conventions

- **UN Convention on the Law of the Sea (UNCLOS)** (1982) provides general obligation to protect the marine environment (see also Chap. 17 in this volume). It does not mention MSP, but its article 123 promotes the cooperation between states bordering enclosed or semi-enclosed seas, to manage, conserve, explore and exploit the living resources of the sea whilst protecting and preserving the marine environment. Whereas its article 192 requires all states to protect and preserve the marine environment (Maes and Cliquet 2015). Coastal states also have full sovereignty over their archipelagic waters, although it should be noted that their sovereignty is "subject to the freedom of innocent passage by foreign vessels and particular rules for certain international straits". This limits MSP "by setting legal requirements for MSP in terms of maritime transportation and navigation" (UNCLOS articles 2 and 17–26) (Hassan and Soininen 2015). Note: UNGA initiated the treaty negotiation for the development of an internationally legally binding instrument on the conservation and sustainable use of marine biological diversity in ABNJ (Fletcher et al. 2017).
- **The International Convention for the Prevention of Pollution from Ships (MARPOL) (1973)** aims to reduce intentional pollution from ships.
- **The Convention on Biological Diversity (CBD)** (1992) mentions that State Parties have the responsibility to ensure that all actions taken within their national jurisdiction shall not have negative impacts on the environment of other states or the environment of ABNJ (Kimball 2005). However, it does not directly apply to the components of biodiversity in ABNJ but instead only to the general impact on biodiversity (Kimball 2005). Each Party to the Convention is responsible for conducting assessments regarding various activities undertaken within their jurisdiction to ensure that they do not have negative impacts on the biodiversity. Moreover, the CBD highlights the need of area-based management approaches and emphasises

the importance that MSP has in promoting the ecosystem-based management approach (Becker-Weinberg 2017).
- **The Regional Sea Conventions.** Some regional seas conventions have a mandate binding on their members for management in ABNJ such as the Barcelona Convention for the Protection of the Marine Environment and the Coastal Region of the Mediterranean and SPA/BD Protocol (Protocol concerning Specially Protected Areas and Biological Diversity in the Mediterranean); the Convention for the Protection of the Marine Environment of the North-East Atlantic (OSPAR Convention); the Convention on the Conservation of Antarctic Marine Living Resources (CAMLR Convention) together with the Antarctic Treaty; the Convention for the Protection of the Natural Resources and Environment of the South Pacific Region (Noumea Convention); and the Convention for the Protection of the Marine Environment and Coastal Area of the South-East Pacific (Lima Convention) (Campbell et al. 2017).

2.2 Agreements and Guidelines

- **UN Fish Stock Agreement (UNFSA)** is an implementing agreement under UNCLOS in force since 2001, which aims to address the problems related to fisheries in high seas (United Nations 2010). The treaty sets forth the principles, legal tools and mechanisms now being employed to maintain sustainable levels of high seas fish stocks, and the RFMOs are one of the primary mechanisms for this (United Nations 2010).
- **Agreement relating to the implementation of Part XI** of the UNCLOS of 10 December 1982—specifically relates to the setting up of the ISA and the context around mining of 'the Area'.
- **FAO International Guidelines on Deep-sea Fisheries on the High Seas.**
- **Agreement to Promote Compliance with International Conservation and Management Measures by Fishing Vessels on the High Seas** (FAO Compliance Agreement).

2.3 Organisations

- **Regional Fisheries Management Organisations (RFMOs)** are intergovernmental organisations (formed by various states) that focus on the implementation of sustainable fishing practices and management measures in the high seas. They play a key role in achieving cooperation between different coastal states regarding the use of fish stocks, although their level of success is uncertain (The Royal Institute of International Affairs 2007).

- **International Maritime Organisation** (IMO) is the global standard-setting authority for the safety, security and environmental performance of international shipping.

2.4 Mechanism for Sustainable Use

- **Ecolabels** (such as Marine Stewardship Council and Friend of the Sea) can be considered as an indirect mechanism for high seas.

3 Identification of Tools to Support MSP in ABNJ

3.1 Can MSP Work in ABNJ?

As discussed earlier, one of the governance challenges present in ABNJ is the lack of a coordinating process or body for the various sectoral management processes. Each sector currently working in ABNJ has its own management process. However, if in the future the number of activities in ABNJ will increase, these sectors will need to better coordinate their actions to avoid incompatible activities occurring in the same spatial location. For example, mining areas being designated across existing deep-sea cables or interacting with vulnerable marine ecosystems (VMEs). The fact that the new implementing agreement for conservation and sustainable use of BBNJ includes area-based management tools provides a potential future mechanism to support improved cross-sectoral coordination.

3.2 Assessment

The use of MSP within national jurisdiction is reasonably common. However, when extending its application in ABNJ, there may be challenges associated with the different governance structures and environment. ABNJ and EEZs have very different physical and ecological characteristics. ABNJ often contain very deep habitats, which are home to slow-growing, potentially fragile ecosystems such as the hydrothermal vent communities (Fisher et al. 2007). Contrastingly, EEZs are characterised by shallower, faster-growing habitats that are often subject to a wider range of human pressures. The contextual differences between EEZs and ABNJ will influence the extent to which a tool is transferable to the high seas. Therefore, it is important to be able to understand the specific characteristics of ABNJ and how they might differ from those found in EEZs, where the tools are typically applied.

Legal framework: The legal and institutional framework in ABNJ is dominated by the high seas provisions of UNCLOS and regional agreements rather than national-level agreements.

Stakeholder engagement: There are a limited number of sectors currently working in ABNJ. However, the connectivity of the ocean and the fact that ABNJ are considered areas where the principle of the common heritage of mankind applies, stakeholders could include the global population.

Pelagic conditions and large size: The greater depth and physical characteristics of water in ABNJ generate distinctive 'oceanic' conditions. Additionally, the habitats and species in ABNJ have evolved to reflect deep cold ecosystems and are generally slow growing. Additionally, the very large size of ABNJ is a unique challenge. In one statistic, 95% of the volume of the ocean is beyond national jurisdiction (Ribeiro 2013).

Data paucity: The distances and costs involved in getting to the high seas and exploring the deep ocean means that there are considerable data gaps in ABNJ.

Management: ABNJ are currently managed in a sectoral way, with individual sectoral-specific management authorities (Gjerde et al. 2016).

Regarding the legal framework, UNCLOS does not specifically mention MSP although it recognises the need to address problems of the ocean space as a whole (Becker-Weinberg 2017). Various articles focus on the preservation of marine ecosystems, *inter alia*, Article 118 on the Cooperation of States in the conservation and management of living resources and Article 194 (para. 5) on the duty of States to protect and preserve fragile ecosystems (UNCLOS 1982). Such provisions provide a legal foundation upon which MSP could be undertaken to achieve the provisions of these articles. They are also particularly relevant to BBNJ discussions. A key characteristic of marine spatial planning is that it is a participatory process. The MSP Guidelines place a strong emphasis on stakeholder engagement, listing mechanisms for enhancing the inclusion of stakeholders. A mechanism for public participation would therefore need to be considered although there would be challenges over this with global population potentially being 'the public'.

The large scale of ABNJ may require that MSP is undertaken over larger areas than currently. This is possible and guidelines for MSP do not specify a limit for the size making it possible. MSP can also be applied to any ecosystems, and guidelines do not limit this; therefore, it could be applied to the variety of ecosystems that exist in ABNJ. There will be a limit to the size of an area that can be planned in relation to the practicality, data and stakeholder inclusion. A specific limitation to a planning process may be the data paucity particularly clear in ABNJ and in deeper waters. Data limitations also apply to waters within national jurisdictions and therefore this situation is not entirely unique to ABNJ. The distances and depths are greater, and therefore the costs would be higher to access some types of data. However, using the precautionary principle, a feature of MSP, it may be possible to undertake initial planning processes, and subsequently modify the measures in an adaptive way, as additional data becomes available.

Potentially one of the major obstacles to the achievement of cross-sectoral planning process in ABNJ is the lack of a clearly mandated leadership organisation or a coordination mechanism. Some coordinating process is needed to undertake marine spatial planning, at both the planning and implementation stages. Currently there is no clear authority in ABNJ with a mandate to lead a cross-sectoral planning process, but it is hoped that the new BBNJ process will result in some organising framework for ABNJ planning.

3.3 Are There Any Existing Tools that Could Be Used in the Different Stages of the MSP Cycle in ABNJ?

The application and effectiveness of MSP are often supported by various processes, approaches and tools, which help to ensure that the most appropriate measures are implemented to meet the agreed upon objectives. Decision-support tools, for example, tend to provide a mechanism for efficient computation or problem-solving in order to support part of an MSP process. Decision-support tools are often designed to perform analyses to support decisions by managers or non-technical people. There are several stages of MSP, where specific decision-supporting tools would be valuable, for example: (1) stocktaking, vision and mapping; (2) development and evaluation of alternative management actions; and (3) monitoring and evaluation. Considering the three stages, it is useful to understand whether it would be feasible to effectively use certain tools for managing these steps in ABNJ. Examples of supporting tools and area-based planning tools that need them are the following:

- Geographic Information Systems (GIS)
- Systematic reserve designing (e.g. Marxan)
- Valuation mapping
- Trade-off analysis
- Cumulative impact assessment

- Future scenario-building
- Enforcement tools

For many of these tools and scientific efforts, the issues related to their use are similar within and beyond national jurisdiction. For example, GIS can be used both in national territories and in ABNJ. The main constraint is the technical capacity of mapping that is needed in order to provide the information into a GIS. The governance organisations around the world, including within ABNJ, all have constraints placed upon them in terms of software, user skills and time. These constraints are not unique in ABNJ but a general issue.

With systematic planning processes, trade-off analysis, cumulative impact assessment and scenario-building, the limits of data availability are a problem, which will likely challenge them all. The process of planning a reserve system, for example, requires the input of a specific set of data to minimise the cost of a reserve system and maximise the benefits. In general, there is no limit to the application of this process in geographic terms, although the data paucity in ABNJ may challenge the application of specific software systems such as Marxan. However, even in data-poor situations, some processes have been undertaken already in ABNJ to better understanding the biodiversity and ecosystem functions. For example, the CBD's Ecologically and Biologically Significant Areas (EBSAs)—ran through expert workshops—can support the identification of important marine areas both within and beyond the limits of national jurisdiction (CBD 2018). Key Biodiversity Areas (KBAs) consider a wider variety of issues and have already been identified in ABNJ, in particular areas of importance to birds, Important Bird and Biodiversity Areas (BirdLife International 2018).

With valuation, one of the common mechanisms to fill data gaps is to undertake benefits transfer, which is the use of values created in one location and extrapolated to another (Richardson et al. 2015). When studies have been undertaken within national jurisdiction, it may not be suitable to use the process of benefits transfer. In addition, how the benefits or costs of the trade-offs or values could be judged in relation to each other at a global scale will require an immense communication effort.

One of the wider challenges, yet to be fully solved, is the enforcement of effective management measures in ABNJ. There are systems capable to recognise ships movements, for example, Vessel Monitoring Systems (VMS), which track ships. These systems can be used but are limited by the challenge of knowing what activities are being undertaken on board. Also, if an infringement is identified, what jurisdiction the infraction is judged through or how the ship is physically intercepted is difficult, given the distances and potential costs involved.

4 Can Marine Spatial Planning Be Effectively Implemented in ABNJ?

4.1 Coordination Process

One of the challenges present in ABNJ is the lack of a coordinating process or body for the various sectoral management processes. Each sector currently working in ABNJ has their own management process. However, as human activities are expected to increase in ABNJ, the involved industries should start coordinating their actions in order to avoid incompatible activities from occurring in the same location. The fact that the new implementing agreement for conservation and sustainable use of BBNJ includes area-based management tools provides a potential future mechanism to support improved cross-sectoral coordination. For example, the following four options could be adopted to properly implement MSP into the agreement:

Option 1: an UNCLOS Implementing Agreement (IA) might establish the common objectives of ensuring the conservation and sustainable use of marine natural resources as well as to develop a network of MPAs in ABNJ, which are effectively managed and represented.

Option 2: an UNCLOS IA might establish a largely regional approach by requiring states and other competent bodies to submit MPAs' proposals for international endorsement. The agreement could define the criteria for submitting proposals, agreeing management measures and procedures for scientific review and endorsement as well as monitoring, control and enforcement measures. Management responsibility could remain at the regional level, operating through regional bodies or through specific collaborations between interested States (i.e. the Sargasso Sea Alliance).

Option 3: an UNCLOS IA might establish a systematic approach in which a global scientific body develops proposals for MPAs, complementary to already existing processes (i.e. at the regional level). Proposals would be based on the results of a scientifically driven process focused on the identification of areas with ecological and cultural significance. Proposals would be submitted to and adopted by the Contracting Parties whilst management responsibility could remain within the regional level and have assistance at the global level.

Option 4: an UNCLOS IA could further initiate a framework for integrated spatial planning and management to facilitate discussions between State Parties and regional and sectoral organisations to ease the coordination of spatial management plans and thus improve the use of marine resources. The agreement could mandate a coordinated process for developing an ecologically and biologically coherent system of MPAs as well as other management measures to achieve the objectives set forth in the agreement and any annexes thereto.

Option 5: in the absence of an agreed UNCLOS IA, the sectors that are currently active in ABNJ could self-organise and mutually agree to a process to identify where potential incompatible activities could occur. Discussions (bilaterally or within a group of existing organisations with mandates) could take place and agreements set up regarding how the various sectors are going to actively engage with other sectors' management designations.

A key characteristic of marine spatial planning is that it is a participatory process. The MSP Guidelines place a strong emphasis on stakeholder engagement, listing a number of mechanisms for including stakeholders and a mechanism for public participation would therefore need to be considered. There would be challenges over this with global population potentially being, 'the public'. Therefore, MSP could support coordination of existing bodies for information exchange about how to involve stakeholder groups related to specific regional and cultural needs.

4.2 Surveillance

Aspects of Surveillance and Implementation of Measures in ABNJ: Is It Feasible to Control Implementation?

Clear legal aspects are needed to ease the enforceability of MSP and facilitate its implementation in ABNJ (UNEP 2017). So far monitoring, control and surveillance systems for high seas fisheries appear to be insufficient (Ardron et al. 2008). For example, the Food and Agriculture Organization of the United Nations (FAO) and RFMO face various challenges in ABNJ. Although RFMOs establish regulations for the management of fisheries, member states are not legally obliged to follow these regulations in the high seas (Ringbom and Henriksen 2017). Moreover, vessels carrying flags of states non-member of the RFMO, cannot be enforced to follow the RFMO's protocol, which may undermine the efforts made by the RFMO in conserving fishing stocks (Ringbom and Henriksen 2017). In 2006, RFMOs—under the call of UNGA—required fishing vessels to stop practising bottom fishing when encountering VMEs and report the encounter (UNGA Resolution 61/105, para 83(d)) (FAO 2015). Most RFMOs with a mandate to regulate bottom fisheries in the ABNJ have responded with some form of encounter protocol. Two distinct approaches have emerged: one primarily for longlining in the

Southern Ocean developed by the Commission for the Conservation of Antarctic Marine Living Resources (CCAMLR) and another for trawl fisheries in the North Atlantic developed by the Northwest Atlantic Fisheries Organization (NAFO) and the North East Atlantic Fisheries Commission (NEAFC) (FAO 2015). Nonetheless, the different VMEs encounter protocols that have been put in place in various fisheries negatively impacted fishers and have resulted in economic losses and increased costs of fishing (FAO 2015).

New Zealand has adopted a unique approach for its vessels fishing in South Pacific ABNJ, while Australia independently developed a protocol similar to the North Atlantic approach. In response to the UNGA resolutions, RFMOs have defined fishery footprints effectively restricting fishing to those areas, and instituted extensive closures, portions of which close parts of each footprint. Those measures are supported by an encounter protocol. As adopted in 2008, the footprint approach was identical across the North Atlantic, but each RFMO has since developed it in regionally specific ways (FAO 2015).

Major difficulties for the industry include the imbalance in the VMEs debate and the challenge the industry faces to comply with strict conservation measures, while also attempting to conduct a sustainable business. From the start, the industry voiced that move-on rules would impact fishing operations. The fishing sector also noted that fishers knew where areas of sensitive habitats were, as well as regional differences with respect to habitats and the types of fisheries that operated in each region. Fishers living on the ocean see a different ocean than policymakers, and a disconnect between fishers and managers was noted. Other challenges faced by RFMOs in managing fisheries in ABNJ are the lower level of data and knowledge (as compared to national areas), the distance, which could negatively affect the costs of assessments and monitoring, as well as control and surveillance (United Nations 2011; Wright et al. 2015).

4.3 Case Studies

Submarine Cable Considerations for Area-Based Planning in ABNJ with Reference to Two Ongoing ISA Processes

Trans-oceanic cables have been deployed in the ocean seabed since 1858 (Carter 2009). Although they are considered to have a minimal environmental impact (Friedman 2017), various uncertainties still exist in relation to the electro-magnetic fields, seabed disturbance and cumulative effect assessment (Johnson 2017). There are currently two main different types of submarine cables: power cables and telecommunication cables. Power cables are larger in size and, compared with telecommunication cables, are less common and have not been placed in ABNJ yet, although the current legislation allows states to freely lay down both types of cables in ABNJ (Art. 87 UNCLOS) (Friedman 2017).

Even though submarine cables are likely to have minimal environmental impacts, the International Cable Protection Committee (ICPC) strongly opposed the idea to use MSP for submarine cables in ABNJ, arguing that it is an unnecessary procedure that would only introduce risks, and that historically, the involved stakeholders have always successfully managed conflicts (ICPC 2016). On the contrary, Johnson (2017) argues that conflict between stakeholders is recognised to be an issue for submarine cable developers, which emphasise the need to improve tools for stakeholder participation, whereas Friedman (2017) notes that excluding cable operations in ABNJ from MSP (or other instruments) would legitimate the request from other human activities to be similarly excepted, which could have negative repercussions.

The implementation of a specific environmental instrument (such as the environmental impact assessment) for submarine cables in ABNJ could be beneficial for the cable industry as it would not directly restrict cable installments, but instead it would allow the sector to be one of the first movers in establishing a fair instrument (Friedman 2017). This is particularly important considering that in the future, conflicts between the submarine cable industry and the seabed mining sector could exacerbate; a scenario that reinforces the idea that the communication between the two sectors would be beneficial (Johnson 2017). In fact, the Secretary General of the ISA has recently announced a workshop with the ICPC to develop guidance for avoiding conflict between the sectors (ENB 2018).

Area-based planning is considered an effective mechanism for design of spatial regulation and for the sustainable use of marine resources as it reduces the risk of possible conflicts between different stakeholders (UNEP-WCMC 2017). Nonetheless, spatial differences exist, and known Best Environmental Practices (BEPs) suitable for national waters are not necessarily appropriate for ABNJ (Johnson 2017). Finally, although the assessment of spatial human activities in ABNJ is a major challenge, the use of MSP for deep-sea environments is increasingly needed to resolve possible space and use conflicts (Johnson 2017).

Area-Based Planning in the Southeast Pacific and Western Indian Ocean Regions

In this area, the Nairobi Convention Contracting Parties expanded the Convention to cover adjacent water in ABNJ to implement an ecosystem-based approach (UNEP-WCMC 2017). It includes the development of ecosystem-based management tools for implementation. The major challenge in implementing activities related to ABNJ in the Western Indian Ocean is the lack of capacity on ABNJ-related issues at the national level (UNEP-WCMC 2017). Here, ongoing research might highlight possible approaches on how to develop a collective governance mechanism, including as many stakeholders as possible and using the Nairobi Convention's Secretariat as coordinator of activities and agreed management approaches.

5 Conclusion

Within the marine environment, a greater number of human activities are taking place and are expected to increase in the future which not only put at risk the availability of many natural resources but also jeopardise the marine biodiversity and thus the benefits people obtain from the services provided by natural ecosystems. Although various legal frameworks exist for the governance of the marine environment, their effectiveness—especially within the ABNJ—in achieving their objectives is questioned. The MSP approach is a valuable tool, which could be used as a framework to achieve better management and spatial use of the marine environment. Although most of the international regimes do not directly deal with MSP, UNCLOS recognises that the activities happening in the oceans are interrelated and should be considered a whole (Becker-Weinberg 2017). Similarly, the UNESCO considers MSP as a 'public process' capable of identifying the different human activities in the marine environment and allocate them in a rational and sustainable manner to reduce negative impacts and possible impacts (Becker-Weinberg 2017, p. 579).

In ABNJ each sector is singularly managed and there is a need for better coordination across the different sectors. MSP could provide a solution to this aspect and not only improve coordination but also deliver a more rational use of the marine environment. However, ABNJ have very different characteristics than areas within national jurisdictions and although MSP has been used in national waters, its use in ABNJ is limited by the fragmented governance framework and by the lack of a coordinating mechanism, or leadership body to facilitate a cross-sectoral planning process. Considering this weakness, it is hoped that the new BBNJ process will discuss MSP together with MPAs and produce a framework for spatial planning in ABNJ, which would facilitate the use of MSP in international waters. Eventually, collaborative actions among states are often the only way to create a legal framework for protecting the marine environment, especially since the oceans do not have physical borders and pollution as well as human pressures do not necessarily stay within designed borders. The MSP approach could facilitate maritime governance and establish new ways of managing the sea that not only takes into consideration the human activities but also considers the interconnections between the marine ecosystems.

Acknowledgements We are grateful to the project 'Economy of maritime space' funded by the Polish National Science Centre for contributing the Open Access fee for this chapter and facilitating our discussions and preparation of the book.

References

Ardron, J., Gjerde, K., Pullen, S., & Tilot, V. (2008). Marine Spatial Planning in the High Seas. *Marine Policy, 32*, 832–839.

Becker-Weinberg, V. (2017). Preliminary Thoughts on Marine Spatial Planning in Areas Beyond National Jurisdiction. *The International Journal of Marine and Coastal Law, 32*, 570–588.

BirdLife International. (2018). Data Zone. Retrieved from http://datazone.birdlife.org/site/mapsearch.

Campbell, D., Hasegawa, K., Lee Long, W., Mongensen, C., Leone, G., Rodriguez-Lucas, L., & Wright, A. (2017). *Regional Seas Programmes Covering Areas Beyond National Jurisdictions*. Nairobi, Kenya: UN Environment Regional Seas Programme.

Carter, L. (2009). Submarine Cables and Other Maritime Activities. In L. Carter, D. Burnett, S. Drew, G. Marle, L. Hagadorn, D. Bartlett-McNeil, & N. Irvine (Eds.), *Submarine Cables and the Oceans: Connecting the World*. UNEP-WCMC Biodiversity Series No. 31 ICPC/UNEP/UNEP-WCMC.

CBD. (2018). Ecologically or Biologically Significant Marine Areas: Special Places in the World's Oceans. Retrieved from https://www.cbd.int/ebsa/.

Druel, E., & Gjerde, K. (2014). Sustaining Marine Life Beyond Boundaries: Options for an Implementing Agreement for Marine Biodiversity Beyond National Jurisdiction Under the United Nations Convention on the Law of the Sea. *Marine Policy, 49*, 90–97.

Earth Negotiations Bulletin. (2018). Summary of the Twenty-Fourth Annual Session of the International Seabed Authority, Vol. 25, No. 157. Retrieved from http://enb.iisd.org/download/pdf/enb25157e.pdf.

EC. (2009). *Legal Aspects of Maritime Spatial Planning: Summary Report*. Luxemburg: Office for Official Publications of the European Communities. Retrieved from https://ec.europa.eu/maritimeaffairs/sites/maritimeaffairs/files/docs/body/legal_aspects_msp_summary_en.pdf.

Ehler, C., & Douvere, F. (2009). *Marine Spatial Planning: A Step-by Step Approach Towards Ecosystem-based Management*. Manual and Guides No 153 ICAM Dossier No 6. Paris: Intergovernmental Oceanographic Commission UNESCO IOC, 99 pp.

FAO. (2007). Report and Documentation of the Expert Consultation on Deep-Sea Fisheries in the High Seas. Bangkok, Thailand, November 21–23, 2006. FAO Fisheries Report no. 829, Rome, Italy (Advance Copy).

FAO. (2015). Report of the FAO Workshop on Encounter Protocols and Impact Assessments for Deep-Sea Fisheries in Areas Beyond National Jurisdiction. FIAF/F1178. Retrieved from http://www.fao.org/3/a-i6452e.pdf.

Fisher, C. R., Takai, K., & Le Bris, N. (2007). Hydrothermal Vent Ecosystem. *Oceanography, 20*(1), 14–23.

Fletcher, R., Scrimgeour, R., Rogalla von Bieberstein, K., Barritt, E., Gjerde, K., Hazin, C., Lascalles, B., Tittensor, D., Vinuales, J. F., & Fletcher, S. (2017). *Biodiversity Beyond National Jurisdiction: Legal Options for a New International Agreement*. Cambridge: UNEP-WCMC.

Friedman, A. (2017). Submarine Telecommunication Cables and a Biodiversity Agreement in ABNJ: Finding New Routes for Cooperation. *The International Journal of Marine and Coastal Law, 32*, 1–35.

Gjerde, K., et al. (2016). Protecting Earth's Last Conservation Frontier: Scientific, Management and Legal Priorities for MPAs Beyond National Boundaries. *Aquatic Conservation: Marine and Freshwater Ecosystems, 26*(Suppl. 2), 45–60.

Halpern, B. S., Walbridge, S., Selkoe, K. A., & Kappel, C. V. (2008). Evaluating and Ranking the Vulnerability of Global Marine Ecosystems to Anthropogenic Threats. *Conservation Biology, 21*, 1301–1315.

Hassan, D., & Soininen, N. (2015). United Nations Convention on the Law of the Sea as a Framework for Marine Spatial Planning. In D. Hassan, T. Kuokkanen, & N. Soininen (Eds.), *Transboundary Marine Spatial Planning and International Law*. Devon: Swales & Willis Ltd.

Johnson, D. (2017). Submarine Cable Considerations for Area-Based Planning in ABNJ with Reference to Two On-Going International Seabed Authority Processes. In press.

Kimball, L. A. (2005). The International Legal Regime of the High Seas and the Seabed Beyond the Limits of National Jurisdiction and Options for Cooperation for the Establishment of Marine Protected Areas (MPAs) in Marine Areas Beyond the Limits of National Jurisdiction. Secretariat of the Convention on Biological Diversity, Montreal, Technical Series no. pp. 19–64.

Maes, F., & Cliquet, A. (2015). Marine Spatial Planning: Global and Regional Conventions and Organizations. In D. Hassan, T. Kuokkanen, & N. Soininen

(Eds.), *Transboundary Marine Spatial Planning and International Law*. Devon: Swales & Willis Ltd.

Merrie, A., Dunn, D. C., Metian, M., Boustany, A. M., Takei, Y., Elferink, A. O., Ota, Y., Christensen, V., Halpin, P. N., & Österblom, H. (2014). An Ocean of Surprises: Trends in Human Use, Unexpected Dynamics and Governance Challenges in Areas Beyond National Jurisdiction. *Global Environmental Change, 27*, 19–31. https://doi.org/10.1016/j.gloenvcha.2014.04.012.

Rayfuse, R. (2012). Precaution and the Protection of Marine Biodiversity in Areas Beyond National Jurisdiction. *The International Journal of Marine and Coastal Law, 27*, 773–781.

Ribeiro, M. (2013). What Is the Area and the International Seabed Authority? Institut Océanographique. Retrieved from http://www.institut-ocean.org/images/articles/documents/1367593542.pdf.

Richardson, L., Loomis, J., Kroeger, T., & Casey, F. (2015). The Role of Benefit Transfer in Ecosystem Service Valuation. *Ecological Economics, 115*, 51–58.

Ringbom, H., & Henriksen, T. (2017). *Governance Challenges, Gaps, and Management Opportunities in Areas Beyond National Jurisdiction*. Washington, DC: Global Environment Facility—Scientific and Technical Advisory Panel.

Soininen, N., & Hassan, D. (2015). Marine Spatial Planning as an Instrument of Sustainable Ocean Governance. In D. Hassan, T. Kuokkanen, & N. Soininen (Eds.), *Transboundary Marine Spatial Planning and International Law*. Devon: Swales & Willis Ltd.

The International Cable Protection Committee. (2016). Submarines Cables and BBNJ. Retrieved from https://perma.cc/NV8W-Q83W.

The Royal Institute of International Affairs. (2007). Recommended Best Practices for Regional Fisheries Management Organizations: Executive Summary. Retrieved from http://www.oecd.org/sd-roundtable/papersandpublications/39374762.pdf.

UNCLOS. (1982). United Nations Convention on the Law of the Sea. Retrieved from http://www.un.org/depts/los/convention_agreements/texts/unclos/unclos_e.pdf.

UNEP. (2017). Thirteenth Meeting of Focal Points for Specifically Protected Areas. Alexandria, Egypt. May 9–12. Retrieved from http://www.rac-spa.org/nfp13/documents/02_information_documents/wg_431_inf_8_msp_and_the_protection_of_bbnj.pdf.

UNEP-WCMC. (2017). *Governance of Areas Beyond National Jurisdiction for Biodiversity Conservation and Sustainable Use: Institutional Arrangements and Cross-Sectoral Cooperation in the Western Indian Ocean and the South East Pacific*. Cambridge: UN Environment World Conservation Monitoring Centre.

United Nations. (2010). Resumed Review Conference on the Agreement Relating to the Conservation and Management of Straddling Fish Stocks and Highly Migratory Fish Stock. United Nation, New Work, May 24–28. Retrieved from http://www.un.org/depts/los/convention_agreements/reviewconf/FishStocks_EN_B.pdf.

United Nations. (2011). Workshop to Discuss Implementation of Paragraphs 80 and 83 to 87 of Resolution 61/105 and Paragraphs 117 and 119 to 127 of Resolution 64/72 on f6/files/A_66_566-EN.pdf.

United Nations. (2012). Resolution Adopted by the General Assembly on 27 July 2012. General Assembly. September 11. Retrieved from http://www.un.org/ga/search/view_doc.asp?symbol=A/RES/66/288&Lang=.E.

United Nations. (2017). Report of the Preparatory Committee Established by General Assembly Resolution 69/292: Development of an International Legally Binding Instrument Under the United Nations Convention of the Law of the Sea on the Conservation and Sustainable Use of Marine Biological Diversity of Areas Beyond National Jurisdiction. A/AC.287/2017/PC.4/2. Retrieved from http://www.un.org/ga/search/view_doc.asp?symbol=A/AC.287/2017/PC.4/2.

Wright, G., Ardron, J., Gjerde, K., Currie, D., & Rochette, J. (2015). Advancing Marine BiodiversityProtection Through Regional Fisheries Management: A Review of Bottom Fisheries Closures in Areas Beyond National Jurisdiction. *Marine Policy, 61*, 134–148.

18

Evaluation of Marine Spatial Planning: Valuing the Process, Knowing the Impacts

Riku Varjopuro

1 Introduction

Marine/Maritime spatial planning (MSP) is an increasingly common approach to manage the use and protection of the resources, the ecosystems and the space of seas (Douvere 2008; Jay et al. 2013). An often-cited definition states that 'Marine [or maritime] Spatial Planning is a public process of analysing and allocating the spatial and temporal distribution of human activities in marine areas to achieve ecological, economic, and social objectives that usually have been specified through a political process' (UNESCO-IOC 2010). A recent European Union directive (EC 2014) states that MSP is 'to promote the sustainable growth of maritime economies, the sustainable development of marine areas and the sustainable use of marine resources' (Directive 2014/89/EU).

The earlier mentioned objectives for MSP are very ambitious. If achieved, the coastal states will gain thriving maritime economies, fewer conflicts at sea and improved environmental status of the marine ecosystems. But knowing whether the plan will help society to achieve all or any of the objectives requires specific attention. This is the evaluative question that is the focus of this chapter.

This chapter applies commonly used approaches in the evaluation of policies and spatial planning on land to the MSP. The emphasis of the chapter is on trying to know the effectiveness of MSP, which brings with it critical

R. Varjopuro (✉)
Marine Research Centre, Finnish Environment Institute, Helsinki, Finland
e-mail: riku.varjopuro@ymparisto.fi

J. Zaucha, K. Gee (eds.), *Maritime Spatial Planning*,
https://doi.org/10.1007/978-3-319-98696-8_18

perspectives on MSP's ability to deliver its objectives. The effectiveness of MSP is largely dependent also on the processes of preparing and implementing MSP. Therefore, characteristics of the process need to be addressed as well.

A distinction from strategic environmental assessment (SEA) helps to explain the consequences of focusing on evaluating effectiveness. In an SEA, which assesses likely environmental impacts of alternative versions of a plan, one should assume that the described alternative is realised. Then likely environmental impacts of each alternative are assessed and mitigation measures are suggested. In an evaluation of effectiveness that focuses on the chosen or proposed plan, one measures or assesses the extent that set targets will be met, any unintended impacts generated and possible obstacles to realising the plan and achieving the desired impacts. Whereas SEA aims to help design a plan that has the least negative environmental impacts, an evaluation of effectiveness aims at improving the implementation of the plan or suggesting improvements for the next versions of the plan. The following figure illustrates the difference (Fig. 18.1).

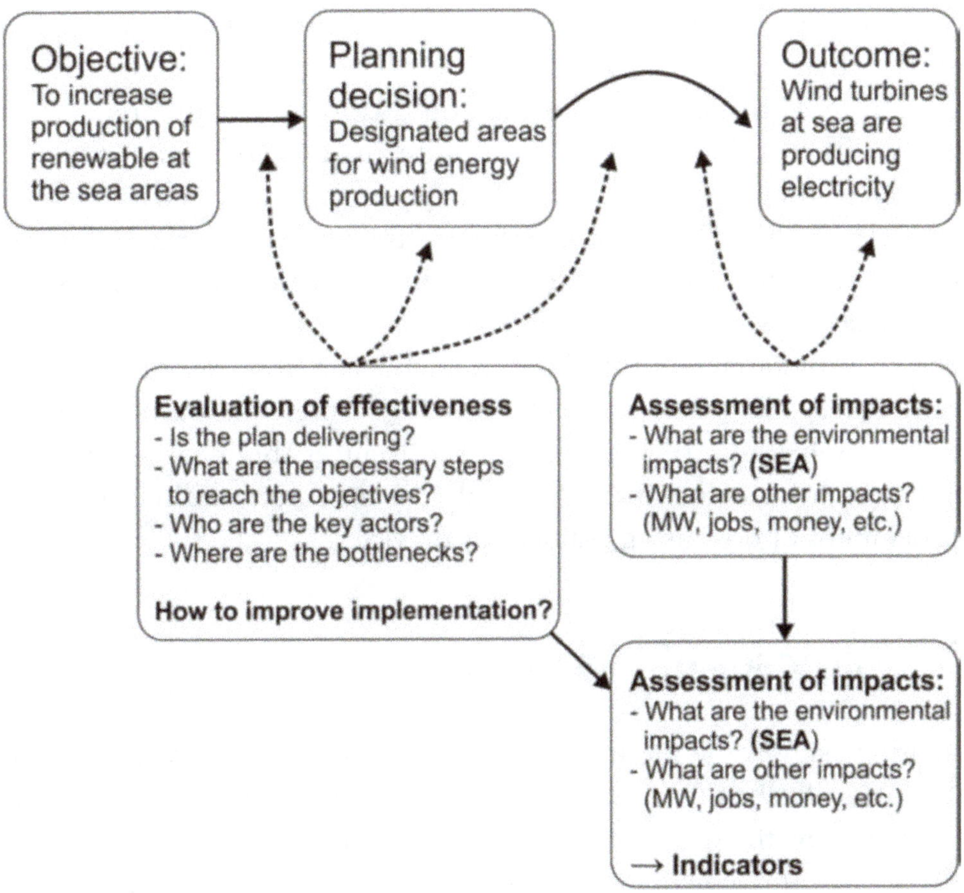

Fig. 18.1 Key questions in an evaluation of effectiveness and SEA. Linkages to indicators

This chapter is based on the work conducted in the Baltic SCOPE project (www.balticscope.eu) in 2015–2017. Within the Baltic SCOPE project the author developed an evaluation and monitoring framework for cross-border collaboration in MSP (Varjopuro 2017). The Baltic SCOPE project brought together MSP authorities from six Baltic Sea Region countries to enhance cross-border integration and coordination of MSP activities in the Baltic Sea. The countries were Sweden, Denmark, Germany, Poland, Latvia and Estonia. Other project partners were the intergovernmental organisations Helsinki Commission (HELCOM) and Vision and Strategies Around the Baltic Sea (VASAB) and the two research organisations NordRegio and the Finnish Environment Institute. For the purpose of preparing the evaluation and monitoring framework, the author and his colleagues followed the Baltic SCOPE process and conducted individual and group interviews to identify factors that influence the success of transboundary collaboration. The collected material also shed light on national MSP processes and practices. One admittedly obvious observation was that MSP is practised in very different ways and with very different objectives. A key conclusion for the prepared evaluation and monitoring framework was that presenting a standard evaluation protocol would not be useful. Instead, it has to be flexible and adaptable for different contexts and cases. The work on developing monitoring and evaluation approaches is now continuing in the Pan Baltic Scope project (http://www.panbalticscope.eu/). In Pan Baltic Scope the focus is on evaluation of national MSP. This chapter describes the methodological findings gained in these two projects.

The chapter starts by presenting approaches and concepts of evaluation of policies and spatial plans. This is followed by a presentation of the progress gained in evaluation of MSP. Section 5 introduces the theory-based evaluation approach for evaluating MSP, while the final Sect. 6 discusses practical considerations of organising evaluations of MSP.

2 Purposes of Evaluation

Evaluation assesses the merit and value of public policies or, as in this case, spatial plans. Evaluation often asks if the set targets are met, but evaluations can also address the processes of policy or plan formation as well as the processes of their implementation (Vedung 1997, 2006). Evert Vedung (2010, 263) justifies the usefulness of evaluating public policies:

> If you carefully examine and assess the results of what you have done and the paths toward them, you will be better able to orient forward. Good intentions, increased funding and exciting visions are not enough; it is real results that count. The public sector must deliver. It must produce value for money.

The main objective of evaluation is thus to improve policies and plans and the processes of producing them. Evaluation should not be seen simply as a judgement of whether or not public authorities have been successful in designing or implementing the policies and plans. Furthermore, it is not enough to determine whether the objectives were or will be met or how satisfied different stakeholder groups are with the process or its outcomes. This is because to improve the policies and plan, it is important to understand why certain elements of the policy or plan work or do not work. In relation to stakeholders, it would also be important to elucidate which of the affected parties have benefitted the most and the least (EC 2013a).

Evaluations that shed light on outcomes as well as on processes of making and implementing policies and plans increase our understanding of various aspects of policies and plans. Such broad evaluations 'provide opportunities to learn about the questions to ask, the goals to set and how to frame the issues as well as the instrumental learning about how to design or implement the policy' (Mickwitz 2006, 18).

A primary purpose of the evaluation process should be to foster learning; but who then learns from the evaluations? The public authorities that commission the evaluation are obviously the ones that learn. This applies to individuals working in such organisations, but learning by an organisation should also be fostered. Evaluation can support both single-loop and double-loop learning (Mickwitz 2006), which are essential for adaptive management cycles (Armitage et al. 2008; Cundill et al. 2012; Pahl-Wostl et al. 2007). Through single-loop learning, one learns how to improve the effectiveness of spatial planning solutions, while double-loop learning helps one to develop the planning system as a whole.

The commissioning public body should not be considered the only one to learn. As pointed out in the quotation from Vedung (2010), public processes should deliver benefits to society at large. It is nowadays the norm that policies and plans are prepared in participatory processes, which considerably enlarge the group of those who could learn from the results of evaluations. It is also suggested that an evaluation itself should be participatory (Carneiro 2013; Hansen and Vedung 2010; Mickwitz 2006). Then persons who are engaged in evaluations can learn from participating in the evaluation process, and when evaluations are conducted as part of participatory planning or policy-making processes, the evaluation process and results reach the widest possible audience.

In addition to fostering learning, evaluation has other important functions. Checking for accountability of public policies is one. Accountability concerns the liability of those who are in charge of and conduct public tasks and spend public resources. Public resources should be used wisely, and it should be ensured that the goals regarding the quality of the process and results are achieved (Mickwitz 2006).

Evaluations can increase trust and the legitimacy of public authorities and processes, as they improve public knowledge and understanding of policies and plans. As evaluations may reveal flaws in processes and unachieved goals, it is important that evaluations contribute to the identification of corrective measures.

3 Alternative Evaluation Approaches

In order for evaluations to produce understandable and justifiable results, they should be done in systematic and rigorous ways (EC 2013a; Mickwitz 2006). This is a generic requirement, as there are many *systematic and rigorous* methods available to be used. A key issue is to choose the right methods for the purpose and scope of the evaluation. Obviously, resources dedicated to evaluation also determine the choice of methods to some extent.

There are several approaches and focuses for the evaluation of spatial planning. According to Terryn et al. (2016), the evaluation of spatial planning has often been based on a linear (or at least cyclical) understanding of planning processes. Consequently the evaluation methods have been structured in simple logical steps to be followed. Terryn et al. (2016, 1085) state that 'most spatial developments do not evolve in a linear, circular or causal way, but rather present themselves more and more in a-linear, pragmatic and adaptive ways'. As this adds a certain level of uncertainty in determining the target of evaluation, they suggest that the evaluation should ideally be conducted as an integrated part of the planning process, as this would allow adjusting the methods to better fit the context of evaluation (Terryn et al. 2016).

The nonlinear and partly unpredictable character of spatial developments is an important point to be taken into account in the evaluation of spatial planning. However, there is also a need to make a distinction between the spatial developments and spatial planning. Spatial developments are outcomes of the combined effects of various processes, while spatial planning is a process that ideally translates the collective aspirations of society into decisions on how to use the area in question. In other words, spatial planning aims to influence, or at least contribute to, the spatial developments.

Even though spatial developments can be nonlinear and unpredictable, the levels of the complexity of the planning context vary—and some planning contexts are not at all complex. The level of complexity of the planning context should be taken into account in choosing the planning methods. Terryn et al. (2016) have studied how different planning approaches would fit different planning contexts. The following matrix presents how different evaluation approaches can fit different planning contexts (Fig. 18.2).

Circular evaluation (lower left in figure) is suitable for simple planning issues and situations when the main focus will be on how the intentions of planning meet the implementation. In such cases it is also well known who the key actors are, what the stakes are and what roles the institutional and non-institutional actors would have in the planning. In other words, the playing field is stable and known. An *adaptive evaluation* (upper left) approach is applicable when the planning issue itself is undefined and possibly changing, but the institutional and societal setting is relatively stable. Adaptive evaluations probe whether the final results meet the needs of changing contexts and various interests. *Participative evaluation* (lower right) is apt when the planning issue is simple, but there are uncertainties regarding the actors, stakes and possible roles of the different types of actors. In such a setting, the evaluation's role as negotiation or dialogue is more important than objective attempts to measure the effectiveness of implementation of the plan. Participatory evaluations review the ability of interest groups to cooperate in a situation of changing playing fields. Finally, the *co-evolutionary evaluation*

		Evaluation approaches	
Planning issue	Highly open, undefined, innovative, new	**Adaptive**	**Co-evolutionary**
	Simple, regular, defined, well-known	**Circular**	**Participative**
		Known, defined, fixed number of agents	Highly dynamic, undefined, volatile
		Playing field	

Fig. 18.2 Evaluation approaches in relation to the degree and reasons of complexity of the planning contexts (Terryn et al. 2016, 1087)

(upper right) approach is needed when both the planning issue and the playing field are not well known or they are in the process of transformation during the planning process or being transformed by the planning. Co-evolutionary evaluation asks if the planning itself is becoming more resilient and adaptive to be able to operate when both planning issues and the playing field are volatile. In such cases it would also be imperative that evaluation is continuous over the process to encourage learning-by-doing and co-evolution (Terryn et al. 2016, 1087–1088).

The importance of evaluations to be close to the evaluated process—or to co-evolve with it—has often been emphasised (Mickwitz 2006; Rae and Wong 2012; Terryn et al. 2016; Vedung 2010). Evaluations need to be sensitive to how the evaluated process unfolds. If need be, the evaluation approach itself should adapt—meaning, for instance, that new evaluation criteria can be learnt during the evaluation (Gomart and Hajer 2003).

Evaluations can have different targets and different timing in relation to the stage of decision-making or planning processes. The selected methodology should respect the nature and complexity of the object of the evaluation. These distinctions are discussed in the following sections.

3.1 Evaluation Can Target Impacts and Processes

Knowing the impacts of policies and plans is essential for evaluating the effectiveness of public policies. The evaluation will try to study to what extent the objectives of the policy or plan have been reached.

Spatial plans are typically aspirational: there are certain goals that are found to be valuable to reach. As pointed out earlier, spatial planning often operates in complex contexts where several factors have an influence at the same time. This happens especially in cases of strategic or general-level planning, which MSP often is. Only some of the factors that generate impacts follow directly from the spatial plan itself. Carneiro (2013) has observed that the current literature on MSP does not pay enough attention to the issue of multi-causality and has not sufficiently discussed the difficulty of isolating the contribution that MSP has or can have on observed changes in the use of sea areas. There is also another important limitation for spatial planning in reaching desired objectives: depending on the plan's legal status, it may have only limited power to directly guide decision-making in other sectors. Then the effectiveness of a spatial plan depends on the other sectors' willingness to follow that spatial plan (Faludi 2000).

How and to what extent are the observed changes attributable to the spatial plan? The notion of causality and attribution then becomes central in the evaluation of effectiveness (EC 2013a). In the evaluation of spatial planning that takes place in complex contexts such as MSP, attribution may become a considerable challenge (Carneiro 2013). When it may be very difficult, if not altogether impossible, to justify causality between the intervention and desired outcomes, that is the *attribution*, one can justify the argument of *contribution* with plausible evidence or a narrative that explains why the evaluated spatial plan can be seen as one of the causes of the observed change (Carneiro 2013; EC 2013a). Even if this cannot reduce uncertainty concerning the effects of plans, it can produce useful findings for improving performance of planning (EC 2013a).

The question of effectiveness is essential for the sake of accountability, but focusing only on the intended goals is often too limited. Identification of unintended consequences is widely recognised as an essential part of the evaluation of spatial planning, especially because spatial planning typically addresses and affects broad areas and a broad spectrum of human activities (Carneiro 2013; Faludi 2000; Mickwitz 2006; Terryn et al. 2016). Therefore, the narratives of MSP's contributions should also pay attention to possible unintended consequences and side effects.

In addition to evaluating the effectiveness of policies and planning, it is also useful to focus on the processes. Evaluation of the process of implementation of a policy or plan not only helps to answer whether the results are met but also helps to understand why it is so. It can also help in explaining the observed results (Carneiro 2013). Furthermore, evaluation of the process of making a policy or a plan gives valuable information for improving the processes in the future, that is, double-loop learning (Mickwitz 2006). Some aspects of policy-making and spatial planning processes as well as processes of implementation have an important intrinsic value, which justifies paying attention to processes in the evaluation. The imperative of public participation in policy-making and spatial planning processes is an example of such intrinsic values. The requirements of transparency and accountability underline the need for focusing on processes (Carneiro 2013; Hansen and Vedung 2010; Mickwitz 2006).

Carneiro (2013) identified several possible foci of evaluations in relation to the planning cycle (Fig. 18.3).

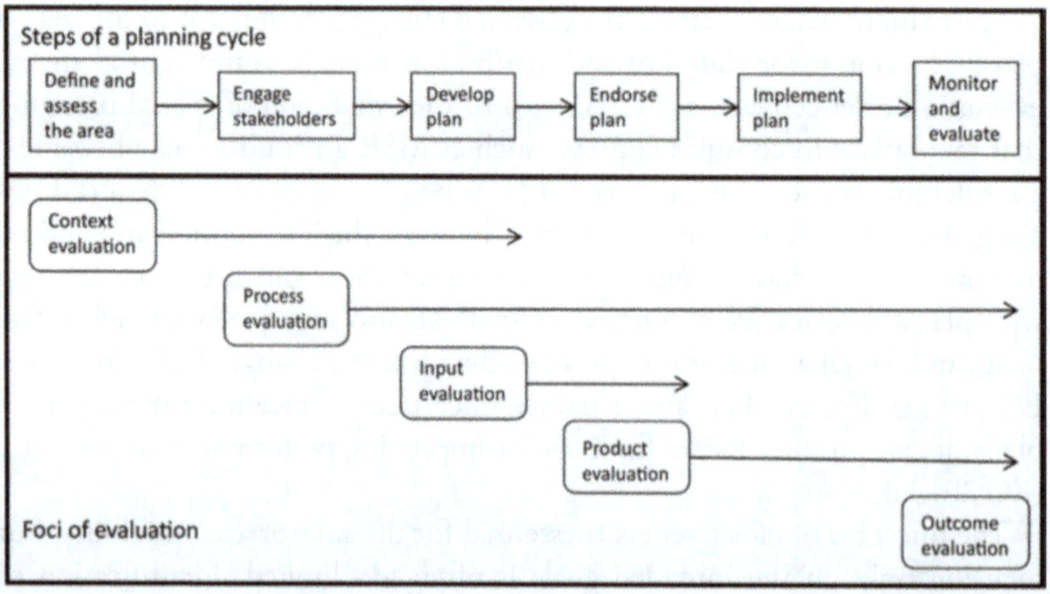

Fig. 18.3 Different foci of evaluation in relation to steps of the spatial planning process (Carneiro 2013, 215)

3.2 Evaluation: Before, During or After the Intervention

Evaluations can be conducted while policies and plans are prepared. Such evaluations give valuable information when designing effective policies and plans. These so-called *ex ante* evaluations anticipate possible future impacts of planned policies. An *ex ante* evaluation should preferably produce results early enough in relation to the policy-making or planning process in order to have a valuable and timely contribution (EC 2013b). SEAs are also done while the policies and plans are prepared.

It is common that policies and plans are evaluated afterwards—*ex post*—or in the late stages of implementation to check whether and to what extent the set results are achieved. *Ex post* evaluations can also study unintended impacts of policies or plans.

Interim evaluations or mid-term reviews generate information that help to assess whether measures are being implemented as planned and whether it seems likely that they will produce the impacts that were anticipated. A more thorough *interim* evaluation can also help to assess whether the assumptions about a policy or plan's effects were correct or not.

4 Evaluation of MSP

The importance of evaluating MSP and different approaches to an evaluation of MSP has already been discussed in early publications on MSP (Carneiro 2013; Day 2008; Douvere and Ehler 2011), but there have been fewer presentations on actual evaluation methods.

EU-funded projects such as TPEA, MASPNOSE, BaltSeaPlan, PlanBothnia and PartiSeaPate have all addressed evaluation of MSP. For instance, TPEA and MASPNOSE produced evaluation frameworks and also tested them to a certain extent during the projects, while PlanBothnia developed approaches to monitor the implementation of MSP. Evaluation approaches of MSP have also been developed in several academic papers (Carneiro 2013; Day 2008; Douvere and Ehler 2011; Fletcher et al. 2013; Kelly et al. 2014; Soma et al. 2014; Stelzenmüller et al. 2013; Vos et al. 2012).

There are also some publications that reviewed the evaluation of MSP (Carneiro 2013; TPEA 2014). The TPEA evaluation report identifies the diversity of evaluation approaches (e.g. the focus can be on ecological or planning aspects; the emphasis is on process or outcomes). And the report concludes that because of the diverse contexts in which MSP is practised, there cannot be a standardised protocol for evaluating MSP. Each evaluation has to be tailored to the context (TPEA 2014).

Even if questioning the usefulness of standardised evaluation approaches, the TPEA evaluation report (TPEA 2014) presents a few general principles:

- Evaluation of MSP should cover all stages of the MSP process, from preparation of planning to implementation.
- Evaluation should be based on a clear understanding of the focus and scope of the evaluation, which helps in defining clear objectives for the evaluation.
- Evaluations should cover context, process, outputs and outcomes.
- Evaluation criteria should be matched by suitable indicators.
- Stakeholder involvement is important for a successful evaluation.

Table 18.1 presents evaluation frameworks that were suggested by the TPEA evaluation report (2014) and Carneiro (2013). The frameworks have many obvious similarities, but they also introduce some unique features. The TPEA approach is more detailed regarding legal, administrative and institutional aspects, which are especially critical for the success of transboundary collaboration in planning—the focus of the TPEA evaluation framework.

Table 18.1 Topics and criteria of two MSP evaluation frameworks (Carneiro 2013; TPEA 2014)

TPEA transboundary evaluation framework (TPEA 2014)	Carneiro's evaluation framework (Carneiro 2013)
Process evaluation: Preparation phase • legal and administrative framework • institutional capacity and cooperation • transboundary MSP area • formulation of strategic objectives **Process evaluation: Diagnosis phase** • area characteristics • uses & activities and cross-border relevance of coastal and maritime issues • governance framework • area of common interest **Process evaluation: Planning phase** • specific objectives • planning alternatives (options and scenarios) • planning documents **Data and information** • data availability and quality **Stakeholder engagement** **Communication** **Implementation** • roles, responsibilities and decision-making • resources • implementation **Outcomes and impact evaluation** • achievement of objectives • wider benefits	**Evaluation of the organisational performance** • planning service quality • organisational quality **Evaluation of the plan-making process** • stakeholder participation • validity of data and analyses • consideration of alternatives • prospective impact assessment • adequacy of resources (for plan-making) **Evaluation of plan contents** • internal coherence • relevance of plan for the region or country • conformance with planning system • external coherence • guidance for implementation • approach, data and methodology • quality of communication • plan format **Evaluation of plan implementation** • prescribed steps and outputs • adequacy of resources (for implementation) • utilisation **Evaluation of plan outcomes and impacts**

The TPEA framework is designed for the evaluation of transboundary processes, while Carneiro's framework is focused more on national MSP

Carneiro's (2013) evaluation framework emphasises the content of the plan itself without neglecting the importance of process evaluation. This framework is mainly meant for the evaluation of national MSP.

The United Nations Educational, Scientific and Cultural Organization (UNESCO)'s International Oceanographic Commission (IOC-UNESCO)

has been promoting MSP and developing methodologies of MSP from very early on (see Ehler and Douvere 2009). IOC-UNESCO produced a guide to evaluate marine spatial plans in 2014 (Ehler 2014). The evaluation guide focuses mainly on outcome evaluation, but it also raises important questions regarding evaluation of the processes. The IOC-UNESCO guide covers the whole sequence of evaluation, from planning the evaluation via the actual evaluation to communicating the evaluation results. These are all also addressed in the TPEA evaluation report (TPEA 2014), but the IOC-UNESCO guide goes one important step further by discussing the use of evaluation results and taking corrective measures.

5 A Theory-Based Evaluation Framework for MSP

Introducing a theory-based evaluation approach to MSP is this chapter's contribution to our common understanding of possibilities and methods for evaluating MSP. There are two main reasons for introducing this approach.

One reason is that as evaluation is a *careful assessment* (EC 2013a; Mickwitz 2006) that requires systematic and rigorous approaches to evaluation in order to produce understandable and justifiable results, systematic and rigorous evaluations are time-consuming. Therefore, those who commission evaluations need to consider the available methods of evaluation in relation to the expected use of the evaluation findings and the available resources (EC 2013b). Thus there is a need to adjust the evaluation approach to the context and knowledge requirements.

The other reason relates to the difficulty in isolating and identifying the actual effects and impacts of MSP from all other factors that influence maritime activities and marine ecosystems. Is MSP contributing to the changes that we can observe? (See Sect. 3.1).

A theory-based approach to evaluation is flexible in the sense that it must always be adjusted to the context and purposes of evaluation. Furthermore, constructing theories of change, which is the key for theory-based evaluation, is also a way to increase knowledge of the possible contributions of MSP.

Theory-based evaluations ask why an intervention—such as a spatial plan—produces intended and unintended effects, for whom and in what contexts and what mechanisms are triggered by the intervention. The goal is to know why an intervention works and whether it would work differently in different localities (Astbury and Leeuw 2010; Coryn et al. 2011; EC 2013a).

5.1 Theories of Change

A theory-based evaluation of spatial planning is based on describing plausible mechanisms through which the plan or the planning process can produce its impacts. The actual evaluation then collects evidence to test whether the implementation of the plan unfolded as anticipated (and why), whether the anticipated results were achieved and whether the implementation of the plan produced any unintended impacts (Coryn et al. 2011; Hansen and Vedung 2010; Mayne 2012; Weiss 1997). A theory-based evaluation does not usually produce numerical results as much as it produces narratives. Its results provide important insights into how spatial planning can work, and later why it worked as it did (EC 2013a).

The key element of theory-based evaluation is the *theory of change*.[1] The term *theory-based* reflects the understanding that all decisions and plans are based explicitly or implicitly on an idea—*a theory*—about how that decision or plan will be implemented and how it will produce results, that is, *a theory of change*. Theories of change are typically described as somewhat simplified, often linear models (Fig. 18.4). Obviously, spatial plans and the generation of their outcomes are not always, or even usually, as linear as depicted in the following figure (see Astbury and Leeuw 2010; Coryn et al. 2011; Hansen and Vedung 2010; Mayne 2012).

An intervention consists of inputs, activities and outputs. Inputs are the required resources (e.g. human, financial, institutional). Activities are the actions taken to define and reach the objectives (e.g. data collection and spatial analyses, production of the plans and planning documents, workshops with stakeholders, consultation). Outputs are the immediate results of action (e.g. the planning decisions).

Impacts of the intervention can be grouped into initial, intermediate and long-term outcomes. Initial outcomes are changes in knowledge, skills and ability of key actors. Intermediate outcomes are typically behavioural changes

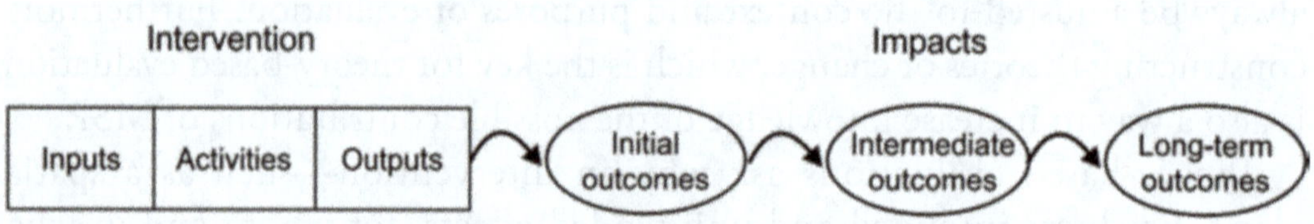

Fig. 18.4 A scheme of a theory of change (Coryn et al. 2011, 201)

[1] Also known as *program theory* or *intervention theory*.

Table 18.2 An example of a plausible theory of change in MSP

Output of transboundary collaboration	Immediate outcome	Intermediate outcome	Long-term outcome
Area designated for wind energy production in MSP	Knowledge of operators and decision-makers increases Interest to build a wind park	Permit applications issued Permit accepted	Turbines are producing renewable electricity

(e.g. a decision to invest in sea areas designated in MSP). And long-term outcomes (sometimes simply called impacts) are a full or partial solution to the perceived problem that the plan was set to address (an outcome corresponding to one of the objectives for MSP) (Coryn et al. 2011, 202). Table 18.2 gives an example of this sequencing.

5.2 Using Theories of Change in Evaluation

Construction of plausible theories of change is a key element of theory-based evaluation, but the actual evaluation focuses on the following if the anticipated steps will or can take place. It is essential to study why the plan produces or could produce the intended and unintended effects; for whom and in which contexts; what mechanisms are triggered by the plan or by the process of producing the plan; how various steps in the theory of change relate to each other and what factors influence the relations (Mayne 2012). These considerations are depicted in Fig. 18.5.

In elaborating the theories of change as step-wise developments, one risk is 'to focus too much on input-output relationships, on linear chains of causality and on building tightly knit models of arrows and boxes' (Weber 2006, 120), which is an important reminder to acknowledge the complexity and situatedness of the planning process. Astbury and Leeuw (2010, 375) suggest that 'a more explicit focus on underlying generative mechanisms might help to counter […] toward oversimplified versions of program theory in the form of linear logical models'.

Theory-based evaluations often rely on participatory approaches to deal with different understandings and preferences. Such processes can help reach jointly agreed theories of change. But Hansen and Vedung (2010) have observed that due to substantive and multilevel complexities and political conflicts, this is not always possible.

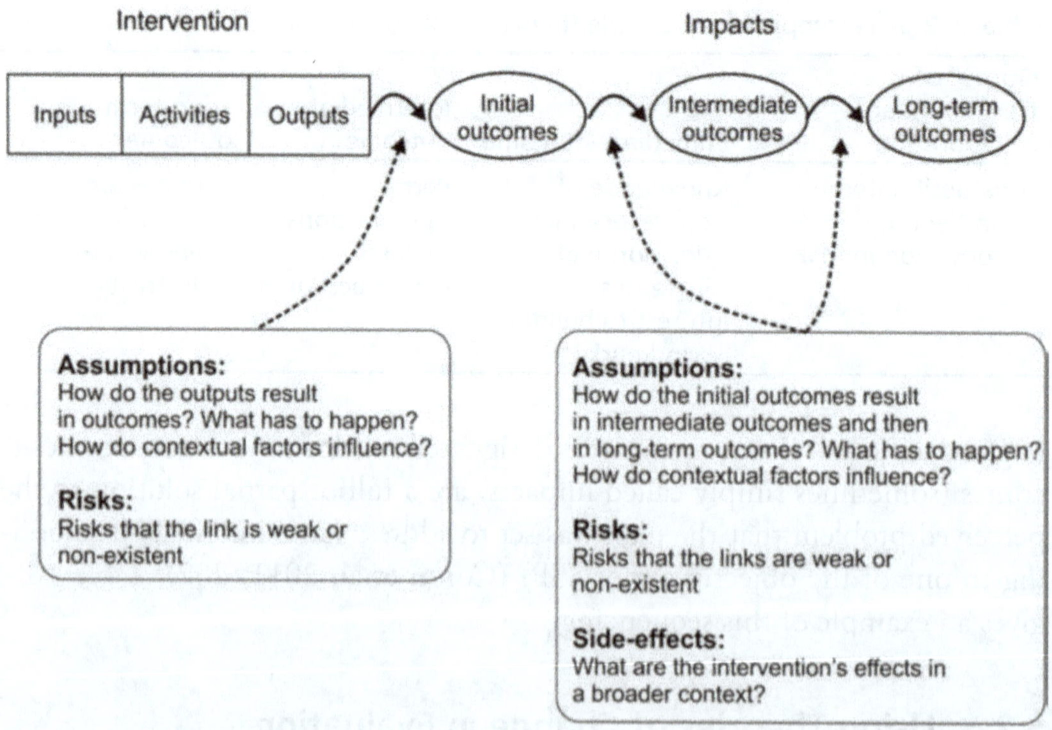

Fig. 18.5 Theory of change, considering factors that influence a logical sequence of events. Modified from Coryn et al. (2011) and Mayne (2012)

It is not always advisable to reduce different perceptions on the intervention to only one theory of change. This is especially important in interventions that 'involve several groups of actors in very different working situations and with very different expectation to the intervention' (Hansen and Vedung 2010, 296). Implementation of MSP typically takes place in situations where success of the plan's implementation is dependent on how the plan changes the behaviour of actors in various marine sectors (Faludi 2000). Approaches that elaborate on alternative, parallel theories of change are needed in complex and conflict-prone interventions that operate nationwide, multisite and multilevel (Hansen and Vedung 2010).

5.3 Testing of Theories of Change

Producing alternative theories of change in collaboration with key actors helps to identify possible impacts and challenges of implementation of MSP in a systematic and transparent way. The actual evaluation tests to what extent the actual cause of events followed or could follow the theories of change and whether or not the goals were reached. Importantly, the evaluation should try and explain why. Theory-based evaluation can be implemented in the following five steps (modified from Coryn et al. 2011, 205; EC 2013a, 56–57):

1. Formulate plausible theories of change to reflect different actors' understandings.
2. Formulate and prioritise evaluation questions around a theory of change:
    a. How can you know that the different steps of the theories of change (will) actually take place? What evidence do you need?
    b. Choose relevant evaluation criteria and indicators.
3. Collect evidence relevant for answering the evaluation questions.
4. Analyse the evidence to test the theories of change:
    a. Which links in the theory of change are strong? Is this conclusion based on strong logic or empirical evidence supporting the assumptions? Is this conclusion widely accepted by relevant actors? And similarly, which links are weak?
    b. Does the observed pattern of outcomes and factors leading to them validate the theory of change? Do or did things unfold as anticipated?
    c. Is it likely that any of the external significant factors had a noteworthy influence on the observed results?
    d. What are the main weaknesses in the descriptions of the theories of change? Would additional data or information be useful?
5. Draw conclusions:
    a. Identify breakdowns (links that did not exist) and respective corrective actions.
    b. Identify side effects and unintended impacts (also identify who was affected).
    c. Determine the effectiveness of implementation. (Were the objectives reached and to what extent?).
    d. Describe and explain cause-effect associations between elements of the theories of change (why things unfolded as they did). Describe also why external factors influence (or influenced) the outcomes.

6 Organisation of Evaluation

Section 5 introduced the theory-based evaluation method. This final section focuses on pragmatic questions on how to organise and manage the evaluation process. There are certain essential steps and challenges that need to be solved. One of the most important issues is to define the scope of the evaluation. One must also decide who is in charge, who performs the evaluation and how many resources will be allocated to the evaluation. Finally, there is a need to decide on the roles stakeholders are given in the evaluation (EC 2013b).

6.1 Targeting the Evaluation: Scope and Purposes

Defining the scope of the evaluation is the most important first decision to be made. The evaluation should be given certain boundaries in terms of institutional, temporal, sectoral and geographical dimensions (EC 2013b). In defining the scope, one asks what exactly will be evaluated and when? It is also essential to consider the expected and possible uses of the evaluation results, which was also raised as an important factor in the IOC-UNESCO evaluation guidance document (Ehler 2014). It is important to understand what future decisions are likely to be informed by the evaluation results (EC 2013b).

For the success of the evaluation, the scope should be defined in a way that gives a clear focus and task for the evaluation. It may turn out, however, that a clear scope for the evaluation cannot be given. This may be the situation especially when a country is producing its first MSP. In such cases the scope of the evaluation needs to be somewhat flexible in the beginning but should be defined more precisely while the process unfolds. This would suggest that the evaluation should be conducted in close cooperation with the planners. However, a flexible scope for evaluation has to be considered against the availability of resources and time for conducting the evaluation. Evaluation questions and the scope should be realistic in relation to the resources, which often require clarity.

6.2 Financial and Institutional Resources for Evaluation

Evaluations should be conducted in systematic and rigorous ways to produce justifiable and relevant results, but the possibilities to live up to this standard can be limited by the availability of resources. However, the most important question for the planning and commissioning of evaluations is to identify the expected use(s) of evaluation findings and fit the resources accordingly.

Resources for evaluation include the financial resources, obviously, but there are other resources available as well. Actors to be involved in the evaluation can give valuable input and information for the evaluation. Also the end users of the evaluation findings are a key resource.

Who is in charge or who is the client for evaluation is critical for the usefulness and actual use of the evaluation findings. It is recommended that the person (or a function in administration) who commissions the evaluation should be in a high enough position to initiate corrective actions to the policies or plans that are being evaluated (EC 2013b). Regarding MSP, this could be the minister in charge of MSP, a representative of the ministry or a high-ranking officer in the spatial planning authority.

Evaluations are often commissioned by consultants who operate under a contract with a public organisation; they monitor or evaluate the policy or planning process being evaluated. If an external evaluator conducts the evaluation, it is important that the evaluation is conducted in close and frequent contact with the client. This ensures that the evaluation results are immediately available, and this will also give an opportunity to adjust the evaluation if new needs emerge or in case the evaluated process is reorganised. Information will then flow in both directions between the evaluator and the client.

In some cases, the public bodies have their own evaluation units. Then it would be advisable that the evaluation is conducted as an in-house service (EC 2013b). It is also possible that the officers who are running or supervising the spatial planning process conduct the evaluation in-house. Such arrangements may create challenges of partiality. In such an arrangement, it is advisable that some of the officers have experience in evaluation methods. If it is decided that the evaluation is conducted in-house and only limited expertise in evaluation methods is available, it is advisable to hire a consultant to facilitate the evaluation process.

The purpose and timing of the evaluation determines to some extent whether the evaluations should be conducted internally or externally. The European Commission's (EC 2013b, 39) evaluation guidance advises that '[i]t may be preferable to rely more on internal resources for formative evaluation inputs or for ex ante exercises but depend more on external resources for the ex post evaluation'. Formative evaluations aim at improving the design and performance of policy-making or spatial planning processes usually while they are conducted. *Ex ante* evaluations have similar objectives, but they are conducted before the processes and also have predictive aims.

The character of issues dealt with in the spatial planning process should be taken into account when deciding whether the evaluation will be conducted internally or externally. If the process will address issues that are known to be controversial, an external evaluation will probably be better received. In such situations, an internal evaluation might appear to be simpler to conduct, but it must be acknowledged that an in-house evaluation of a controversial process will shed doubts on the reliability and impartiality of the evaluation.

An assessment of the financial resources needed for conducting or commissioning evaluations needs to be carefully considered against the anticipated purposes and expectations for the evaluation (EC 2013b). It is generally estimated that an evaluation of a rather routine policy or planning process would require a relatively small amount of money in proportion to the resources for the whole process—normally less than 1%. For evaluations of extensive and new types of policies or spatial planning processes, and if there are high-learning expectations and substantial investment in stakeholder participation, the costs are likely to be relatively high in proportion to the overall programme costs—up to 10% (EC 2013b). The EU Commission's (DG REGIO) guidance document points out that *ex ante* evaluations usually have a rather limited time and limited possibility of acquiring data for the evaluation. Then the required resource is not that high in comparison to evaluations that come in later stages. Especially *interim* evaluations, if they have strong formative ambitions, may require a lot of evidence and extensive stakeholder engagement, which increases both the costs and time needed. *Ex post* evaluations do not necessarily require substantial resources, depending on the scope given for the evaluation. In conclusion, the most important factor that determines the required budget is the nature and scope of the evaluation (EC 2013b).

6.3 Stakeholder Engagement in Evaluations

It is common and even recommended that evaluations engage stakeholders at different stages (EC 2013b; Hansen and Vedung 2010). It is important to notice that the earlier discussion about whether to conduct the evaluation in-house or by external consultation is not related to the need for engaging the stakeholders—the evaluations conducted as an in-house service should also aim at engaging the stakeholders. In fact, due to the possible risk that in-house evaluations are perceived as not being transparent and neutral, it is even more important for such evaluations to be inclusive.

Stakeholder involvement is advisable for two broad reasons. First, stakeholders possess expertise, knowledge and information that can be an invaluable resource for the evaluation. For instance, in evaluations that are run within a limited time and with limited resources, well-organised stakeholder involvement can be decisive for the success of the evaluation. The second reason follows from an often-cited definition of a stakeholder: 'a stakeholder is an individual or group influenced by—and with an ability to significantly impact (either directly or indirectly)—the topical area of interest' (Glicken 2000, 1). Various individuals, communities and organisations are affected, positively or negatively, by the spatial plan that is being evaluated. Therefore, they have an

Steps	**Possible methods**	**Outputs**
1. Define the scope and purpose(s) of the evaluation and define evaluation questions (by the public body that commissions the evaluation)		Terms or Reference for the evaluation
Ex ante		
2. Get familiar with the context and objectives of the spatial planning programme	Desk study, meetings with the planning authority reps	Detailed evaluation plan, identification of key actors and stakeholders
3. Formulate theories of change to reach the objectives in collaboration with the planning authority reps	Desk study + a workshop	Draft theories of change
4. Test the theories of change with other actors (e.g. stakeholders and sector authorities)	Workshop and/or interviews	Theories of change (joint understanding of possible results and impacts of the evaluated intervention and understanding of differences among the actors)
5. Define evidence and indicators for follow-up programmes (need to match with evaluation criteria as defined in the Terms of Reference	Desk study + (possibly) a workshop or focus group with key actors	Set of indicators and identified sources of information (evidence)
During the processes of planning and implementation		
6. Monitor the evaluated process and its outputs	Desk studies to analyse documents, observation of the planning process, interviews of key actors, workshops or focus groups to collect evidence	Evidence for the process evaluation (and outcome evaluation for *interim* outcomes)
Ex post (or during the processes, if there are interim evaluations)		
7. Monitor impacts	Desk studies to analyse documents and evidence that was collected, interviews of key actors, workshops or focus groups to collect evidence	Evidence for the outcome evaluation
8. Assess the theories of change against the evidence	Desk studies, workshops and focus groups	Updated understanding of how the plan produces impacts and what impacts, who are affected
9. Draw draft conclusions and recommendations	Desk studies, workshops and focus groups	Draft results of the evaluation and feedback on them
10. Communicate evaluation results	Reporting and dissemination to decision-makers, planners and key actors	Final results are communicated to decision-makers and planners
11. Decide and implement corrective actions (by the public body that decides about spatial planning)		Improved planning process and plan

Table 18.3 Steps of an evaluation process

interest in the evaluation results and outcomes. Stakeholders' willingness is also important for successful implementation of the spatial plans, especially if the plans are nonbinding (Faludi 2000).

Stakeholders are a resource for evaluations in their capacity to provide information and insights that help design and implement the evaluation. It has been suggested that stakeholders should be involved at all stages of evaluation processes (Carneiro 2013; EC 2013b). In the early stages, their input can be valuable in defining the scope of the evaluation and in outlining the key evaluation questions.

It has been emphasised that theories of change should ideally be constructed together with the stakeholders, since in complex environments it is likely that there are well-justified alternative understandings of possible impacts and how they might be generated by the planning decisions (Hansen and Vedung 2010; Mayne 2012). Hansen and Vedung (2010) even point out that elucidation of different understandings of how interventions might play out and different valuations of the impacts is often one of the most important results of theory-based evaluations.

Finally, the stakeholders should be given an opportunity to comment on the evaluation results (Carneiro 2013). Participation of the stakeholders at different stages of evaluation aims at ensuring that there is ownership of the evaluation findings (EC 2013b).

Table 18.3 presents a summary of Sects. 5 and 6 as practical steps of the evaluation.

Acknowledgements The author is grateful for EU co-funding to the Baltic SCOPE and Pan Baltic Scope projects. The Open Access fee of this chapter was provided by the Pan Baltic Scope project. The Baltic SCOPE project was funded from the European Maritime and Fisheries Fund under the agreement EASME/EMFF/2014/1.2.1.5/SI2.703760. The Pan Baltic Scope is funded from the European Maritime and Fisheries Fund under the agreement EASME/EMFF/2016/1.2.1.6/01/SI2.763067.

References

Armitage, D., Marschke, M., & Plummer, R. (2008). Adaptive Co-management and the Paradox of Learning. *Global Environmental Change, 18*(1), 86–98. https://doi.org/10.1016/j.gloenvcha.2007.07.002.

Astbury, B., & Leeuw, F. L. (2010). Unpacking Black Boxes: Mechanisms and Theory Building in Evaluation. *American Journal of Evaluation, 31*(3), 363–381. https://doi.org/10.1177/1098214010371972.

Carneiro, G. (2013). Evaluation of Marine Spatial Planning. *Marine Policy, 37*, 214–229. https://doi.org/10.1016/j.marpol.2012.05.003.

Coryn, C. L. S., Noakes, L. A., Westine, C. D., & Schröter, D. C. (2011). A Systematic Review of Theory-Driven Evaluation Practice from 1990 to 2009. *American Journal of Evaluation, 32*(2), 199–226. https://doi.org/10.1177/1098214010389321.

Cundill, G., Cumming, G. S., Biggs, D., & Fabricius, C. (2012). Soft Systems Thinking and Social Learning for Adaptive Management. *Conservation Biology, 26*, 13–20. https://doi.org/10.1111/j.1523-1739.2011.01755.x.

Day, J. (2008). The Need and Practice of Monitoring, Evaluating and Adapting Marine Planning and Management—Lessons from the Great Barrier Reef. *Marine Policy, 32*(5), 823–831. https://doi.org/10.1016/j.marpol.2008.03.023.

Douvere, F. (2008). The Importance of Marine Spatial Planning in Advancing Ecosystem-Based Sea Use Management. *Marine Policy, 32*(5), 762–771.

Douvere, F., & Ehler, C. (2011). The Importance of Monitoring and Evaluation in Adaptive Maritime Spatial Planning. *Journal of Coastal Conservation, 15*(2), 305–311. https://doi.org/10.1007/s11852-010-0100-9.

EC. (2013a). *EVALSED Sourcebook: Method and Techniques* (p. 165). Brussels: European Commission/DG REGIO.

EC. (2013b). *EVALSED: The Resource for the Evaluation of Socio-Economic Development* (p. 119). Brussels: European Commission/DG REGIO.

EC. (2014). Directive 2014/89/EU of the European Parliament and of the Council of 23 July 2014 Establishing a Framework for Maritime Spatial Planning, OJ L 257, 28.8.2014, pp. 135–145.

Ehler, C. (2014). *A Guide to Evaluating Marine Spatial Plans*. IOC Manuals and Guides No. 70, ICAM Dossier 8. Paris: UNESCO, Intergovernmental Oceanographic Commission UNESCO IOC, 96 pp.

Ehler, C., & Douvere, F. (2009). *Marine Spatial Planning: A Step-by Step Approach Towards Ecosystem-based Management*. Manual and Guides No 153 ICAM Dossier No 6. Paris: Intergovernmental Oceanographic Commission UNESCO IOC, 99 pp.

Faludi, A. (2000). The Performance of Spatial Planning. *Planning Practice & Research, 15*(4), 299–318.

Fletcher, S., McKinley, E., Buchan, K. C., Smith, N., & McHugh, K. (2013). Effective Practice in Marine Spatial Planning: A Participatory Evaluation of Experience in Southern England. *Marine Policy, 39*, 341–348. https://doi.org/10.1016/j.marpol.2012.09.003.

Glicken, J. (2000). Getting Stakeholder Participation "Right": A Discussion of Participatory Processes and Possible Pitfalls. *Environmental Science & Policy, 3*, 305–310.

Gomart, E., & Hajer, M. (2003). Is that Politics? For an Inquiry into Forms in Contemporary Politics. In B. Jorgens & H. Nowotny (Eds.), *Social Studies of Science and Technology: Looking Back Ahead* (pp. 33–61). Dordrecht, Boston: Kluwer Academic Publishers.

Hansen, M. B., & Vedung, E. (2010). Theory-Based Stakeholder Evaluation. *American Journal of Evaluation, 31*(3), 295–313. https://doi.org/10.1177/1098214010366174.

Jay, S., Flannery, W., Vince, J., Liu, W.-H., Xue, J. G., Matczak, M., Zaucha, J., Janssen, H., van Tatenhove, J., Toonen, H., Morf, A., Olsen, E., de Vivero, J. L. S., Mateos, J. C. R., Calado, H., Duff, J., & Dean, H. (2013). International Progress in Marine Spatial Planning. In A. Chircop, S. Coffen-Smout, & M. McConnell (Eds.), *Ocean Yearbook* (Vol. 27, pp. 171–212). Leiden: Martinus Nijhoff.

Kelly, C., Gray, L., Shucksmith, R., & Tweddle, J. F. (2014). Review and Evaluation of Marine Spatial Planning in the Shetland Islands. *Marine Policy, 46*, 152–160. https://doi.org/10.1016/j.marpol.2014.01.017.

Mayne, J. (2012). Contribution Analysis: Coming of Age? *Evaluation, 18*(3), 270–280. https://doi.org/10.1177/1356389012451663.

Mickwitz, P. (2006). *Environmental Policy Evaluation: Concepts and Practice*. Helsinki: The Finnish Society of Sciences and Letters.

Pahl-Wostl, C., Sendzimir, J., Jeffrey, P., Aerts, J., Berkamp, G., & Cross, K. (2007). Managing Change Toward Adaptive Water Management Through Social Learning. *Ecology and Society, 12*(2), 30.

Rae, A., & Wong, C. (2012). Monitoring Spatial Planning Policies: Towards an Analytical, Adaptive, and Spatial Approach to a 'Wicked Problem'. *Environment and Planning B: Planning and Design, 39*(5), 880–896.

Soma, K., Ramos, J., Bergh, Ø., Schulze, T., van Oostenbrugge, H., van Duijn, A. P., Kopke, K., Stelzenmüller, V., Grati, F., Mäkinen, T., Stenberg, C., & Buisman, E. (2014). The "Mapping Out" Approach: Effectiveness of Marine Spatial Management Options in European Coastal Waters. *ICES Journal of Marine Science: Journal du Conseil, 71*(9), 2630–2642. https://doi.org/10.1093/icesjms/fst193.

Stelzenmüller, V., Breen, P., Stamford, T., Thomsen, F., Badalamenti, F., Borja, Á., Buhl-Mortensen, L., Carlstöm, J., D'Anna, G., Dankers, N., Degraer, S., Dujin, M., Fiorentino, F., Galparsoro, I., Giakoumi, S., Gristina, M., Johnson, K., Jones, P. J. S., Katsanevakis, S., Knittweis, L., Kyriazi, Z., Pipitone, C., Piwowarczyk, J., Rabaut, M., Sørensen, T. K., van Dalfsen, J., Vassilopoulou, V., Vega Fernández, T., Vincx, M., Vöge, S., Weber, A., Wijkmark, N., Jak, R., Qiu, W., ter Hofstede, R., & Urquhart, J. (2013). Monitoring and Evaluation of Spatially Managed Areas: A Generic Framework for Implementation of Ecosystem Based Marine Management and Its Application. *Marine Policy, 37*, 149–164. https://doi.org/10.1016/j.marpol.2012.04.012.

Terryn, E., Boelens, L., & Pisman, A. (2016). Beyond the Divide: Evaluation in Co-evolutionary Spatial Planning. *European Planning Studies, 24*(6), 1079–1097. https://doi.org/10.1080/09654313.2016.1154019.

TPEA. (2014). Evaluation Process Report (pp. 42): Transboundary Planning in the European Atlantic Project. Retrieved from http://www.tpeamaritime.eu/wp/wp-content/uploads/2013/09/TPEA-Evaluation-Report.pdf.

UNESCO-IOC. (2010). Marine Spatial Planning (MSP). Retrieved July 4, 2012, from http://www.unesco-ioc-marinesp.be/marine_spatial_planning_msp?PHPSESSID=v0mrck0t5csledjmhqhkn4kjo3.

Varjopuro, R. (2017). Evaluation and Monitoring of Transboundary Aspects of Maritime Spatial Planning: A Methodological Guidance. Baltic SCOPE Report. Retrieved from http://www.balticscope.eu/content/uploads/2015/07/BalticScope_EvaluationMonitoring_WWW.pdf.

Vedung, E. (1997). *Public Policy and Program Evaluation*. New Brunswick, NJ and London: Transaction.

Vedung, E. (2006). Evaluation Research. In B. G. Peters & J. Pierre (Eds.), *Handbook of Public Policy* (pp. 397–416). London: SAGE.

Vedung, E. (2010). Four Waves of Evaluation Diffusion. *Evaluation, 16*, 263–277.

Vos, B. d., Stuiver, M., & Pastoors, M. (2012). Review and Assessment of the Cross-Border MSP Processes in 2 Case Studies. MASPNOSE project Deliverable D1.3.2. (pp. 49). Retrieved from https://www.wur.nl/upload_mm/0/b/5/67c197e9-377f4572a497a64e3c7bc679_MASPNOSE%20D1.3.2%20Evaluation%20of%20MSP%20in%20case%20studies.pdf.

Weber, K. (2006). From Nuts and Bolts to Toolkits: Theorizing with Mechanisms. *Journal of Management Inquiry, 15*(2), 119–123. https://doi.org/10.1177/1056492605280237.

Weiss, C. H. (1997). Theory-Based Evaluation: Past, Present, and Future. *New Directions for Evaluation, 1997*(76), 41–55. https://doi.org/10.1002/ev.1086.

19

Education and Training for Maritime Spatial Planners

Helena Calado, Catarina Fonseca, Joseph Onwona Ansong, Manuel Frias, and Marta Vergílio

H. Calado (✉)
MARE—Marine and Environmental Sciences Centre, Lisboa, Portugal

Faculty of Sciences and Technology, Department of Biology, University of the Azores, Ponta Delgada, Portugal
e-mail: helena.mg.calado@uac.pt

C. Fonseca
CICS.NOVA—Interdisciplinary Center of Social Sciences, Faculty of Social Sciences and Humanities, University Nova de Lisboa, Lisboa, Portugal

CIBIO—Research Center in Biodiversity and Genetic Resources/InBIO—Associate Laboratory, University of the Azores, Ponta Delgada, Portugal

J. O. Ansong
MaREI Centre for Marine and Renewable Energy, Environmental Research Institute (ERI), University College Cork, Cork, Ireland

M. Frias
HELCOM (Helsinki Commission) Secretariat, Helsinki, Finland

M. Vergílio
CIBIO—Research Center in Biodiversity and Genetic Resources/InBIO—Associate Laboratory, University of the Azores, Ponta Delgada, Portugal

J. Zaucha, K. Gee (eds.), *Maritime Spatial Planning*,
https://doi.org/10.1007/978-3-319-98696-8_19

1 Introduction

The idea that became Maritime Spatial Planning (MSP) was initially proposed in 1976 by international and national interests in developing marine protected areas (e.g. the Great Barrier Reef Marine Park) as a response to the environmental degradation of marine areas caused by human activities (Olsson et al. 2008). The MSP concept then evolved in the North Sea, a sea basin under high human pressure, where attention was placed on managing the multiple use of marine space driven by new maritime uses, such as offshore wind, especially in areas with conflicts amongst users and conflicts between users and the environment (Olsen et al. 2014; Douvere 2008). It has developed and been implemented in about 20 countries over the past decade (Ehler et al. 2018), and a community of MSP disciplines is developing that calls for more specifically qualified professionals (Ansong et al. 2018). This development offers new opportunities for practitioners with different marine backgrounds, including planning officers, planning/policy consultants, and planning policymakers, that are emerging with the new demands of the field. MSP professionals and sectoral agencies therefore need a comprehensive understanding of the process of MSP and a range of competences and skills, including not just scientific, oceanographic, ecosystem functioning, and geospatial analytical aspects but also planning, programme management, and stakeholder engagement, amongst others (McCann et al. 2014). There is the opportunity to define and explore theories, research, and concepts in such a new field but not much practical experience or real successful cases of MSP from which to learn and to inform training materials and courses that can address training needs. Specific teaching materials and practical manuals are therefore limited. Education and training in MSP, however, must respond to challenges associated with the complexity of this particular planning process, different competence/skill needs, limitations of resources, and the transdisciplinary nature of MSP by a joint effort of disciplines (Ansong et al. 2018; Gissi and de Vivero 2016).

This chapter explores the various training needs and the competences, skills, and backgrounds needed to achieve a successful professional practice. A brief analysis of the existing MSP training resources and a more detailed analysis of an Erasmus Mundus programme specific to MSP are presented. This chapter further contributes to answering the critical questions of the most effective approaches for MSP training by gathering information from various professionals in the MSP community, including agency MSP officers, academics, policymakers, consultants, and other professionals.

2 Evolving Maritime Spatial Planning and Training

The multiple aims and objectives of MSP call for different competences, including those related to environmental, economic and social disciplines, sciences, and skills. MSP is currently a process to foster and encourage spatial efficiency, by promoting coexistence and synergies between maritime use, whilst other cross-border advantages may also have been considered.

Glegg (2014), in answer to the question "who needs training?", suggested: the planners who will be responsible for creating the plans, those with a mainly statutory responsibility to participate in the planning, and those representing a particular interest, whether commercial (e.g. marina owners or fishers) or interest-based (e.g. sailing or non-governmental conservation organisations). Individuals should be trained/educated in MSP to improve their skills, knowledge, and behaviour for a successful MSP process. Increasing MSP capacity is required at all levels and should strengthen legal, administrative, financial, technical, and human resources to address various multifaceted issues that complicate the MSP process. MSP professionals with broad skill sets and knowledge beyond traditional disciplines are urgently needed to include understanding of the legal frameworks, programme management, and social skills involved in working on interdisciplinary teams.

This very broad spectrum of educational and training targets represents different dimensions. Each target group needs a different answer and design programme, but the terms "education" and "training" in this chapter are used regardless of the public or group to be targeted but mainly referring to those directly involved in the process making of the plan.

Initial approaches to MSP implementation and training were largely focused on the consolidation of concepts, mainly because few practical experiences were available from which to draw. Progress in the practical implementation of MSP, however, has been made, and recent approaches to education and training should reflect the practical aspects of MSP to ensure that the multiple objectives of MSP are achieved (Ansong et al. 2018; Jones et al. 2016). Training and education must address these objectives to ensure that the gap between concept and practice is bridged and to emphasise the development of practical skills.

As Gissi and de Vivero (2016) stated that education/training offers a clear dominance of contents for "environmental analysis/assessment" and "maritime uses", whilst "experiences in MSP" and "planning theory" are the least represented. MSP courses may be continuous with education in the management of coastal zones, where ecological science, applied geography, and physical science are the most emphasised theoretical subjects, along with assessment and monitoring. Another reason for the dominance is that environmental studies and marine science have a long tradition in managing marine resources from the perspective of conservation. These courses appear to be reorientations of already established programmes towards new questions and demands inherent to MSP instead of being established to cover specific areas of MSP. A new orientation associated with the combination of "environmental analysis and assessment" with "maritime sectors" has been introduced in Germany and Norway, which may have originated in the drivers for MSP involving the support of blue growth and key maritime sectors, such as renewable energy.

The different drivers for MSP, such as historical, cultural, and geographical contexts, and governance approaches based on certain marine industries or sectors in place inform the various planning approaches and the implementation of MSP (Kidd and Shaw 2014). The challenges for marine planners (such as engaging with relevant parties; assimilating marine plans into existing management; understanding the current framework; willingness to negotiate a role for marine plans; integrating science and policy to support appropriate decisions; awareness of available and important information; and identifying, involving, and maintaining the commitment of individuals in the planning process (Glegg 2014)) shape new demands on education and training with a stronger emphasis on the social dimensions.

In summary, Ansong et al. (2018) examined the existing educational materials, such as those supported by Intergovernmental Oceanographic Commission—United Nations Educational, Scientific and Cultural Organization (IOC-UNESCO), on the educational dimension of increasing MSP capacity, mostly through its web platform, sharing experiences from all over the world, and the publication of MSP manuals and guides. The European MSP Platform provides in-depth information on specific aspects of MSP and complementary information on MSP processes and projects of European Union (EU) member states. OpenChannels is another important platform due to its role in disseminating MSP initiatives, tools, and literature and for promoting and supporting debate. IOC-UNESCO and the EU are the main international actors in increasing MSP educational capacity, with initiatives such as Erasmus Mundus Master Course in Maritime Spatial Planning (EMMCMSP), the ERASMUS+ Strategic Partnership for MSP, and the Marie Skłodowska-Curie Action Planning in a liquid world with tropical stakes: solutions from an EU-Africa-Brazil perspective (PADDLE project). Some public administrations have also held MSP training projects for MSP and sectoral officers such as in the case of Poland and the Marine Management Organisation (MMO).

Practical teaching approaches in MSP education are still being developed, but experience in terrestrial planning can be instructive for MSP training and education. Ritchie et al. (2015) outlined some practice-oriented approaches to planning education that can be used in MSP education. Mock inquiry, one of these approaches, can be transferred to MSP training. This model uses a role-play format to develop student understanding of the framework that enables planning decisions. Student teams study a real planning application that has been refused but not appealed. The module culminates in a live simulated mock planning inquiry, chaired by practicing planners, for students to obtain a practical understanding of the issues and challenges of planning. Experiential approaches to learning that give prominence to soft skills, such as the ability to collaborate, work in groups, read social cues, and respond adaptively, are also needed.

3 Europe and Maritime Spatial Planning Education

The education offered is characterised by multiple combinations of contents and methods generated by different interpretations of what MSP has been and what is now being practised in Europe and other countries. New educational initiatives in MSP should respond to this complexity by (1) developing the transdisciplinary approach that began to be adopted at the beginning of this process and (2) including the successive environmental- (ecosystem-based management) and economics-based (blue growth) foci that dominate the present approach to marine planning (Gissi and de Vivero 2016).

Ansong et al. (2018) stressed the need for a process approach to cover the entire MSP cycle by referring to the paradigm shift in recent teaching experiences, from MSP being used as an environmental approach and managing conflicts amongst uses to a more holistic approach for coordinating sectoral policies, facilitating transboundary cooperation, and planning advantages. Flannery et al. (2018) argue that MSP negotiations examined as a boundary object ("something which brings diverse stakeholders together, which each view from their own perspective, yet negotiate a common understanding of—provides a theoretically driven analysis of the processes through which actors collaborate or act so as to deny the actions of others") facilitate a greater understanding and explanation of the negotiation process, co-option, and domination that occur within MSP initiatives, balancing powers, invested interests, and conflicts. Increasing MSP capacity was therefore defined as a process by which the abilities of individuals, institutions, and their networks are developed and enhanced for making effective and sustainable decisions about the temporal and spatial ordering of human activities in the marine space.

EMMCMSP is a two-year advanced professional Master's degree programme designed within the Erasmus Mundus 2013–2019 programme, with the participation of three European universities: Università Iuav di Venezia, University of Seville, and University of the Azores. This particular Joint Master's degree course intends to prepare students to become specialists in MSP, operating both in public institutions and as independent professionals or researchers. The course was designed to provide students with skills to plan, design, and evaluate projects and policies, which consider terrestrial, coastal, and marine dimensions, and to develop their ability to manage decision processes towards an adaptive and integrated approach. The course familiarises students with key issues involved in policy formulation and planning strategies for maritime space to improve the management of resources from environmental, economic, social, and legal perspectives within the framework of maritime policies. The EMMCMSP degree course is organised into four terms, including practice-oriented classrooms, a period of internship in different countries, and the development of a final thesis.

This Erasmus Mundus programme in MSP, integrated in the cooperation and mobility programme in the field of higher education that aimed to enhance the quality of European higher education and to promote dialogue and understanding between people and cultures through cooperation with third countries (Decision No 1298/2008/EC (EC 2008)), has currently completed its 5th edition cycle. Considering its EU funding support, its international dimension, the successful employment of former students, and the tripartite academia as its basis (Portugal, Spain, and Italy), this Master's degree course is used as an example to support a brief analysis of the present educational interests. It has also to be stressed that this course was designed during the discussion of the EU Directive on MSP (Directive 2014/89/EU (EC 2014)), and great effort was dedicated to incorporating the various demands of the Directive published at the end of the first year of the Erasmus Mundus Master course on MSP.

The five editions of the EMMCMSP have already welcomed about 60 students from several countries. More than 50% of the students come from Asia or Africa, and the number of students has been stable throughout the editions. Seven students in the 2nd edition were from Asia, which is the highest number of students from a particular region in all editions. The popularity of the EMMCMSP in these regions is probably due to the aim of the Erasmus Mundus programme to improve cooperation with Third World countries and to the extensive connections with these regions.

The number of students (six) from the EU peaked in the 2nd edition but decreased to only two in the 5th edition, which was surprising because MSP has become very popular due, for example, to the MSP Directive and the efforts towards a blue-growth economy encouraged by the Commission. The programme, however, provides only two or three grants per edition for EU students, which is probably mainly responsible for the level of bias towards these countries. This gap in MSP EU training strongly needs to be addressed. South America has provided the largest increase in the number of students, from only one student in the 1st edition to four students in the last edition. Is it just chance or has the coordination office been more active in advertising the EMMCMSP in South America because MSP is not yet developed enough to justify the increasing interest in the EMMCMSP?

Almost one-third of all students from all editions have had a background in architecture or engineering (Fig. 19.1). The number of students with a background in environmental science and geography is slightly lower but is also close to one-third. The difference in the trend of these two categories, however, is notable; the number of students with backgrounds in architecture and engineering has been steadily decreasing over all editions, but the number of students with backgrounds in environmental science and geography has remained the same. Justification for the background in architecture may be because architecture is an extension of land-use planning, where many architects are specialists. The number of students with backgrounds in marine biol-

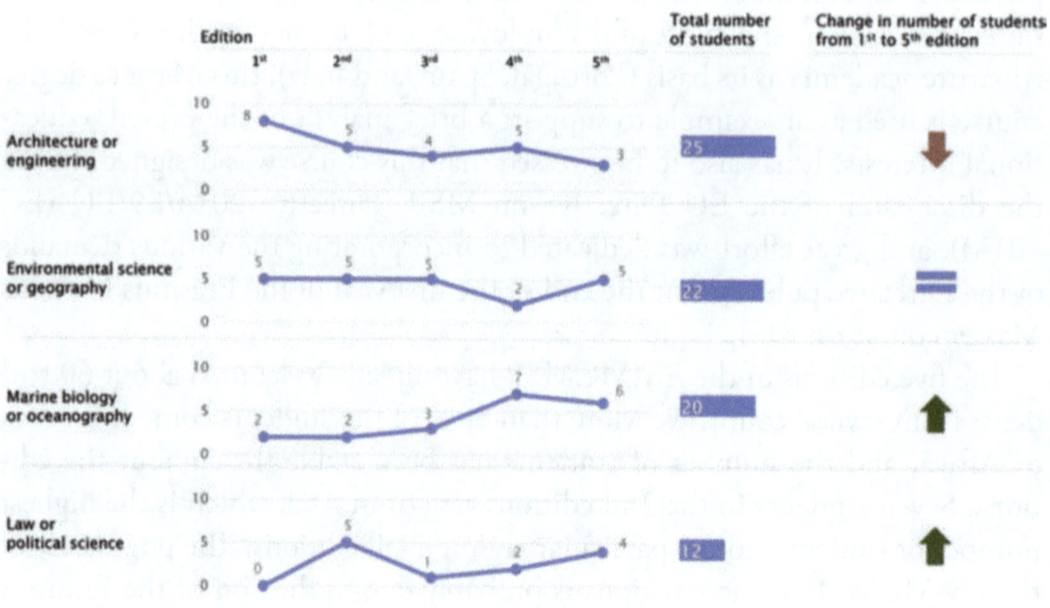

Fig. 19.1 Number of students in the Erasmus Mundus Master Course on Maritime Spatial Planning per background and edition

ogy and oceanography, in contrast, has increased remarkably since the 1st edition, which is perhaps logical because MSP has a strong environmental component, at least in Europe. What is interesting, however, is that the number of students with this background has not been high from the beginning.

The fewest students across all editions had backgrounds in law and political science.

In addition to the geographical origin of the students, the kind of issues looked for when deciding to integrate EMMCMSP is also worth analysing. Most theses are about governance and ecosystem-based management. More than one-third of students decide to write their theses on the issue of governance, and the number of theses about this topic has been slightly increasing since the 1st edition (Fig. 19.2), which is not surprising because governance is such a broad topic. The number of theses about other topics has remained stable across all years.

Almost half of the students of all editions decided to focus their theses as a recommendation to a certain topic (Fig. 19.3). Writing recommendations has been the favourite output of all editions except the 2nd. This kind of output decreased remarkably in the 2nd edition in favour of "lessons learnt". Writing about supports to decision-making has not been popular in any edition. In fact, the number of this kind of output topic is decreasing drastically.

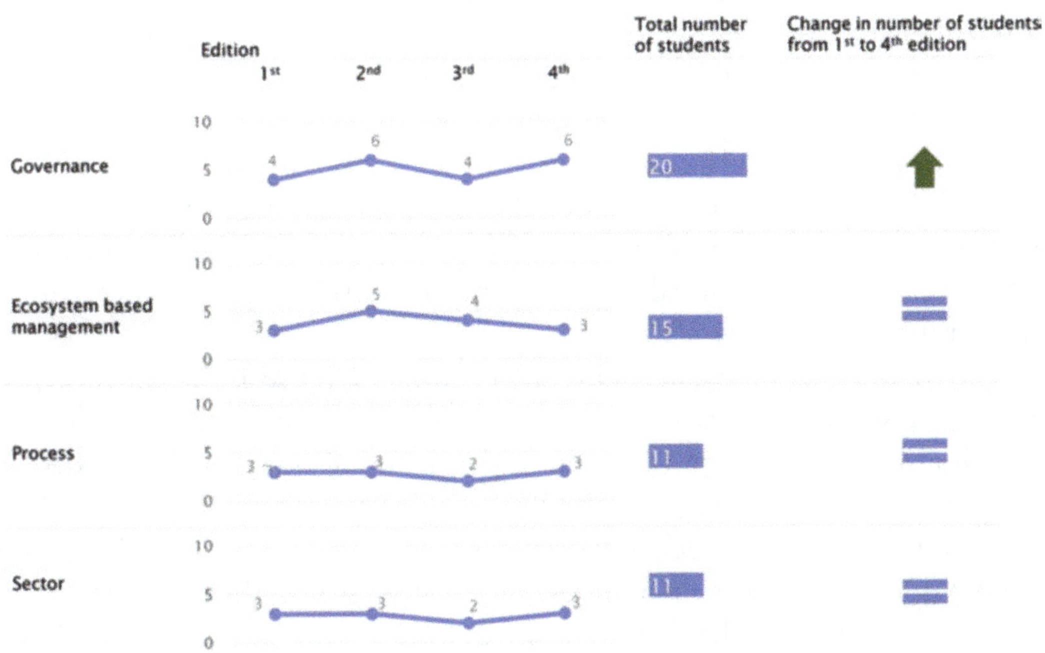

Fig. 19.2 Number of theses from the Erasmus Mundus Master Course on Maritime Spatial Planning per theme and edition

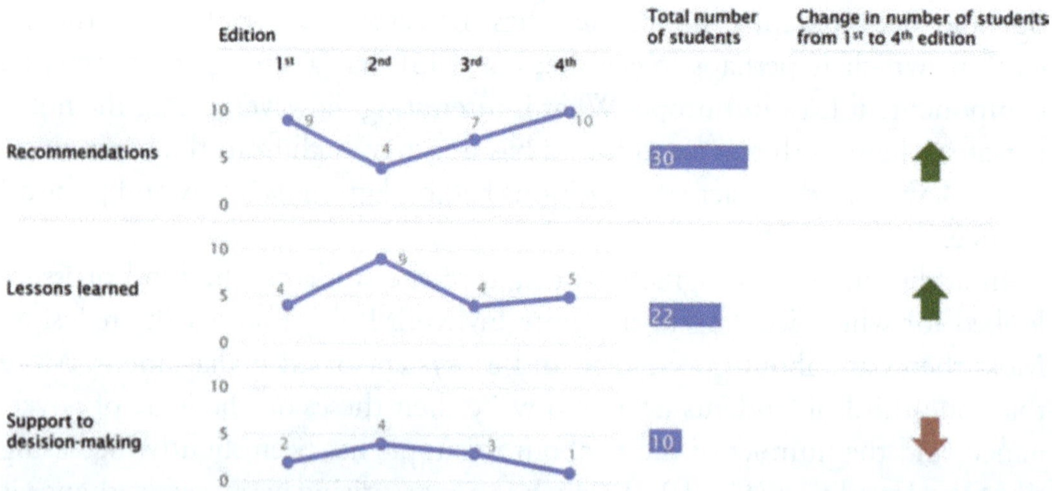

Fig. 19.3 Number of theses from the Erasmus Mundus Master Course on Maritime Spatial Planning per topic and edition

Almost 40% of all these deal with Step 3 in Fig. 19.4 "Organising the MSP Process" from the UNESCO "A Step-by-Step Approach toward Ecosystem-based Management" (Ehler and Douvere 2009). Steps 8 "Implementing the plan" and 5 "Analysing existing conditions" are the foci of one-third of the theses. In addition to the personal preferences of the students, the available time for the elaboration of the thesis may have influence on the chosen topic. Steps in the process involving stakeholders' engagement will be more difficult to develop in a short period of time. It is also worth mentioning that steps are sequential and, for example, future condition cannot be addressed without a proper analysis of the current conditions. In what concerns the step for evaluation, as not many plans are implemented, it is expected that not many theses are developed under this topic.

EMMCMSP has reached its initial programming, and new editions are dependent on the approval of a new cycle of studies by the Erasmus programme. The educational offer, however, still exists under other opportunities.

The Marine Planning and Management Master Course at the University of Liverpool is another initiative oriented towards MSP. This course is a full-time (12 months) or part-time (24 months) programme open to all first-degree subjects. The MSc in Marine Planning and Management is also designed on a multidisciplinary approach and provides graduates with the knowledge and skills required to meet the job opportunities arising from the recent adoption of MSP and related developments in marine conservation and maritime industries. Some topics selected by students for development in their dissertations include implementation of MSP in Portugal, global food security, find-

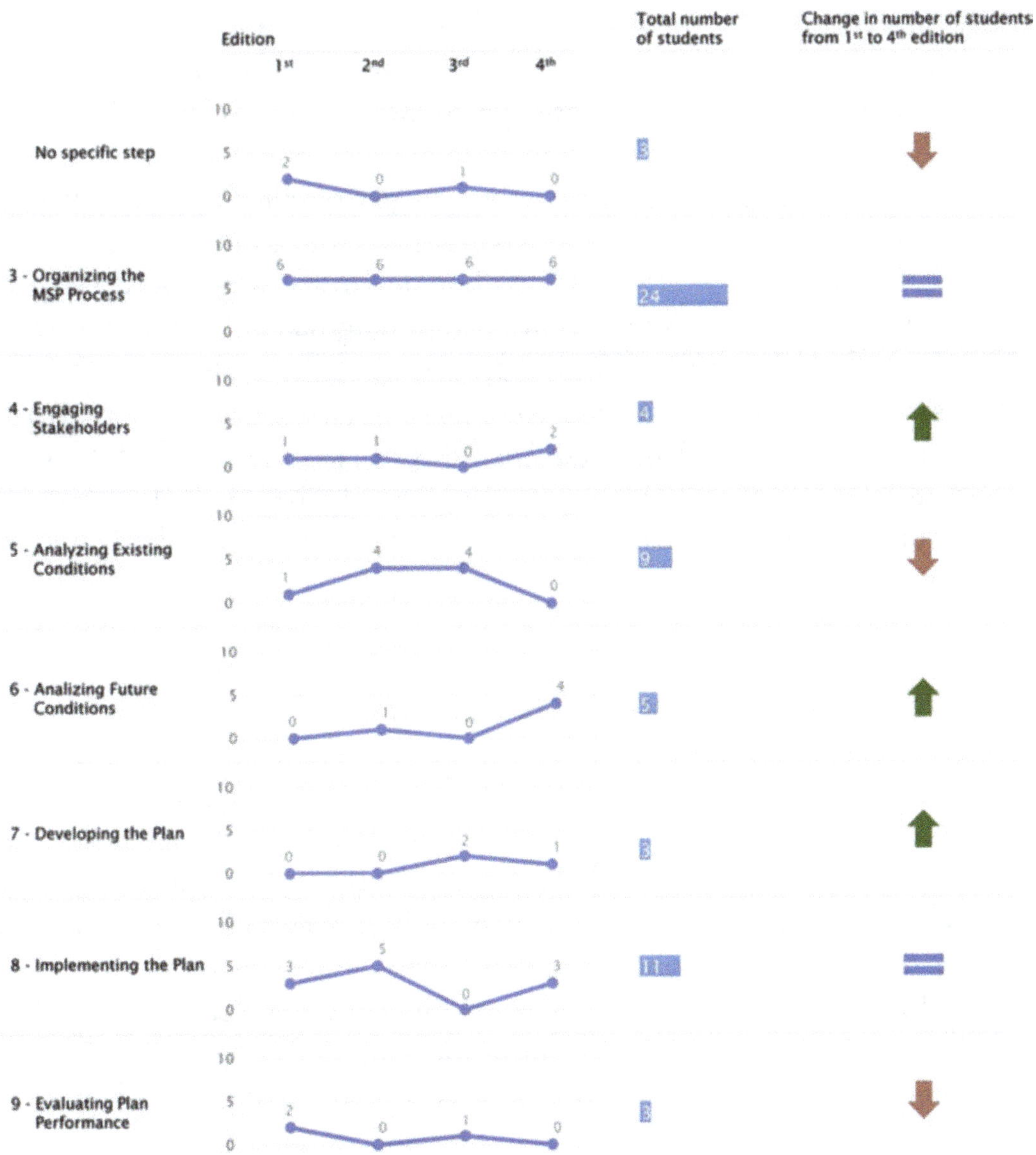

Fig. 19.4 Number of theses per MSP step and edition (MSP steps from the UNESCO "A Step-by-Step Approach toward Ecosystem-based Management" (Ehler and Douvere 2009))

ing space for aquaculture, stakeholder participation in marine planning in the UK, success factors for offshore wind energy, China's system of marine functional zoning, assessing the value of sand-dune systems in north-western England, stakeholder involvement in the Irish Sea Conservation Zone project, reducing the impact of offshore wind farms on seabirds, and mitigating the impacts of tidal barrages (Marine Planning and Management MSc 2018). This analysis, however, focused only on a programme designed from the beginning to cover the needs of the MSP process based on EU trends.

The new Strategic Partnership for MSP is a cooperation to address current issues in the emerging field of MSP, with the overall aim to reach a common understanding and transdisciplinary approaches on a transnational level for higher education. The partnership will present a common European educational agenda and is expected to have some degree of tailoring and downscaling of proposals to the different MSP training and educational needs covering the EU. Other initiatives, including ad hoc initiatives, were also developed, namely SeaPlanSpace (a project aiming to strengthen the competences of employees of administration and the private sector, as well as students and university graduates, in the area of MSP and sustainable marine governance (University of Gdańsk 2018)), BONUS BaltSpace (a summer school for early career professionals and PhD students (The Baltic University 2018)), and the MSP Course for professionals (Maritime Spatial Planning in the Baltic Sea Region 2018).

4 Skill and Competence Needs

As a discipline or planning process such as MSP evolves, the skills and competences necessary for its proper development and implementation also evolve. A bipartite approach was developed to identify the upcoming needs for skills and competences in MSP. Firstly, skill and competence needs were identified based on "Marine spatial planning—A Step-by-Step Approach toward Ecosystem-based Management" by Ehler and Douvere (2009) and on the Davies et al. (2011) study on "future working skills". Secondly, skill and competence needs for MSP were discussed by consulting a pool of MSP professionals.

4.1 Skill and Competence Needs Based on the Maritime Spatial Planning Process

Davies et al. (2011) proposed ten key working skills needed by professionals to be active and effective for the next ten years. The study concluded that skills for rationality (the ability to determine the deeper meaning or importance of what is being expressed), social intelligence, novel and adaptive thinking, cross-cultural competency, computational skills, new-media literacy, transdisciplinarity, design mindsets, cognitive-load management, and virtual collaboration are relevant across a range of professions. These skills are summarised in Table 19.1. Marine spatial planners and professionals need to be

Table 19.1 Competences of a marine planner/marine planning team based on the MSP process

Maritime planning process and activities	Skills and knowledge	Background and expertise
Defining/selecting the planning area *Discussion on boundary of the planning area based on jurisdictional boundary, patterns of maritime activities and bio-regions.* Collecting and understanding existing information *and plans available in the planning area*	Existing jurisdictional boundaries, bioregions, dialogue, intuitive reasoning, rationality	Legal and policy expertise, ecologists, marine geographers, spatial planners, social scientists
Stakeholder engagement/ consultations *Identification of stakeholders, when & how to engage them*	Stakeholder analysis, stakeholder engagement tools, facilitation, negotiation, communication	Social scientists, specialists in graphics and geographic information systems (GISs), communication experts
Gathering data & evidence/ stock-taking *Collecting, storing, and managing scientific data and information for the planning area*	Data-collection methods, spatial-database management, existing governance system, digital thinking, intuitive reasoning, rationality	GIS, specialists in data and information technologies, statisticians
Definition of visions and objectives *Definition of visions and objectives of the planning area are based on evidence and engagement*	Policy analysis, logical framework analysis, scenarios, facilitation, negotiation	Legal and policy expertise, terrestrial planning, communication expert, social scientists
Analysis of current and future conditions *Issue identification, spatial conflicts, options/ alternatives, scenarios* *Analysing current and future spatial/temporal trends and requirements*	GIS, scenario analysis, sector assessment, synthesising information, spatial analysis, socio-economic analysis, environmental analysis, intuitive reasoning, rationality, spatial planning	Marine scientists and technical specialists, for example, oceanographers, ecologists, surveyors, statisticians, economists, terrestrial planners, environmentalists, heritage and cultural specialists, GIS specialists, specialists in data and information technologies, social scientists Sectoral interests such as fisheries and marine industries

(*continued*)

Table 19.1 (continued)

Maritime planning process and activities	Skills and knowledge	Background and expertise
Development of plan policies/measures *Measures and alternatives to achieve planning objectives and visions*	Existing sectoral policies, activity planning, analysis of existing governance system, communication, facilitation, negotiation	Sustainability appraisal, legal and policy expertise, sectoral interests, social scientists, growth strategies and regeneration
Plan approval and adoption *Review of draft plan to include comments and inputs from consultations with necessary arrangements for approval*	Policies and legislation	Legal and policy expertise, social scientists
Plan implementation *Ensuring coordination, compliance, and enforcement of measures and policies defined by the plan*	Project/organisational management	Sectoral interests and agencies
Monitoring and evaluation *Reporting and monitoring the progress of the plan and necessary planning reviews*	Understanding a "logic model" and indicators, existing monitoring programmes	Statistical and reporting experts, social scientists
Management of the planning process/other competences *Coordination and organisation of the various activities and processes*	Programme/project management, systems thinking, and management processes	Terrestrial planning, project managers, social scientists
Additional skills	Politics and legislation	

Adapted based on Ansong et al. (2018)

trained to deal with current and future challenges and requirements. Ansong et al. (2018) also proposed a set of skills and backgrounds needed for MSP based on MSP practice and informed by inputs from MSP professionals. Table 19.1 lists the steps identified by Ehler and Douvere (2009) in the maritime planning process and is therefore a combination where the skills developed by Davies et al. (2011) were juxtaposed against the background and skill set for a marine spatial planner/team by Ansong et al. (2018) to demonstrate that rationality, social intelligence, and computational thinking skills are largely covered and emphasised in MSP practice according to professionals. Transdisciplinarity, novel and adaptive thinking, design mindsets, and cognitive-load management are implied in these MSP skills and backgrounds, whilst cross-cultural competency, virtual collaboration, and new-media literacy are not referred to or indicated.

Cross-cultural competency is not emphasised probably because statutory MSP processes are normally within national jurisdictions. Stakeholder engagement, however, is expected to cater to different cultures and understand how they can be brought into the MSP process. MSP is changing, and international cooperation will become increasingly important in the future. MSP education and training should therefore cater to training professionals to deal with different cultural settings. Virtual collaboration is an important skill for MSP professionals, which has been noticed in MSP projects and partnerships between institutions in different countries, due to upcoming technologies, which allow for hosting audio-visual meetings between professionals and stakeholders and amongst the wider MSP community. Integrating new-media literacy into MSP educational programmes is also closely linked to developing virtual collaboration skills. Using new-media platforms and presenting visually stimulating information is also a critical skill to be developed. Experience has demonstrated that more effort is needed in such an area to engage more with marine stakeholders in understanding MSP issues.

Additional emphasis on developing practical skills such as critical thinking, insight, and analytical capabilities in MSP education and training is needed.

4.2 Skill and Competence Needs: Expert Consultation

Professionals who have worked in MSP and active professionals currently developing activities in the MSP sphere have the most experience in the field and can probably best illustrate the skills and backgrounds needed by future MSP professionals. Active professionals also constitute potential employers, and many represent institutions in charge of or associated with MSP processes that have already been or will be developed. A pool of professionals, not intended to be representative of MSP practitioners but rather representative of the different dimensions in need of MSP specific skills, was invited to collaborate in this study, contributing their knowledge of the current needs for MSP skills and competences. This pool was organised to include professional categories: policymakers/governmental agencies, scientific representatives, industry/sectoral professionals, consultants, and professionals with MSP experience (Fig. 19.5). The pool was drawn from different countries/regions in the EU where MSP processes have already been implemented, plans have already been developed, and agencies have legitimacy in the MSP process (namely England, Scotland, and Germany). The majority of the respondent

Fig. 19.5 Professional categories included in the pool of professionals

professionals have been teachers, tutors, and/or facilitators in any type of specific training in MSP. Masters courses and other types of training (e.g. short courses) are the most common training format in which the respondents have taught, followed by seminars and summer schools. The questions were organised into three sections: the first focused mostly on the background and past training of the professionals; the second focused mostly on the skills and backgrounds currently needed in MSP; and the third focused mostly on the issues that MSP training should include.

Training Attended by Respondent Professionals

More than half of the respondents had attended at least one type of training in the last five years. The most common type of training was games, followed by seminars, workshops, and other types of training, all at the same level (Fig. 19.6a). Master courses, summer schools, and fieldwork were the least attended types of training. Games and workshops seem to be suitable for professionals to become updated on MSP evolution. Opinions about the benefits of the training to the professionals on a scale of 0 to 5 were divided, but the professionals generally considered their training to have been beneficial (Fig. 19.6b). None of the respondents considered that the training was not at all beneficial (value zero for benefit). The most selected levels of benefit, quantified by the number of answers, were medium (level 3) and high (level 5).

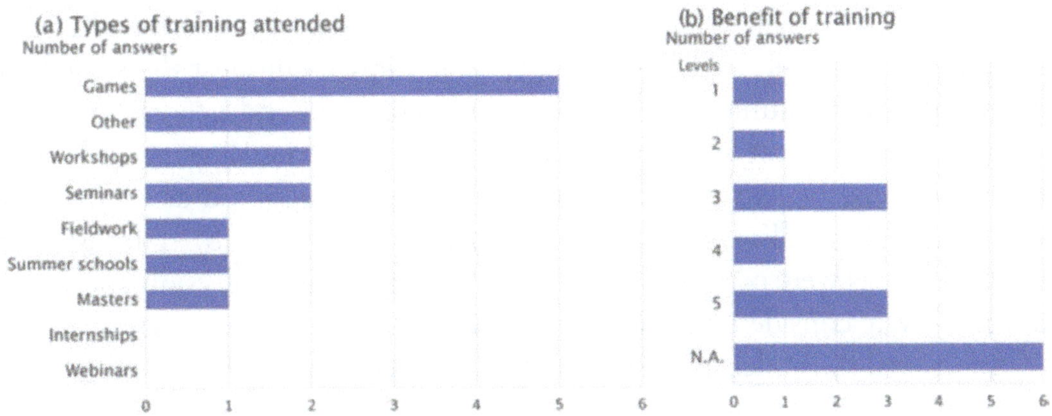

Fig. 19.6 (a) Main types of training attended by professionals in the last five years and (b) level of benefit of the training to the professionals

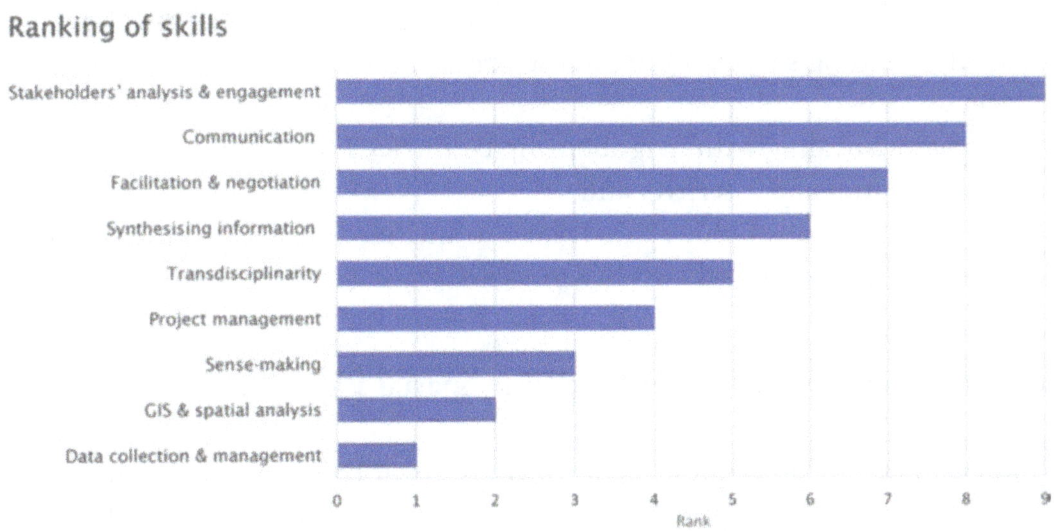

Fig. 19.7 Ranking of the importance of skills needed by an MSP practitioner, according to the expert respondents; highest values indicate more important skills

Benefit levels 1, 2, and 4 were also selected, demonstrating that the professionals recognised the benefit of their training.

Skills Needed by an Maritime Spatial Planning Practitioner

A set of skills needed by an MSP practitioner was proposed and presented to the professionals, who were asked to rank the skills in order of importance. They were also asked if other skills were needed by MSP practitioners. The results were compiled to represent skills of higher importance with higher values and skills of lower importance with lower values (Fig. 19.7).

The ability to analyse and engage with stakeholders was identified as the most important skill needed by an MSP practitioner, amongst the suggested set of skills. Communication and skills for facilitation and negotiation were identified as the next most important. Skills for synthesising information, transdisciplinarity, and project management were categorised as moderately important. Rationality, geographic information system (GIS) training, and spatial analysis were considered less important, and data collection and management were considered the least important.

Additional skills identified by the professionals were:

- Political skills and navigating politics, such as understanding processes of law, policy, and decision-making and understanding the legislative framework of MSP and its constraints,
- Spatial-planning skills (itself a set of skills including some of the above),
- Understanding social and natural sciences and the ability to take a holistic view,
- Capacities dealing with neighbouring countries and cultures,
- Being neutral and assertive, and
- Presentation skills (which might be also included in the suggested skill of communication) and the ability to write clearly (particularly policies).

Backgrounds Needed by an Maritime Spatial Planning Team

Similar to the previous question, a set of proposed backgrounds needed by an MSP practitioner was proposed and presented to professionals, who were asked to rank the backgrounds by order of importance. The professionals were also asked if other backgrounds were needed by MSP practitioners. The results were compiled to represent backgrounds of higher importance with higher values and backgrounds of lower importance with lower values (Fig. 19.8).

Spatial planning was considered the most important background, followed by marine sciences, legal framework, and social sciences. Industry and technology, GIS, and other environmental sciences were ranked as moderately important. In descending order, economics, political science, communication sciences, and transportation and statistics were considered the least important.

Additional backgrounds identified by respondent professionals included

- Administration,
- Modelling marine physical processes,
- Facilitation and moderation, and
- Conflict management.

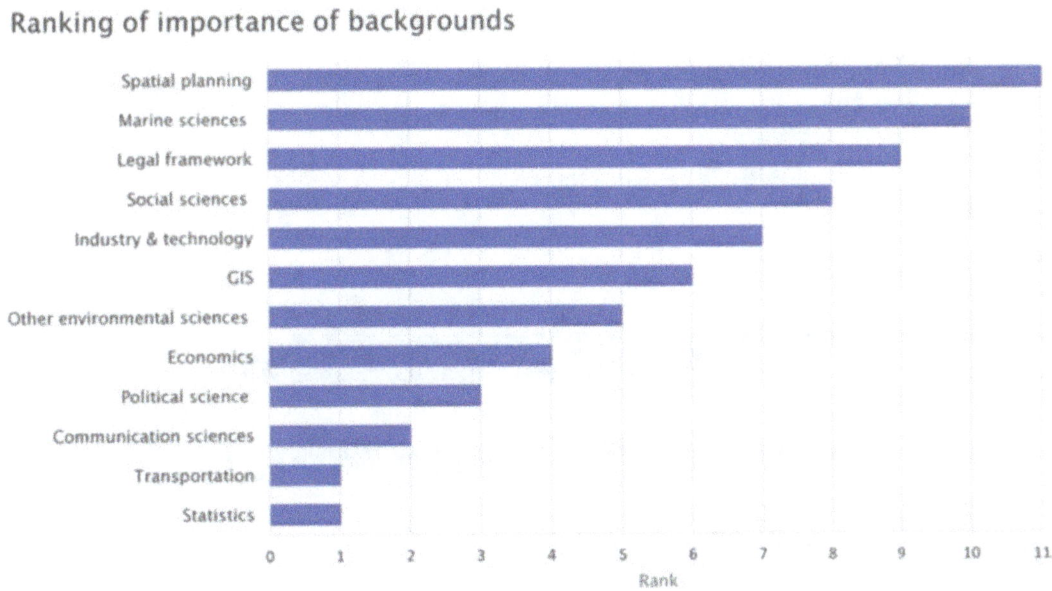

Fig. 19.8 Ranking of importance of backgrounds needed by an MSP practitioner, according to the expert respondents; highest values indicate more important backgrounds

Importance of Specific Training and the Most Important Areas of Knowledge

On a scale from 0 (level of no benefit) to 5 (level of highest benefit), none of the respondents considered that training in MSP was not important to practitioners (value 0 for importance) or selected importance level 2. Importance level 4 was the most selected (Fig. 19.9), followed by levels 3, 5, and 1, demonstrating that the professionals considered specific training to be important or very important for MSP practitioners.

Several areas of knowledge were considered by professionals as important to be covered by specific training in MSP. Similar answers were grouped to shorten the list and identify the most consensual areas of knowledge (Fig. 19.10). This question was an open question, and the results overlapped to some degree with the results for the *ranking of skills* and *ranking of backgrounds*. Most of the professionals agreed that stakeholders' identification and engagement, governance, legal and political frameworks, and spatial-planning theory and practices should be covered by specific MSP training. Stakeholders' analysis and engagement was the highest ranked skill, whilst spatial planning was the highest ranked background. For governance, legal and political frameworks, and administrative and economic frameworks, MSP training should focus on country- or region-specific contexts, and training, including real-world examples, was considered of utmost importance.

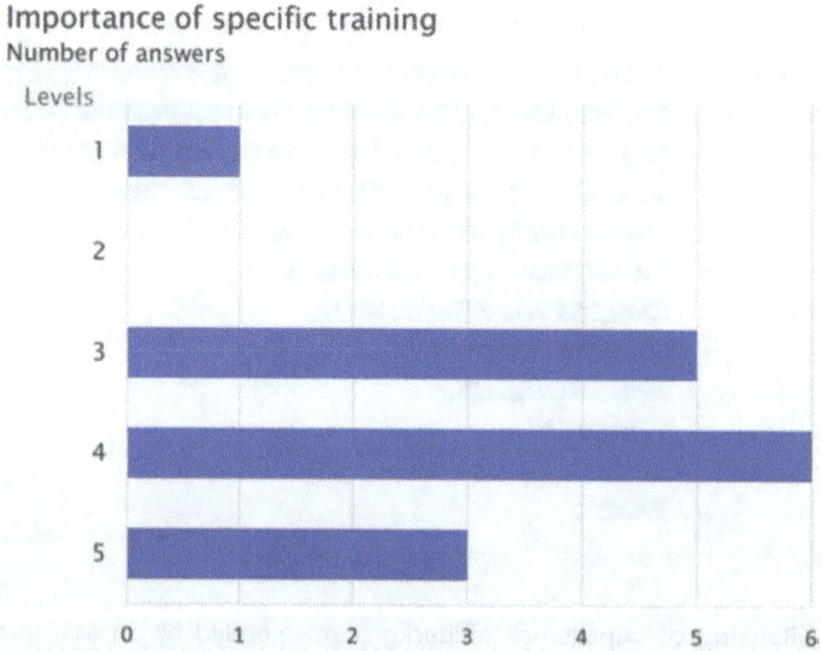

Fig. 19.9 Levels of importance of specific training for MSP practitioners

About half of the professionals considered that "hands-on" training (e.g. fieldwork and internships) was the most effective type of training and preparation for developing and implementing MSP processes (Fig. 19.11). Masters, summer schools, and workshops were also effective but with less consensus amongst the professionals. The category "others" included, for example, the professionals' references to short courses and continuous professional development. Networking was also considered a good opportunity to increase skills and backgrounds in MSP. Real-world cases and examples, problem-solving, and practical training were again considered highly effective.

Additional comments of the professionals included the reiteration of some training needs for MSP staff, namely communication (written, oral, digital media, press, public speaking), moderation and negotiation skills, knowledge of the socio-economic and cultural aspects of MSP, provision of real-world cases and examples, increased knowledge about the implementation of the ecosystem-based approach, development of innovative resource-efficient methods (e.g. online) for engaging stakeholders and increasing participation, exploring the links between MSP and terrestrial planning (land/sea interface), and a Bayesian MSP (progression from Boolean thresholds (yes/no) of area suitability to a spectrum of values of area suitability and integration of data uncertainty).

Fig. 19.10 Areas of knowledge that the respondent professionals considered should be covered by specific MSP training

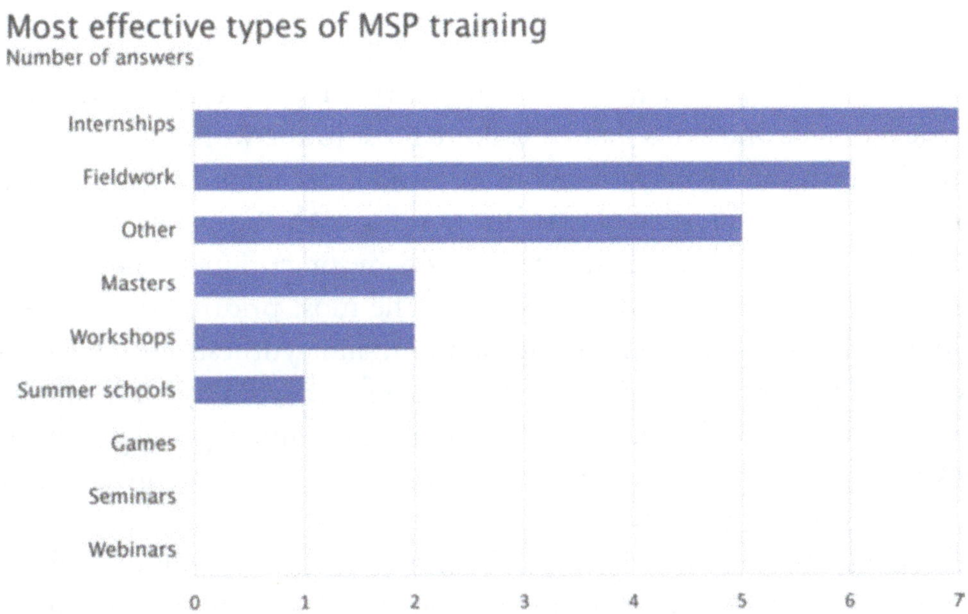

Fig. 19.11 Most effective types of MSP training according to the respondent professionals

5 Discussion and Recommendations

This chapter intended to initiate a discussion of the basic and most important skills and backgrounds important to achieve a successful professional MSP practice and how specific education and training can contribute to that end. As the practice of MSP matures, the educational supply must evolve to respond (or even anticipate) the expectations of both students and potential employers. These three dimensions (educational offer, students' expectations, and market requirements), however, often evolve at different rates creating mismatches that need to be identified and corrected.

The discussion and results indicate that skills requested by MSP practitioners are sometimes common to other practitioners or are at least considered to be general skills, because they are not specific to the MSP process and rationale. Some examples of such skills are rationality, digital thinking, and new-media literacy, including communication and writing competences (Table 19.2) and those considered future working skills identified by Davies et al. (2011). Providing future professionals with these skills will be a challenge not only for MSP but also for employers, as they are also part of a live learning process. The curriculum/syllabus of programmes in MSP, however, must reflect this challenge, and new contents in digital literacy and expression must be added.

Even though covering all subjects and disciplines is important, resources (time and funding) for MSP education and training are limited. Identifying the backgrounds and skills that contribute most to the working environment of MSP practitioners may therefore be needed to identify the issues that should be given the most attention. The ranking of the skills and backgrounds needed by marine planners and teams and by practitioners indicated that stakeholder engagement and analysis were the most prioritised, followed by communication, facilitation and negotiation, and synthesising information. These rankings support the earlier analysis that future training for MSP should focus on new-media literacy and communication and especially how these skills help to engage stakeholders, both in person and virtually. Additional skills suggested by the MSP professionals as also important, including presentation skills, ability to deal with neighbouring countries, political skills, and ability to understand various legislative frameworks, reaffirmed the need for MSP education and training to focus more on communication, legislation, and cross-border engagement skills.

The professionals suggested that spatial planning, marine sciences, legal frameworks, and social sciences were the most important backgrounds for MSP practitioners. Most students accepted into EMMCMSP have a back-

Table 19.2 MSP competences and knowledge and related skills

Skill/key drivers	MSP competences and knowledge	Definition	Summary
Rationality	Dialogue skills, intuitive reasoning Policy analysis, logical framework analysis, scenarios, understanding a "logic model" and indicators, knowledge of existing monitoring programmes	Ability to determine the deeper meaning or importance of what is being expressed	As smart machines take over rote routine manufacturing and services jobs, critical thinking or rationality will emerge as a skill on which workers will increasingly need to capitalise
Social intelligence	Stakeholder analysis, stakeholder engagement tools, facilitation, negotiation skills, communication skills	Ability to deeply and directly connect to others to sense and stimulate reactions and desired interactions	As we collaborate with larger groups of people in different settings, we need socially intelligent employees able to quickly assess the emotions of those around them and adapt their words, tones, and gestures accordingly
Novel & adaptive thinking		Proficiency at thinking and generating solutions and responses beyond those that are rote or rule-based	As automation and offshoring continue, job opportunities will require the ability to respond to unique unexpected circumstances of the moment
Cross-cultural competency	Sectors, policy, environment, socio-economic integration	Ability to operate in different cultural settings	In a truly globally connected world, the ability to adapt to changing circumstances and to sense and respond to new contexts will be necessary for all workers who operate in diverse geographical environments

(*continued*)

Table 19.2 (continued)

Skill/key drivers	MSP competences and knowledge	Definition	Summary
Digital thinking	Data-collection methods, spatial-database management, existing governance system, digital thinking	Ability to translate vast amounts of data into abstract concepts and to understand data-based reasoning	As the amount of data that we have at our disposal increases exponentially, fundamentals of programming virtual and physical worlds will enable us to manipulate our environments and enhance our interactions. Workers must nevertheless remain able to act in the absence of data or algorithms to guide decision-making
New-media literacy	Communication skills, stakeholder engagement	Ability to critically assess and develop content that uses new forms of media and to leverage these media for persuasive communication	As immersive and visually stimulating presentations of information become the norm, workers will need more sophisticated skills to critically assess these kinds of media and to use these tools to engage and persuade their audiences
Transdisciplinarity	Economic analysis, environmental analysis, intuitive reasoning, rationality	Literacy in and ability to understand concepts across multiple disciplines	The ideal worker of the next decade should be able to converse in the language of a broader range of disciplines, which requires a sense of curiosity and a willingness to learn long after their formal education

(continued)

Table 19.2 (continued)

Skill/key drivers	MSP competences and knowledge	Definition	Summary
Design mindset	Project/ organisational management Programme/project management, systems thinking, and management processes	Ability to represent and develop tasks and work processes for desired outcomes	Workers of the future will need to become adept at reorganising the kind of thinking that different tasks require and making adjustments to their work environments that enhance their ability to accomplish these tasks
Cognitive-load management	Policy analysis, logical framework analysis, scenarios, understanding a "logic model" and indicators, knowledge of existing monitoring programmes	Ability to discriminate and filter information for importance and to understand how to maximise cognitive functioning using a variety of tools and techniques	The next generation of workers will have to develop their own techniques for solving the problem of cognitive overload, which is due to a world rich in information streams in multiple formats and from multiple devices
Virtual collaboration	Digital thinking	Ability to work productively, drive engagement, and demonstrate presence as a member of a virtual team	Connective technologies allow us to work, share, and be productive despite physical separation, but leaders need to develop strategies for engaging and motivating a dispersed group, whilst team members need to find environments (either physical or virtual) that promote productivity and well-being

Source: Experts inputs, authors construction on Davies et al. (2011)

ground in architecture or engineering (but which has been decreasing), environmental science and geography, and marine biology and oceanography (which has been increasing). The difference between what is perceived as necessary and what is actually happening is decreasing. The reorientation of background areas of knowledge, if needed, can be achieved by the demand for particular areas as an admission requirement. This demand would also allow for shorter introduction courses, thus freeing time for some specialisation within the general MSP skills and competences. As one of the professionals said, "specific expertise is not so important, it's possible to get specialists to cover this (such as environmental science, social science, economics), but there must be awareness of the basics of each field (e.g. the difference between qualitative and quantitative data, different research methods) and the importance of all these fields to MSP".

The professionals also suggested that the most important skills for MSP practitioners to acquire were stakeholders' analysis and engagement, communication, facilitation, negotiation, and synthesising information. Most EMMCMSP theses' themes are about governance and ecosystem-based approach, with recommendations and lessons learnt as the most common outputs from the theses. The students are free to choose topics, but their judgement is noticeably disconnected from the needs of and recommendations by the professionals. Therefore, guidance must be provided. Greater contact between students and the reality of the labour market should also be promoted from an early stage to adjust the expectations of students and potential employers. These efforts need to monitor MSP evolution and evolve with it, always adapting to current needs.

On a final note, attention must be drawn to the fact that teaching and training for MSP has been analysed here assuming a certain degree of knowledge on basic concepts and skills of spatial analysis. However, this might not be the reality in some particular areas, remote and isolated communities, or some underdeveloped countries where focus sessions using friend mapping tools might be the ideal first step.

As MSP is context specific evolving with science and technology but also with market demands, one of the identified actions for training and education initiatives is to permanently update the scope of modern and emerging skills and competences in order to efficiently design curricula and syllabus.

The purpose of this chapter was not to present a detailed analysis of the existing education and training offer/methods but rather to provide a glimpse of the experience of professionals on what is needed for future MSP practice and start discussion on how this should shape a new stage on MSP training.

Training must also diversify and adapt to serve different needs and specific features and contexts: short courses for professional adaptation and evolution, postgraduate and Master courses reactive to new trends in employment, and training for stakeholders and representatives of organisations and agencies.

In conclusion, more training in specific skills and competences is need. The emphasis at all levels of training and education must be placed on skills (especially those connected to facilitation/negotiation, communication, and digital and media literacy) and not only on scientific backgrounds and support. This challenge is even more important at the level of higher education (degrees and masters which are naturally the first choice for specialisation of new professionals), traditionally designed more to provide deep knowledge on specific topics than to develop technical and social competences in a practical context.

However, since most students accepted into the EMMCMSP were coming from architecture and engineering backgrounds, it is still important that environmental sciences and ecosystem aspects remain in the core of skills to develop into future professionals in this particular case. As MSP develops, it is important that MSP education and research takes account of how plans are implemented and related skills to achieve it. Finally, interdisciplinary and transdisciplinary training allows students to develop skills and acquire knowledge in the entire range of subjects underlying MSP practice.

Acknowledgements The authors thank all the collaborating professionals who willingly answered the questions and improved the discussion with additional comments. We also thank Daniela Gabriel, who was an important source of help with the preparation of different materials, the GPSAzores project (Ref. ACORES-01-0145-FEDER-00002 GPS Azores) funded by structural funds ERDF and ESF (Azores 2020 Operational Programme, and FEDER funding from the Operational Programme for Competitiveness Factors) COMPETE, and national funding from FCT (Foundation for Science and Technology) under the UID/BIA/50027/2013 and POCI-01-0145-FEDER-006821 grants. The Open Access fee of this chapter was provided by GPS Azores project (Ref. ACORES-01-0145-FEDER-00002 GPS Azores—funded by structural funds ERDF and ESF—Azores 2020 Operational Programme).

References

Ansong, J., Calado, H., & Gilliland, P. (2018). A Multifaceted Approach to Building Capacity for Marine/Maritime Spatial Planning: Review and Lessons from Recent Initiatives. *Marine Policy* (Submitted).

Davies, A., Fidler, D., & Gorbis, M. (2011). *Future Work Skills 2020*. Palo Alto, CA: Institute for the Future for University of Phoenix Research Institute. Retrieved from http://www.iftf.org/uploads/media/SR-1382A_UPRI_future_work_skills_sm.pdf.

Douvere, F. (2008). The Importance of Marine Spatial Planning in Advancing Ecosystem-Based Sea Use Management. *Marine Policy, 32*(5), 762–771. https://doi.org/10.1016/j.marpol.2008.03.021.

EC. (2008). Decision No 1298/2008/EC of the European Parliament and of the Council of 16 December 2008, Establishing the Erasmus Mundus 2009–2013 Action Programme for the Enhancement of Quality in Higher Education and the Promotion of Intercultural Understanding Through Cooperation with Third Countries. Retrieved from https://eur-lex.europa.eu/LexUriServ/LexUriServ.do?uri=OJ:L:2008:340:0083:0098:EN:PDF.

EC. (2014). Directive 2014/89/EU of the European Parliament and of the Council of 23 July 2014 Establishing a Framework for Maritime Spatial Planning, OJ L 257, 28.8.2014, pp. 135–145.

Ehler, C., & Douvere, F. (2009). *Marine Spatial Planning: A Step-by Step Approach Towards Ecosystem-based Management*. Manual and Guides No 153 ICAM Dossier No 6. Paris: Intergovernmental Oceanographic Commission UNESCO IOC, 99 pp.

Ehler, C., Zaucha, J., & Gee, K. (2018). Marine Spatial Planning at the Interface of Planning, Geography and Social Sciences. In J. Zaucha & K. Gee (Eds.), *Marine Spatial Planning—Past, Present, Future*. Palgrave Macmillan.

Flannery, W., Clarke, J., & McAteer, B. (2018). Politics and Power in Marine Spatial Planning. In J. Zaucha & K. Gee (Eds.), *Marine Spatial Planning—Past, Present, Future*. Palgrave Macmillan. (Submitted).

Gissi, E., & de Vivero, J. L. S. (2016). Exploring Marine Spatial Planning Education: Challenges in Structuring Transdisciplinarity. *Marine Policy, 74*, 43–57. https://doi.org/10.1016/j.marpol.2016.09.016.

Glegg, G. (2014). Training for Marine Planners: Present and Future Needs. *Marine Policy, 43*, 13–20. https://doi.org/10.1016/j.marpol.2013.03.011.

Jones, P., Lieberknecht, L. M., & Qiu, W. (2016). Marine Spatial Planning in Reality: Introduction to Case Studies and Discussion of Findings. *Marine Policy, 71*, 256–264. https://doi.org/10.1016/j.marpol.2016.04.026.

Kidd, S., & Shaw, D. (2014). The Social and Political Realities of Marine Spatial Planning: Some Land-Based Reflections. *ICES Journal of Marine Science, 71*(7), 1535–1541. https://doi.org/10.1093/icesjms/fsu006.

Marine Planning and Management MSc. (2018). UK: University of Liverpool. Retrieved from https://www.liverpool.ac.uk/study/postgraduate-taught/taught/marine-planning-and-management-msc/overview/.

Maritime Spatial Planning in the Baltic Sea Region. (2018). *Maritime Spatial Planning in the Baltic Sea Region*. Retrieved from http://tuba.bth.se/lo/msp/index.asp.

McCann, J., Smythe, T., Fugate, G., Mulvaney, K., & Turek, D. (2014). *Identifying Marine Spatial Planning Gaps, Opportunities, and Partners: An Assessment.* Narragansett, RI: Coastal Resources Center and Rhode Island Sea Grant College Program. Retrieved from http://seagrant.gso.uri.edu/wp-content/uploads/2014/10/CRC_MSPassessmentreport_FINAL.pdf.

Olsen, E., Fluharty, D., Hoel, A. H., Hostens, K., Maes, F., & Pecceu, E. (2014). Integration at the Round Table: Marine Spatial Planning in Multi-Stakeholder Settings. *PLoS One, 9*(10), e109964. https://doi.org/10.1371/journal.pone.0109964.

Olsson, P., Folke, C., & Hughes, T. P. (2008). Navigating the Transition to Ecosystem-Based Management of the Great Barrier Reef, Australia. *Proceedings of the National Academy of Sciences of the United States of America, 105*(28), 9489–9494. Retrieved from http://www.pnas.org/content/pnas/105/28/9489.full.pdf.

Ritchie, H., Sheppard, A., Croft, N., & Peel, D. (2015). Planning Education: Exchanging Approaches to Teaching Practice-Based Skills. *Innovations in Education and Teaching International, 54*(1), 3–11. https://doi.org/10.1080/14703297.2015.1095645.

The Baltic University. (2018). *Maritime Spatial Planning Summer School.* Retrieved from http://www.bup.fi/index.php/645-maritime-spatial-planning-summer-school.

University of Gdańsk. (2018). *Faculty of Law and Administration Project 'SEAPLANSPACE' Selected for Funding by the Monitoring Committee of the EU's Interreg South Baltic Programme 2014–2020.* Retrieved from https://en.ug.edu.pl/media/aktualnosci/72514/faculty_law_and_administration_project_seaplanspace_selected_funding_monitoring_committee_eus.

Index[1]

A

Acceptance, 16, 124, 229, 301, 383
Accountability, 295, 421, 424
Agenda 21, 220, 298, 384
Agnostic planning, 181
Aichi Target 11, 74, 76, 78, 86, 89
Alternative Futures for Marine Ecosystems (AFMEC), 328, 345
Approaches
 cross-sectoral, 384
 ecosystem, 8, 11, 37, 49, 51, 55, 64, 155, 220, 267, 275, 296, 298, 300, 301, 357, 367, 384–386, 388, 389
 holistic, 33, 152, 167, 391, 392, 445
 integrated, 14, 16, 37, 247, 258, 385, 391, 392, 446
 multidisciplinary, 2–4, 74, 449
 sectoral, 384, 385, 387, 392
 transboundary, 4, 54
Aquaculture, 49, 57, 58, 60, 64, 124, 127, 129, 131, 137, 140, 142, 143, 146, 201, 207, 251, 253, 263, 273, 279, 284, 287, 296, 312, 314, 315, 330, 334, 344, 345, 391, 450
Area, 380
Areas Beyond National Jurisdiction (ABNJ), 16, 30, 75, 76, 377, 379, 397–412
Autonomy, 329, 335, 345

B

Balance, 8, 54, 74, 91, 176, 187, 189, 211, 318, 320, 381
Baltic Marine Environment Protection Commission (HELCOM), 8, 8n6, 228, 228n4, 260, 263, 266, 388, 419
Baltic SCOPE, 137, 221, 231, 419
BaltSpace, 14, 137, 221, 231, 451
Barcelona Convention, 258, 273–278, 282, 289, 290, 402
Baseline/sea baseline, 53, 378, 379
Behaviour change, 153, 162
Benguela Current Large Marine Ecosystem (BCLME), 79–89, 91

[1] Note: Page numbers followed by 'n' refer to notes.

J. Zaucha, K. Gee (eds.), *Maritime Spatial Planning*,
https://doi.org/10.1007/978-3-319-98696-8

Biodiversity Beyond National Jurisdiction (BBNJ), 399, 400, 403–405, 407, 412
Biodiversity/biological diversity, 5, 54, 72–80, 82–91, 107, 127, 157, 166, 262, 272, 273, 283, 286, 295, 298, 312, 381, 385, 389, 391, 397–401, 406, 411
Blue economy, 4, 106, 125–127, 135, 137, 301
Blue growth
  and MSP, 122, 125, 126, 135, 137–145, 179, 207, 208, 211, 444
  policy, 36, 37, 121–146
  strategy, 126, 127, 132, 143–144, 255
  studies, 123, 128–132, 134, 135, 144
  sustainable, 132, 137, 255, 383
Blue space, 162–164, 166
Boundary
  object, 3, 202, 209–210, 212, 445
  spanning, 210

C

Cables, 32, 105, 106, 141, 143, 249, 251, 255, 284, 334, 380, 397, 403, 409–410
Capacity, 5, 15, 39, 51, 58, 74, 75, 82, 108, 111, 131, 133, 139, 154, 161, 162, 179, 180, 186, 189, 194, 204, 205, 207, 208, 210, 226, 227, 230–232, 234, 239, 275, 283, 291, 302, 312, 320, 331, 380, 383, 392, 398, 400, 406, 411, 437, 443–445, 457
Celtic Seas Partnership, 55, 63, 252, 257, 329, 330
Citizen science, 202, 209–211
Civil society, 179, 220, 286, 299, 301, 305, 309, 315–318
Classical location theory, 97–116
Clearly defined areas, 381
Climate change, 16, 35, 43, 49, 73, 91, 127, 144, 253, 281, 328, 331, 380, 385, 391, 399
Coastal state, 29, 31, 245, 258, 272, 376–379, 381, 382, 387, 389, 392, 401, 402, 417
Collaboration, 14, 134, 144, 209, 210, 222, 228–230, 237, 238, 260, 267, 273, 275, 299, 355, 407, 419, 426, 431, 451, 454
Common Fisheries Policy (CFP), 51, 54, 58, 63, 110
Common heritage of mankind (CHM), 30, 31, 380, 391, 398, 404
Competences, 126, 282, 360, 364, 387, 389, 390, 442, 443, 451–459
Conflicting
  synergies, 287, 288
  urgencies, 176
Consultation, 6, 179, 222, 225, 229, 237, 239, 284, 295, 298, 301, 311–313, 315, 317, 319–321, 355, 356, 359–361, 365, 382, 388, 429, 435, 454–459
Convention on Biological Diversity (CBD), 53, 74–76, 82, 87, 90, 298, 400, 401, 406
Cooperation, 6, 13, 53–56, 58, 62–65, 121, 127, 142, 143, 146, 259–261, 263–265, 272–275, 279, 283, 284, 289, 290, 304, 313, 327, 330, 333, 335–337, 343–347, 357, 375, 382–386, 388, 390, 391, 400–402, 433, 445, 446, 451, 454
  cross-border, 65, 261, 284, 343, 384
Cultural
  heritage, 112, 133, 153, 154, 190, 263, 398
  landscape, 41–42, 140

D

Deep habitats, 403
Degree course, 446
Deliberation, 156, 160, 181–183, 220, 235–237, 360
Democracy, 179, 181, 182n5, 194
Democratic
  decision-making, 5, 108, 114, 176, 202
  deficits, 202, 206, 208
Development
  economic, 4, 80, 114, 144, 176, 178, 194, 334, 385
  future, 138, 144, 146, 167, 239, 263, 333
  marine/maritime spatial, 101, 104, 106, 110–114, 251
  spatial, 6, 12, 14, 98, 101, 111, 327, 330, 333, 334, 421, 422
  sustainable, 1, 3, 7, 8, 49, 51, 62, 74, 75, 79, 82, 88, 109, 110, 112, 121, 144, 161, 162, 175, 206, 207, 246, 258, 260, 274, 276, 284, 287, 298, 330, 362, 376, 385, 386, 417
Discourse, 3, 4, 37, 153, 176, 181, 203, 204, 207, 211, 225, 227, 228
Doughnut economics, 188

E

Ecologically and Biologically Significant Areas (EBSAs), 72, 75–77, 79, 80, 82–92, 406
Ecological Support Area (ESA), 73, 74, 77, 86, 88–90
Economic growth, 176, 177, 177n3, 179, 188, 193, 195, 312, 334
Economies of agglomeration, 98–101, 106, 114, 115
Ecosystem
  approach, 49, 220, 267, 275, 278–280, 282, 283, 290, 296, 298, 300, 301, 357, 367, 385, 388
  approach to management, 47, 51
  based management, 37, 47–65, 74, 82, 88, 158, 247, 267, 335, 386, 387, 401–402, 411, 445, 448–451
  types, 77, 78, 84–88
Ecosystem services, 49, 51–53, 61, 63, 65, 73–75, 80, 104, 115, 153, 155–160, 164, 253, 257, 329, 362, 389
  cultural, 63, 64, 152, 153, 155–160, 183–185, 183n6
Education, 155, 162, 185, 187, 193, 234, 257, 442–466
Effectiveness (of MSP), 260, 282, 405, 417, 418
Environmental
  degradation, 5, 35, 442
  impact assessment, 13, 140–141, 357, 381, 400, 410
  protection, 11, 36, 54, 63, 64, 113, 114, 176, 179, 193–195, 260, 287, 362, 376, 380–383, 386, 387, 389–391, 412
Equality, 180, 181, 186, 187n7, 189, 190
Equity, 155, 178, 179, 186–189, 189n8, 194, 195, 375
  planner, 195
Erasmus Mundus programme, 442, 446
European Atlantic, 262, 295–321
European Commission (EC), 4–8, 11, 36, 37, 54–56, 109, 121, 123–127, 131, 132, 134–137, 143, 155, 207, 220n3, 248, 249, 251, 255, 256, 275, 279, 290, 295, 296, 300, 302, 330, 383, 400, 420, 421, 424, 425, 428, 429, 432–435, 437, 446
European MSP Platform, 137, 204, 251, 252, 259, 261, 444

European Observation Network for Territorial Development and Cohesion (ESPON), 249, 257, 281, 327
Evaluation, 10, 15, 126, 144, 157, 177, 185, 205, 206, 208, 209, 221, 234, 235, 286, 302, 365, 367, 405, 417–437, 449
Exploitation, 16, 24, 29, 31–33, 35, 36, 64, 65, 101, 273, 284, 290, 301, 335, 382, 398, 399
Exploration, 279
Exploratory scenarios, 331
Externalities, 64, 97, 99, 100, 107, 108, 110, 112–114

F

Fishers, 27, 58, 110, 114, 164, 186, 227, 233, 409, 443
Fishery, 28, 49, 52, 54, 58, 60, 63, 64, 102, 104, 108, 125, 127, 128, 131, 139, 140, 143, 146, 180, 183, 190, 231, 251, 272–274, 279, 281, 284, 287, 290, 311, 313, 315, 329, 342, 391, 402, 408, 409
Forecasts, 255, 327, 328, 331, 332
Fragmented governance, 411
Freedom of the seas, 29, 378, 379

G

Global ocean governance (GOG), 375, 377, 384, 385, 390, 391
Good environmental/ecological status (GES), 11, 37, 54, 116, 187, 252, 253, 273, 275, 277, 314, 330
Governance failure, 111

H

HELCOM-VASAB joint group, 388
Higher education, 446, 451, 466
Human activities, 1, 4–8, 51, 101, 155, 159, 245, 249, 251, 260, 262, 263, 275, 283, 300, 318, 353–355, 367, 384, 386, 387, 389, 392, 398, 399, 401, 407, 410–412, 417, 424, 442, 446

I

Impacts (of MSP), 143, 428
Inclusion, 49, 87, 167, 177–179, 183–186, 190, 193, 195, 208, 220, 222, 302, 386, 404–405
Indicators, 39, 87, 88, 157, 159, 164, 275, 283, 329, 360, 418, 426, 432, 453, 462, 464
Influence, 10, 24, 29, 41, 42, 105, 106, 111–114, 137, 139, 141, 155, 160–162, 178–180, 182, 186, 193, 208, 222, 232–235, 238, 262, 267, 302, 313, 316, 321, 344, 392, 403, 419, 422, 423, 428, 430–432, 449
Integrated coastal zone management (ICZM)/integrated coastal management (ICM), 6, 124, 145, 220, 220n3, 247–249, 258, 260, 264, 265, 271–291, 385, 386
Integrated management, 62, 258, 276, 384, 386–388, 391, 392
Integrated maritime policy (IMP), 6, 37, 47, 121, 122, 124, 132, 134, 145, 355, 388–391, 398
Integration practices, 265
Intergenerational equity, 176, 187, 194
Intergovernmental Oceanographic Commission (IOC), 245, 290, 376, 382, 383, 386, 387
International Cable Protection Committee (ICPC), 410
International Maritime Organization (IMO), 138, 382, 383, 397, 399, 403

International Organization for Standardization (ISO), 354, 363–365, 367
International Seabed Authority (ISA), 380, 397–399, 402, 409–410
International Union for Conservation of Nature (IUCN), 72, 88, 385
Irish Sea, 47–65, 304

K

Knowledge, 2, 3, 14, 33, 35, 37, 39, 40, 100, 109, 111, 113, 121, 123, 125, 126, 128, 133, 134, 137, 154, 156, 157, 167, 177, 180, 181, 183–186, 190, 210–212, 220, 225, 227, 228, 230, 231, 234, 235, 262, 266, 277, 282, 296, 301, 302, 308, 318, 321, 355, 357–361, 367, 389, 409, 421, 428, 429, 436, 443, 449, 454, 458–459, 462–466
  production, 210–212

L

Land-sea interaction (LSI), 6, 140, 273, 278–282, 289, 330
Land-Oceans Interactions in the Coastal Zone (LOICZ) Programme, 249
Law of the Sea, 27, 375–392
Legitimacy, 15, 101, 179, 181, 184, 193, 208, 210, 211, 220, 228, 232, 233, 421, 454

M

Manado Ocean Declaration, 385
Mare
  clausum, 26–29
  liberum, 26–29, 376
Marine areas
  archipelagic waters, 378, 401
  the area, 380
  contiguous zone, 271, 376, 378, 379
  continental shelf, 271–272, 377, 380
  exclusive economic zone (EEZ), 1, 13, 16, 30, 48, 53–55, 58, 72, 76, 79, 81, 83, 90, 201, 239, 247, 271, 272, 279, 290, 304, 311, 314, 363, 377–379, 381, 382, 398, 399, 403, 404
  high seas, 30, 83, 272, 378–382, 386, 398, 399, 401–404, 408
  internal waters, 30, 271, 378
  territorial sea, 5, 30, 48, 53, 58, 201, 247, 271, 272, 276, 278, 279, 312, 313, 376, 378, 379, 381, 398
Marine citizenship, 152, 153, 160–162
Marine ecosystem conservation, 383
Marine environment, 6, 8n6, 13, 36, 37, 39, 40, 51, 54–56, 63, 79, 153, 154, 159–162, 163n1, 166, 178, 183, 201, 202, 205, 207, 219, 245, 247, 252, 253, 259, 273–275, 281, 283, 284, 288, 295, 304, 314, 328, 367, 368, 376, 377, 379–390, 392, 401, 411, 412
Marine genetic resources, 400
Marine governance, 4, 15, 107–113, 151, 167, 176, 177, 179, 201–203, 206, 207, 209, 211, 296, 297, 375, 451
Marine management, 24, 47, 102n6, 155, 166, 202, 206, 208, 287, 302, 327, 330, 388
Marine Management Organisation (MMO), 329, 364, 444
Marine/maritime/sea space, 1, 30, 49, 74, 97, 121, 175, 201, 219–240, 245, 271, 295, 327, 354, 375, 398, 417, 442

Marine/maritime spatial planning (MSP), vii, 1, 24, 47, 71, 97, 121, 151, 175, 201, 219, 271, 295, 327, 353, 375, 397, 417, 442
Marine protected areas (MPAs), 5, 75, 76, 82, 88–90, 92, 159, 166, 288, 314, 343, 381, 386, 389, 400, 407, 408, 412, 442
Marine Spatial Management and Governance Project (MARISMA), 82, 89
Marine Strategy Framework Directive (MSFD), 37, 47, 54–56, 62, 109, 114, 121, 122, 124, 252, 253, 266, 275, 314, 344, 354, 355, 398
Marine sustainable governance, 375
Marine users, 219, 231, 335
Maritime sectors, 7, 122, 133, 134, 138, 142, 249, 256, 273, 279, 284, 313, 330, 333, 334, 346, 444
Maritime spatial plans
  government-approved, 1, 16
  indicative, 5
  nonbinding, 437
  strategic, 16, 132, 143, 179, 182
  visionary, 328
Market failure, 97, 107, 108, 113
MARPOL (International Convention for the Prevention of Pollution from Ships), 383, 399, 401
MSP Directive/Maritime Spatial Planning Directive (Directive 2014/89/EU), 175, 277, 278, 300, 330, 387, 417, 446
MSP Expert Group, 251
MSP principles, 8, 9, 229, 279, 388
  management principles in MSP, 360–361

N

Natural landscape, 41
Nature conservation, 35, 121, 220
Neoliberalism
  free-market neoliberalism, 204, 205
Neoliberal logic, 204–206
Non-utilitarian perspective, 35–36, 38
Normative scenarios, 331

O

Ocean
  energy, 129, 131, 140–141, 143, 144, 146, 207, 296, 314
  literacy, 153, 154, 160–162
  mining, 80
Offshore wind, 3, 60, 62, 102, 104, 107, 122, 127–129, 131, 140–141, 143, 188, 256, 312, 334, 337, 343–345, 442, 450
Oslo–Paris Convention (OSPAR), 52, 56, 252, 402
Ownership, 26–29, 31, 35, 89, 107, 301, 302, 437

P

Pan Baltic Scope, 419
Participation
  centralised decision-making, 205
  consensus, 227, 235
  ladders, ix, 223, 235
  public participation, 208, 220, 229, 235, 283, 300, 310, 311, 404, 408, 424
  tokenistic, 204
Participatory mechanisms, 298, 309, 317
Participatory planning, 204, 206, 208, 301, 420
Partnerships Involving Stakeholders in the Celtic Sea Ecosystem (PISCES), 55

Phronetical evaluation, 15
Pipelines, 4, 85, 105, 141, 143, 249, 251, 255, 284, 334, 380
Place, 25, 32, 34, 38–43, 47, 73, 77, 100, 102, 144, 153, 156, 158, 183, 193, 267, 444
Planning
 criteria, 364
 cultures, 14, 16, 357
 process, 3, 23, 47–65, 72, 113, 122, 151, 175, 204, 220, 245, 273, 301, 327, 353, 377, 399, 417, 442
Policymaking, 36, 176, 227, 353, 355, 357, 359, 361, 386, 421, 424, 425, 434
Political mobilization, 233
Political system, 353–356
Possibility space, 328, 329, 333–338, 343–346, 348
Post-political processes/planning, 177, 203–206, 211, 212
Power
 and politics, ix, 201–212
 relations, 43, 177, 180, 181, 194, 202, 206, 211
 sharing, ix, 221, 226, 228, 235
 of stakeholders, 177, 181, 187, 237
Pressures map, 84
Process responsibility, 237
Process-based planning/MSP, 11
Prospective, 31, 327, 427
Protection and preservation of the marine environment, 247, 376, 380–383, 386, 389, 390
Public
 choice, 5, 8, 97, 98, 107–114, 116
 perceptions, 154, 161

Q
Quality of life, 4, 110, 154, 163, 164, 188

R
Rationality, 202, 206, 207, 210, 211, 321, 451–453, 457, 461–463
 and power, 202, 206–209
Rational organization, rational use, 74, 207, 376
Regional development, 143, 266
Regulatory frameworks, 166, 274, 354, 356, 386
Rent spatial/location, 98, 102–105, 107, 113, 116
Representation, 25, 49, 73, 76–78, 80, 86, 90, 179, 181, 184, 185, 190, 220, 233, 301, 309, 310, 316, 317, 319
Resilience, 49, 73, 74, 109, 110, 283, 399
Res usus publicum, 380
Risk
 analysis, 12, 365, 386
 assessment, 357, 358, 364
 identification, 364, 365
 management, 354, 355, 358, 363–367
 treatment, 364, 365
Roadmaps, 125, 137, 279, 311

S
Scenario
 predictive, 331
 scenario-building, x, 327–348, 406
Scientific discovery, 32–37
Sea basin, viii, 4, 6, 10, 122, 124, 125, 127–132, 140–141, 258–261, 266, 267, 273, 289, 303, 314, 315, 442
 differences, 122, 128–133
Seabed, 142, 207, 273, 279, 296, 377, 378, 380, 385, 398, 409, 410
Seascape, 39, 42, 74, 75, 84, 153, 155, 185

Shipping, 3, 4, 50, 56, 57, 64, 72, 81, 85, 105, 113, 124–126, 128, 129, 131, 138–139, 143, 144, 146, 246, 251, 256, 263, 264, 273, 279, 284, 288, 303, 311, 334, 342, 344–347, 354, 363, 383, 385, 397–399, 403
Supporting Implementation of Maritime Spatial Planning in the Celtic Seas (SIMCelt), 52, 62, 64, 311, 313, 336
SIMWESTMED, 274, 278, 287, 314
Skills, x, 100, 124–126, 133, 134, 187, 321, 368, 406, 429, 442, 443, 445, 446, 449, 451–459, 461–466
Social
  cohesion, ix, 123, 178, 189–190, 193
  inclusion, 178, 190, 193
  justice, 5, 107, 108, 155, 176, 185, 186
Societal values, 5, 108, 116, 328
Socio-cultural benefits, 185
  values, ix, 160, 183–186
socio-economic, 78, 81, 82, 86, 88, 91, 146, 177, 184, 186, 251, 253, 255–257, 260, 262, 267, 281, 282, 303, 316, 320, 329, 331, 353, 354, 356, 376, 385, 386, 459
Socio-political spaces, 208
Solway Firth (SWF) Partnership, 61–63
Sovereignty/sovereign rights, 30, 311, 335, 377–380, 382, 401
Spatial efficiency, 8, 334–337, 343, 346, 347, 443
Spatial footprint, 334, 345
Stakeholders
  clearly defined roles, 230, 238, 433
  definition, 297, 321, 436
  elite, 204, 209
  engagement, 85, 164, 179, 180, 194, 282–284, 306, 308–309, 313, 404, 408, 427, 435–437, 442, 454, 461–463
  interactions, 202, 316
  involvement, 62, 220, 221, 229, 314, 359, 361, 364, 426, 436, 450
  mechanisms, 310
  participation continuum, 305, 316, 317
  powerful, 205, 318
  processes, 134, 295–321
State jurisdiction, 29, 378, 379
Strategic environmental assessment (SEA), 140, 141, 357, 418, 425
  SEA Directive, 330
Strategies, 7, 36, 73, 81, 82, 108, 121, 123–126, 132, 144, 145, 194, 258, 259, 261–266, 274, 278, 279, 284, 296, 306, 309, 314, 318, 321, 327, 328, 332, 362, 385, 446, 453, 464
Substantive political equality, 180, 181, 187n7
SUPREME, 274, 278
Sustainability
  goal 14, 109
  hard, 187
  social, 175–195
  soft, 187
Systematic Conservation Planning (SCP)
  tools, 71–91

T

Targets, 16, 58, 75, 77–80, 84–87, 89, 90, 121, 122, 133, 144, 233, 238, 275, 296, 314, 319, 331, 346, 362, 418, 419, 421, 423–425, 433, 443
Territorial cohesion, 110, 123

Territorial Spatial Planning (TSP), 246, 267
Thünen model, 102
Training, 133, 257, 276, 367, 368, 442–466
  resources, 442
Transboundary
  MSP, 228–231, 234–236, 239, 296, 303, 305, 308, 327, 333, 398
  partnership, 55
  pollution, 381
Transparency, 9, 87, 228, 295, 319, 359, 363, 364, 367, 388, 424
Transboundary Planning in the European Atlantic (TPEA), 55, 304, 305, 308–311, 313–316, 319, 426–428
Trust/mistrust, 15, 49, 181, 190, 226–228, 230, 231, 275, 302, 309, 354, 359, 367, 421

U

Uncertainties, 14, 111, 114, 135, 221, 273, 343, 354, 358, 359, 361, 362, 367, 398, 409, 421, 422, 424, 459
UN Environment World Conservation Monitoring Centre (UNEP-WCMC), 86, 87, 398, 410, 411
UN General Assembly (UNGA), 76, 246, 400, 401, 408, 409
United Nations Agenda for Sustainable Development 2030, 386
United Nations Convention on the Law of the Sea (UNCLOS), 29–31, 36, 38, 53, 113, 246, 247, 271, 272, 279, 355, 376–383, 387, 391, 398–402, 404, 407, 410, 411
United Nations Environmental Programme (UNEP), 47, 122, 265, 271–275, 277, 279, 280, 282, 284, 286, 383, 408
United Nations Millennium Declaration, 385
Utilitarian perspective, 38

V

Value chain analysis, 249, 256, 257
Value conflicts, 221, 227, 228
Vision and Strategies around the Baltic Sea (VASAB), 5–9, 228, 229, 260, 419
Visions, x, 7, 11, 123, 137, 143–145, 205, 259, 275, 277, 282, 287, 305, 327, 329, 332, 348, 386, 405, 420
Vulnerable social groups, 179, 186, 188

W

Water Framework Directive (WFD), 54–56, 354
Well-being
  and quality of life, 4, 163, 188
  three dimensions, 91, 178
World Ocean, 375–377, 380, 382, 384, 387, 389, 392

Z

Zero-sum game, 208

www.ingramcontent.com/pod-product-compliance
Lightning Source LLC
LaVergne TN
LVHW062156200926
848484LV00007B/1923